Atlanta's Fighting Forty-Second

MERCER UNIVERSITY

MERCER UNIVERSITY PRESS

Endowed by

TOM WATSON BROWN

and

THE WATSON-BROWN FOUNDATION, INC.

ATLANTA'S FIGHTING FORTY-SECOND

Joseph Johnston's "Old Guard"

W. Clifford Roberts, Jr.
Frank E. Clark

MERCER UNIVERSITY PRESS
Macon, Georgia
2020

MUP/ H985

Published by Mercer University Press
1501 Mercer University Drive
Macon, Georgia 31207

9 8 7 6 5 4 3 2 1

Books published by Mercer University Press are printed on acid-free paper that meets the requirements of the American National Standard for Information Sciences—Permanence of Paper for Printed Library Materials.

Printed and bound in Canada.

This book is set in Adobe Garamond Pro.

Cover/jacket design by Burt&Burt

ISBN 978-0-88146-741-3
Cataloging-in-Publication Data is available from the Library of Congress

Contents

Acknowledgments

In 2011, my coauthor, Frank Clark, began an appeal to raise the funds necessary to bring the personal and military papers and artifacts of Lovick Pierce Thomas, lieutenant colonel of the 42nd Georgia Volunteer Infantry, back to the home state of this noteworthy Southern soldier. An antiquarian dealer in Williamsburg, Virginia, had begun to offer the collection, which centered around two large scrapbooks, to discerning collectors and institutions. As the founder and curator of the Bell Research Center in Cumming, Georgia, Frank believed that the collection belonged back in the red hills of the Georgia Piedmont, and, more specifically, at his research facility. It was a daunting task for a small nonprofit to raise the money necessary to make such a purchase, and nobody except Frank thought it could be done. What followed was a fundraising drive that brought in contributions from six heritage groups, thirteen individuals, and the Brady Foundation of Alpharetta, Georgia. When the scrapbooks arrived in 2012, the Bell Center held a well-attended open house displaying featured items from the collection.

Present that day was Frederick Huff, a Thomas descendant and author of a detailed 1993 family history *Four Families: Winn, Thomas, Ware, Garrett of the Southern United States.* Mr. Huff had spent years researching his family and transcribing the letters and diaries of L. P. and wife Jennie Thomas. Mr. Huff's excellent book gave us a human perspective to our newly acquired collection. In 2015, Dr. Lovick Pierce Thomas VI visited us from his home in Santa Fe, New Mexico. "Vick" brought with him his enthusiasm, a chest of family letters, photographs, and records, and the sword carried by his great-great-grandfather and namesake into the fray at the head of the 42nd Georgia in the Battle of Atlanta.

As our project expanded beyond L. P. Thomas to the men who had served alongside him, I sought out war-time letters, diaries, and correspondence from members of the 42nd in the Atlanta History Center, the Georgia Archives, the National Archives in Washington, the Georgia Historical Society, the Gwinnett Historical Society, the Tennessee Archives, our own Bell Research Center, and the university libraries at Alabama, Emory, Georgia, LSU, and North Carolina. As a result, our narrative is filled with firsthand accounts from these soldiers describing how they endured military service in some of the most difficult campaigns and fights in the western theater of the war. Steve Zerby of Cherry Hill, New Jersey, has added his professional research at the United States Army War College Library in Carlisle, Pennsylvania. Mr. Zerby contributed recollections and reports from Northern soldiers whose battle lines were directly across from the 42nd Georgia during their many engagements. The eight excellent maps in this book were drawn by Frank from my rudimentary sketches. Mike Cunningham did the detailed lettering for each map.

We are indebted to several great people who have shared important artifacts from their private collections for publication in this book. Joseph Casea of Franklin, Tennessee, and Georgians Robert Elliott of Grayson, Dr. Jim Latimer of Jonesboro, Bill Erquitt of Atlanta, Kerry Elliott of Newnan, Lisa Yarn of Athens, and Annette Wade of Monroe have all made this a much better book. Jill Powell of the Confederate Museum in Charleston gave me access to Colonel Irving Walker's war memorabilia. Several important people have added their expertise to this project. Joe Bailey of Norcross, Georgia, has shared his burial database on the members of the Georgia Brigade. Tyler Holman, of the Gwinnett Historical Society, shared his research on the Lawrenceville Manufacturing Company. Dr. Gordon Jones, of the Atlanta History Center, has helped us with the history and details of the Cyclorama Painting and the positioning of the DeGress Battery in the Battle of Atlanta. Timothy Burgess of White House, Tennessee, shared the unpublished battlefield casualty reports from Clayton's Division assembled by chief surgeon Gustavus Thornton.

Visiting each battlefield where the 42nd Georgia stood and fought has made the descriptions of these battles in this book more accurate and precise. Charles Lunsford and Mark Pollard took Frank, Vick Thomas, and me on a tour of Jonesboro, Nash Farms, and Lee's Mill, the location of Lovick's brief capture. Historians Perry Bennett and Bill Erquitt showed me the trench lines outside of Atlanta that were held by the 42nd Georgia during the grim summer of 1864. Jim Kay of the Nashville Battlefield Trust gave me an insightful tour of Overton Hill and the trench lines of December 16, 1864. I have also been fortunate to participate in several of the yearly Congresses held by the General Barton Stovall Historical Association. This group of descendants from the famous brigade, led by founder Gary Goodson and President Mike Griggs, visits a different area of the South each year and explores where the brigade marched, camped, and fought. I have joined them for gatherings in Atlanta, Eastern Tennessee, Kentucky, Vicksburg, and Dalton. The collective knowledge of their members is extremely impressive. Finally, my band of reenactor compatriots—Jerry Bryan, Mike Couch, Mike McAlpin, Brett Martin, Ralph Mills, and Chad White—have taken me to historical sites in Alabama, Georgia, Kentucky, Mississippi, Tennessee, and Virginia.

A final and forever thank you to family and friends who have encouraged Frank and me to persevere and see this labor of love across the finish line.

Cliff Roberts
Charleston, South Carolina
September 2019

MAPS

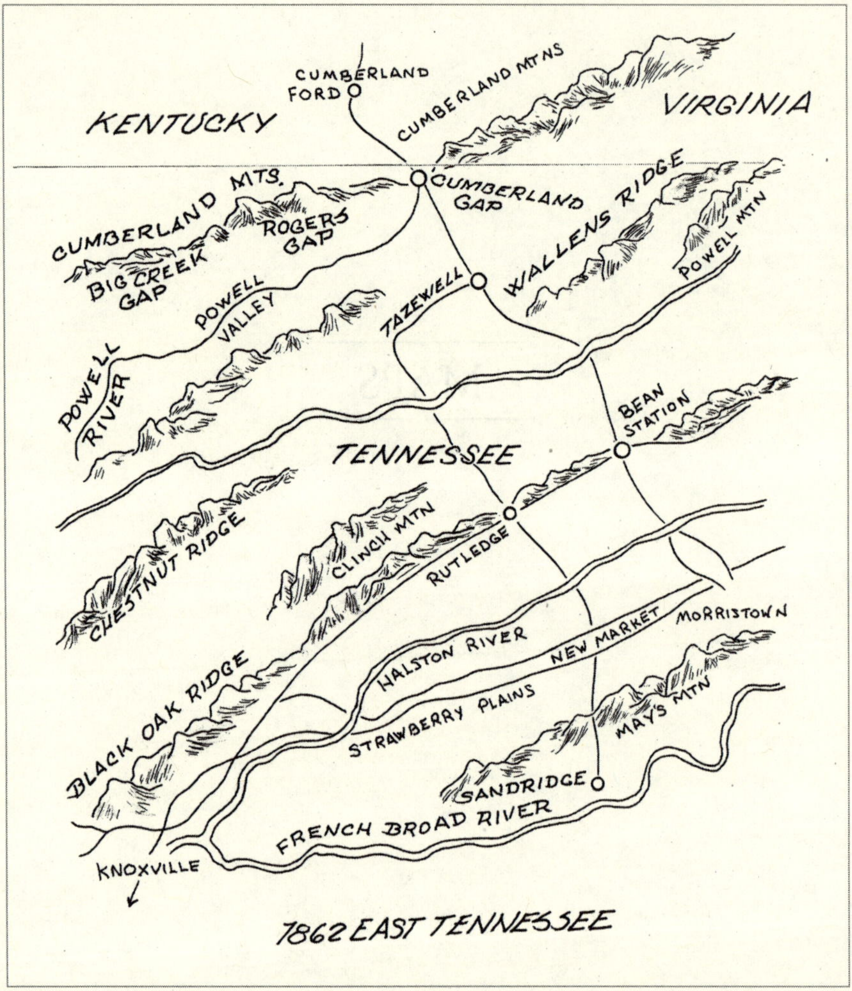
CUMBERLAND FORD
KENTUCKY
CUMBERLAND MTNS
VIRGINIA
CUMBERLAND MTS.
CUMBERLAND GAP
ROGERS GAP
CUMBERLAND MTS.
BIG CREEK GAP
WALLENS RIDGE
POWELL MTN
TAZEWELL
POWELL VALLEY
POWELL RIVER
BEAN STATION
TENNESSEE
CHESTNUT RIDGE
CLINCH MTN
RUTLEDGE
MORRISTOWN
NEW MARKET
HALSTON RIVER
BLACK OAK RIDGE
STRAWBERRY PLAINS
MAY'S MTN
SANDRIDGE
FRENCH BROAD RIVER
KNOXVILLE
1862 EAST TENNESSEE

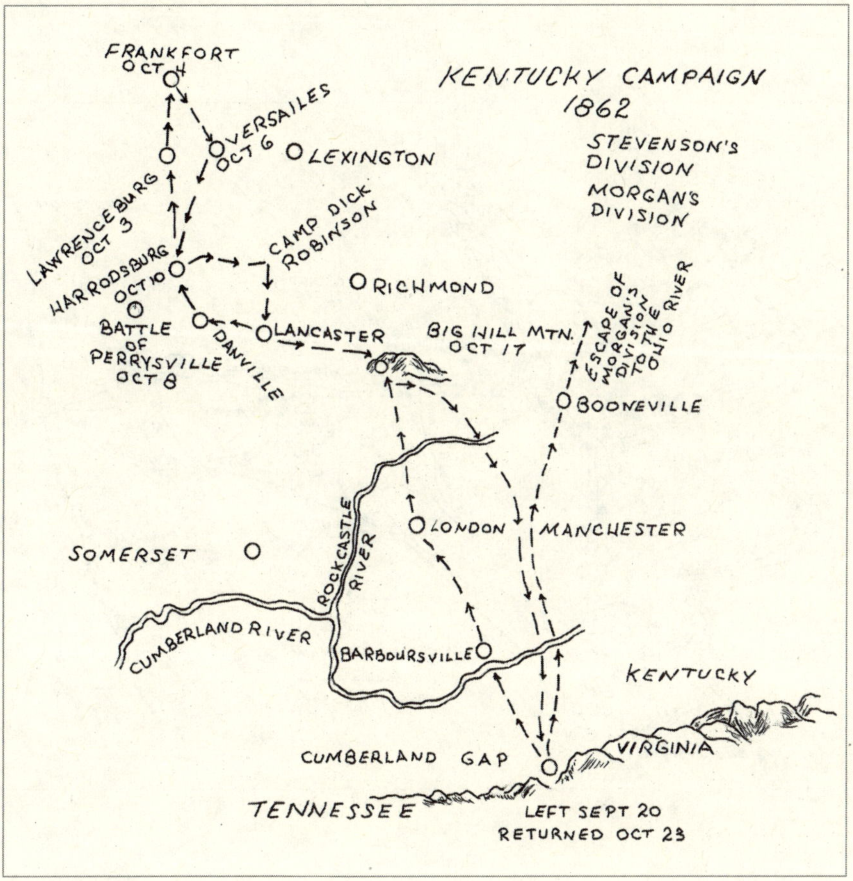
KENTUCKY CAMPAIGN
1862
STEVENSON'S
DIVISION
MORGAN'S
DIVISION
FRANKFORT
OCT 4
VERSAILES
OCT 6
LEXINGTON
LAWRENCEBURG
OCT 3
HARRODSBURG
OCT 10
CAMP DICK
ROBINSON
RICHMOND
BATTLE
OF
PERRYSVILLE
OCT 8
DANVILLE
LANCASTER
BIG HILL MTN.
OCT 17
ESCAPE OF
MORGAN'S
DIVISION
TO THE
OHIO RIVER
BOONEVILLE
ROCKCASTLE
RIVER
LONDON
MANCHESTER
SOMERSET
CUMBERLAND RIVER
BARBOURSVILLE
KENTUCKY
VIRGINIA
CUMBERLAND GAP
TENNESSEE
LEFT SEPT 20
RETURNED OCT 23

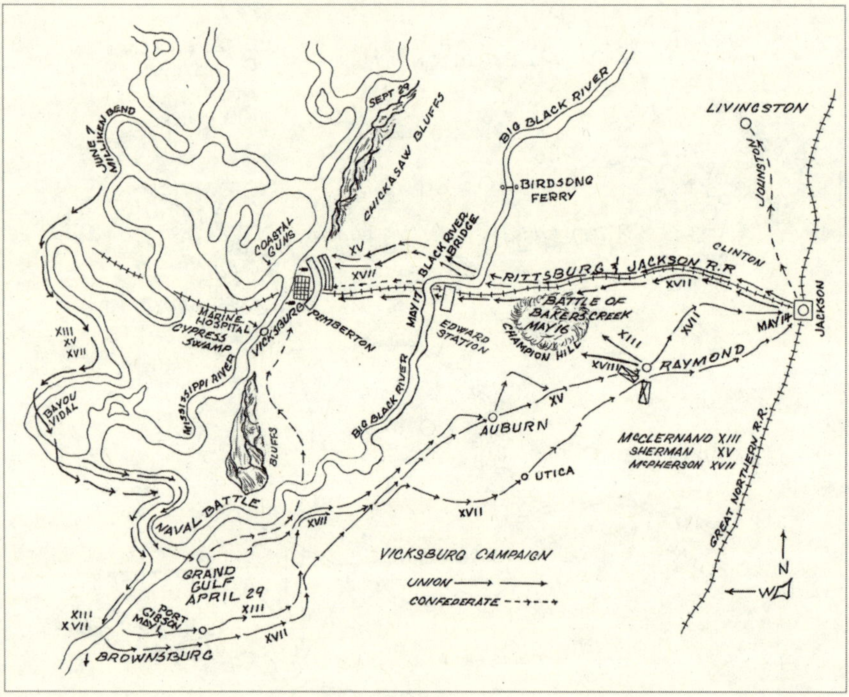

LIVINGSTON
JOHNSTON
CLINTON
JACKSON
BIG BLACK RIVER
BIRDSONG FERRY
SEPT 29
CHICKASAW BLUFFS
COASTAL GUNS
BLACK RIVER BRIDGE
XV
XVII
RITTSBURG & JACKSON R.R
XVII
BATTLE OF BAKERS CREEK MAY 16
CHAMPION HILL
MAY 17
EDWARD STATION
XIII
XVIII
XVII
RAYMOND
MAY 14
JUNE 7 MILLIKEN BEND
MARINE HOSPITAL
CYPRESS SWAMP
VICKSBURG
PEMBERTON
XIII XV XVII
BAYOU VIDAL
MISSISSIPPI RIVER
BLUFFS
BIG BLACK RIVER
AUBURN
XV
UTICA
McCLERNAND XIII
SHERMAN XV
McPHERSON XVII
XVII
GREAT NORTHERN R.R.
NAVAL BATTLE
XVII
GRAND GULF APRIL 29
VICKSBURG CAMPAIGN
UNION
CONFEDERATE
N
W
XIII XVII
PORT GIBSON MAY 1
XIII
XVII
BROWNSBURG

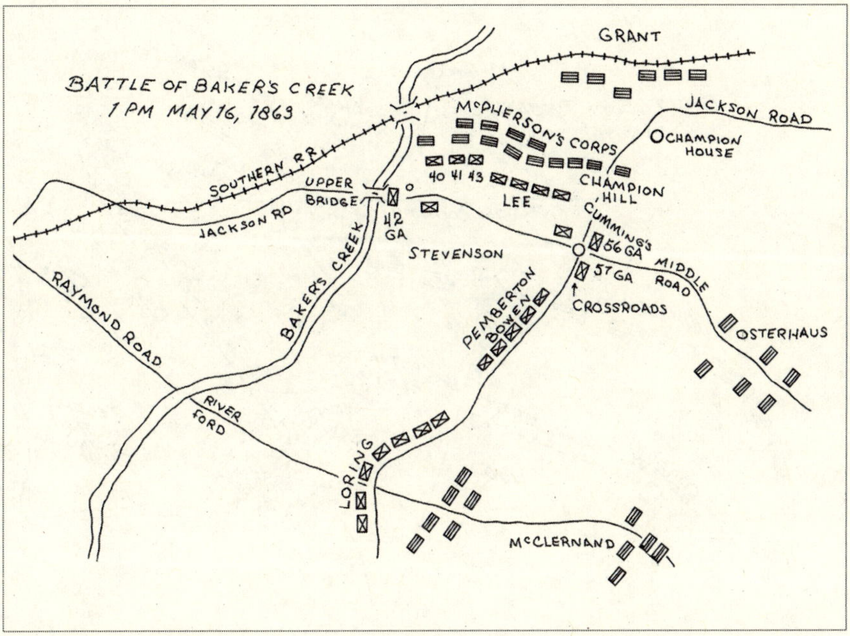
BATTLE OF BAKER'S CREEK
1 PM MAY 16, 1863
GRANT
JACKSON ROAD
McPHERSON'S CORPS
CHAMPION HOUSE
SOUTHERN RR
UPPER BRIDGE
40 41 43
CHAMPION HILL
JACKSON RD
42 GA
LEE
CUMMING'S
56 GA
STEVENSON
57 GA
MIDDLE ROAD
CROSSROADS
RAYMOND ROAD
BAKER'S CREEK
PEMBERTON
BOWEN
OSTERHAUS
RIVER FORD
LORING
McCLERNAND

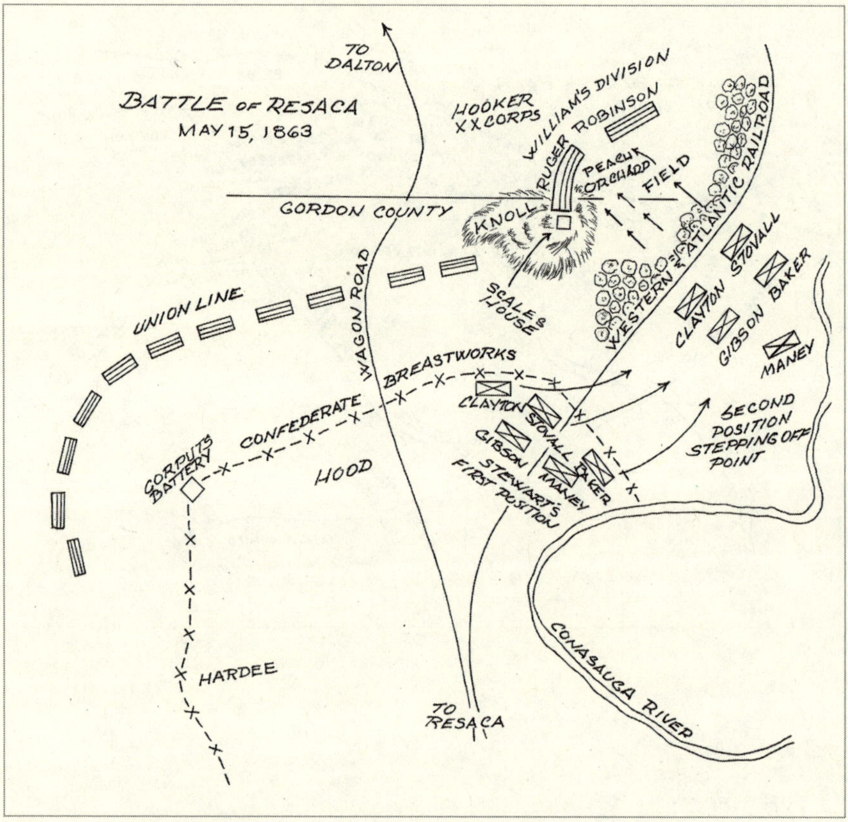
BATTLE OF RESACA
MAY 15, 1863
TO DALTON
HOOKER XX CORPS
WILLIAM'S DIVISION
ROBINSON
RUGER
PEACH ORCHARD
FIELD
KNOLL
GORDON COUNTY
SCALES HOUSE
WESTERN & ATLANTIC RAILROAD
WAGON ROAD
UNION LINE
STOVALL
BAKER
CLAYTON
GIBSON
MANEY
BREASTWORKS
CONFEDERATE
CORPUTS BATTERY
HOOD
CLAYTON
STOVALL
GIBSON
MANEY
BAKER
STEWART'S FIRST POSITION
SECOND POSITION STEPPING OFF POINT
HARDEE
TO RESACA
CONASAUGA RIVER

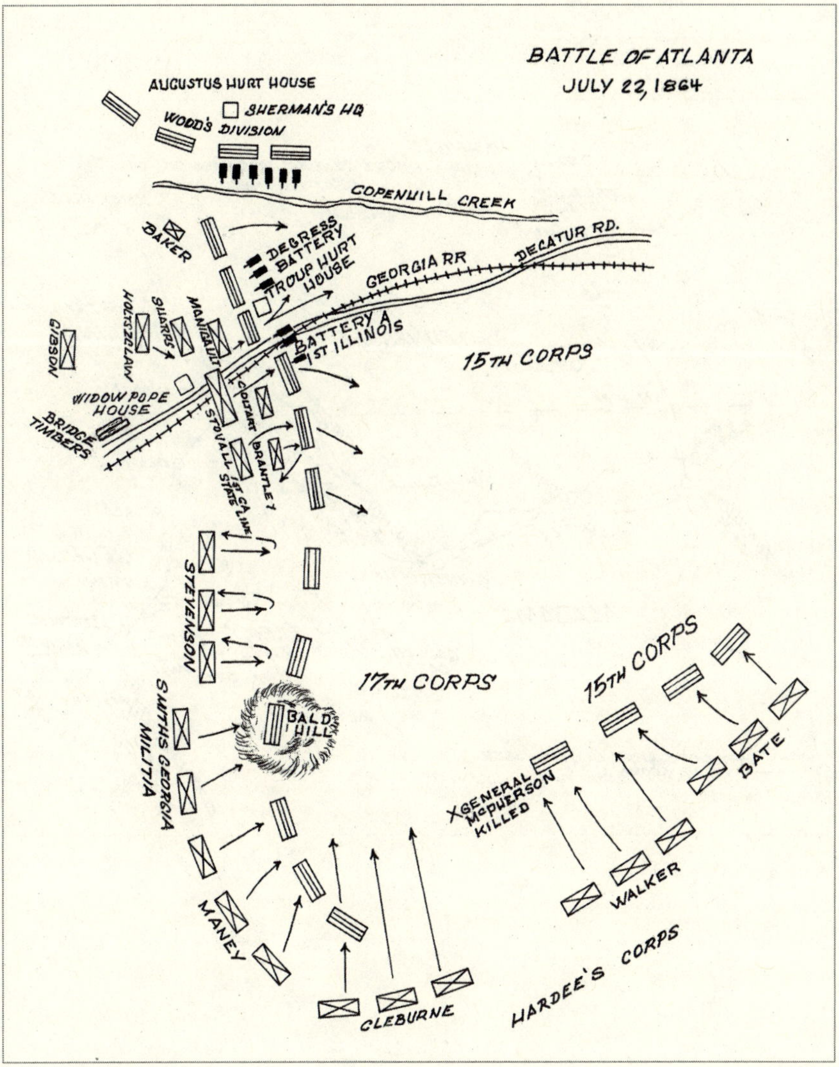
BATTLE OF ATLANTA
JULY 22, 1864
AUGUSTUS HURT HOUSE
SHERMAN'S HQ
WOOD'S DIVISION
COPENHILL CREEK
BAKER
DEGRESS BATTERY
TROUP HURT HOUSE
GEORGIA RR
DECATUR RD.
GIBSON
HOLTSCLAW
SHARPS
MANIGAULT
BATTERY A 1ST ILLINOIS
15TH CORPS
WIDOW POPE HOUSE
BRIDGE TIMBERS
COLTART
STOVALL
1ST GA STATE LINE
BRANTLEY
STEVENSON
17TH CORPS
BALD HILL
SMITHS GEORGIA MILITIA
15TH CORPS
BATE
X GENERAL McPHERSON KILLED
WALKER
MANEY
HARDEE'S CORPS
CLEBURNE

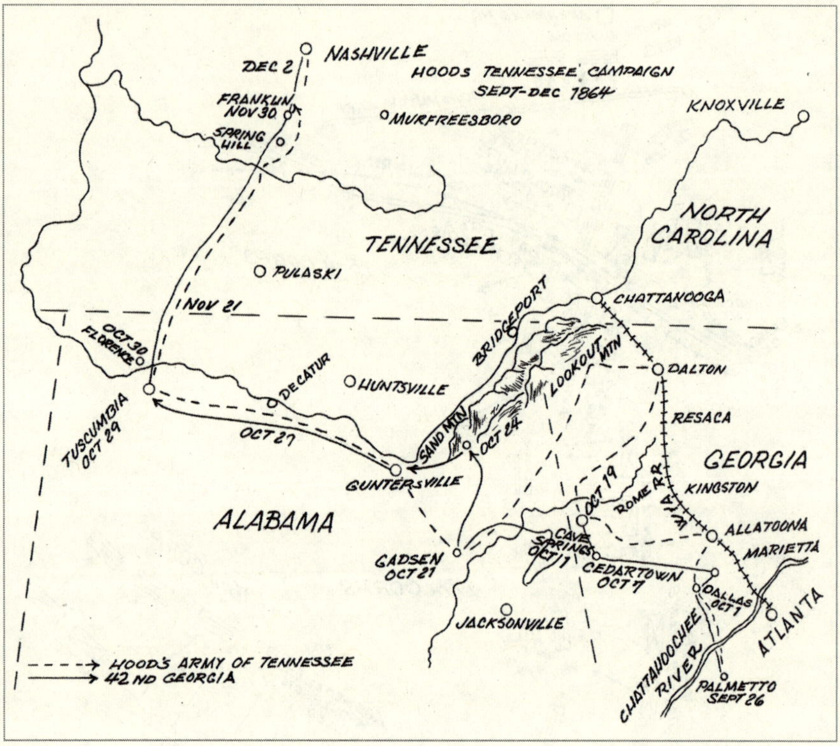
HOODS TENNESSEE CAMPAIGN
SEPT-DEC 1864
NASHVILLE
DEC 2
FRANKLIN
NOV 30
SPRING
HILL
MURFREESBORO
KNOXVILLE
NORTH
CAROLINA
TENNESSEE
PULASKI
CHATTANOOGA
NOV 21
BRIDGEPORT
OCT 30
FLORENCE
LOOKOUT MTN
DALTON
DECATUR
HUNTSVILLE
TUSCUMBIA
OCT 29
OCT 27
SAND MTN
OCT 24
RESACA
GEORGIA
GUNTERSVILLE
KINGSTON
OCT 19
ROME RR
ALABAMA
ALLATOONA
CAVE
SPRING
MARIETTA
GADSEN
OCT 21
CEDARTOWN
OCT 7
DALLAS
ATLANTA
JACKSONVILLE
CHATTAHOOCHEE
RIVER
PALMETTO
SEPT 26
HOOD'S ARMY OF TENNESSEE
42 ND GEORGIA

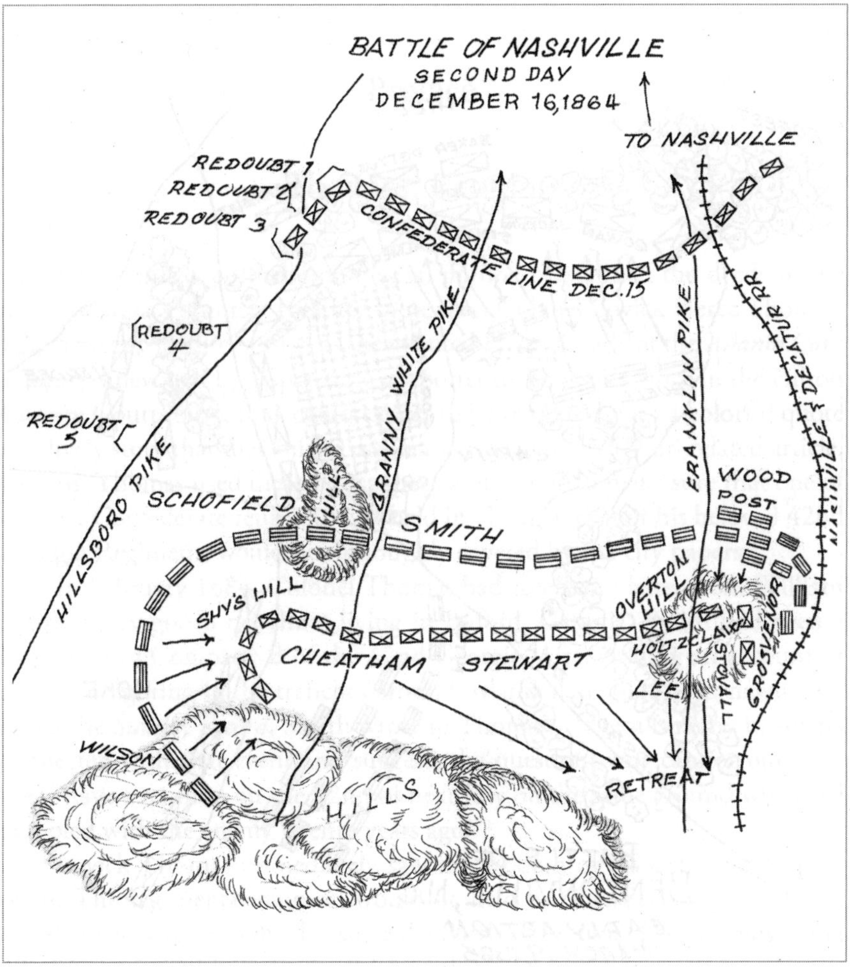
BATTLE OF NASHVILLE
SECOND DAY
DECEMBER 16, 1864
TO NASHVILLE
REDOUBT 1
REDOUBT 2
REDOUBT 3
REDOUBT 4
REDOUBT 5
CONFEDERATE LINE DEC. 15
GRANNY WHITE PIKE
FRANKLIN PIKE
NASHVILLE & DECATUR RR
HILLSBORO PIKE
SCHOFIELD
HILL
SMITH
WOOD
POST
OVERTON HILL
SHY'S HILL
CHEATHAM
STEWART
HOLTZCLAW
STOVALL
GROSVENOR
LEE
WILSON
HILLS
RETREAT

remaining years of his life. The stroll was pleasant and, apart from crossing the busy tracks of the West Point & Central Railroad, allowed him to entertain thoughts of what the day would bring.

Despite the promise of another hot and humid summer day in Atlanta, Lovick wore a formal suit with a starched white shirt. Jennie seemed as excited as he was about the first reunion of the 42nd Georgia, and she assured Bud, as she called him, that she would bring their eight children to Ponce de Leon Park to see their father on the pavilion. As the deputy sheriff of Fulton County, the forty-nine-year-old Thomas made the trip to the courthouse nearly every day of the week. Soon, he hoped to play an even larger role in city and county affairs; he had entered his name for the office of high sheriff of Fulton County in the upcoming fall election. Politics was not on L. P.'s mind that Tuesday, however, as he had taken the day off from work, something he rarely did. Twenty years earlier to the day, L. P. Thomas had drawn his sword and led the 42nd Georgia Infantry on a momentous charge along the tracks of the Georgia Railroad in the Battle of Atlanta. His regiment had scaled and captured a section of the Federal breastworks, thereby securing possession of two Napoleon howitzer cannons. It was one of the few highpoints for him and his regiment in three years of constant and brutal sectional war.

With its prominent clock tower, the new two-story red brick courthouse had already become something of a city landmark. Two blocks south of the Union Depot, this seat of county affairs had been selected by Thomas as an easy rendezvous point for the initial reunion of members of the 42nd Georgia Volunteer Infantry Regiment. By 9:00 that July morning, some 250 veterans had assembled near the front steps of the courthouse. Some were friends, neighbors, and work colleagues, but many others Lovick had not seen since the final surrender of the Army of Tennessee in late April 1865. That day, the depleted regiment stood in an old farm field near Hillsboro, North Carolina, and stacked their Enfield rifles, furled their battle flag for the last time, and accepted their written Federal pardons. Now two decades later, these "old veterans" were finally gathered in one place again to remember their bonds of comradery in time of war.

As chairman of the reunion organizing committee, Thomas was grateful that so many of the veterans had turned out for the day's events. The former members of the 42nd Georgia had traveled from at least eight counties of the Georgia Piedmont to be a part of this reunion and re-forge the bonds that forever bind old soldiers. Most were about fifty years of age, and each accepted with pride the blue reunion badge that was carefully pinned to their chest. Among the recipients was Henry Wilson and Abram Hardeman, two of a handful of former black servants who had gone to the front with their white masters. Like the men they now stood with, Hardeman and Wilson were still Confederates, bonded by friendship and mutual experience, and not by current social position.

Led by a brass band, the former members of the 42nd Georgia moved as a group up Pryor Street toward the new buildings of a resurgent Atlanta. At the railroad depot, the Southern veterans boarded special horse-drawn streetcars that carried them out of the city limits to the traditionally "whites-only" Ponce de Leon Park. The large crowd that had been gathering in the park cheered at the arrival of the old soldiers. The program for the day soon began from a crowded pavilion.

Master of ceremonies Robert F. Maddox, who now represented Atlanta in the state General Assembly and had been the first lieutenant-colonel of the 42nd Georgia, called the reunion to order. Reverend T. R. Kendall, the pastor of Trinity, the city's largest Methodist church, invoked divine blessings on the assembled men. He was followed by the mayor of Atlanta, John Goodwin. The stocky politician announced that "the hearths of the Gate City would always glow the warmest for the veterans of this historic regiment." Cheers then greeted Judge Lowndes Calhoun, who had led Company K, a group of Atlanta men better known as the "Calhoun Guards," through the years of conflict. William's father, James Calhoun, had been the wartime mayor of Atlanta. As one of the city fathers, the elder Calhoun performed the sad task of surrendering his shell-scarred city to Tecumseh Sherman after Hood had withdrawn the Southern army. Lowndes Calhoun had followed in his father's footsteps, serving a term as mayor of the Gate City from 1879 to 1881. Still serving the people of Atlanta, he was now judge of the Ordinary Court for Fulton County.[3]

The former captain and mayor delivered an address on the history and services of the regiment. Calhoun pointed out that the 42nd was the largest regiment in the Army of Tennessee, a fact he attributed to the energy of its officers and the popularity of its first commander, Robert Henderson. Twenty-one battles were briefly described, with particular emphasis given to the regimental charges at Resaca and Atlanta. In May 1864, 113 men from

---

[3] "Around the Old Flag: Reunion of the Forty-Second Georgia Yesterday," *Atlanta Constitution*, July 23, 1884, 7.

the 42nd Georgia fell near the tracks of the Western & Atlantic Railroad by the hamlet of Resaca. The horrific casualties included both the colonel and major of the regiment, requiring command of the 42nd Georgia to devolve to the senior company captain, Lovick P. Thomas of Gwinnett County. Fortunately for the men of the 42nd, Thomas proved to be a capable leader during the last year of the War of Southern Independence. Lowndes Calhoun remembered July 22 as the day the Georgia boys, despite fearful odds, had driven the Federals back and "captured some of the most famous guns in Sherman's army." Calhoun then praised the 1st Georgia State Line, which had charged next to the 42nd that day in Atlanta. Colonel Albert Howell, now the owner of the *Atlanta Constitution*, had led that regiment in their brave dash, and his presence on the podium received a warm ovation.

It was at this moment that the veterans jumped to their feet, saluting and giving their high-pitched Rebel yell. General Robert Henderson of Covington, their first regimental commander, had come to the front of the pavilion. This Newton County plantation owner and politician had turned the men before him from civilians to soldiers on the training fields of Camp McDonald near Big Shanty. Henderson had held them together on the windswept mountain above Cumberland Gap in eastern Tennessee, through the long march into Kentucky, and again through forty-seven days of half-rations at the Siege of Vicksburg. Henderson's head wound at Resaca had taken him from the regiment for most of the Atlanta campaign, but he had returned to hold Stovall's Georgia Brigade together as they constituted Hood's rear guard in the desperate retreat from Nashville. The members of Henderson's former regiment gave him a wild reception, unambiguous testimony of their continued high regard for him. Next to speak was the popular Colonel William Hulsey, another casualty of Resaca. Hulsey had been the chivalrous "boy" major of the regiment. He was now a state politician and another former mayor of Atlanta. Loud calls from the veterans came for two more men on the pavilion to speak a few words. Governor Henry McDaniel, a former Confederate major who had been wounded and captured in Maryland on the retreat from Gettysburg, complimented the veterans for being "as true in peace as they had been in war." The governor was only a few months away from starting the construction of a new state capitol on a prominent hill looking down on the Union Depot. Next stood former governor Alfred Colquitt, now first-term United States senator. The former Confederate brigadier-general was part of the "Bourbon Triumvirate" that seemed to control all aspects of Georgia politics. Senator Colquitt, ever sensitive to the fact that Georgia was only five years removed from Federal military rule, remarked that this occasion should not be interpreted as an effort to "revive the war spirit in the South," but as a simple reunion of old comrades. The proceedings were closed by the reading of a poem, and the governor and senator led the veterans to dinner tables, where a colossal barbecue awaited the hungry former soldiers.[4]

A two-hour break allowed the veterans and guests to stroll the grounds, patronize the rides, and exchange stories on park benches. The afternoon proceedings had a different feeling to them. A representative from each of the ten companies gave a short extemporaneous speech. Many of their remarks were noted for their wit and humor. Lovick Thomas was kidded for leading five of the companies in a spirited charge through the woods near Orangeburg, South Carolina, only to find that they had managed to capture every member of the 40th Georgia but not a single Yankee. Having properly acknowledged past adventure, hardship, and peril, the group needed to resolve the business of the future.

Should these reunions continue? This was not an easy question in that the Federal government viewed these gatherings with suspicion. Ladies' Memorial Association had honored the Confederate dead in Southern cemeteries without criticism, but the formal reunions of Confederate veterans were new territory. Among the attendees, there was a unanimous consensus to "perpetuate these anniversaries." Lovick Thomas was declared president of the new Survivors of the 42nd Organization, and ten men, one from each company, became vice presidents. It was resolved that each veteran should contribute a dollar a year to a relief fund for any survivor who might fall into destitute circumstances.

At 5:00, the men were surprised by the arrival of the governor's horse guard. The horsemen, in their red dress uniforms and polished brass, drew their steeds up in front of the pavilion. Rising from his stirrups, Captain John Milledge tendered a ringing address to the veterans, with a tribute of honor and respect from "the volunteer soldiery of the New South." Responding from the podium in fitting terms was a 42nd veteran Leonidas Felix Livingston of Newton County. Now an aspiring politician, Livingston would soon represent Atlanta and the 5th District in the House of Representatives in a political career that would span ten congressional terms. With the speeches concluded, the Guard enthusiastically cheered the former Confederates and

[4] Ibid.

rode away. Lovick P. Thomas would, until his death in 1910, remain president of the 42nd Infantry Survivor Association. He considered it one of his favorite and most important responsibilities. Reunions of the regiment would be held every July 22 without exception until 1912 when the ranks finally dwindled to a handful of veterans and Lovick Thomas had been buried at Oakland Cemetery.

As dusk set on the first reunion of the 42nd Georgia Infantry, Colonel Thomas stepped off the pavilion. The veterans, with an arm on the shoulder of the man next to them, had begun to gather around the old regimental flag. Thomas shook hands with William F. Edwards, who held the revered standard. Edwards had been the third man to grasp the regimental colors on that bloody day in Resaca, and he continued, until the final surrender in North Carolina, to serve as the brave color bearer of the 42nd. Somehow, at the Greensboro parole ceremony, Edwards had managed to smuggle the battle flag away tucked in his boot. The old flag had survived, but it was torn and riddled with bullet holes. When it had draped over the speaker's podium, some of its stars had hung by a few threads. Nothing needed to be said. The regimental flag "in silent eloquence told the tale of the heroic courage of the men who surrounded it."[5]

[5] Ibid.

Principal Figures

Zachariah Jabez Armistead (1844–1877)—Teenage corporal in the Walton Tigers (H).

Brigadier General Seth Maxwell Barton (1829–1900)—West-Point graduate from Virginia who led the Georgia Brigade at the battles of Chickasaw Bayou and Baker's Creek and at the Siege of Vicksburg.

Ethelbert Pendleton Bedel (1831–1865)—Brother-in-law of L. P. Thomas who accompanied him to California. Killed leading a Georgia militia unit late in the war.

Captain William Lowndes Calhoun (1837–1908)—Captain of the Calhoun Guards (K). Mayor of Atlanta in 1878. Judge of the Fulton County Ordinary Court from 1881 to 1897. Historian of the 42nd Georgia Survivors Association. Commander of the Fulton County Confederate Veterans Association.

Captain Elijah Henry Clarke (1835–1898)—Elected captain of the Fowler Guards (D) after the resignation of Nathan Clay in February 1863. Methodist minister, sawmill owner, and DeKalb County politician.

Major-General Henry De Lamar Clayton (1827–1889)—Division commander in 1864 and 1865. Became president of the University of Alabama in 1886.

Major Maximilian Van Den Corput (1824–1911)—Belgium-born commander of the Georgia Cherokee Artillery. His guns were captured at both the battles of Baker's Creek and Resaca. A prominent architect after the war, Corput designed the second Union Station in Atlanta. His post-war home was on the site where the 42nd Georgia had seized the Federal trenches in the Battle of Atlanta.

George Hewitt Daniel (1816–1864)—Englishman who became a Covington merchant. Visited the 42nd Georgia outside of Vicksburg in 1863. Killed by Gerrard's cavalry in Covington raid.

Captain Francis DeGress (1840–1883)—German immigrant who commanded Company H of the 1st Illinois Artillery at the Battle of Atlanta. Sold firearms in Latin America after the war.

Lieutenant Colonel Christopher John Dickerson (1828–1872)—Colonel of the 10th Michigan at the Battle of Buzzard's Roost on February 25, 1864. Highest ranking Federal officer captured by the 42nd Georgia.

Dr. William Meigs Durham (1846–1919)—Enlisted as a teenager in the

Calhoun Guards (K). Rose to "Boy Captain" adjutant of the 42nd Georgia. Decatur physician after the war. Commander of Atlanta UCV Camp # 159.

Sergeant William Franklin Edwards (1844–1926)—Regimental color bearer from the Battle of Resaca until the end of the war. Carried the banner to UCV reunions and parades for more than thirty years.

Brigadier General Clement Anselm Evans (1833–1911)—Wounded five times as commander of the 31st Georgia. A prominent Methodist minister and Confederate historian, Evans headed the Georgia UCV for twelve years.

Major General John Brown Gordon (1832–1904)—Lee's trusted general was the long-serving commander of the UCV. Part of the Bourbon Triumvirate, Gordon served as senator from Georgia.

Master Sergeant James Henry Griffin (1840–1911)—Born to a wealthy Newton County family, Jim Griffin worked closely with the officers of the 42nd Georgia to direct the actions of the regiment.

Lieutenant James Monroe Gresham (1840–1913)—Colorful 1st lieutenant of the Walton Tigers (G). Captured at Battle of Nashville and left an account of his time as a prisoner at Johnson's Island. School teacher, farmer, and politician after the war.

Dr. Ira Lawson Gunter (1830–1904)—Herbal doctor from Henry County assigned to Division hospital at Bean Station. Passed his Confederate medical boards in March 1863 and served as assistant surgeon in army hospitals in Atlanta. His recently discovered wartime letters are detailed, opinionated, and thoughtful. Physician, politician, and Baptist minister after the war.

Brigadier General Robert Johnson Henderson (1822–1891)—Covington planter and politician before the war. Competent commander of the 42nd Georgia until he was wounded at Battle of Resaca. Led Cumming's Brigade on daring charge at Bentonville in 1865.

Lieutenant General Daniel Harvey Hill (1821–1889)—Outspoken South Carolinian who fell out of favor with Jefferson Davis. Led Stephen D. Lee's Corps at Battle of Bentonville in March 1865.

Colonel William Arnold Hemphill (1842–1902)—Wealthy business manager of the *Atlanta Constitution*. Instrumental in organizing the Fulton County Confederate Association. Mayor of Atlanta in 1890.

Lieutenant General John Bell Hood (1831–1879)—The 42nd Georgia served in the corps of this controversial general in the Atlanta Campaign. Hood took over the Army of Tennessee in July 1864 and led it through the disastrous Tennessee campaign. President of New Orleans insurance company after the war.

Colonel Albert Howell (1843–1927)—Promoted to colonel of 1st Georgia State Militia that served in Stovall's Brigade during the Atlanta Campaign. Held the controlling interest in the *Atlanta Constitution*.

Captain Evan Howell (1839–1905)—Commanded Howell's battery at Battle of Atlanta. Editor of his brother's *Atlanta Constitution*. Influential Atlanta politician.

Lieutenant Colonel William Henry Hulsey (1838–1909)—Dashing young major of the 42nd Georgia for most of the war. Wounded at Battle of Resaca. Mayor of Atlanta in 1869. Prominent Atlanta lawyer and politician.

Colonel Abda Johnson (1826–1881)—Prominent Bartow County lawyer before the war. Commanded the 40th Georgia and often the Georgia Brigade. Mayor of Cartersville in 1872 and delegate to 1877 Georgia Constitutional Convention.

General Joseph Eggleston Johnston (1807–1891)—"Old Joe" assumed command of the Army of Tennessee in December 1863 and defended Georgia during the ninety-day Atlanta Campaign. One of only five full generals in the Confederate military.

Lieutenant-General Stephen Dill Lee (1833–1908)—Hero of Chickasaw Bayou and youngest lieutenant-general in the Confederacy. Became corps commander in Army of Tennessee in 1864 with mixed success. Second commander of the United Confederate Veterans. First president of the Mississippi Agriculture and Mechanical College.

Congressman Leonidas Felix Livingston (1832–1912)—The quartermaster of the 42nd Georgia during the Cumberland Gap campaign, "Uncle Lon" won ten terms in the U.S. House of Representatives as a populist Democrat.

Colonel Robert Flourney Maddox (1829–1899)—Sheriff of Troup County before the war. Second-in-command of the 42nd Georgia in 1862 and 1863. Left to organize and command the 2nd Georgia State Militia. Atlanta politician and businessman after the war. Founder of the Maddox-Rucker Bank. His son became mayor of Atlanta in 1908.

Brigadier General Arthur Middleton Manigault (1824–1886)—Wealthy Charlestonian and Mexican War veteran. Brigade commander who captured the DeGress Battery at the Battle of Atlanta. Severely wounded at Battle of Franklin. His vivid narrative of the war was not published until 1983.

Sergeant Moses Martin (1831–1898)—Served in the Gwinnett Beauregards (A) and lost a leg at the Battle of Bentonville in 1865. Served twelve years as doorkeeper of the Georgia Assembly.

Captain Singleton Allen Maxwell (1832–1864)—Elected captain of the Milton Tigers in December 1863 after the resignation of Henry Paris. Kept a wartime journal. Killed leading his troops at the Battle of Atlanta.

Captain Thomas Jefferson Mercer (1826–1886)—Captain of Harper Guards (E) from Newton County. His company served as color company of the 42nd Georgia. Operated mill near Decatur after the war.

Captain Lovick Pierce Thomas, Sr. (1812–1878)—The "Fighting Quartermaster" of the 35th Georgia, Thomas Sr. became Atlanta's police chief immediately following the war. Over the course of his life, Lovick ran hotels in Athens, Gainesville, Newnan, and Rome.

Lieutenant Colonel Cornelius Irvine Walker (1842–1927)—As commander of the South Carolina Division of the UDC, Walker contested the claim that the 42nd Georgia had captured the DeGress Battery in the Battle of Atlanta. A Citadel graduate, Walker was the adjutant and later commander of the 10th South Carolina. He was a successful author and book publisher in Charleston after the war.

Captain Benjamin Walker (1831–1864)—Popular captain of the Independent Rebels (B) killed at the Battle of Franklin. Union soldiers spared his home after seeing his Mason ribbons in the front hall.

Captain Andrew Jackson West (1844–1917)—A former captain in the 41st Georgia, A. J. West hired L. P. Thomas as a traveling salesman for his wholesale grocery business. Destroyed by the Panic of 1873, West rebuilt his fortune in Atlanta real estate. He was director of the 1895 Cotton States Exposition and held numerous positions in the UCV.

Courtland Simmons Winn (1863–1940)—Son-in-law to L. P. Thomas, Winn was elected mayor of Lawrenceville in 1884 at age twenty-four. He later moved to Atlanta and practiced law. In 1911 he was elected mayor of Atlanta.

Chapter 1

Cumberland Gap

"Although late entering service, the 42nd regiment was made up of excellent material, men who knew that the time had come to fight for their homes and loved ones and that whipping Yankees had ceased to be talked of as child's play."

—Lieutenant James Monroe Gresham, Walton Tigers[1]

Enlistment

The April 1861 bombardment and surrender of Fort Sumter in Charleston harbor was soon followed by predictions, in both the North and South, that the war would be over in six weeks. By spring 1862, there was a realization everywhere that this would, in fact, be a long, cruel affair. Rebel victories at First Manassas, Virginia, and Wilson's Creek, Missouri, had given the citizens of the new Confederate States of America some degree of optimism that their independence could be sustained. This was a tall order for an agrarian nation of fewer than ten million people, of whom only six million were white citizens, and, of these, roughly one million were eligible for military service. To defend their new Richmond, Virginia, capital, Southern troops would have to hold a 160-mile front from Fredericksburg to the Allegheny Mountains. The western front was even more imposing, a 600-mile land border from the Appalachian Mountains to the Ozarks. The resourceful President Abraham Lincoln was a formidable opponent, determined to crush what he termed the "rebellion," no matter the cost. By the beginning of 1862, the numerous factories in the North—the Yankee Leviathan—were in full war production and large Federal armies were being mustered, trained, and organized. Southern military and political leaders were desperate to raise more soldiers to defend their vast rural nation that covered more than 750,000 square miles.[2]

On February 2, 1862, the Confederate War Department in Richmond

---

[1] Gresham, "A Partial History of the Walton Tigers," 660.

[2] McCurry, *Two Great Rebel Armies*, 14.

sent Georgia governor Joseph Brown a request for twelve regiments that would serve for three years or the duration of the war. Georgia had already raised some thirty-eight infantry regiments for the Cause, which, almost to a man, had been sent to the Virginia front. These "Boys of '61" had enlisted when passions were at their fullest, the first wave being after Fort Sumter and a second wave of recruits joining after the Battle of First Manassas. Most of these volunteers were in their mid-twenties, and almost three of every four were single. With fewer familial responsibilities and business commitments, they were, in the words of historian Joseph Glatthaar, the "Minutemen" of the Confederacy.[3]

Governor Brown was now calling for the "Boys of '62" to step forward. He issued a proclamation calling on all able-bodied men between the ages of eighteen and forty-five to appear on March 4 at their respective county militia parade grounds. Those volunteering on that date would receive a $50 bonus and the right to elect their own company officers. It was understood that there would eventually be conscription for those not present and that these late "volunteers" would not receive the promised bonus upon enlistment.[4]

On the appointed day of March 4, 1862, men from the countryside of Georgia appeared on the grounds near their local county courthouses. Most had come at the urging of a prominent local citizen. Stirring speeches, patriotic music, and good food created the proper mood, and soon men stepped forward to enroll their names in the muster books. For all the patriotic spirit that spring day, there was also an omen of what lay ahead. Northern general Ulysses S. Grant had captured Fort Donelson on the Cumberland River only two weeks earlier, and the latest news was that Nashville, the capital of Tennessee, had fallen to Yankee invaders. With the stakes thus raised, these new volunteers, these "Boys of '62," put pen to paper thereby committing their lives, fortunes, and sacred honor to Southern independence. They were on average slightly older than the recruits of 1861, more likely to be married, with a handful of children, and more likely to own the land they worked. The objective on that spring day was to gain at least 100 signatures on the muster roll, as this would constitute a company at full strength. With the names of each volunteer duly recorded, elections were held to select a captain, junior

---

[3] Glatthaar, *General Lee's Army*, 18, 203.

[4] Bohannon, "The Northeast Georgia Mountains during the Secession Crisis and Civil War," 131–32. On April 16, 1862, the Confederate Congress passed a national conscription act for Southerners between eighteen and thirty-five, followed by a September act for men between thirty-five and forty-five.

officers, and noncommissioned sergeants and corporals. Finally, the men of the newly formed company would choose a colorful name for their unit that would convey their spirit and patriotism.[5]

Many of the men from Gwinnett County who enlisted that day at the Lawrenceville Courthouse came at the behest of a local shopkeeper named Lovick Thomas, who had actively recruited men for the new company in the town and neighboring countryside. His efforts were rewarded when, on March 4, enough men stepped forward to form two complete companies. One hundred fourteen men from Lawrenceville and farms to the west and north of the county seat proudly designated themselves the "Gwinnett Beauregards." Not surprisingly, twenty-seven-year-old Lovick Thomas was elected captain of the company, with Darling McDaniel, Noah Brogdon, and John Medlock gaining the rank of lieutenant. A second company made up of men with homes east and south of Lawrenceville formed the "Independent Rebels." They elected Benjamin Weaver, a prominent farmer living near the Yellow River, as their captain. Sergeant George Hoyle Jones of the Gwinnett Beauregards proudly stated that his company, "was composed almost entirely of young, strong farmers, country boys, all known to each other, and all ready and eager for the front."[6]

Similar enlistment ceremonies took place in small towns across Georgia on this designated enlistment day. The volunteers from each company would elect one captain, three lieutenants, five sergeants, and five corporals from among their peers. Near the town of Decatur, in DeKalb County, the "Fowler Guards" came together for the first time. Most of these men arrived from farms located in the Indian Creek District of the county, and this new company would serve alongside the Gwinnett Beauregards in the soon-to-be-formed 42nd Georgia Regiment. On the piazza of mill owner Ezekiel Mason, located near the DeKalb courthouse, Miss Anna Davis presented the men of the Fowler Guards with their own "Stars and Bars" company flag that she had lovingly sown. It had a circle of white stars upon a field of blue and the Confederate bars of red and white. Miss Georgia Hoyle, wearing a homespun dress to show her Southern allegiance, followed Miss Davis with a patriotic

---

[5] McCurry, *Two Great Rebel Armies*, 91; Glatthaar, *General Lee's Army*, 203.

[6] Jones, "George H. Jones Recollections," 116. As well respected as L. P. Thomas was in Lawrenceville, even more people were familiar with his uncle, Henry P. Thomas. A wealthy plantation owner and politician, Henry had raised the first military company in Gwinnett County a month before Fort Sumter had been fired upon. By summer 1862, the fifty-two-year-old Henry Thomas was serving in Virginia as the colonel of the 16th Georgia Infantry.

County. Thomas's "Gwinnett Beauregards" were christened Company A, and Weaver's "Independent Rebels," would be Company B. Thirty-one-year-old Benjamin Putnam Weaver was the son-in-law of Thomas Maguire, a wealthy Irishman who owned a 1,000-acre plantation known as the "Promised Land." C Company had been organized in the county seat of Alpharetta, and they called themselves the "Milton Tigers." The Alpharetta men had elected their 6'4" county sheriff Henry Paris to be their captain. D Company was the seventh company to be organized in DeKalb County and was named the "Fowler Guards" in honor of Captain John Fowler, who had organized the first military company in the county, the Dekalb Light Infantry, back in May 1861. Companies E and F were from Newton and Jasper Counties, and G and H were from Walton County. Only Company K, from Atlanta, could say they were predominately town people.[12]

The new regiment would have three senior officers in all. Reporting to Colonel Robert Henderson would be a lieutenant colonel and a major. These two executive positions were selected by the ten captains in an election on March 20. Robert Flourney Maddox, an Atlanta store owner and former sheriff of Putnam County, was elected lieutenant colonel and quartermaster. Captain Robert Maddox's immediate promotion meant that Company K, the "Calhoun Guards," would be commanded by a twenty-four-year-old lawyer, William Lowndes Calhoun, the son of Atlanta's wartime mayor James Calhoun. The major of the new regiment would be the young and dashing William Henry Hulsey, a twenty-four-year-old lawyer from Monroe County. Since the start of the war, Hulsey had been the 1st lieutenant of the Mitchell County Independents, later Company F of the 6th Regiment. With "an ugly wound disabling him," Hulsey was given an 1861 Christmas furlough back to Monroe County. He used the opportunity to gain a transfer to the newly formed 42nd Georgia. The success of any regiment would fall to the officers and their ability to lead and communicate with one another. Fortunately, Robert Henderson had selected companies with talented and energetic leadership. These captains, in turn, bore complete respect for their new colonel.

[12] Henry W. Robinson to home, March 25, 1862, Robinson letters. Robinson maintains that Weaver made this gesture despite the wishes of his men. Margaret Mitchell made several trips to Thomas Maguire's Gwinnett County property to review his extensive plantation records from the Civil War period as research for *Gone with the Wind*; Chelsea Thomas, "Plantation and Its Owners Give Glimpse into Gwinnett's Remarkable History," *Gwinnett Daily Press*, January 5, 2012, 1.

Captain Calhoun would later describe Henderson as "a strict disciplinarian...a fine drill master...was brave but not rash, and was remarkably expert in managing his command in the field of battle."[13]

Other companies in Camp McDonald were also being organized into regiments. At the same time that the 42nd was in camp, other officers were organizing their men into the 34th, 39th, 40th, 41st, 43rd, 52nd, and 56th Georgia infantry regiments. "We are all formed into Ridgements," wrote Henry Robinson from Camp McDonald on March 25, 1862. Robinson was a thirty-two-year-old blacksmith from Jackson County and a member of Ben Weaver's Independent Rebels of Gwinnett County. Like many of the men joining the Confederate cause, he was a prolific writer whose insights and wit shine through his phonetically spelled prose.

Organizing a regiment also meant requisitioning supplies. On March 24, 1862 Captain Weaver acknowledged receiving for his ninety-five men: two wall tents, thirteen axes, nine camp kettles, nine buckets, nineteen cooking pans, four spades, two hatchets, nine ovens with spider lids, six frying pans, and ninety-five tin cups. A March requisition for cords of firewood made by each company put the overall size of the regiment at 920 officers and men, not counting Henderson's small regimental staff.[14]

The mixing of so many young men in one place led to the outbreak of infectious diseases in March and April. Most of these new volunteers came from isolated communities and had never been exposed to smallpox, measles, typhoid fever, or malaria. "The men in my regiment are all passing the ordeal of Camp sickness," wrote Weir Boyd, a Dahlonega attorney who had become the colonel of the 52nd Georgia. "Seven men have died in my regiment here," he wrote his wife, and "over a dozen have died in Col. Henderson's (42nd) regiment." Henry Robinson wrote his wife, Elizabeth, from his tent that, "Dr. OCelley [Francis O'Kelley] in our Compney ses I have the nettle rash. I have been taken powders and pills from him ever cence I have been hear, but it does me no good." On April 10, 1862, Hiram Edgar, a thirty-two-year-old sergeant from the Walton Blues (G) wrote home from "Camp Macdolar" that "our company has the Most sickness of any company in the Regiment." Edgar

---

[13] Allardice, *More Generals in Gray*, 127; "Thirty Years Ago, and Now: The Strange Changes Which Take Place in Men," *Atlanta Constitution*, October 16, 1892, 21.

[14] Request for firewood by company broke down to (A) 113, (B) 95, (C) 91, (D) 106, (E) 79, (F) 91, (G) 103, (H) 77, (I) 91, (K) 74; Special Requisition, March 24, 1862, Confederate service records of Captain Benjamin Weaver, NARA, M266, roll 63.

was "Expecting to Take the Measles any day." Service records show that examining physicians dismissed sixteen men from the 42nd Georgia for various disabilities. More importantly, some two dozen men in the regiment died from disease in their first few weeks of Confederate service. On April 10, Colonel Henderson resorted to giving nearly half the men in his new regiment fifteen day passes to return home and regain their health. A handful more died at home in the following weeks.[15]

Pay for being a Confederate soldier would vary by rank. Company officers received between $80 and $130 a month. A private earned only $11 a month, a musician $12, and a 1st sergeant $20 a month. On March 26, the ten captains signed their Captain's Commissions, which had been issued under the signature of Governor Joseph Brown from his executive mansion in Milledgeville. As deployment seemed imminent, the men realized that it would be months before they would see home again. Henry Robinson wrote his wife, Elizabeth, on March 26, "I had Ruther be at home & stay one month then to have this world and all its worth."[16]

Cumberland Gap

On March 28, 1862, General Robert E. Lee, who was serving as the Confederate military chief of staff to President Jefferson Davis, informed General Kirby Smith in Knoxville that he had ordered the 39th, 40th, 41st, 42nd, 43rd, 52nd, and 9th Battalion of Georgia infantry to East Tennessee. Back at Camp McDonald, there was a growing discontent among the new soldiers as the promised $50 recruitment bonus had not yet been paid. On April 8, Colonel Henderson telegraphed Governor Joseph Brown in Milledgeville: "Sir, Shall I order the Regiments now here to move to Tennessee forthwith under order from Confederate Authorities before they are mustered into service and bounty money paid?" The Georgia governor replied the next day that he had, "no authority to countermand the order given by the president." Fortunately for Colonel Henderson, the $50 bounty money was distributed on April 13, 1862.[17]

---

[15] Kinsland, "A Band of Brothers," 10; Henry Robinson to home, March 25, 1862, Robinson letters; April 10, 1862_letter written from Camp McDonald to wife Mary Edgar from Hiram Edgar.

[16] O'Kelley, *Dahlonega's Blue Ridge Rangers*, 15; Henry Robinson to home, March 26, 1862, Camp McDonald, Robinson letters.

[17] *OR*, ser. 1, vol. 10, part 2, 370; Governor's log, April 8, 1862; telegrams between Colonel Robert Henderson and Governor Joseph E. Brown.

Excitement had erupted on the previous morning of April 12 when Union raiders, dressed in civilian clothes and under the direction of James Andrews, stole the locomotive "the General" at Big Shanty only yards from the unsuspecting guards of Camp McDonald. The wild chase northward on the Western & Atlantic was led by the conductor William Fuller and resulted in the eventual capture of all nineteen of the raiders. Four days later, just as news of the Confederate defeat at the bloody Battle of Shiloh was reaching Atlanta, the newly trained members of 42nd Georgia boarded a train, arriving in Knoxville the next day.[18]

Upon disembarking at the Knoxville train depot, the men marched west from town for two miles, to a broad valley that had been taken over by Confederate authorities and designated Fort Van Dorn. The 40th and 52nd Georgia would establish a camp here for the next two months along the banks of Third Creek, which ran through the valley. The creek water, it turned out, was not healthy for the men, and disease—smallpox, typhoid fever, measles—ran rampant through the camp. Some 150 men would perish from just these two regiments in the coming weeks. Private Henry Robinson of the Independent Rebels (B) called it, "the damndest place you ever saw." He had talked with members of the 40th Georgia who told him that some 100 of their regiment were dead and that another 50 were in the large brick house used for a hospital that, according to Robinson, "out stinks a dead horse." Colonel Weir Boyd of the 52nd Georgia, wrote home, "Measles, mumps and fever are the prevailing diseases…a man of sympathy would be sick to see so much suffering as is seen here."[19]

The overall Rebel commander in East Tennessee, Major-General Kirby Smith, was not confident that his forces could defend this portion of the new Confederate nation. Writing from Fincastle, Tennessee, on May 3, he informed his Richmond superiors, "I shall resist the enemy's entrance into East Tennessee with all the means at my disposal, but with the people in my midst enlisted against me, and with a force of at least four to one, more efficient and better equipped, it will be alone assistance from on High that enables us to maintain possession of this department." Though pleased to have the newly trained regiments sent from Camp McDonald, he wrote, "The Georgia regiments ordered to this department were weak when reported; they have

[18] An initial suspicion was that the train had been brazenly stolen by recruits upset that they had not been paid their promised $50 bonus.

[19] Henry Robinson to home, April 17, 1862, Knox County, Robinson letters; Kinsland, "A Band of Brothers," 12.

of an Ohio volunteer regiment at the start of the Mexican War. While south of the border, he was brought into the regular army as a colonel and, because of his bravery at the Battle of Churubusco, the twenty-six-year-old Morgan was made the youngest brevetted brigadier general in Winfield Scott's American army. In April 1862, Morgan's 7th Division, made up of four green brigades of infantry was fourteen miles from the Cumberland Gap, camped at Cumberland Ford in Kentucky.[25]

As commander of the 42nd Georgia, Robert Henderson would report to Colonel James Rains. A graduate of Yale University, the thirty-nine-year-old Rains was a former Nashville lawyer and newspaper editor. He had skillfully defended Cumberland Gap over the previous winter and knew the topography of the area well. With the arrival of Stevenson in March, Rains became one of his two brigade commanders. In all, Carter Stevenson, as the new overall commander, would have 6,700 untried Southerners to defend this strategic outpost.

It was Rains's new brigade, defending the eastern end of the Confederate mountain entrenchments some 600 yards from Fort Pitts, that the Federals first assaulted on May 7, 1862. Only one day after completing a march of seventy-five miles from Knoxville to the Cumberland Gap, the men of the 42nd Georgia were under attack. Private Robinson wrote that "the enemy attacked us and we kept up the fire for 2 to 5 hours with only three of our men being wounded." Here, on a Tuesday, on the Tennessee-Kentucky border, the men of the 42nd "saw the elephant," a euphemism for their first occasion under live fire. The same could be said for the blue-clad soldiers who were advancing toward them. The armed Federal reconnaissance was made by the brigades of Samuel Carter and John DeCourcy. The stout Rebel resistance convinced the overall Federal commander, George Morgan, "that the position could only be carried with immense loss of life," and he quickly ordered a withdrawal to the Cumberland Valley.[26]

Lieutenant Thomas Wright, a staff officer under Colonel Marcellus Stovall in the 3rd Georgia Battalion, was a convalescent in the local field hospital that morning recovering from pneumonia. Wright was also the special correspondent of Atlanta's *Southern Confederacy* newspaper, formerly the *Gate City Guardian*, and wrote under the pen name "T. D. W." His May 8

[25] Ibid., 333–34; W. C. Davis, "Cumberland Gap, 1862," 26.

[26] Henry Robinson to home, May 7, 1862, Cumberland Gap, Robinson letters; field report of General George W. Morgan, 7th Division, Army of the Ohio, June 22, 1862, *OR*, ser. 1, vol. 10, part 1, 58–62.

dispatch reported that the Federals, "had possession of a skirt of timber some 800 yards from our main battery." Detachments from the 29th North Carolina and the 42nd Georgia were posted to check their advance. That afternoon, Lieutenant Wright could hear the intensity of fire pick up as artillery joined the skirmish. Minie balls seemed to whistle about everywhere. Henry Robinson wrote that, "one ball fell a foot or two from me. It came through Captain Thomas' tent and Struck at the dore of Capt. Weaver's."[27]

The Federal soldiers had better rifles, but that did not stop the men of the 42nd from "peppering away" with their older muskets. Nightfall brought a close to the exchange and the Federal brigade finally withdrew. Rumors of a "big battle" on Pinnacle Mountain above the Cumberland Gap soon filled the newspapers. T. D. W. wrote, "They had it reported in Knoxville that we repulsed the enemy eleven times, and that the dead were laying around in heaps, when in fact we only repulsed them once." Like the Confederate defenders, the Yankees were untried and new to the region, and they did not try to press the well-entrenched Rebel troops. Private Robinson only saw the Yankees "carry off 15 wounded or dead."[28]

Captain L. P. Thomas would remember after the war, "how white and clean our new army tents looked, high up on the mountainside." Each officer had their name stenciled in bold Roman letters on the side of their tent. "We all had well-filled trunks, too," remembered Thomas, "and our mess boxes were models of beauty, and they were generally well filled." The Rebels had cut down all the trees on this mountain for a mile in front of their defenses. "You can look as fur as your eyes can see," wrote Henry Robinson from his picket post on Pinnacle Mountain, 900 feet above Cumberland Gap, "and you cant see nothing but mountains." Standing near one bastioned fort atop Pinnacle Mountain was a stone marking the intersection of three states. Robinson marveled, "I can be walking Round one tree and walk in three states—Caintuckey, Virginia, and Tennessee."[29]

---

[27] Thomas D. Wright, "Our Special Cumberland Gap Correspondence," *Atlanta Southern Confederacy*, May 8, 1862, in Stovall scrapbook; Henry Robinson to home, May 7, 1862, Cumberland Gap, Robinson letters.

[28] Wright, "Our Special Cumberland Gap Correspondence," *Atlanta Southern Confederacy*, May 8, 1862; Henry Robinson to home, May 7, 1862, Cumberland Gap, Robinson letters.

[29] L. P. Thomas, "The Battle of Resaca: The Part the Forty-Second Georgia Took in It," *Atlanta Constitution*, February 2, 1890, 12; Henry Robinson to home, May 7, 1862, Cumberland Gap, Robinson letters.

Both commanders found it difficult to supply their soldiers in this remote part of Appalachia, and soldiers on either side had no choice but to forage the fields of the local farmers. E. T. Petty of the 42nd Ohio described the country around Cumberland Gap as "genuine backwoods" that was, "well adapted to moonshine enterprises." He believed that the poorly educated citizens "were about equally divided on the question of the war, and each side was extremely zealous in its views." Owen Hopkins, also of the 42nd Ohio, recalled, "At times, our men would be gathering green corn from one side of a field while the Rebels were at the other side, and skirmishes occurred daily." Stevenson's Confederate soldiers spent their time atop Pinnacle Mountain extending the log-and-earth field fortifications under the direction of veteran engineer Albert Miller Lea.[30]

Fortunately for the Georgia regiments assigned to East Tennessee, a bounty of new recruits enlisted in their respective Georgia communities on May 12, 1862. Most of these men were meeting the final deadline that allowed them to both avoid conscription and to join a regiment of their choosing. By missing the previous March 4 deadline, however, these Georgians had forsaken a $50 bonus and the right to participate in the election of company officers. A persuasive recruiter was 2nd Lieutenant Elias Woolley, who signed up thirty-seven men in Atlanta for his Newton Rifles (F). A handful of these new recruits possessed military experience, having recently completed their six-month obligations in regiments belonging to Governor Joseph Brown's Georgia State Troops. In October 1861, these early volunteers had marched off with shotguns and hunting rifles to defend the coastline and, once in Savannah, had been given rudimentary military training. With their terms of service now expired, most of these militia soldiers were looking for Confederate regiments to join for the duration of the war. The 42nd Georgia drew several soldiers from the "Sunny South Guard," formerly Company H of the 6th Regiment of Georgia State Troops.[31]

Back in Tennessee, supplies for the ill-equipped regiment began to slowly arrive. On June 5, 1862, Lieutenant H. B. Mikel, the acting commander of the Walton Blues (G), signed for 46 percussion muskets, 113 out-of-date flintlock muskets, 159 cartridge boxes and waist belts, 44 cap

---

[30] Petty, "The Cumberland Gap Campaign," 120; Hopkins, *Under the Flag of the Nation*, 31; Stevenson, *Cumberland Gap*, 10.

[31] In October 1861, 8,000 men had volunteered for a six-month term in the Georgia militia. They had been organized into ten companies in three brigades (Bragg and Scaife, *Joe Brown's Army*, 2–3).

pouches, 135 haversacks, and 15 bayonets with scabbards. Good news came to the Milton Tigers (C) with the arrival of new uniforms that they had ordered from the Roswell Mill factory back in their home county. Private Robinson considered them well made and was fortunate to be able to buy a pair of pants and a frock tail coat for himself. He explained, "I give $8.25 cents fur it Redy made & you cant buy the cloth fur that money." Henry Robinson also attempted in June to transfer to Captain Yeizer's artillery company. Robinson thought there were several advantages to artillery service: "I git 13 dollars per month and doant have to Stand gard & can get everything hauled and Ride my self." Unfortunately for the irascible infantryman, neither Captain Weaver nor Colonel Henderson would approve his request.[32]

Life was never easy for the defenders on the windswept mountains above Cumberland Gap. "We have drawn up a piece of Riting and all of the company signed it," wrote Henry Robinson of the Independent Rebels (B), "that if one in the company dies or gets kild, to send him home & let him be buried in his native soil. Let it cost what it will." The two Georgia regiments, the 42nd Georgia and the 3rd Georgia Battalion, constituted half of Rains's Second Brigade. Lieutenant Thomas Wright of the 3rd wrote of the 42nd, "I met many familiar faces and, in my judgment, the 42nd Georgia is an ornament to their State." Wright asked Colonel Robert Maddox of the 42nd what he thought of the Confederate camps at Cumberland Gap. Wright reported, "the colonel [Maddox] says the culinary department is not as good as expected, but he makes himself comfortable."[33]

Picket duty was hazardous for both sides. The Federals contended with nighttime bushwhackers who crept up close enough to take shots at the posted guards. On May 21, Private Robinson's picket duties were interrupted by "the hardest hail storm I ever saw." With trees falling all around them, the sentries, "got as wet as rats, but we Stoode it without a word." On June 1, Henry Robinson reported that "General Stevenson says that the 42nd is a scandal to this Brigade becas we will Shoot at a Fuss," rather than first identifying what in the darkness was creating the noise. Robinson speculated that he and his fellow soldiers were shooting because "it is So dark that We cant see our hands before us." Captain L. P. Thomas of the Gwinnett Beauregards

[32] "Requisition for Ordnances and Ordnance Stores for the use of detached Company," Confederate service records of Henry B. Mikel, NARA, M226, roll 42; Henry Robinson to home, June 28, 1862, Bean Station, Robinson letters.

[33] Henry Robinson to home, May 7, 1862, Cumberland Gap, Robinson letters; Wright, "Our Special Cumberland Gap Correspondence," *Atlanta Southern Confederacy*, May 8, 1862.

(A) remembered that Stevenson may have had a point. As he recollected, in 1890, his greenhorn soldiers, in their first nights on picket duty, fired at advancing lightning bugs. The men, it seems, imagined themselves under attack from Yankees advancing with the help of lighted torches. "One gun was fired and then another, and soon the entire line was letting fly the missels of war into the imaginary enemy," chuckled Thomas. "We fell back, not whipped exactly; but badly demoralized."[34]

Henry Robinson lamented that the guardhouse had thirty or forty Georgians under guard every day, "fur steeling & Running out in the country & giting drunk." The blacksmith from Gwinnett County was also convinced that, at the end of the day, the Georgians were better fighters than the "G.D. Tennesseans." This was debatable as the 11th Tennessee, under the capable leadership of Colonel George Gordon, was an excellent regiment raised from Nashville and counties to the west of the state capital. The same could not be said for the 36th Tennessee, which was made up of native East Tennessee troops. Desertions to the Federal army and unauthorized trips home plagued this unit to the point that, in May, an exasperated Kirby Smith ordered the regiment off Cumberland Gap with orders to go by train to Savannah. This did not happen because a few weeks later all Tennessee regiments were reorganized under the Conscript Act. The 36th was disbanded, and the soldiers sought other Tennessee regiments.[35]

This was the backdrop, when, on June 7, four companies of the Georgia 42nd and some 3,000 men of the brigade were drawn up on three sides by the headquarters of General Stevenson. A cavalry unit closed the square. The Southerners watched as six deserters were tied to a stack of arms and each given thirty-nine lashes with cowhide on their bare backs. They were then drummed out of the Southern camp under the tune of "Yankee Doodle."[36]

On June 9, Ira Lawson Gunter, one of Elias Woolly's late recruits, wrote his wife, Ann, from Knoxville. When the war had begun, the thirty-two-year-old Gunter was practicing medicine in McDonough, Georgia. He had formerly been apprenticed to Dr. Levi Brantley, an expert in using medicinal

---

[34] Henry Robinson to home, May 25, 1862, Cumberland Gap, Robinson letters; L. P. Thomas, "The Battle of Resaca: The Part the Forty-Second Georgia Took in It," *Atlanta Constitution*, February 2, 1890, 12.

[35] Henry Robinson to home, June 1, 1862, Cumberland Gap, Robinson letters. Gordon would soon be a general and command his own brigade in Cheatham's Division of the Army of Tennessee.

[36] Deuteronomy 25:3 states that no more than forty "stripes" are to be administered as punishment. Jewish Talmudic law later reduced flogging to thirty-nine lashes.

plants and herbs. After Fort Sumter had fallen, the young physician had written his brother-in-law about the upcoming war. Gunter was angry with the "Chicanery of Southern hot heads and fanatics," who had driven his region to declare for secession from the Union long before the citizens were financially or militarily ready to defend their new nation. In the end, though, the doctor declared, "I am Southern born, Southern raised and none but Southern blood courses my veins, and every impulse of my poor heart is with my brethren of the South." Now he was in a camp waiting to join his regiment with the other new recruits. He wrote Ann, "I am confined here as a volunteer and cannot go on to the Regiment until we all go." Sixty-nine men from his group had been detached the previous day to guard Loudon Bridge or serve fatigue duty in Morristown. Ira Gunter would prove to be a welcome addition to the 42nd in eastern Tennessee, serving principally as a hospital nurse helping the many desperately sick soldiers in the regiment.[37]

In the middle of June, Federal commander George Morgan began executing his well-conceived strategy of flanking Carter Stevenson's defenses at Cumberland Gap. Morgan, acting something like Hannibal of Carthage, personally led two of his four brigades along a narrow bridle path thru a small notch or defile in the Cumberland Range at Rogers's Gap. Moving the artillery train became a herculean task. According to Captain Jacob Foster of the 1st Wisconsin Battery, "Two hundred men from the infantry were detailed to assist in overcoming the steep ascents and descents, which was to be done by ropes and pulleys." By June 15, the rear-guard of Morgan's force, the 33rd Indiana, was marching into Powell Valley "to the air of 'Dixie.'"[38]

The Federal commander had managed to mystify the Confederates atop Pinnacle Mountain by getting 8,000 of his troops within a dozen miles of their undefended rear. Carter Stevenson's outnumbered defenders now faced two Federal brigades to their front and two more behind them. "Had Kirby Smith been personally in command we should have had a battle," wrote George Morgan, "but it was evident to me that the general in command felt uncertain of the ground upon which he stood." The Federal commander believed, "[James] Rains, a gallant and dashing officer, was in our immediate

[37] Ira Lawson Gunter to home, June 9, 1862, Knoxville, and undated (May–June 1862) letter possibly to his brother-in-law, J. M. Williamson, Latimer collection.

[38] Field report of Captain Jacob Foster, 1st Wisconsin Battery, *OR*, ser. 1, vol. 10, part 1, 64–67; W. C. Davis, "Cumberland Gap," 26–27; Field report of General Morgan, June 22, 1862, *OR*, ser. 1., vol. 10, part 1, 59.

front, and [Seth] Barton was a few miles from our right flank, while Stevenson was in supporting distance." To make matters worse for the Confederates, a Union division under General James Negley had punched into central Tennessee and was threatening Chattanooga. Fearful that the Yankees would soon cut Stevenson's supply line, Kirby Smith ordered his division commander to make a tactical withdrawal from the mountain fortress and retreat toward Morristown, Tennessee.[39]

Colonel Rains's 2nd Brigade was the last to leave the defenses of Cumberland Gap and was given the assignment of covering the hasty withdrawal. According to Private Robinson, the men cut their tents to pieces and destroyed their cooking vessels before departing. They would be traveling as light as possible. Each man was allowed to carry two pairs of pants, two slips, two shirts, and one blanket. "Old Tom," their heavy 32-pounder smoothbore gun, said to have served in the Mexican War and later brought to the Gap from Manassas, Virginia, was among the cannons spiked and thrown down the mountain. Captain Lovick Thomas was not there to witness the withdrawal because he had left Pinnacle Mountain on June 12, leading fifty men from companies A and B of the 42nd on a scouting expedition to Big Creek Gap. "We had to make his company up to 50 & it took 14 of us to do it," wrote Henry Robinson of Company B, "If I hadent Volenteard, I would had to went anyway." Thomas and his detachment camped at the base of the natural opening thirty miles southwest of their fortifications on Pinnacle Mountain. They stood watch for Union cavalry until June 17 when they heard news that Stevenson was abandoning the main Confederate position above Cumberland Gap.[40]

On June 22, a jubilant George Morgan reported to Union General Don Carlos Buell that "the Gap is ours, and without the loss of a single life." The Stars and Stripes now floated atop the "American Gibraltar." Construction immediately began on seven Union forts that faced south toward the retreating Confederates. Soon thereafter, hundreds of men from East Tennessee, bringing their wives and children with them, headed to the Gap to declare their loyalty to the Union. These mountain men were organized into regiments and mustered into the Federal Army. According to Private Owen Johnston Hopkins, these new recruits were awkward and uncouth in appear-

[39] Field report of General Morgan, June 22, 1862, *OR*, ser. 1, vol. 10, part 1, 60–61.

[40] Provine, "The Legend of 'Long Tom' at Cumberland Gap," 260; Henry Robinson to home, June 28, 1862, Bean Station, Robinson letters.

ance and proved difficult to drill and train into soldiers. It was not long, however, before Morgan faced the same issues that had plagued Stevenson when he controlled the Gap; namely, how to supply his men with provisions in such a remote and barren region. "The citizens brot large quantities of peaches and other fruits to the picket posts for sale," recounted E. T. Petty of the 42nd Ohio, "but could not be prevailed upon to accept greenbacks in Payment."[41]

Carter Stevenson would stop his command on the other side of the Clinch Mountain Range, in the village of Bean Station, a valley town about ten miles west of Morristown. Seth Barton's brigade was placed within supporting distance at the village of Rutledge. Here, the division was close enough to the rail line to come to the defense of Knoxville if necessary. After marching all night and wading across two rivers, Thomas's small contingent reached the new Confederate lines at Bean Station and rejoined their brigade. They had sent the company wagon with their knapsacks and supplies on to Morristown but, to their consternation, would later learn that the wagon had broken down en route and the escort soldiers had chosen to burn it to the ground. "We have but two frying pans and one skillet to our compney," wrote Henry Robinson. The soldiers in Stevenson's Division worried about the implications of their retreat, but they were hopeful that a pitched battle with Morgan's Yankee Division in the open fields of Powell's Valley was soon in the cards. On June 19, Captain Temple Cooper of the 52nd Georgia wrote home, "We thought here a while that we should fall back to Chattanooga and perhaps further south, but now the impression is that we will die or hold the T. and Va. Railroad which I fear will cost us the loss of many lives in battle with disease."[42]

A Visit from Jennie Thomas

On July 14, 1862, Jennie Thomas of Lawrenceville received a letter from her husband, Captain Lovick Thomas, asking her to visit him in Tennessee. The decision to make the trip would be hers. Two days later Jennie wrote in her diary, "I am at a loss what to do. There are various reasons for and against.

---

[41] Field report of General Morgan, June 22, 1862, *OR*, ser. 1, vol. 10, part 1, 61; Hopkins, *Under the Flag of the Nation*, 30; Petty, "The Cumberland Gap Campaign," 120.

[42] Henry Robinson to home, June 22 and June 28, 1862, Bean Station, Robinson letters; Temple Cooper to home, Bean Station, TN, June 19, 1862, Winsor, letters of Temple Franklin Cooper Family, 6.

a grocery store in the county seat of Eatonton.[46]

On January 22, 1857, Lovick Thomas married Jemima Jane "Jennie" Peeples in Lawrenceville, Georgia. They were a striking couple. He was tall, lean, and outgoing, and she was lovely to the eye, possessed a quick wit, and was deeply rooted in her Methodist faith. Their temperaments, intelligence, and work ethic were suited to the mercantile trade, and it was not long before Lovick owned a half-interest in a general store. In a 1904 *Gwinnett Journal* newspaper story, Lovick wrote that in 1856, "I married one of Gwinnett's purest and brightest daughters (Jennie Peeples), and soon thereafter at the brick store, on account of the great manufacturing institution then in your enterprising little city, it came to be known as the factory store. At this place, I commenced a mercantile business under the style of Wilson & Thomas." His business partner was Joshua S. Wilson, a South Carolinian who was ten years his senior. Lovick's financial stake in the new store had come from his "his western gains," and he saw his investment as "the nucleus of a fortune."[47]

Their mercantile destiny depended upon the economic success of the Lawrenceville Manufacturing Company, one of the South's most substantial cotton spinning concerns. Located in a four-story, 40,000 square foot building, it employed, in 1854, eighty operatives who worked 30,050 spindles and eighty looms. By 1857, there were twenty-seven worker homes built around the mill, giving it the feel of a factory town and the impression that Lovick & Wilson were operating a factory store. The mill, unfortunately, struggled financially. Lawrenceville was not an ideal location for a large factory. The enterprise was sixteen miles from the closest railroad in Stone Mountain, and the spindles and looms had to rely on steam power because no river ran through Lawrenceville. Burdened with debt, the mill fell into bankruptcy in 1857. A local lawyer named James Simmons bought the operation in 1858 and renamed it the Gwinnett Manufacturing Company. Simmons had the foresight to hire Enoch Steadman, an experienced New England manufacturer, to run the mill. Steadman, who would soon become a vocal secessionist, proved to be an inspired choice as he turned the mill into a profitable and growing enterprise.[48]

---

[46] Huff, *Four Families*, 1, 62; Louisiana, New Orleans Passenger Lists, 1820–1945; "Col. Lovick P. Thomas," in *Memoirs of Georgia*, 1:943; "List of all passengers taken on board the Steam Ship *Danl Webster*," June 1855, NARA, M259, roll 42.

[47] Thomas, "Reminiscences of Forty-Second Georgia," 14; "Col. Lovick P. Thomas," in *Memoirs of Georgia*, 1:943.

[48] Holman, "Factory Town," 58–61.

In the Federal Census for Gwinnett County, taken in June 1860, L. P. Thomas is listed as a twenty-five-year-old merchant. He owned two slaves, probably wedding gifts from the Peeples family, and recorded his personal property at $3,500. At home was twenty-four-year-old Jennie, a two-year-old daughter, and an eight-month-old son. Lovick Thomas had much to be proud of, and, as he later recalled about his factory store, "I was enjoying prosperity in this business when the tocsin of war was first sounded."[49]

At the end of a long deliberation, Jennie chose to make the journey to East Tennessee to visit her husband. With daughter Mat, age four, and son Bud, age three, she reached Bean Station on August 3. Lovick could only stay one night with his family before he was called back. Jennie "found him much fleshier than when he left home. The children seemed to know him." She would have to wait five days for her soldier to return from Clinch Mountain. He came back on the eighth and they spent time together until August 15. There was unwelcome news to share as Lovick's brother-in-law, Colonel Archibald Hunt, husband of Lovick's sister Mattie, had been killed during a cavalry raid in Kentucky. The dashing Arch Hunt was the colonel of the 16th Georgia Cavalry Battalion and was riding under the command of John Hunt Morgan when he was mortally wounded on July 7 in a raid on Tompkinsville, Kentucky.[50]

T. D. W., writing in the *Southern Confederacy* from Morristown, warned families not to visit the men on the front. "I have seen the wives of soldiers arrive here by the cars," he wrote in an August 8 dispatch, "some of whom I know, and here [Morristown] they found the only consolation, that conveyance to the camp could not be had—the distance too far to walk, and hardly a house to shelter them in if they did go." Jennie Thomas had somehow managed, but it was a difficult journey back to Georgia. She and the children reached Lawrenceville by way of Morristown, Strawberry Fields, Knoxville, and Atlanta. Mindful of every war-time expense, Jennie wrote, "My trip cost about $82.00."[51]

---

[49] The 1860 Federal Census lists Jennie as "Jenna," daughter Martha Caroline as "Caroline," and son Lovick III as "William" Thomas family members called Martha "Mattie," and Lovick III was known as "Bud."

[50] Allardice, *Confederate Colonels*, 204.

[51] Wright, "Our Special Cumberland Gap Correspondence," *Atlanta Southern Confederacy*, August 8, 1862; Jennie P. Thomas, diary, August 8, 1862, in Huff, *Four Families*, 75.

Battle of Tazewell, Tennessee

On July 14, 1862, General Kirby Smith informed "His Excellency Jefferson Davis" of the military situation in East Tennessee. "I have organized my command into two divisions," he wrote the Confederate president. "General Stevenson, with four brigades of infantry and one of cavalry, occupies the line of the Clinch. His force of 10,000 effective is mobilized, well equipped, and efficient. General Stevenson is an officer of worth and merit, and has been a treasure to me in this department." In Chattanooga, Smith had placed another 10,000 men under General Henry Heth to keep an eye on the Federal Corps of Don Carlos Buell. Smith also ordered John Hunt Morgan with 1,300 troopers into Kentucky, and another three regiments of cavalry under Colonel Nathan Bedford Forrest into Middle Tennessee. Smith's Confederates were beginning to move to the offensive.[52]

On July 17, Ira Gunter headed his letter to his wife, Ann, "In the hospital at Bean Station, TN." Knowing that she would be worried about his health, he quickly added that he was not a hospital patient but was working as a nurse, with a salary of $17.50 per month. "I have charge of one room, and have under my care at present, nine patients all from our regiment." The former McDonough physician was not only serving the sick but also exempt from guard service, out of the weather, and able to "escape those rough marches that we sometimes take over these mountains." He had a high opinion of the two surgeons serving the 42nd Georgia, Dr. William S. R. Hardeman and Dr. John S. Wilson, but a poor opinion of the other brigade surgeons. Gunter explained, "There are four of them and I could be busy only half my time and do more than every one of them." Not only were the surgeons lazy, unfeeling, and incompetent, the four "jackanapes" were drunks. Gunter wrote with disgust, "I saw them drink four bottles of fine brandy that was intended for the sick yesterday morning before breakfast."[53]

Ira Gunter had hopes he would soon be appointed a permanent medical steward, but his regiment was also on the move and he might soon be called to the front. The 42nd was being temporarily reassigned to the command of General Seth Barton, "one of the most inhuman and unfeeling wretches ever born." Ira wrote Ann that he dreamed of her and "being at home pursuing my old avocation. I dreamed that peace and plenty was reigning in our land." The morning dawn brought back the reality of war, but Dr. Gunter hoped

[52] Kirby Smith to Jefferson Davis, July 14, 1862, *OR*, ser. 1, vol. 16, part 2, 727.

[53] Ira Gunter to Ann Gunter, July 17, 1862, Bean Station, in Latimer collection.

that one day his dream would be realized and prayed that until that day, "may we all be preserved from dangers, sickness, and death, by the righteous hand of Providence."[54]

"We have tite Rools over us now," wrote Henry Robinson on July 28. Furloughs were canceled as the expectation of combat increased. After six months of Confederate service, what had been cordial relationships between longtime neighbors were now firmly replaced by military discipline between officers and soldiers. "The order was Red out to us on dress parade the other day that we will all have to pull off our hats when we go to the colonel or general," an exasperated Robinson explained, "and you know that is one thing I won't do. I will see you in hell before I pull my hat off to any man."[55]

On August 2, Rains's Brigade was drawn up in a three-sided square to witness the execution of one of its Tennessee soldiers. The condemned man arrived in a wagon and was riding on top of his coffin. "To see a man a seting on his own coffin, it was awful," Henry Robinson wrote. Colonel Rains read the charges; the actions of the soldier had caused the death of a lieutenant. "He was ordered to set down on his coffin & was blindfolded," wrote Private Robinson, "and twelve men to shoot him but onley six to fier at a time & half of them was blank cartridges." Every man present must have had a different set of emotions. "I do hope and trust that Georgia will never have to Regret the deth of a soldier in that way," wrote Henry. He also could not help but wonder why the condemned soldier had made no attempt to run. "If it had been a case of mine," he speculated, "they would a had to a Shot me on the wing."[56]

On a much lighter note, the thirty-six-year-old Robinson believed it was time for his regiment to depart Bean Station. On August 3, he declared, "If we stay heare much longer, in about nine months from now, thare will be more little Georgians a squalling then you can shake a stick at." He comforted his wife that "I doant intend to be cald Pap so fare from home, but thare is a great many a pitching in."[57]

Cumberland Gap's strength was also its weakness. Situated in a "bleak, barren, untilled mountain region," George Morgan's Federal Division was 130 miles from their nearest supply base in Kentucky. Short on provisions, the Yankee soldiers were soon on half rations. In the first week in August,

---

[54] Ibid.

[55] Henry Robinson to home, July 28, 1862, Bean Station, Robinson letters.

[56] Ibid., August 2, 1862.

[57] Ibid., August 3, 1862.

Kentucky and 16th Ohio. "Most of the boys in the 40th Georgia procured a No. 1 overcoat and blanket apiece, and a large number of knapsacks," reported a member of the 40th Georgia, writing under the pen name "Stilesboro" in the *Southern Confederacy.* "It is not a very pleasant reflection to know," wrote Captain Davidson of the 14th Kentucky, "that they got several blankets, knapsacks, canteens, haversacks, &c., which formerly belonged to Uncle Sam's boys."[63]

J. F. Belton, Kirby Smith's assistant adjutant-general, wrote Carter Stevenson from Knoxville that the general "instructs me to say that the good conduct displayed on that occasion (Tazewell) was only what he had reason to expect from troops in the state of discipline to which you had brought them." An exchange of prisoners soon commenced between generals Morgan and Stevenson. The Confederates returned one captain and fifty privates in exchange for one lieutenant-colonel (George Washington Gordon of the 11th Tennessee), one lieutenant, and nine privates.[64]

Death and Disease

The men of the 42nd Georgia were now veterans of sorts. Their experiences in the mountains of East Tennessee in 1862 had been difficult, but nothing like those of Georgia soldiers who were fighting in the Army of Northern Virginia. The Georgians in Robert E. Lee's army had spent 1862 fighting the Army of the Potomac in many large-scale battles in Virginia and Maryland. The men of the 42nd Georgia, by contrast, were pawns in a cat and mouse game of who could maneuver best to control the strategic Cumberland Gap. Neither side felt they had sufficient manpower to aggressively confront their opponent nor hold Pinnacle Mountain in the face of a determined siege.

It would be measles, not battle casualties, that had from the beginning, reaped a significant toll on the readiness of the 42nd Georgia to fight. On June 5, Lovick Thomas's Gwinnett men lost their thirty-two-year-old lieutenant Noah Brogdon, who died of the mumps in a Knoxville hospital. Between May and August, Thomas lost six privates to disease, mostly at the Fair

[63] Stilesboro, "Battle of Tazewell," *Atlanta Southern Confederacy*, August 17, 1862, 1; Davidson, "Battle of Tazewell."

[64] J. F. Belton to C. L. Stevenson, August 9, 1862, *OR*, ser. 1, vol. 16, part 2, 748; George Morgan (at Cumberland Gap) to J. B. Fry, August 8, 1862, *OR*, ser. 1, vol. 16, part 2, 295.

Grounds Hospital in Knoxville. It was a similar story in the other nine companies of the 42nd Georgia. "Thare is 2 or 3 dying every weak out of the regiment," wrote Henry Robinson on August 3. Officers were also afflicted with disease; Captain Enoch McCollum of the Walton Blues (G) died of measles at Bean Station on July 7, and Captain Jett Mitchell of the Walton Tigers (H) resigned on July 21 being no longer able to take the field.[65]

Colonel Hiram Parks Bell of the 43rd Georgia described a difficult transition from "the habits and comforts of home-life, to the exposure, privation and duties of life in the camp." The 43rd had departed from Camp McDonald for Tennessee at the same time as the 42nd Georgia. As Bell wrote in his 1907 autobiography, "Measles, flux, dysentery and brain fever attacked the troops; some died and nearly all, were more or less, sick." On June 20, Kirby Smith had written Joseph E. Brown, governor of Georgia, that the Georgia regiments under his command "have now been reduced, chiefly by sickness, to an average of about 400 men. Can you not take measures for adding to them from conscripts and by calling upon all absentees who are now able for duty to rejoin their regiments, so that the muskets be kept constantly in use."[66]

Poorly supplied, the Georgia men stationed in remote East Tennessee had difficulty maintaining their good health. William Looper of Dawson County, a member of the 52nd Georgia, wrote home, "We have been fed very poorly during the last month. Sometimes we have been without food for three days at a time and hardly ever have half enough to eat. Part of the time we have bread and no meat, then meat and no bred, then neither."[67]

In a letter dated July 19 from Upson Spring, Tennessee, Private Francis O'Kelley of the Independent Rebels (B) wrote the wealthy plantation owner Thomas Maguire, back in Gwinnett County, about the health of Maguire's son-in-law, Captain Benjamin Weaver. O'Kelley, who had a medical background, wrote,

> The Captain thinks he is doing very well although his bowells—they are still running off at times and the Surgeons have nothing to give him for it only opium and that only checks them for a few hours at a

---

[65] Confederate service records of Enoch McCollum, NARA, M226, roll 40; Confederate service records of Jett T. Mitchell, NARA, M226, roll 43; Henry Robinson to home, August 2, 1862, Bean Station, Robinson letters.

[66] Bell, *Men & Things*, 83; General Kirby Smith to Governor Joseph Brown, June 20, 1862, *OR*, ser. 1, vol. 16, part 2, 694.

[67] Lane, *"Dear Mother: Don't Grieve about Me,"* 179.

> time. They sent him up a little port wine last night and I am giving him some of that and hope it may have a good effect on his bowels. He is taking some elixir vitriol as he is inclined to swet too much at knights and he has some relish for light diet. And he wishes you to send Dr. Wm. Bond to see him just as soon as you receive this and tell him to bring what medicine he thinks necessary in his case and lemons and loaf sugar and one quart of good french brandy, for such things can't be had here.

Mr. Maguire immediately dispatched Dr. Bond to East Tennessee.[68]

In fall 1862, "camp fever" swept through the 42nd Georgia, resulting in the deaths of twenty-seven men from the regiment in the Confederate hospital at Bean Station. On September 12, Private Isiah Woolley of the Newton Rifles (F) wrote Martha Trimble from a camp on the Powell River. Martha was the mother of George Trimble, another private in the same company. "Mr. W. R. Hick, a member of our company, has just returned from Tazewell this evening," he wrote, "and informed me he has seen your son George." Dr. Richardson ran the hospital and told Hicks that there "is not any chance for him to get well or live but a few days." Richardson was correct in his diagnosis; George Trimble died three days later. This was one of many sad letters coming from East Tennessee to the homes of Georgia soldiers. The 42nd Georgia would lose an average of one soldier a day to disease in 1862.[69]

---

[68] F. C. O'Kelley to Ben Weaver, July 19, 1862, Upson Springs, Tennessee, Weaver letters.

[69] Isiah Woolley to Martha Trimble, September 12, 1862, in Bond, ed., *Civil War Letters*.

Chapter 2

Kentucky Heartland Offensive

"We have had the hardest time I reckon of any army in the Confederacy. We marched 800 miles or more through heat and dust, with the worst water ever was drunk. Part of the time on half rations, and sometimes not time to cook it."
—Lewis Stovall, a sergeant in the 43rd Georgia[1]

Heck Thomas in Virginia

On August 8, 1862, Jennie Thomas wrote in her diary that she had returned to Lawrenceville from her Tennessee trip only to find "Pa and Henry gone to Virginia and Ma speaking of moving to Oxford." In this wartime setting, Jennie would have her hands full with the strong personalities in her husband's family.[2]

"Pa" was Lovick Thomas, Sr., who had returned from the gold fields of California to run the Etowah House, a hotel in the steamboat town of Rome, which sat on the Coosa River near the border with Alabama. When the Civil War began, the fifty-year-old Lovick was recruited by his younger brother, Edward Lloyd Thomas, to be quartermaster of the 35th Georgia. Edward was the colonel of the 35th Georgia and on his way to becoming a brigadier general, leading a division in Robert E. Lee's army.

By June 1862, the 35th Georgia was a proud part of A. P. Hill's Light Division. Captain Lovick Thomas Sr. had, a month earlier, helped carry General Joseph E. Johnston from the field after he was severely wounded at the Battle of Seven Pines. Seeing his regiment deficient in field officers, Lovick volunteered to help lead the men in a frontal charge at the Battle of Mechanicsville, the first major engagement of the Seven Days Battle. The elder Lovick earned the moniker the "Fighting Quartermaster" for his actions that day, but he also took a minie ball through the right lung. He would have died

[1] Lewis Stovall to home, October 25, 1862, Forsyth Historical Society collection.
[2] Jennie P. Thomas, diary, August 8, 1862, in Huff, *Four Families*, 75.

where he fell had he not been assisted off the field by Chaplain George Yarborough, another man who shared deep connections to Oxford, Georgia.

"Ma" was Martha Ann Fullwood Bedell Thomas, mother of Lovick, and wife of Lovick, Sr. She was the younger sister of Sarah Vivian Fullwood, the grandmother of the famous Sidney Clopton Lanier. This relationship made the younger Lovick Thomas a first cousin once removed to the famous Georgia poet. Twenty-four-year-old Martha had married Lovick, Sr. in Watkinsville in 1834, two years after the death of her first husband, Pendleton Bedell. She brought land, fifteen slaves, and three young sons to the new marriage. Martha and Lovick would have five children of their own, with fifteen years separating their first-born, Lovick, from the last child, Henry.[3]

"Oxford" was the little village of Oxford, located a few miles north of Covington. In 1837, Lovick's grandfather, Edward Lloyd Thomas, had marked off 125 heavily wooded town lots for the new town of Oxford. The village was to be the home to the new Methodist college of Emory. Edward was the son of a Revolutionary War soldier and had come to Georgia from Virginia when he was nineteen. In his lifetime, he was a successful surveyor, land speculator, farmer, and earnest Methodist minister. As a surveyor, Edward Thomas had earned praise for his precise survey that established the Alabama-Georgia border north of the Chattahoochee River in 1826 and, two years later, for planning and laying out the new town of Columbus on 1,200 acres by the fall line of the Chattahoochee River.[4]

As payment for his services in laying out the town of Oxford, Edward Thomas was awarded four town lots. The ethos of the place appealed to the master surveyor, and he selected one particular lot on the highest hill in Oxford, at the corner of Wesley Street and Watson Street, to build a substantial house. Oxford soon became a civilized village in the wilderness, one that drew educated Methodist preachers and teachers, and where it was possible each evening to hear Latin and Greek being spoken on the open house porches. By 1862, Edward and Mary Thomas were dead, but to their children and many grandchildren, the Thomas home had become a place of refuge during the War Between the States.[5]

---

[3] Frances Winn Smith, "Ancestral Lineage of Lovick Pierce Thomas, II," Thomas Family folder, GHS.

[4] Goff, "Edward Lloyd Thomas, Surveyor"; Newton County Historical Society, *History of Newton County*, 410.

[5] Dorward, Deep Running Roots, 4.

Lovick, Sr., the wounded quartermaster, was sent home to Lawrenceville, Georgia, where his family had relocated at the beginning of the war. After a few weeks on the mend, he returned to Virginia, bringing the family slaves Riley and Johnson as his camp servants. He also brought his twelve-year-old son, Henry Thomas, to act as a courier and drummer boy. It was a rash act to bring a boy into the vortex of war, and he likely accomplished it only because Jennie was not there to stop him. Young Henry Thomas would grow up to become the famous Wild West Marshall "Heck" Thomas, but that would be some two decades in the future.

In a 1910 letter to his niece Fannie Courtland, Heck Thomas recalled his impressions some forty-two years later. "I was there on a picnic," he wrote, "and was in everybody's way." Young Heck arrived in Virginia in time to witness the Confederate victory at Second Manassas. On September 1, 1862, the retreating Yankees made a stand at the Battle of Chantilly. At the end of the day, as rain poured down and darkness began to set in, Major General Philip Kearny dashed up to the men of Thomas's Georgia Brigade. He rode a black stallion with much gold lace on his uniform and one sleeve pinned back (he had lost an arm in the Mexican War). "I was nearby," recalled Heck, "and it never occurred to me that the man on that horse was a Federal soldier, and I don't reckon anyone who saw him for a second even suspected the real situation." Heck wrote, "That horse was a picture, a perfect picture. It reminded me of book pictures of Bonaparte. With head high in the air, champing the bit, prancing and stamping restlessly and apparently hard to handle." The general demanded to know what regiment the men belonged to. A ragged and hungry Confederate saluted the general and replied that he was now in front of the 45th Georgia. Realizing his peril, Kearny "wheeled his steed and dashing spurs into his sides darted off." Rifles cracked from the Rebel line, and the general fell to the ground dead some 200 yards from their position. Holding a lantern, Major General A. P. Hill was one of the first Rebel officers to reach the body of the illustrious soldier. He exclaimed, "You've killed Phil Kearny. He deserved a better fate than to die in the mud."[6]

Robert E. Lee held a profound respect for Philip Kearny because they had shared a long history in the United States Army, having served together in the same Mexican campaign. Twelve-year-old Heck Thomas was an eyewitness to both Kearny's death and General Lee's kind treatment of his old friend's body, horse, and accouterments.

---

[6] Edward C. Bruffey, "Bruff's Column," *Atlanta Constitution*, August 24, 1910, 8; Werstein, *Kearny the Magnificent*, 235–36.

The 35th Georgia missed the Battle of Antietam as they were shepherding captured supplies and prisoners from Harper's Ferry to Virginia. Antietam remains the bloodiest single day in American history. Both armies were too weary and shocked to resume fighting the next day. During the lull, General Lee called for Heck Thomas, who, under a flag of truce, crossed the battlefield strewn with Confederate and Union dead to the Federal lines. The boy, in his homemade Rebel uniform, led the black stallion with the body of General Kearny and a condolence note from Lee all the way through the Yankee regiments to Kearny's widow. Agnes Kearny thanked the young lad for his care of the horse and her admiration for him as a young soldier. Heck's Confederate military career proved to be a brief one because he soon fell sick with typhoid fever, "and was dead to the world when I got back to old good Georgia."[7]

Thirty-three Days in Kentucky

In August 1862, as Lee's Army of Northern Virginia crossed the Potomac River into western Maryland, the Confederate troops in Tennessee launched an aggressive campaign northward into Kentucky. Two Rebel columns, one under General Braxton Bragg, who moved toward Louisville on the Ohio River, and the other under General Kirby Smith, who marched on Frankfurt, the Kentucky state capital. Historians call this the Confederate Heartland Offensive. Bragg hoped that a move into Kentucky would avert Union attention away from Vicksburg and from a possible attack on Chattanooga by a large Union force in Nashville under Don Carlos Buell.

On August 16, 1862, Kirby Smith, leading the divisions of Patrick Cleburne and Thomas Churchill, crossed through Roger's Gap into Kentucky with 6,000 soldiers. Marching sixty miles in fifty hours, his forces captured Federal supply trains in Barboursville and London. From the perspective of Federal commander George Morgan atop Pinnacle Mountain, the arrival of Kirby Smith north of his position meant that his tenuous communication and supply line had been cut off. To Morgan's south, Carter Stevenson and his 9,000 Rebel troops had left Bean Station on August 3 and pushed their way up to Jones Station, about sixteen miles south of the Gap. On August 17, Stevenson pushed forward again, effectively investing Morgan's position from the east, south, and southwest. The Confederates drew themselves up, three miles from Morgan's defenses, in a semi-circle stretching from

[7] Edward C. Bruffey, "Bruff's Column," *Atlanta Constitution*, August 24, 1910, 8.

mountain to mountain. "Looking from our camp down the valley," wrote Owen Hopkins of the 42nd Ohio, "the sun shone on thousands of white tents which had sprung up like mushrooms in every direction in Powell's Valley, while upon the north side John [Hunt] Morgan's cavalry pickets could be seen upon the road leading to the Cumberland Ford." Union deserters were soon informing Stevenson that Morgan and his 7,000 Cumberland Gap defenders were rapidly running out of supplies.[8]

The problem for Kirby Smith, who was now north of the Gap, was that he found this remote part of Kentucky "completely drained of all supplies" and "the roads between here and East Tennessee being much worse than I had supposed." On August 20, General Smith wrote his counterpart, General Braxton Bragg, that he could not supply his modest army in their present position near Cumberland Ford. Smith could fall back to East Tennessee and be closer to his supply base, or he could advance toward Lexington in the Bluegrass region of central Kentucky, where he believed supplies would be found in abundance. The next day, Kirby Smith wrote Jefferson Davis that he had chosen the bolder choice of marching northward toward Lexington, as it was "the time to strike a decisive blow in the West." He instructed Carter Stevenson to send him Stovall's 3rd Georgia Brigade from Rains's Division to bolster his force, and the order was given to advance farther into Kentucky. Smith expounded on his reasoning to Richmond authorities on August 24, writing, "I find the people in this section openly and bitterly opposed to us. Unless the people on the other side of the mountain are very different from those here, nothing whatsoever can be expected from them. I shall, however, push on and endeavor to test the question of the loyalty of the blue-grass region." In other words, could Kentucky, with the Southern armies of Bragg and Smith in their midst, be successfully coaxed into joining the new Confederate nation?[9]

Dr. Ira Gunter had stayed back in Bean Station nursing Lieutenant Newton Hicks, at a private residence after the officer had contracted what proved to be a fatal case of camp fever. Gunter viewed Hicks as "one of the most exemplary young men I ever saw," and viewed his death as an irreplaceable loss to the Walton Blues (G). On August 16 the doctor, freed of his nursing duty, caught up with his company at Jones Station. In a letter to his

---

[8] Cathey and Waddey, *"Forward My Brave Boys!,"* 92; Hopkins, *Under the Flag of the Nation*, 34; Stevenson, Cumberland Gap, 14–15.

[9] Kirby Smith to Braxton Bragg, August 20, 1862, *OR*, ser. 1, vol. 16, part 2, 767; Kirby Smith to Jefferson Davis, August 21, 1862, *OR*, ser. 1, vol. 16, part 2, 769.

bands. The 42nd Georgia had at least a half dozen musicians in their regimental band, which was led by twenty-four-year-old George Washington Lyon, a superb musician and composer. Lyon even wrote original music for the marching band.[19]

"I thought we made a splendid spectacle," recalled Joseph Bogle of the 40th Georgia. Ladies waved handkerchiefs and small Confederate flags as the soldiers passed, and, according to Douthit, "Frequently a citizen would feed 50 to 100 men without price or charge." The Southern soldiers, however, also found the stores closed as shopkeepers did not want to accept Confederate currency. The rank and file, according to Bogle, were not above slipping out of line to forage for food. Officers were secretly compliant with this undertaking if they too were included in the subsequent distribution of spoils. The Confederate soldier, recalled Bogle, has never been "'adequately illustrated" as an expert forager when rations were scarce.[20]

On October 1, Stevenson's Division passed through Lawrenceburg headed for Frankfurt, a town of 8,000 inhabitants and the capital of Kentucky. They were among the Confederate soldiers present on October 4 for the inauguration of Richard Hawes as the new Confederate governor of Kentucky. The ceremony had to be cut short when uninvited Federal artillery opened up on the town. That night, Stevenson's Division marched fourteen miles through the darkness to the town of Versailles. They camped there for three days as Kirby Smith concentrated his brigades as the Federal forces drew closer. An appreciative John Douthit wrote, "We got as good water here as there is in Kentucky and were happy of the privilege of resting." Forty miles to the south, the wing of Bragg's army under General Leonidas Polk fought a Federal corps under General Alexander McCook to a tactical draw at the Battle of Perryville. The 41st Georgia took a big part in the battle, losing all of its field officers while attacking Union artillery positions. Even at a great distance, the Georgians in Stevenson's Division could hear the sounds of battle. "The heavy booming of artillery greeted our ears nearly all day," wrote Douthit, as "the frightened citizens were leaving their homes all along the

---

[19] "Professor G. W. Lyon Dead," *Atlanta Constitution*, April 9, 1903. Professor Lyon went on to teach music for forty-eight years. He was the author of fifteen sacred songbooks including *The New Song Waves* (1903). The musicians in the 42nd Georgia included William C. Cockrell (K), George W. Lyon (D), W. M. Lyon (H), Benjamin F. Moore (G), Warren H. Williams (H).

[20] Douthit, "Johnny Reb's Impressions of Kentucky in the Fall of 1862," 207; Bogle, "Some Recollections of the 40th Regiment, C.S.A., in the Civil War," in Bogle and Calhoun, *Historical Sketches*, 13.

road not knowing when or where the fury of battle might rage."[21]

Braxton Bragg now faced an advancing Federal army while his forces remained scattered about Kentucky. The Confederate commander ordered his various Confederate columns to reunite at the nearby town of Harrodsburg. Carter Stevenson's division reached the rendezvous point at dawn on October 10, and his men were soon drawn up in a line of battle across a series of hills two miles southwest of town. John Douthit wrote, "We marched out meeting them [the Union Army] and took up position for battle and unpacked knapsacks ready to do or die for liberty." The men of the 42nd Georgia, like all of the soldiers in the division, prepared themselves for their first experience in a large-scale Civil War battle. "We had about ten thousand men in my line of sight," marveled Douthit. Despite orders to stay in place and be ready for battle, Douthit remarked that "in five minutes the boys were out cracking walnuts and getting corn and pumpkins for supper and Starting an occasional Rabbit. The men would join in the shout as though no enemy was near." Despite a busy signal corps waving their flags up and down the line of gray infantry, no advance of the Federal army came that day. That evening the men stacked the rifles as heavy rain began to fall.[22]

At a council of war held that soggy night, the Confederate generals present voted to retreat. News had reached them that, in recent weeks, large Confederate armies had been defeated at Antietam and Corinth. Staying in Kentucky at the start of winter without tents or proper clothing would have been a bold move, one not in Braxton Bragg's conservative nature. The Confederates under Bragg and Smith were tired, outnumbered two-to-one, and low on ammunition. The citizens of Kentucky had not rallied to the Southern banner as had been expected at the start of the campaign. Losing a major battle deep in the Bluegrass state would be catastrophic, resulting in a desperate and long retreat through the Kentucky hills. Even if, in the coming days, the Confederates prevailed in a battle with Buell's army, the Southerners would be in no position to follow up their victory. It was better, thought Bragg and his high command, to consider the Kentucky campaign a success and return to Tennessee. The prickly "Old Porcupine" ordered his army home to Middle Tennessee for the winter.[23]

General Bragg's decision to retreat was not well received by his troops.

---

[21] Douthit, "Johnny Reb's Impressions of Kentucky in the Fall of 1862," 208.

[22] Ibid.

[23] Hafendorfer, *Perryville*, 424–27.

Nashville. "This whole country is full of soldiers," noted Benjamin Weaver. "An old citizen told me today that he thought there was not less than 100 thousand in Middle Tenn. I think we will have a big fight or a big backout soon. This is too fine a country for both armies to winter in."[32]

On December 10, 1862, Confederate president Jefferson Davis met with generals Braxton Bragg and the new "Theater Commander" Joseph E. Johnston in Murfreesboro. The conversation became heated over how to deploy their limited forces. With Vicksburg, Mississippi, under increasing pressure, Davis, who was a resident of Vicksburg's Warren County, ordered Bragg to send his largest division, that of Carter Stevenson to the "Gibraltar of the Confederacy."

At this time, the 42nd was camped near Readyville, on the Stones River forty miles southeast of Nashville. Ino reported on the sixteenth that the men were receiving, "Sugar, molasses, soap, and other acceptable rations." On the seventeenth, Captain Weaver wrote his parents, "We are drilling now with a vim—have company drill at 9—Brigade drill at 11 and Battalion drill at three—so we have but little spare time during the day. The weather is fine now though rather cold. We are in a fine wooded country." The official order was given to Stevenson on December 18. That morning Captain Weaver's breath had frozen on his beard as he gave morning roll call. He wrote home, "Reported this morning that we go to Mississippi—transferred to another brigade. I don't know the truth of this report." Indeed, changes were afoot. The 41st and 42nd, which had been serving in Rains's Brigade, was reassigned to the First Brigade of Carter Stevenson's division. This brigade of only Georgia regiments, the 40th, 41st, 42nd, 43rd, and 52nd, would be commanded by the handsome Brigadier General Seth Barton of Fredericksburg, Virginia. Barton lost the 30th and 31st Alabama regiments in the exchange.[33]

As part of these changes, General Kirby Smith promoted Lieutenant Thomas Hill of Stovall's 3rd Georgia Battalion to captain and gave him command of the Fulton Blues (I) of the 42nd Georgia. Hill had been born and raised in Bastrop County, 100 miles west of Houston, Texas. When the Civil

---

32 B. P. Weaver to wife, Mary MaGuire Weaver, December 9, 1862, Readyville, TN, Weaver letters.

33 Wright, "Our Special Cumberland Gap Correspondence," *Atlanta Southern Confederacy*, December 16, 1862, dispatch from Readyville; Benjamin Weaver to home, December 17, 1862, Readyville, TN, Weaver letters; Herring, *Company A of the Fortieth Georgia Infantry in the Confederate Service*, 28.

War began, the twenty-one-year-old Texan was an instructor at the Georgia Military College in Marietta, Georgia. After instructing raw recruits at Camp McDonald, Hill was commissioned a lieutenant and given a six-gun battery of artillery. He was assigned to Stovall's 3rd Georgia Battalion and sent to Cumberland Gap. During the Kentucky campaign, Lieutenant Hill had been detailed to serve as provost marshal for the city of Lexington. Captain Hill was a welcome addition to the officer ranks of the 42nd Georgia.[34]

The regiment set out for Vicksburg on trains reaching Atlanta on December 21. Captain Lovick Thomas of the Gwinnett Beauregards (A) was sick with jaundice and gained permission to stay home in Lawrenceville for one month to regain his health. Jennie wrote in her diary, "He looked very yellow and had it pretty bad though did not keep his bed." Even in a weakened state, there were pressing business matters, and Lovick pushed himself to complete these transactions. With a long war ahead, it was appropriate to end his business partnership with Joshua Wilson. Lovick sold his stock of goods and interest in their Lawrenceville general store to Joshua's younger brother, Jim Wilson, for $3,000.[35]

On January 20, 1863, Lovick went to Atlanta to take a train for Vicksburg. Before leaving, he purchased $2,000 worth of tobacco with Confederate money from his brother-in-law Joseph T. Porter, who owned a business on Marietta Street. The boxes of tobacco were stored in an Atlanta warehouse, with some boxes going to his wife, Jennie Thomas, in Lawrenceville. Porter noticed that Lovick still had a fair amount of money left and persuaded the young officer to buy a town lot on Crap's Alley in the growing city of Atlanta for $800. Lovick purchased the 100' by 200' lot sight unseen. Still very weak, Thomas went to his mother in Oxford and from there gained a one-month medical extension on his furlough. Jennie wrote that he "soon commenced improving and got fleshier than he has been since we were married. I believe weight 152. His complexion looks much better too."[36]

---

[34] "Death of Honored, Good, True Citizen Captain T. A. Hill," *Weimar (Texas) Mercury*, October 10, 1924, 1.

[35] Jennie P. Thomas, diary, December 23, 1862, in Huff, *Four Families*, 77.

[36] Winn, "A Sketch in the Life and War Record of Col. Lovick Pierce Thomas II," 50; Jennie P. Thomas, diary, December 23, 1862, in Huff, *Four Families*, 77.

Captain Calhoun's Detachment

Throughout the campaigns in East Tennessee and Kentucky, the 42nd Georgia had been without the services of fifty of their men. In April 1862, Captain Lowndes Calhoun's detachment had been sent back to Georgia to guard the 476 political prisoners rounded up near Knoxville. Calhoun had initially been under the impression that his assignment would be a brief one: escort the prisoners to Milledgeville, the capital of Georgia, and then quickly return to his regiment. It turned out, however, that Georgia state officials were utterly unprepared to accept the prisoners; therefore, Calhoun and his men would be needed as prison guards until other arrangements could be made.

The Tennessee prisoners were ultimately assigned to the abandoned buildings of the Madison Steam Factory Cotton Mill by the side of the Georgia Railroad in Morgan County. The large building was surrounded by outbuildings and enclosed by a tall plank fence complete with gates. It made for a suitable prison. The twenty-five-year-old Captain Calhoun soon found himself the officer of the guards for the new prison. Local resident Isaac Walton, a man of Unionist sympathies, was appointed the camp superintendent.

Further complicating Calhoun's responsibilities was the decision to deposit Union officers and soldiers captured at the Battle of Shiloh at the Cotton Mill Prison. A thirty-six-page handwritten registry, now at the University of Georgia, holds the names of 891 prisoners held at the mill from April to December 1862. Thirteen of the prisoners had died that year. Generals Benjamin Prentiss and Thomas Crittenden were among the Federal officers held there. Members of the Andrews Raiders are listed under "Bridge Burners & Engine Thieves."[37]

Ira Ford, a member of the 18th Wisconsin, recorded a favorable impression of Captain Calhoun, "a nephew of Senator Calhoun of South Carolina." He probably did not know that the young Confederate officer was also the son of James M. Calhoun, the wartime mayor of Atlanta. Ford recounted that twenty-four members of the 18th Wisconsin spent three weeks digging an eighty-four-foot clandestine tunnel from the cellar basement to the outside. Before an escape could be attempted, however, Ford and his messmates were put on boxcars and transported to Augusta for an exchange.[38]

After eight months of duty as prison guards, a frustrated Calhoun made a personal appeal to Henry Wirz (later the commander of the Andersonville

---

[37] Confederate States of America prison (Madison, Ga.) records.

[38] Erquitt, "Cotton Mill Prison Canteen Revisited," 26–37.

Prison) that he and his men be returned to their regiment. What the impatient young captain may not have fully realized was that central Georgia was a far safer and healthier assignment than windswept Cumberland Gap. Only one of Calhoun's men would die from disease in 1862 while the 42nd Georgia was, on average, losing one man a day campaigning in Tennessee and Kentucky. Calhoun's contingent of men finally rejoined the 42nd in December 1862. The empty Cotton Mill Prison was then turned into a major hospital by Confederate authorities.[39]

Traveling to the Mississippi River

The 42nd was now a member of a new all-Georgia brigade. The regiments in this brigade, the 40th, 41st, 42nd, 43rd, and 52nd Georgia, would remain together for the rest of the long, cruel war. Having trained with these regiments at Camp McDonald and having served alongside the 40th and the 52nd in the Cumberland Gap campaign, the men of the 42nd were already thoroughly familiar with the soldiers in their new Cracker brigade.

The slight geographic differences from where these regiments drew their men from did, in fact, create minor cultural differences among the rank and file. The men of the 42nd were from Piedmont counties around Atlanta, and most were literate and had accumulated some degree of land ownership and taxable wealth. A handful even brought a black slave with them as their camp servant. L. P. Thomas described the 42nd as "hardy young men from the best families of our farming communities." The 52nd, by contrast, came from the thinly populated mountain counties of northeast Georgia. Their ranks were filled with rough hardscrabble men with little education. The 43rd drew from the hilly counties between the Piedmont and the mountains, the 40th from the Ridge and Valley counties stretching from Dalton down past Rome, and the 41st were from Piedmont counties north and west of Atlanta.[40]

The 41st was, at this point in the war, the only regiment in the new brigade to have participated in a significant battle, having fought well at Perryville on October 8, 1862. As part of Maney's Division in Braxton Bragg's army, they had helped to capture federal artillery on both Parson's Ridge and Starkweather's Hill, before failing in an assault on a third knob. Three of their

---

[39] Ibid. Slocum's XX Corps ultimately burned the hospital during Sherman's March to the Sea.

[40] L. P. Thomas, "The Battle of Resaca: The Part the Forty-Second Georgia Took in It," *Atlanta Constitution*, February 2, 1890, 12.

color bearers had fallen in the charges as well as their respected commander, Colonel Charles McDaniel. In recognition of their gallantry in this battle, the cannon cross was engraved across their regimental flag "in token thereof." In the Battle of Perryville, the 41st Georgia suffered 23 killed, 125 wounded, and 3 missing, or a loss of 30 percent of the regiment.[41]

The new brigade commander was the handsome thirty-four-year-old Seth Maxwell Barton. An 1849 graduate of the U.S. Military Academy, Barton had been Stonewall Jackson's chief engineer during the storied Valley Campaign. Friction between Jackson and his engineer, however, had led to Barton's promotion to brigadier general and transfer to Tennessee. General Barton was given a brigade command in Stevenson's Division. The new general would soon earn a reputation for gallantry on the battlefield but not one of perceptive decision making under fire. Barton would also be labeled by historians, in subsequent campaigns, as uncooperative with his fellow Confederate officers, especially after his failure to follow General George Pickett's orders during the North Carolina campaign of 1864. With President Davis's directive, Stevenson's entire 12,000-man division, including Barton's new Georgia Brigade, was on the move to the defense of Vicksburg.[42]

Captain Lovick Thomas made the trip only as far as Atlanta. His health had collapsed after the Kentucky campaign, and he would need time to recover. Company A would now be under the temporary command of thirty-seven-year-old Lieutenant Darling Peeples McDaniel. Named for a Baptist minister in Barnwell District, South Carolina, Darling McDaniel left a large family and seven slaves on a farm near Rockbridge in Gwinnett County.

Getting to Vicksburg by rail would require a long, circuitous route. According to the travel log of Private Francis Nash, the 42nd traveled from Chattanooga to Atlanta and on through West Point, camping on the twenty-second in Montgomery, Alabama. Here they had to disembark, help the teamsters load up their baggage on drays, and then march across town to the Mobile railroad terminal. Moving an entire division and their artillery by rail was a massive undertaking requiring several trains. The agrarian soldiers, riding in cold boxcars, were keen observers of the land they crossed. Lieutenant Calvin Smith, a Mexican War veteran in the 31st Tennessee, wrote in his

---

[41] Hafendorfer, *Perryville*; 454; M. Stovall, unpublished autobiography, Stovall papers, ASU, 7. McDaniel was an Emory College graduate and the first president of Bowden College in Carroll County.

[42] Warner, *Generals in Gray*, 18–19.

diary, "The best farms I saw in Georgia or Alabama is after you leave Montgomery. You pass some beautiful little towns."[43]

Northwest of Mobile, the soldiers disembarked near the Tensaw River and boarded steamboats. Most of the soldiers had never seen vast swamps filled with giant cypress trees. The mile-wide river emptied them into the giant Mobile Bay, which, according to Smith, was filled with, "several war vessels, rams, many steamer boats and transports of every form and size." Landing in Mobile, they were again required to transport themselves and their baggage across town to the Mobile & Ohio Railroad. Henry Robinson embraced it all. He wrote home, "I saw some of the largest fish in Mobiel I ever saw. The scales was as large as a half dollar." At 5:00 P.M. on Christmas Eve, after eating vast quantities of fish and drinking heavily, they departed Mobile.[44]

On Christmas Day 1862, they reached the railroad center of Meridian, Mississippi, after a journey of 134 miles. "You pass through extensive pine barrens, some large level track of prairie-looking country," wrote Calvin Smith. Henry Robinson wrote that his comrades looked for barrels of sugar at each train depot. "If a shugar barrel hadn't the head out," he wrote, "we would Soon have it out & eat it up." As a result of this urban foraging, many of the men got sick. Advance elements of Stevenson's Division reached Vicksburg at midnight on December 26 by way of Jackson, Mississippi. In all, it was a taxing trip of roughly 830 miles. Captain Lowndes Calhoun of the Calhoun Guards (K) would later recall at a Confederate reunion that upon his first view of the Mississippi River, the "father of waters" as he called it, "I took my first and last drink of Louisiana Rum."[45]

---

[43] C. M. Smith, *A Smith in Service*, 37.

[44] Ibid.; Henry Robinson to home, January 2, 1863, Vicksburg, MS, Robinson letters.

[45] C. M. Smith, *A Smith in Service*, 38; Henry Robinson to home, January 6, 1863, Vicksburg, MS, Robinson letters; Calhoun, "Siege of Vicksburg; Historical Address Delivered by Judge W. L. Calhoun," in Bogle and Calhoun, *Historical Sketches*, 39.

Chapter 3

Defending Mississippi

"Every man stood like a lion."
—Private Henry Robinson, describing his fellow soldiers at the Battle of Chickasaw Bayou.

"As the Yankees came from out the timbered land, our boys from their rifle pits let loose the dogs of war and down went the blue coats."
—George Hewitt Daniel of Covington, describing the 42nd Georgia at the Battle of Chickasaw Bayou[1]

Chickasaw Bayou

With Vicksburg in mortal danger, the men of Seth Barton's brigade were the first of Stevenson's reinforcements to arrive to defend the river port. They were met at the train station and directed on a five-mile march north of the city to prepared rifle pits at the base of bluffs that looked down on Chickasaw Bayou. The Georgians filed into the trenches in the darkness but awoke to Federal artillery and a view of frost on the open field before them. The lakes and bayous that lay beyond the grass shimmered in ice. The newly arrived brigade had a commanding view, as their trenches, elevated on a promontory, were about fifteen to twenty feet above the swampy lowland. The Federal capture of Vicksburg would mean Union control of the Mississippi River and the splitting of the Confederacy in two. The strategic importance of this citadel, sitting above a great elbow in the river on 200-foot high bluffs, was clearly understood by both sides. Since fall 1862, General Ulysses S. Grant had envisioned a three-pronged approach. General Nathaniel Banks would push his command northward from New Orleans while Grant's army would drive south from Tennessee into central Mississippi. After capturing Grenada, Grant would turn west and threaten Vicksburg.[2]

Meanwhile, General William Tecumseh Sherman would take a third

---

[1] Alexander and Bogle, "George Hewitt Daniel (1817–1864)," 26.

[2] McDonough, *William Tecumseh Sherman*, 363.

army by transport from Memphis and, in an amphibious assault, attempt to capture Vicksburg from the north. Grant's master plan soon unraveled as neither Banks nor Grant could protect their long supply lines from Confederate cavalry raids. Their respective armies were compelled to retreat to more secure positions. In late December 1862, it fell to Sherman to capture the river city. On the 26th of the month, Sherman's men began disembarking at Johnston's Plantation, a few miles above Vicksburg, on the edge of a triangular piece of low and marshy land between the Yahoo and Mississippi Rivers and the Chickasaw bluffs. The area was cut by lakes and bayous, the largest being Chickasaw Bayou, and in many places deeply forested with water oaks and cypress trees. Despite these natural obstacles, Sherman hoped to move swiftly through the low land of the Yazoo valley and gain control of the bluffs that looked down on this spit of land. From the higher dry ground, it was only a few miles of marching to reach the relatively undefended rear of Vicksburg.

Unfortunately for Sherman, overseeing the defense of Vicksburg was a twenty-nine-year-old South Carolinian artillery general named Stephen Dill Lee. Upon the request of President Davis for an accomplished artillerist to take charge of the defenses around Vicksburg, Robert E. Lee had recommended S. D. Lee. At Second Manassas and Antietam, young Lee had shown himself to be "brave, soundly competent, energetic, and resourceful." Promoted to brigadier general, Lee arrived in Vicksburg in mid-November to a warm welcome from the citizens of the river town. He found Mississippi to be "one of the strangest places I have ever seen—nothing but ravines and bluffs—it is difficult to find sufficient level ground in this vicinity even to encamp the troops." Lee placed his artillery in a strong defensive position—atop the Walnut Hills Line, looking down at the flooded and muddy terrain of the bayou—but he only had 2,700 men to stop Sherman's army of 30,000. The arrival of each train filled with infantry reinforcements was critical to any chance of Southern victory. By sending forth a strong skirmish line, General Lee managed to bluff and delay the Yankee advance throughout the 26th and 27th.[3]

By December 28, the Confederate infantry works, built along a shelf of land that extended along the base of the bluffs, had been heavily reinforced with Maxey Vaughn's Tennessee brigade holding the left flank, Stewart's Brigade moving to the right, and Seth Barton's Georgia troops occupying the

[3] Hattaway, *General Stephen D. Lee*, 64–65.

center trenches. Opposing the Rebels on December 28 was the division of George Washington Morgan, the very men that the Georgians had confronted for months at Cumberland Gap. The fiercest fighting at Chickasaw Bayou, however, was reserved for December 29. Sherman thought a powerful thrust into the Confederate center would allow him to take the elevated base of Walnut Hill and might, with a bit of good fortune, carry his advance to the tops of the bluffs beyond the Rebel position. Late on the afternoon of December 28, the 42nd Georgia was sent to Lee's Brigade to reinforce his left flank. Colonel Robert Henderson wrote that he reported to General Lee that evening and "was immediately assigned a position in the rifle pits on his extreme left with orders to hold it at all hazards, it being of the greatest importance."[4]

General Frank Blair, leading Sherman's First Brigade in Frederick Steele's Fourth Division, was not confident of Union victory. He wrote, "These formidable works, defended by a strong force of desperate men such as held them on the 29th, would seem to require almost superhuman effort to affect their capture." General Morgan planned to gain more punch in his attack on Stephen Lee's position by assembling a pontoon bridge on his right flank, which would require crossing the widest and deepest portion of the bayou. At daylight, Rebel artillery spotted the Yankee pioneers at work and began shelling the bridge builders. Realizing there was no infantry in that sector of his line, General Lee dispatched Colonel Allen Thomas with the 42nd Georgia and the 29th Louisiana to stop these efforts to build a usable pontoon bridge across the 80-foot waterway. Thomas reported that the men took their positions "under a very galling fire from the enemy's artillery and sharpshooters."[5]

Colonel Robert Henderson wrote that the 42nd moved under "a raking fire of shell and shot." Thomas deployed four Georgia companies and two Louisiana companies "on a mound opposite to and commanding the pontoon." A spirited firefight took place between the Butternut defenders and the men of the 69th Indiana who had taken up a position in the woods behind the bayou and were attempting to provide cover for the Union pioneers. Henderson wrote, "My four rifle companies opened a vigorous fire upon

[4] Henderson, "Chickasaw Bayou Field Report"; Alexander and Bogle, "George Hewitt Daniel (1817–1864)," 29.

[5] Military pioneers were trained to perform engineering and construction tasks for the army. They principally built bridges and fortifications and improved roads (Bearss, *The Campaign for Vicksburg*, 1:196; C. A. Evans, *Confederate Military History*, 17:695).

them which was vigorously returned for a short while by the enemy, many of whom were seen to fall and lie on the banks of the bayou." Ultimately, efforts to latch the eight pontoon boats together in the icy wind never came to fruition.[6]

About the same time, shortly before noon, eighteen Union cannons signaled the frontal assault of Morgan's 6,000-man division, made up of the brigades of Blair, Thayer, and DeCourcey. The Bluecoats emerged from the woods and took frightful casualties as they crossed the bayou at two thin points, worked their way through the abatements, waded the stream, and scaled the steep bank. They finally reached the open ground in front of Stephen Lee's Tennessee and Louisiana regiments. While the beleaguered Yankees managed to capture the advanced rifle pits of the Confederates, little more could be done. Most of the Federal troops fell to the ground to avoid certain death. Captain Joseph Woodward, commanding Company D of the 52nd Georgia, watched the charge from his position along Barton's line. He wrote, "From the position we occupied on the left, we could see them fighting on our right, and saw the enemy charge and saw them fall by the hundreds, and routed them in perfect confusion—saw shell after shell burst in their ranks, fired from our batteries above, thinning their ranks at every fire—while the bullets were whizzing over us all the time from the Yankee lines in front."[7]

The 4th Iowa, under Colonel James Williamson, came behind DeCourcey's troops and, after crossing the levee, swung to the right, crossed a rail fence, and closed on Lee's two left flank regiments, the 42nd Georgia and the 29th Louisiana. Colonel Henderson believed the attack might have succeeded "had it not been for the gallant conduct of the four Cos which I left along the road to my right. They met the enemy in the field and drove him with great slaughter." Private Henry Robinson wrote, "The yankeys come like tigers at companies B & H. The balls come as thick as hail. We pord it in to them for some time." It was an impressive stand by these two companies standing in the open without protection. Robinson believed, "Every man stood like a lion. I stood in an open place with Sevrel Round me & while I

---

[6] Henderson, "Chickasaw Bayou Field Report," 29. Robert Henderson's field report on the actions of the 42nd Georgia does not appear in the *OR*. It was found among the papers of George Hewitt Daniel and reprinted int eh *Atlanta Historical Bulletin*, September 1968. Henderson had four "rifle" companies, which were armed with better rifles. Armed with axes, "pioneers" were designated soldiers who cleared and improved roads and bridges.

[7] Bearss, *The Campaign for Vicksburg*, 1:200; Captain Joseph W. Woodward, "Account from Captain Joseph W. Woodward," in Hewitt, et al., eds., *Supplement to the Official Records*, part 1, vol. 3, 340.

that the men had rented a comfortable room for their fellow soldier. Upon James's death, "we managed to bury him very decently. We got a nice coffin and the Lady of the house where he staid let me have a nice sheat and gave us the privilege of burying him in the family grave yard." According to Angus, James "was thought a heap of by the whole company," and "his death caused a gloom over the whole company but we will try to console ourselves with the thought that he is better off than we that are permitted to live."[18]

In late January, the 42nd was visited by George Hewitt Daniel, a forty-six-year-old merchant from Covington. Daniel was an Englishman from London, having settled in Newton County in the 1830s, where he established himself as a successful businessman and merchant. He was a good friend of Robert Henderson and knew many of the men in the Newton County companies. He wrote his daughter on January 24 that, "I wandered over interminable hills, lofty peaks and craggy sides in search of the 42nd Geo. & after a tiresome & weary walk of near 4 miles I found Col. Henderson & his regt. encamped in a corn field forming a sort of cove amid the hills." Colonel Henderson and the Newton men were surprised to see someone from Covington so far away from home. "They greeted me warmly," he wrote, "and I was really gratified to meet them and find them *in* such excellent spirits and *with* such *excellent spirits,* a portion of which offered me, I *imbibed*—of course." There was much to catch up on. Daniel described his arduous eight-day trip to Vicksburg by way of Mobile. He was not impressed with Mississippi. "I have travelled a great deal," he told his daughter, "but never have I seen such miserable roads, absolutely dangerous and oftimes nothing to eat, mud, rain, & cold prevailing." He was disappointed to find few goods in Vicksburg, and "articles of all kinds are cheap in Georgia as compared with prices here."[19]

Daniel had walked the battlefield at Chickasaw Bayou that day and had seen the trees, "riddled with balls, limbs struck off and some few dead Yankees lying stiff & stark, food for the vultures—a fit repast." The men showed him their regimental flag, which "had ten bullet holes thru it—plain evidence that the bearer and our chivalrouse boys were at the front when duty called them." Colonel Robert Maddox, Major William Hulsey, and the surgeon of the regiment all had bullet holes in their clothing and hats to show him. "I wandered about the camp shaking hands & talking with everybody," recalled Daniel. He found, "A good many of the Co. were sick—near 100—occasioned principally from the bad camp & worse water but now I learn are improving."

[18] Angus McLeod to Griffin Vaughn, January 20, 1863, Elliott collection.

[19] Alexander and Bogle, "George Hewitt Daniel (1817–1864)," 25.

With nightfall, he hiked the four miles back to Vicksburg.[20]

Ira Lawson Gunter of the Newton Rifles (F) never went to Vicksburg but had stayed in Atlanta working as a hospital steward in the Fair Grounds Hospital. Located a block south of Oakland Cemetery and convenient to the Georgia Railroad, his hospital was a collection of buildings and tents in the former fairgrounds of Atlanta. In early January 1863, the facility had received hundreds of casualties from the Battle of Stones River in Murfreesboro, Tennessee. As he wrote to Ann, "The poor fellows are wounded in every way and almost in every portion of the body and limbs. And taken collectively exhibit an amount of physical suffering which is heart rendering to contemplate." By February 1863, Gunter was hopeful that he could continue his medical work in Atlanta and not be sent to his regiment in Vicksburg. "So here I am yet," he wrote Ann on February 22, "and probably will stay as long as my company and Regimental officers are willing or can be induced to let me stay. I am subject however to [be] ordered to my regiment at any time, though I believe Captain [James] Summers will let me stay as long as he can, consistent with his duties as an officer." Gunter would ultimately remain in the medical service for the duration of the war, passing his Confederate medical boards in Atlanta on March 12, 1863.[21]

On February 10, Regimental Musician Benjamin Franklin Moore of the Walton Blues (G) wrote home from Vicksburg that Captain William Malcom had died from his wounds received at Chickasaw Bayou. It was the second time that the Walton Blues had lost their leader. Enoch McCollum, the first captain of the company, had died of measles at Bean Station on July 7, 1862. Thomas Davis, 2nd Lieutenant of Company K, was elected the new captain of the Walton Blues. On February 18, Henry Robinson reported that nine members of his company were already dead and feared that "if we stay heare til Sumer we will (all) dye." Benjamin Moore went on, "The weather here is as bad here as in any place I've ever been; and we do have 11 tents now to our camp, so when it rains we can all get in but we are badly arranged in them." On March 25, John Nash of the Independent Rebels (B) wrote that "The weather is getting verry warm hear and the woods wears a green appearance." Vicksburg was a difficult place to defend because "[t]he mosquitoes hum their solemn tune for us to sleep and plenty of wiggle tails are in our

[20] Ibid., 26–27.

[21] Ira Gunter, January 6, 1863, and February 22, 1863, letters from Fair Grounds Hospital, Atlanta, Latimer collection; Welsh, *Two Confederate Hospitals and Their Patients*, 12–15.

water we have to drink." Henry Robinson, the irascible chronicler from the Independent Rebels (B), perished on April 12. He died in Hospital #2 in Vicksburg, which was reserved for smallpox patients. Lieutenant William Williams wrote Robinson's wife to resolve his affairs and send his effects home.[22]

In January 1863, Grant withdrew his army to Memphis where steamers were being assembled for a move on Vicksburg. Between February and April, Grant attempted four Bayou Expeditions north of Vicksburg. All were repulsed through a combination of skillful Confederate responses, difficult terrain, and severe weather. At the end of April, a frustrated Grant attempted, according to historian Timothy B. Smith, "as close to an all-or-nothing gamble as any Union commander would ever make during the entire war." The Union commander had to get his army south of Vicksburg where rolling hills and a network of roads made an offensive operation possible. He would risk everything by landing his army south of Vicksburg and use maneuver and guile to defeat the dispersed Rebels. Grant recognized that he would be cut off from help and would probably be unable to retreat in the event his army lost a significant battle. There were roughly 50,000 Confederate soldiers in the Port Gibson-Jackson-Vicksburg triangle. In a rare reversal of troop strength, Grant would only have 40,000 Federal soldiers at his disposal.[23]

Grant's Vicksburg campaign would begin with the ironclads of Admiral David Dixon's Brown Water Navy challenging Vicksburg's extensive collection of river batteries. The newly built ironclads would have to escort a flotilla of barges and transports carrying provisions and ammunition past a furious bombardment of emplaced Confederate artillery. The men of the 42nd, it turned out, would be witnesses to it all. Lowndes Calhoun, captain of the Calhoun Guards (K), wrote that the 42nd had, in January, left the trenches for the swamps below Vicksburg, "when from exposure, malaria, and bad and insufficient food, the men became walking skeletons." They were returned to Vicksburg in February and placed as infantry support behind the lower water batteries.[24]

As rations improved, with ham and flour superseding blue beef and sour

---

[22] Goodson, *Letters and Diaries*, 49; John T. Nash to Robert Bolt Nash, March 25, 1863, Elliott collection; Henry Robinson to home, February 18, 1863, Vicksburg, MS, Robinson letters.

[23] T. B. Smith, *Champion Hill*, 15, 106.

[24] Calhoun, "Siege of Vicksburg; Historical Address Delivered by Judge W. L. Calhoun," in Bogle and Calhoun, *Historical Sketches*, 36.

cornmeal, so did the men's morale. "We had a system of signals," writes Calhoun, "consisting of the firing of picket guns from above the upper batteries to give notice of the approach of the vessels, at which the infantry supporting them was required to take position in line and in their rear. The gunboats sometimes opened upon us, but being dark, generally missed their mark." Sergeant Tim Worley of the 52nd Georgia wrote that when the gunboats approached Vicksburg the long drum roll would be beaten, and "we would be ordered into line and double quicked down to the wharf about half asleep the most of the time on a false alarm."[25]

On March 25, John Nash of the Independent Rebels (B) wrote his uncle Robert Bolt Nash about events on the river. The 42nd was quartered in the brick Marine Hospital building, just at the lower end of town on the bank of the river. The nearby Marine Hospital Battery was a formidable river defense, mounting three heavy 42-pounder smoothbores, two 32-pounder smoothbores, and two 32-pounder rifles. Nash wrote that he was awoken that morning by calls to "fall in" and the roar of the coastal cannons. Standing in a line of battle behind the Marine Hospital Battery, he "looked up the river and...saw one of the monster Iron clads come steaming along down the river and closely behind it fobled by another." These were the rams *Switzerland* and *Lancaster* attempting to run the Vicksburg gauntlet en route to Port Gibson. "I tell you things roared and the ground trembled," wrote Private Nash as the Confederate batteries fired on the approaching ships. According to *Harper's Weekly*, the trailing *Lancaster* was hit thirty times. With the entire bow shot away, she sank quickly, "turning a complete summerset as she went down." Nash thought, "the crew was mostley lost. I saw many of the blue coats loope off into the water, but I think it was so coald and swift that but few of them ever landed." The *Switzerland* was disabled after a 64-pound ball penetrated the steam-drum. The side-wheeler floated helplessly down river until saved by the *Albatross*, which ran up beside her and towed her to safety. Nash could, "plainly see that the first one was badley wounded and it is reported that it sunk just below [Vicksburg]." The Georgian believed the Yankees "will get tired of trying to run their iron clads by our batteries as they have all met disaster that has made the trip."[26]

April 16 was a moonless night as Admiral Porter sent a dozen Union boats toward Vicksburg. The Confederates used barrels of tar and cotton

---

[25] Ibid., 37; Kinsland, "A Band of Brothers," 18.

[26] "Running the Vicksburg Blockade," *Harper's Weekly*, April 18, 1863, 1; John Nash to R. B. Nash, March 25, 1863, Elliott collection.

soaked in turpentine to set ablaze buildings on the opposite bank at De Soto Point. Despite being backlit on the river, the Union ships survived the gauntlet of fire from the Confederate gunners. Some 1,562 shells were fired at the passing ships. Losses were small again when a second convoy attempted the passage on April 22. Captain Calhoun was transfixed by April 16, the night of the first convoy. He wrote, "A constant and terrific fire was kept up on them, commencing at the upper batteries and continuing for three miles along the riverfront, making the very earth tremble. It was a scene that I, and those who witnessed it, will never forget."[27]

According to Jennie Thomas's April 14 diary entry, Lovick Thomas had made it back to Vicksburg despite the ubiquitous rain and washed away bridges. He reported to her that his company was barracked in an old hospital (Marine Hospital) and that he and Lieutenant Darling McDaniel shared a room. She wrote, "They had to live very hard for a while, corn bread & molasses, no meat & they commenced eating rats in the place of meat, but they got plenty now." Billington Wilson of the Walton Blues (G) wrote his sister on April 22 to "tell mother and Martha that I am well and as fat as two hogs." On April 25 and 26, General Barton marched his Georgia Brigade out of Vicksburg down to the river bottoms near Warrenton where large-scale drills were conducted. Lieutenant Calvin Smith of the 31st Tennessee watched during the warm afternoons as "they formed line, wheeled into columns of Companies, and passed the General and Staff." Brass bands played as the soldiers marched back to their camps in Vicksburg.[28]

Soldiers in the XIII Corps under Major John McClernand were the first Yankee infantry to establish an eastern bridgehead at Bruinsburg, some forty-four miles south of Vicksburg. They were followed by the XVII Corps under General James Birdseye McPherson, and would eventually be joined by Grant's third portion of his army, the XV Corps, under his favorite subordinate General William Sherman. After a march inland of ten miles, McClernand's 17,000 men approached the important crossroads town of Port Gibson, where a major battle for control of the town broke out on May 1, 1863. Carter Stevenson sent Edward Tracy's Alabama Brigade to reinforce the outmanned Confederates defenders. Events did not go well for the Southerners

---

[27] Calhoun, "Siege of Vicksburg; Historical Address Delivered by Judge W. L. Calhoun," in Bogle and Calhoun, *Historical Sketches*, 37.

[28] Billington A. Wilson to his sister, April 22, 1863, UDC collection, 6:365; Jennie P. Thomas, diary, April 14, 1863, in Huff, *Four Families*, 78; Carnes, "'We Can Hold Our Ground,'" 26.

as General Tracy was mortally wounded by a Union sharpshooter, as he organized his initial battle lines. The Botetourt Artillery of Virginia would be overrun in the engagement and lose six of its Napoleon cannons, fifty-six horses, and thirty-seven of its officers and men. After a stern but unsuccessful fight, the outnumbered Confederates retreated to a defensive line behind the Big Black River.[29]

With the Federal army now firmly established south of Vicksburg, the rest of Stevenson's Division left Vicksburg on May 1 and marched ten miles south to Warrenton, where they camped. From May 4 until May 12, the 42nd Georgia defended the most direct roads from Port Gibson to Vicksburg, holding the various fords and ferries near Hankinson's Ferry on the Big Black River. Pemberton had very little cavalry and was therefore handicapped in gaining a true picture of Grant's troop dispositions. The Confederate commander chose to concentrate his army between Grant and Vicksburg, a citadel he had committed to President Davis that he would defend to the end. Grant remained a step ahead of the Confederates and did not move directly on Vicksburg but instead struck eastward toward Jackson, the capital of the Magnolia state. On May 12, McPherson's men brushed aside John Gregg's Brigade at the Battle of Raymond and moved toward Jackson. Joseph Eggleston Johnston, Pemberton's superior and head of the Department of the West, raced to the state capital and important railroad crossroads. Despite the impending arrival of Rebel reinforcements by rail, Joe Johnston believed that the wisest course was to evacuate the city and hope to unite with Pemberton's army somewhere to the north of Jackson. It was a serious mistake on Johnston's part, as historian Ed Bearss points out, for it gave the audacious Grant a free hand to turn his attention away from Johnston's forces and march his Union army swiftly westward and deal with an isolated Pemberton in the open field.[30]

On May 14, John Pemberton moved 23,000 of his Vicksburg defenders eastward, following the tracks of the Southern Railroad of Mississippi. He had left the divisions of Smith and Forney in Vicksburg and had come forward with the divisions of Stevenson, Loring, and Bowen. Pemberton foresaw attacking Grant's army from the rear as Johnston provided a stout defense of

---

[29] Johnston, "Joseph W. Anderson, B. L.," 392. Upon the death of General Tracey, General Stephen D. Lee assumed command of the Alabama Brigade.

[30] According to Lieutenant Singleton Maxwell of the Milton Tigers (C), the 42nd set up their camp on the Big Black River "near A Cuff's farm" (Maxwell journal, 7); Bearss, *The Campaign for Vicksburg* 2:567–68.

Jackson. After a day's march, an uncertain Pemberton held a council of war with his generals at Edward's Station. Pemberton was dealing with a presidential directive from Jefferson Davis to hold Vicksburg at all costs, but he was also receiving dispatches from Joe Johnston to abandon Vicksburg and to concentrate his forces with those of Johnston somewhere near Clinton. Doing both at the same time seemed impossible. Pemberton had compromised and left two of his divisions behind in Vicksburg.

Confederate generals William Loring and Carter Stevenson urged Pemberton to move his expeditionary force south and east and hit Grant's supply line running from Grand Gulf to Jackson. There were reports of a Yankee supply train at Dillon's Plantation, west of Raymond. This third course of action was soon adopted by the council and orders were given for a march on Raymond. Well before dawn on the morning of May 15, the column set off, led by Loring's Division, followed by Bowen, and then Stevenson's large division. The army's supply train of 400 wagons, each pulled by four mules, trailed the procession. The infantrymen had left Vicksburg with three days of cooked rations in their haversacks. Almon Phillips, a private in the Milton Tigers (C), wrote his brother that it was a confident Rebel army that was marching to capture Grant's supply train. Seth Barton wrote home to Virginia that his dear friend Joseph Anderson rode a part of the way with him. The artillery major from Virginia, according to Barton, "told me that he had a presentiment that he should not survive the impending engagement; that the feeling was new to him, but did not at all depress him; that he could laugh at it, and if it had to be, he believed he could meet it as became him, but that he could not shake it off."[31]

Lieutenant William Drennan, an ordinance officer in Featherston's Mississippi Brigade, wrote his wife that, "An army in motion is a grand sight, with its long lines of bayonets glistening and flashing in the sun—the rumbling of the artillery and the noise of the trains—all conspire to throw over one a feeling of the greatness and magnificence of War." The march in muddy conditions along primitive roads, however, did not go well. As the army was marching eastward along the Raymond Road, they reached an impassable ford at Baker's Creek. The preceding week's rain had brought the swirling brown Baker's Creek to a flood stage. The perplexed officers ordered the men to turn northward along the western banks of the forty-foot-wide creek,

[31] Almon Phillips to his brother and sister, May 7, 1863, Fowler papers; Johnston, "Joseph W. Anderson, B. L.," 396.

marching for two miles to reach a bridge that crossed the angry water.[32]

After reaching and crossing the wooden bridge, the army marched east for two miles along the Jackson Road until they reached a country crossroads. Champion Hill, which had been a tall constant on their left, dominated the landscape. The lower slopes of this small mountain, marked with dense wood and undergrowth, were crisscrossed with ravines and gullies. At this juncture of dirt roads, the Jackson Road turned north and disappeared over Champion Hill. The Rebel column instead turned south on Ratliff Road, a small plantation trail that would reconnect them to the Raymond Road. It was 3:00 A.M., and after some twenty hours of marching, Stevenson's men were finally allowed to collapse by the roadside. "I shall never forget this tiresome march," recalled Captain Lowndes Calhoun of the Calhoun Guards (K). At dawn, Pemberton's army was strung out with Loring's Division having regained the Raymond Road and Bowen's and Stevenson's divisions still on the plantation trail. The wagon train, at the end of the column, was still on the Jackson Road, having not reached the crossroads yet. No one suspected that Grant's large army was rapidly approaching the sleeping and exhausted Confederates. Years later, Lieutenant James Monroe Gresham of the Walton Tigers (H) summed up the subsequent battle when he wrote, "Grant attacked Pemberton before he got up in the morning, and routed him before he could form a line of fight—a complete surprise."[33]

The Battle of Baker's Creek or Champion Hill

John Pemberton had never commanded troops in a pitched battle, and for that matter, neither had Carter Stevenson, who commanded Pemberton's largest division. Young Lieutenant Drennan did not think much of Stevenson's men. He wrote two weeks after the battle, "Stevenson's Division had never been under fire, and was composed of the last regiments from Georgia and Alabama—very ignorant men—far away from their homes, and had been marched heavily for the past few days." May 16 would mark the fourth straight day of marching for the men of the 42nd Georgia. Feeling both tired and sore, they started what promised to be another hot day as the sun came

[32] William Drennan to Georgia Torrey Drennan, May 15, 1863, in Edwards, ed., "Lt. William Drennan, Ordinance Officer."

[33] Calhoun, "Siege of Vicksburg; Historical Address Delivered by Judge W. L. Calhoun," in Bogle and Calhoun, *Historical Sketches*, 39; Gresham, "A Partial History of the Walton Tigers," 661.

up.

General Pemberton received startling news in the early morning that the capital of Jackson had fallen to Grant's men and that Johnston was in full retreat with his Confederate forces toward Canton. Old Joe still wanted to unify the two Rebel armies near Clinton, northwest of Jackson. To accomplish this directive, Pemberton would have to countermarch his command back to Edward's Depot. At this moment, though, his army was strung out for four miles on small country roads in a very remote portion of Mississippi. The logistics of turning around his force would not be easy. Carter Stevenson's Division, which had been bringing up the rear of the long column, would now become the vanguard for the march back to the depot. The brigade of Alexander Reynolds's Tennessee soldiers, the last of Stevenson's four brigades, was ordered to take the supply wagons back across Baker's Creek and head west for Edward's Depot. To do this meant turning 400 heavily-laden wagons around on a narrow road. Reynolds would be followed by the brigades of Stephen Lee, Alfred Cumming, and, finally, Seth Barton.[34]

It was quickly apparent to all, however, that Grant's much larger Army of the Tennessee had seized the initiative and was rapidly approaching the strung-out Confederate line of march from the west. At 7:00 A.M. Rebel officers began organizing their soldiers into battle lines. According to twenty-one-year-old E. M. Underhill, a former English army officer assigned to Stephen Lee's staff, "The men were completely broken down by hard marching and none had had anything to eat for nearly 36 hours. Under these auspices we awaited battle on a glorious May morning." Anticipating trouble, Stevenson ordered Henderson's 42nd Georgia, along with a section of the 1st Mississippi Light Artillery, to march the two miles back to the country bridge on Baker's Creek. This strategic water crossing had to be held if Pemberton ever needed to order a retreat.[35]

General Stevenson spent the next two hours getting the 400 wagons of munitions turned around and then moved across the bridge and out of danger. Alexander Reynolds and his brigade from East Tennessee covered the retreating train from the bridge to the Southern Railroad and then for another five miles above Edward's Ferry. Having gotten the wagon train dispatched toward Vicksburg, Carter Stevenson finally focused his attention on securing

---

[34] William Drennan to Georgia Torrey Drennan, May 15, 1863, in Edwards, ed., "Lt. William Drennan, Ordinance Officer."

[35] Stephen E. M. Underhill to his mother, June 28, 1863, in "Lieutenant S. E. M. Underhill…Account of Vicksburg Campaign."

the crossroads at the base of Champion Hill. If the Confederate army was to retrace their steps and make a successful escape across the bridge over Baker's Creek, this road juncture had to be held. The division commander could see that coming toward the crossroads from the east were 2,700 Federals under Peter Osterhaus. Three miles away at the southern end of Pemberton's strung-out army, Loring's Division drew up on a timber ridge line and prepared to defend the Raymond Road as Andrew Jackson Smith's Federal Division cautiously approached.

At 9:00 A.M. all seemed to be going well for the Confederates as they were aligned to defend their position against the two approaching Federal columns. This situation deteriorated when news reached General Stephen Dill Lee, who was holding the crossroads with his brigade of five Alabama regiments, that a large Federal force had been detected to his left. Thousands of Bluecoats were climbing the other side of Champion Hill and heading directly for the crossroads. Lee, the hero of Chickasaw Bayou, ordered his men from the crossroads and, by "files left," up the Jackson Road to the bald crest atop the seventy-five-foot high Champion Hill. According to General Manning Force of the 20th Ohio, Champion Hill was "a considerable eminence about a mile across." It was more imposing from the Union side, viewed as a steep hill, with knobs and ravines on all sides, and covered in large trees and underbrush, with the one exception of a cornfield on the very top. The battle would start here on Champion Hill, without the benefit of trenches or prepared positions for either side.[36]

Upon reaching the crest of Champion Hill, Stephen Lee was stunned by what lay in front of him. McPherson's entire Corps, 16,000 Bluecoats with several artillery batteries, led by the division of Alvin Hovey, was moving directly up the hill toward the Confederates in three long battle lines, with the clear intention of crushing Pemberton's left flank. John "Black Jack" Logan's 3rd Division was setting up to the right of Hovey and extending Grant's line toward the flat Baker Creek valley and overlapping Lee's own left flank by a wide margin. Lee sent his courier to Carter Stevenson requesting immediate help.

General Alfred Cummings, a native of Augusta, left two regiments of his brigade at the crossroads, to defend an advance from the east, and rushed three Georgia regiments, the 34th, 36th, and 39th, to line up to the right of Lee's Alabama regiments. As the Georgians filed into position and attempted

---

[36] Force, "Personal Recollections of the Vicksburg Campaign," 1:302.

to catch their breath, two divisions of Federal troops, some 10,000 men, swept forward into the maze of ravines and ridges of Champion Hill. "We met a stubborn resistance from the start," remembered Thomas Williams from the 56th Ohio. While the Alabama regiments held their position, Cumming's Georgia regiments were overwhelmed by Federal numbers, and it was not long before they began a pell-mell retreat back to the crossroads. "It was like a bank crumbling away before the action of a torrent," wrote the Englishman Stephen E. M. Underhill. The charging Yankees came down the hill and captured the Virginia artillery at the crossroads, but not before the cannoneers had fought for their guns until most of them were killed or wounded.[37]

Seth Barton's Brigade, meanwhile, was moving at a double-quick to reinforce the two besieged Rebel brigades atop Champion Hill. The 1,000 men from Georgia had a mile and a half of pastures, fences, and undergrowth to cross if they were to reach Lee's extreme left in time to stem the blue tide. Upon arriving on the field, Seth Barton did not like what he saw. "The position was not a good one," he wrote, "the country much broken and covered in most parts with dense woods." Through the timber, he could see that two Union brigades were moving unchecked past Lee's left flank. Barton posted the Georgia Cherokee Battery, under Captain Maximilian Van Den Corput, on his extreme left. The four 10-pounder Parrott rifles were positioned on a ridge next to the Roberts's farm house just north of the Jackson Road, about 600 yards short of the bridge over Baker's Creek held by the 42nd Georgia. From the Confederate perspective, there was simply not enough infantry to extend Barton's line from Lee's formation all the way to the Baker's Creek bridge. Division commander Carter Stevenson lamented, "My line, as will thus appear, was necessarily single, irregular, divided, and without reserves." According to Wesley Olin Connor, a twenty-two-year-old Rebel artilleryman, Corput's guns immediately came under fire from a battery of Yankee Napoleons about 600 yards away. "We fired fifteen or twenty rounds from each gun, but it was hot work," wrote Connor in his dairy, "Shot, shell, and shrapnel flew thick and fast around us." Artillerymen began to fall dead and wounded around the sweating Connor. On the left of Corput's four guns,

---

[37] Williams, *An Historical Sketch of the 56th Ohio Volunteer Infantry*, 44, 45; Stephen E. M. Underhill to his mother, June 28, 1863, in "Lieutenant S. E. M. Underhill...Account of Vicksburg Campaign."

the second section of the 1st Mississippi Light Artillery under Captain Samuel Ridley unlimbered and joined the battle.[38]

Meanwhile, General Barton was putting his three infantry regiments into a battle line, from left to right: the 40th, 41st, and 43rd Georgia regiments. With no time to spare, the Virginian pulled his sword high into the air and ordered the three regiments forward into the smoke-filled woods. Only the 52nd Georgia was held in reserve to protect the six artillery pieces by the Jackson Road. Federals from three Illinois regiments and the 20th Ohio were advancing in three lines when the surging Georgians hit their center. A year later in a private letter, Barton recalled, "Our little band charged with fury, broke through the first line, throwing it back in confusion on the second, which in turn fell on the third." Artillery Major Joseph Anderson, with cap in hand, led the 40th Georgia on the left flank into the fray. On the right side, Colonel Skidmore Harris was mortally wounded as he bravely led the 43rd Georgia into a volley of fire from the 20th Ohio. Unfortunately for Barton and his Georgia regiments, their advance of 300 yards pushed them well beyond Stephen Lee's beleaguered line and then into formations of Bluecoats and shell-fire from at least six Federal batteries. As Lee's overmatched Alabama troops began a slow withdrawal, it only made the situation worse for Barton by exposing his right flank to Union fire.[39]

To complete this perfect storm, John Stevenson's Union Brigade was steadily working their way unchecked through the forest and ravines pushing themselves well past Barton's left flank. When Stevenson's four regiments emerged in the open, they were three hundred yards from the half-dozen Rebel artillery pieces firing from the Roberts farmhouse. The overmatched Confederate general ordered his last reserves, the 52nd Georgia under twenty-seven-year-old Colonel Charles Phillips, to counterattack these oncoming Federals. One lone regiment, even moving in "handsome style," could not stop a brigade. The 52nd Georgia ran into hundreds of leveled Yankee muskets and were soon driven from the field, with Colonel Phillips lying wounded in four places. Eighteen-year-old Captain Augusta Boyd from Dahlonega fell dead in the fight after a minie ball struck him in his forehead.[40]

---

[38] T. B. Smith, *Champion Hill*, 214; *OR*, ser. 1, v. 24, part 2, 100. The Cherokee Artillery had been organized in Rome, Georgia, in 1861. It was assigned to Barton's Brigade at the Battle of Baker's Creek; Connor, Cherokee Artillery war diary, May 16, 1863; field report of Carter Stevenson, July 29, 1863, *OR*, ser. 1, vol. 24, part 2, 95–96.

[39] Johnston, "Joseph W. Anderson, B. L.," 396; Grabau, *Ninety-Eight Days*, 287.

[40] T. B. Smith, *Champion Hill*, 227.

The 81st Illinois and the 32nd Ohio charged Captain Ridley's 1st Mississippi Artillery guns on the far left. With shell and double-cannister ripping their ranks, the Bluecoats pressed forward across Austin Creek and deep ravines until they were close enough to level their rifles and kill men and horses alike. Carter Stevenson wrote of Ridley's two besieged guns, "Double-shotted, they were fired until, in many instances, the swarms of the enemy were in amongst them, the officers and men standing by them to the very last moment that they could be served." The section was captured despite the heroism of "the gallant Ridley, refusing to leave his gun, single-handed and alone fought until he fell, pierced with six shots." Captain Corput, to Ridley's right, had managed to get his four guns hitched and ready to escape, but, according to Private Connor, "a line of advancing infantry emerged from the woods about two hundred yards in our front, and as our infantry had given way, and the battery in that direction already taken, no obstacle to their progress was presented." Two guns were unlimbered, but it was too late for Corput to make an effective stand. He ordered his men to run for the rear. "I felt so mortified that I stopped once or twice to make sure that I was not the first to leave the guns," wrote Wesley Connor in his diary that night.[41]

Enveloped from both the left and right, Seth Barton ordered Robert Henderson, via courier, to advance the 42nd Georgia from the bridge at Baker's Creek. Henderson's men charged blindly forward, hoping to prevent the capture of the six artillery pieces. How much ground they covered in the 600 yards between the bridge and the Roberts house is uncertain, but their regimental banner was wrestled away from the unfortunate color bearer by Joseph R. Fitch of Company D of the 11th Indiana. With the previous regimental battle flag shot to pieces at Chickasaw Bayou, the 42nd was following the handsome flag of the Fowler Guards (D) that day. The 42nd Georgia must have gotten reasonably close to the guns as General Alvin Hovey, commander of the 12th Division, would mistakenly report that, "Colonel Macauley [of the 11th Indiana] has the battle flag of the Fowler battery."[42]

Henderson's 42nd soon found itself in a no man's land, unable to reach

---

[41] Field report of Carter Stevenson, July 29, 1863, *OR*, ser. 1, vol. 24, part 2, 95–96; Connor, Cherokee Artillery war diary, May 16, 1863.

[42] The 11th Indiana found itself out front of the other Federal units for much of the Battle of Champion Hill. The regiment reported losing 28 killed, 126 wounded, and 13 missing. The Fowler Flag ended the war in the possession of Captain John P. McGrew, the adjutant of the 11th Indiana. McGrew served as chief of the United States Capitol Police Force from 1900 to 1913 (Scribner, *Indiana's Roll of Honor*, 2:271–74; field report of Brigadier General Alvin Hovey, May 25, 1863, *OR*, ser. 1, vol. 24, part 2, 39–44).

Corput's guns, secure Barton's left flank, or protect the bridge they had left behind. The Georgia colonel elected to have his regiment fall back to the bridge in decent order. According to James Madison Brown of the Walton Tigers (G), Colonel Henderson said to his men, "We are between the enemy and the creek and they have got the bridge. Follow me and take care of yourselves." Mounted on his horse Old Roan, Henderson led a hasty retreat to Baker's Creek. Private Brown followed his colonel, "and when Rhone got out of the creek I, by the help of his tail, was at the bank of the creek." The bridge was soon secure and became a rallying point for Seth Barton, the 42nd Georgia, and several companies of the 40th Georgia. Only the 40th and 42nd, according to Barton's official report of the battle, came out of the disaster with "unbroken ranks."[43]

Barton's remaining shattered regiments were outnumbered and nearly surrounded. "Having vainly endeavored to cover the left with the Forty-second Regiment, brought forward for that purpose," reported the general of the brigade, "I was compelled to fall back." With hope lost, Barton's men broke and ran for the rear, not stopping as they passed the Jackson Road. The 124th Illinois, who would lose sixty-three men killed and wounded on Champion Hill, followed in pursuit and would later proudly claim that they captured more men from the 43rd Georgia than numbered in their own ranks. Major Joseph Anderson, who had led the 40th Georgia into the battle, was shot off his horse and mortally wounded near the dirt road as he attempted to re-form Barton's shattered ranks. Carter Stevenson, in command of one-third of the Confederate army, had, in less than two hours, been soundly beaten, and McPherson's Federals had gained complete control of Champion Hill and the crossroads below it. General Grant had successfully shattered Pemberton's left flank.[44]

At 1:00 P.M., General Pemberton ordered a dramatic counterattack by the troops that were holding his center. John Bowen's Division of 4,500 veterans from Arkansas and Missouri were the best combat troops in Pemberton's army. The movement began with Francis Cockrell forming his Missouri Brigade just below the crossroads. Colonel Cockrell rode down his line, holding a large magnolia flower in one hand and waving a sword in the other. Martin Green's Brigade followed on Cockrell's right a few minutes later. This elite line of Confederates slammed into the winded and disorganized Federals

[43] Brown, "James Madison Brown's 1925 Recollections," 438.

[44] Field report of Seth Barton, June 18, 1863, *OR*, ser. 1, vol. 24, part 2, 100; Records of regiments in the Union Army, MOLLUS, 3:335.

under James McPherson and drove them back over Champion Hill. Joseph Bogle of the 40th Georgia was with a handful of Georgians who had rallied near General Stephen Lee back on the Confederate side of the hill. They offered to follow the young general back into the fray, but, with a sad smile, the Charlestonian, who had lost three horses in the fight, replied, "Boys, there are not enough of you; if there were, we would redeem the day yet; as it is, you might as well go [retreat] with the rest."[45]

Ulysses Grant finally organized a successful stand at the bottom of the Federal side of Champion Hill with fresh troops under Marcellus Crocker. Low on ammunition, Bowen's Rebels began a slow fighting retreat back across Champion Hill and back to the crossroads. The English observer Underhill wrote, "Barton's brigade had been demolished and Green's wild Missourians, after having completely routed one Yankee division, retaken all our artillery and made 560 Yankees prisoners, had been surrounded by two other divisions and had only cut its way out with frightful loss."[46]

It was now time for the Confederate commander to extradite his outnumbered army. Fortunately for the Rebels, John Pemberton had had the foresight in the morning to order his engineers to construct a second wooden bridge across the lower ford of Baker's Creek. This new structure was several miles below the bridge that Seth Barton, and the men of the 42nd Georgia, still held. From Pemberton's perspective, the Federals were now between Barton and the demoralized Southern army. Leading the withdrawal over the second bridge were the shattered remnants of Barton and Cumming's Georgia brigades from Stevenson's Division, described by Underhill as "scarcely more than an armed mob." Only Stephen Lee's Yellow Hammer Alabama regiments retreated with any sense of order.[47]

Seth Barton, with several hundred men, still held the bridge at the upper crossing. By late afternoon, Barton could see that it was only the lower crossing that the Rebels were using for their retreat. When Federal artillery began to unlimber to shell his position, it was clear to the young general that it was time for his Georgia troops to abandon the upper bridge before their own avenue of escape was closed. The 42nd Georgia and remnants of the 40th Georgia were alone on their two-mile retreat up the Jackson Road north to

---

[45] T. B. Smith, *Champion Hill*, 240; Bogle, "Some Recollections of the 40th Georgia Regiment, C.S.A. in the Civil War," in Bogle and Calhoun, *Historical Sketches*, 16.

[46] Grabau, *Ninety-Eight Days*, 308.

[47] Stephen E. M. Underhill to his mother, June 28, 1863, in "Lieutenant S. E. M. Underhill...Account of Vicksburg Campaign"; Grabau, *Ninety-Eight Days*, 308.

Edward's Depot. It was a fighting withdrawal as they continually fended off sorties from Federal cavalry. Barton would stop in front of Edward's Depot and hold this position until nearly dark. Stragglers from Cumming's Brigade rallied to Barton's colors. Unfortunately, Barton's withdrawal from the upper bridge, at the first sign of trouble, had allowed Federal artillery to move quickly across the creek, unlimber, and begin shelling the departing Confederates who had just crossed the lower bridge. Loring's Division, which had not yet reached the lower crossing, could not cross Baker's Creek and continue west along the Raymond Road toward the Big Black River. Out of good options, General Loring abandoned his artillery and turned left, marching his entire division through the woods, and away to the south. His men escaped their Yankee pursuers and would eventually join Johnston's army in Jackson, but they would not be a part of the siege of Vicksburg.

Blame for the Confederate defeat at Baker's Creek was directed at Carter Stevenson and the collapse of his division atop Champion Hill. Particularly harsh words were directed at the Georgians in Barton and Cumming's brigades. Seth Barton argued that his troops had fought with courage, pointing out that his brigade had taken 42 percent casualties—58 killed, 106 wounded, and 737 captured. Numbers were not broken down by regiments, but the 43rd Georgia probably lost close to 60 percent of its soldiers in the fight. The 42nd Georgia, which had held the bridge for most of the battle, had at least 5 killed, 12 wounded, and 38 captured. Lieutenant Singleton Maxwell of the Milton Tigers (C) estimated in his journal that the regiment, "lost about 60 in killed, wounded and prisoners." It would take many more campaigns for the Georgia Brigade to overcome the stigma of its inferior performance at Baker's Creek.[48]

The Debacle at Big Black River

To be safe from Grant's pursuing army, General Pemberton needed to get his Confederate army over to the western side of the unfordable Big Black River using the bridge belonging to the Southern Railroad. Alexander Reynolds had, as the sun went down, saved Pemberton's supply train by crossing the river two miles further north using a pontoon bridge. After crossing the broad river, the Tennesseans destroyed the temporary bridge just as Federal infantry arrived. The disheveled brigades of Lee, Barton, and Cumming were

[48] T. B. Smith, *Champion Hill*, 383; Maxwell journal, 7; Barton, June 18, 1863, *OR*, ser. 1, vol. 24, part 2, 99–100.

the first to reach the giant railroad bridge at 8:00 P.M. on May 16. They were ordered to push on another five miles, before collapsing in exhaustion just short of the village of Mt. Alban. The observant Lieutenant Underhill was at the bridge and, "from a gentle eminence I could see Edwards' Depot and the fine plantations and country seats in a blaze, showing too plainly the advance of the pursuing foe."[49]

John Bowen's Division of Arkansas and Missouri troops followed the Georgians from the lower crossing, and they reached the bridge at midnight. Tired but still in good spirits, they were instructed to fill the trenches defending the eastern bridgehead, on either side of a brigade of untested East Tennessee transcripts under John C. Vaughn. Coming by way of Edward's Depot, General Barton with the 42nd Georgia, finally reached the Big Black River bridge at 1:00 A.M. They were ordered to cross the bridge and continue toward Vicksburg. Hoping that William Loring, with the last of the three Confederate divisions, would soon arrive, General Pemberton continued to hold the bridge through the night. The Confederate commander had no idea that General Loring was retreating southward and away from Pemberton's position.[50]

The next morning, John McClernand's XIII Corps attacked the 5,000 overstretched bridge defenders. Breaking through at Vaughn's trench line, the Federals rushed through the gap in the center of the Confederate's crescent line and toward the bridge. It quickly became a debacle for Bowen's Confederates as they gave up their exposed positions and rushed madly for the safety of the bridge. Before they could reach the bridge, 1,751 Southerners were cut off and became prisoners of war. The Federal spoils at the brief Battle of Big Black River Bridge included eighteen cannons and five Rebel battle flags. For the survivors on the western bank, there was little to do but burn the bridge and trudge the ten miles back to Vicksburg.[51]

---

[49] Stephen E. M. Underhill to his mother, June 28, 1863, in "Lieutenant S. E. M. Underhill…Account of Vicksburg Campaign."

[50] *OR*, ser. 1, vol. 24, part 2, 98.

[51] Bearss, *The Campaign for Vicksburg*, 2:680.

Chapter 4

The Forty-Seven-Day Siege of Vicksburg

"Rations were growing small, and at last were reduced to peameal bread and no meat, except mule meat—said by the men who ate it to taste very well."
—Captain Lowndes Calhoun,
describing conditions along the Vicksburg siege line.[1]

Rifle Pits at South Fort

Seth Barton's Georgia Brigade did not participate in the disaster at the Big Black River Bridge on Sunday, May 17. Already past the temporary defenses at the bridge, they led the procession of defeated Confederates down the Jackson Road on the six-mile march back to Vicksburg. The citizens of the river town met them with buckets of water and food to refresh the tired soldiers. Emma Balfour wrote in her diary, "I hope never to witness again such a scene as the return of our routed army." Colonel Winchester Hall of the 26th Louisiana, watching from a boarding house window, lamented, "I saw at once that it was our army in retreat and in utter confusion—a long line of stragglers. There would be a squad of infantry, a horseman—a gun—a few more infantry, and so on; with no more order than travelers on a highway." That evening the men set up a temporary camp and spent the night in the fields and hollows between the town and the empty rifle pits that formed a jagged nine-mile half-crescent around Vicksburg.[2]

On the morning of May 18, General Pemberton gave instructions to Carter Stevenson to have his division establish a defensive line from the tracks of the Arkansas & Vicksburg Railroad to the South Fort on the Mississippi River—a distance of four miles that covered the southern half of the Rebel defenses. North of the railroad tracks, where much of the future fighting was expected, the divisions of John Forney and Martin Smith held the Rebel line

---

[1] Calhoun, "Siege of Vicksburg; Historical Address Delivered by Judge W. L. Calhoun," in Bogle and Calhoun, *Historical Sketches*, 43.

[2] Hoehling, *Vicksburg*, 20, 22; Bearss, *The Campaign for Vicksburg*, 3:737.

with John Bowen's veteran Division serving as their reserve force. To conserve food supplies in Vicksburg, the Rebel cavalry was sent away with orders to do what they could to torment Grant's supply line.

After the Battle of Baker's Creek, John Pemberton had little trust in many of Stevenson's brigades. Historian Edwin Bearss maintains that Barton's Georgians, "the worst offenders in the Champion Hill debacle," were assigned the farthest post from impending danger to the rifle pits closest to South Fort and overlooking the river. Pemberton believed, correctly as it would turn out, that this would be the quietest sector during the impending siege. To the left of Barton would be Reynolds's Tennesseans and then Cumming's shaky Georgia Brigade. Champion Hill had shown that Lee's Alabamians were Stevenson's best fighters, so they were entrusted with the line from Square Fort to the railroad tracks. This piece of land was reasonably flat, and, as such, one of the few points where the Yankees could mount an infantry assault. Eleven Texas infantry companies under Colonel Thomas Waul and one company of Louisiana Zouaves were placed in a hollow south of the railroad and constituted a ready reserve.[3]

Large slave work crews had constructed the trench lines around Vicksburg in fall 1862 under the intelligent direction of Pemberton's chief engineer, Major Samuel Lockett. Consisting of redoubts, redans, and lunettes all connected by rifle pits, they formed a crescent a mile or so from the outskirts of Vicksburg. Substantial earthen and timber stockades commanded the six roads that entered Vicksburg and the tracks of the Southern Railroad that headed out of town east toward Jackson. As formidable as it all looked from a distance, the intervening winter and spring rains had eroded and weakened much of the earthen defenses. The Georgians immediately set to work with entrenchment tools to fortify their new position of irregular trench lines that held the commanding ridges. All the timber and cane in front of the lines were felled to create clear firing lines, and the fallen trees were lined up to construct a formidable abatis. The rifle pits were generally five feet deep with firing steps. The 42nd Georgia occupied a bluff that faced more south than east. To their left was the 43rd Georgia, where the land fell away to Stout's Bayou. Between the regiments was the Pointe Coupee Artillery from Louisiana. To the right of the 42nd was the trench line held by the 41st Georgia, and a quarter-mile further down the line rose the imposing South Fort, where

[3] Bearss, *The Campaign for Vicksburg*, 3:737. The Battle of Baker's Creek is referred to in the North as the Battle of Champion Hill.

the 40th Georgia protected the heavy artillery, under William Capers of Louisiana. Sitting on a giant bluff that fell off some forty feet into a swampy marsh that bordered the Mississippi River, the fort anchored the Confederate defenses south of the city. The primary responsibility for Barton's Georgia Brigade would be to stop any Federal advance up the Warrenton Road, a wagon trail that paralleled the river and ran north into the streets of Vicksburg.

May 1863, Siege of Vicksburg

The confident Yankees were not long in arriving, with Sherman's XV Corps forming Grant's right, McPherson's XVII Corps his center, and McClernand's XIII Corps the left flank. McClernand's four brigades moved west along the A & V Railroad directly toward the position of Stephen Lee's Alabama troops. Initially, there were no Federal troops in front of Stevenson's other three brigades. This was fortunate for the men in the 42nd Georgia because it allowed them to rebuild and strengthen their earthworks unmolested by Yankee sniper fire.

In the eighteen days since landing below Vicksburg, Grant's modest army had boldly marched 200 miles, won five battles, and managed to inflict some 5,000 casualties on the scattered Confederate defenders. Believing that his veterans could storm the Confederate positions around Vicksburg with a determined assault, General Grant ordered his entire army forward on May 19, 1863. The Bluecoats were repulsed at every point that day, as they had underestimated the will of the Confederate soldiers fighting from entrenched positions. Grant lost nearly 1,000 casualties on the nineteenth and gained little except favorable ground to place forward artillery emplacements. Lee's Alabama troops from Stevenson's Division again earned accolades for the stubborn defense of their position just south of the railroad.

Worried that Joseph Johnston in Jackson, Mississippi was organizing a new Southern army to come to the relief of Vicksburg, Grant decided that time was not his friend and that he needed to redouble his effort to break Pemberton's defensive line and seize the prize that he and his army had invested the last year in pursuit of. In the following days, supplies were stockpiled for Grant's army, and the well-placed Federal artillery gained clear superiority over the battlefield. Mud splattered Bluecoats constructed wooden scaling-ladders and pushed their trench lines to within 300 yards of the Rebel lines.

On the third day of the siege, the mortar fleet of Admiral David Porter anchored on the Arkansas shore just south of Vicksburg and began its bombardment of the besieged city with 13-inch mortars. Each big gun was mounted on its own flatboat, and cranes were used to lower the 350-pound shells into the mortar. According to Edward Reynolds of the 46th Illinois, "About a dozen men are required to man each boat or gun, and when ready to fire, all step on shore, raise on tiptoes, cotton in ears and a piece of lead between the teeth, the lanyard is pulled and the shell is sent on its way." To escape the exploding missiles, huge caves were excavated into the steep hillsides by terrified civilians. Being two miles south of Vicksburg, Barton's Georgians were the closest Rebels to the river and Commodore Porter's fleet. They soon realized that they were in more danger from mortar fire coming from the river behind them, than they were from infantry fire to their front. Bomb-proofs had to be dug even deeper into the yellow clay-like loess soil. In the trenches next to the Georgians, Calvin Smith of the 31st Tennessee wrote in his diary that each night he could, "see the course of the shells marked out by the fire tail bomb ejecting a star like tail and whisteling noise passing high in the heavens seemingly among the starey orbs of the heavenly firmament."[4]

On the night of May 21, the Georgians were surprised to see campfires burning in their front. They were soon under skirmish fire from the 2nd Illinois Cavalry. All of this was a Union ruse as General John McClernand was protecting his left flank as his Corps began, the following morning, a massive assault on Lee's Alabamians holding the Railroad Redoubt. To McClernand's right, Sherman's Corps was given the task of attacking the Stockade Redan Complex, which was the highest and most heavily defended Confederate position. Under the cover of a furious artillery barrage, thousands of Federal soldiers, from late morning to early afternoon, assaulted the Rebel lines in separate and disjointed attacks. In almost every case, they were thrown back by determined Confederate soldiers. Only at the Railroad Redoubt did a partial Federal breakthrough even begin to threaten Pemberton's defense of Vicksburg. The redoubt was soon recaptured after a brave counterattack by a contingent of Waul's Texans. In all, Grant's Army of the Tennessee took more than 3,000 casualties on May 22 with little to show for their sacrifice.

Barton's Georgia Division on the far right of the Confederate line had a miserable time the night before the main Federal assault. In addition to Un-

---

[4] Hoehling, *Vicksburg*, 31; Reynolds diary, 17; Carnes, "'We Can Hold Our Ground,'" 30.

ion cavalry in their front, three ironclads—*Benton*, *Mound City*, and *Carondelet*—steamed up the Mississippi that night to shell their entrenchments around South Fort. Early on the morning of May 22, Admiral Porter again sailed up the Mississippi with the three ironclads plus a fourth, the *Tuscumbia.* In one hour, the firepower of the Federal ships made short work of the outmatched Rebel guns at South Fort, as well as the next Southern battery located a half mile up the river. With the lower batteries silenced, Porter sailed north toward Vicksburg with the three ironclads of his original force. The *Tuscumbia* was left behind to watch over South Fort and to encourage Barton into believing his position was untenable. A mile farther up the Mississippi River, the Marine Hospital Battery had seven big coastal guns and seven smaller guns to contest Porter's progress. For the next two hours, the determined Louisiana artillerists dueled with Porter's forty-six naval guns at close range. Their accurate fire on the ships did enough damage to keep Porter from sailing any closer to Vicksburg.[5]

The men of the Georgia Brigade had not been quiet onlookers as Grant assaulted the Confederate lines on May 22. At 9:00 A.M. Barton's forward outpost, which was three miles down the Warrenton Road from South Fort, began firing on advancing Federal troops. The Bluecoats were the vanguard of General John McArthur's 3rd Brigade, six regiments from Illinois and Iowa, who had marched north from the Federal supply base near Port Gibson. One piece of Grant's master plan for the general assault on May 22 was to have McArthur approach the Confederate line by following the Warrenton Road along the Mississippi River. Any Federal breakthrough along the general Rebel line, thought Grant, might compel Pemberton to pull reinforcements northward from Stevenson's idle brigades to plug the breach. McArthur's men, who could by now read the signs painted on the buildings of downtown Vicksburg, might be in an ideal position to capture South Fort and then, in quick fashion, follow the unguarded Warrenton Road north, straight into the streets of the "Gibraltar of the Confederacy." It would be up to the men of the 40th Georgia to hold South Fort against any attack from McArthur's Hawkeye Brigade.[6]

The Midwesterners moved slowly as they were under converging fire from four well-placed Rebel batteries, but, by 11:00 A.M., they were closing in on South Fort. The men in the fort attempted to mount two 12-pound howitzers, probably belonging to Company A of the 14th Mississippi Light

[5] Bearss, *The Campaign for Vicksburg*, 3:855–56.

[6] Bearss, *Campaign for Vicksburg*, 3:856–57.

artillery, but accurate shell-fire from the *Mound City* and *Carondelet* forced them to abandon the effort and return to the shelter of the fort's bomb-proofs. Sharpshooters from the 11th, 13th, and 16th Iowa were brought forward, and by midafternoon, they had worked themselves to within forty yards of the ditch fronting the contested fort. McArthur issued orders up and down his battle line to prepare to storm the enemy's rifle pits. Within the fort and along the trenches, the officers of Barton's Brigade held to their resolve and told their men that there would be no retreat or surrender. They vowed to one another that here in these Mississippi trenches, looking across a level cleared plateau broken up by small ravines, was their moment of redemption—the opportunity to avenge their catastrophic defeat at Champion Hill only six days earlier.

It was at this extraordinary moment that everything suddenly changed. Two staff officers from General McClernand reached McArthur in quick succession. General McArthur ignored the first dispatch, but the second dispatch was signed by Ulysses S. Grant. Both messages called for McArthur to disengage and march to the right in support of McClernand's troops, who were presently assaulting the Alabama troops holding the railroad redoubt. It was a foolish order as the topography of the land meant that it would take McArthur the better part of a day to reach McClernand's position. John McArthur, a brigade commander, knew this but also felt that he had no choice but to obey the orders of his major general. The Scottish native recalled his sharpshooters and withdrew his brigade from their advanced position. The circuitous route around Stouts Bayou meant that McArthur's men did not reach the XIII Corps sector until noon the following day. One can only speculate what would have happened if the brigades of Barton and McArthur had been allowed to square off on the afternoon of May 22, 1863.[7]

With the complete failure of the May 22 general assault, Grant and his generals were now content to employ a classic siege of Vicksburg and its defenders. Without the open back door that later kept Atlanta alive in 1864, it would only be a question of time before Pemberton's 28,000 defenders would face starvation and be forced to surrender. With 50,000 troops, and reinforcements arriving each week by way of Memphis, Grant wanted his engineers to add to the pressure on the Rebels by steadily pushing trenches and mines closer to the Confederate defenses. The twelve miles of Federal lines—camps, trenches, and gun emplacements—ran from the Yazoo River north of

[7] Bearss, *Campaign for Vicksburg*, 3:857.

Vicksburg across hills and ridges all the way around to the Warrenton Road south of the entrapped city. Historian Stephen Ambrose believed that "The whole thing anticipated the more sophisticated trench warfare of the Western Front in World War I."[8]

The wildcard for Grant was Joseph E. Johnston, who by May 24 had gathered 22,500 Southern soldiers for his Army of Relief, which he divided into two wings. General William Loring in Jackson had roughly half the Rebel infantry with him, while General William H. T. Walker, a Georgia fire-eater, commanded the northern wing based in Canton, Mississippi. Another veteran division under General John C. Breckinridge, some 5,500 troops, was headed for Jackson as well. It was a large enough force to give General Grant pause as Johnston and Pemberton together would have roughly the same number of soldiers as the Federals currently besieging Vicksburg. Grant reacted by organizing a Federal Expeditionary Corps, some 12,000 Yankees under the aggressive Frank Blair, to keep Johnston and his Confederates away from Vicksburg.

On the evening of May 25, John McArthur's Hawkeye Brigade, which had returned to the Warrenton Road sector, was ordered away again, this time as part of Blair's new Expeditionary Corps. Replacing them was a brigade of four Illinois regiments under Colonel Cyrus Hall. After completing a dusty two-day march up the Warrenton Road from Snyder's Bluff, the newly arrived Federals established their camp 600 yards south of South Fort. According to the account of Colonel Benjamin Dornblaser, five companies of the 46th Illinois were placed in a forward position, "from and near a cotton gin on the Warrenton road westward to a swamp which we supposed and had been informed was impassable." As Edward Reynolds of the 46th Illinois wrote in his diary, "We were so near the enemy that we could hear the noises of work in their works."[9]

Confederate chief engineer Samuel Lockett interpreted the arrival of Hall's Illinois regiments that evening as a Federal attempt to seize the muddy floodplain that extended between the bluff and down to the Mississippi River. With summer approaching, the mighty river had been steadily receding, and more of the marshy ground was being exposed. Lovick Thomas, who was

---

[8] Ambrose, "Struggle for Vicksburg," 54.

[9] Bearss, *The Campaign for Vicksburg*, 3:886. The Hawkeye Brigade included the 14th, 15th, 46th, and 76th Illinois Infantry regiments (B. Dornblaser to S. P. Bourquin, June 7, 1963, in Woodbury, *Complete History of the 46th Illinois Veteran Volunteer Infantry*, 54; Reynolds diary, 18).

captain of the Gwinnett Beauregards (A) at the time, shared Lockett's belief that this spit of land could unlock the back door to Vicksburg. Thomas recalled to an Atlanta newspaper reporter in 1900 that the floodplain had become contested territory. He loosely explained that at some point during the siege, "I was selected to command a brigade skirmish line which extended from the banks of the Mississippi River through a dense cane break to a point on the Warrenton road below Vicksburg." He remembered that his instructions were "to hold it at the sacrifice of every man under me." Standing at an elevated vidette post closest to the river, Thomas was startled to spot Yankees disembarking from gunboats and working their way toward his position. "As soon as they saw us, and the surprise seemed mutual, they opened fire," he went on, "and the fight which began at that point was one of the most thrilling and interesting I witnessed during the entire siege." Thomas found himself in the awkward position of commanding a line of Georgia men, none of whom were members of his own regiment, fighting for control of a swamp. He marveled more than thirty years later, "I have never seen any mention anywhere of this particular engagement, which to my mind was as fierce as any that occurred during the war."[10]

On May 25, General Seth Barton was ordered to extend his perimeter by felling trees and constructing an abatis across the floodplain all the way to the banks of the river. That evening under the cover of darkness, Lieutenant George Donnelly led his fatigue party into the muddy swamp. They were protected by a strong patrol of riflemen under Colonel William Ezra Curtiss of the 41st Georgia. Observing the Union campfires of the newly arrived 46th Illinois, Curtiss from Carroll County saw an opportunity to create a bit of chaos that evening, and he ordered Donnelly's work party back to South Fort. Following a little-used trail through the thick cane break, the Georgians quietly passed the flank of the Federal pickets and circled behind them. With more shouts than shots fired, the men from the 41st sprang their trap on the unsuspecting Midwesterners. In a manner of minutes, 113 Yankees were prisoners of William Curtiss's sortie.

"A rebel yell broke out directly in our rear," wrote Edward Reynolds of the 46th Illinois, "between us and our reserve camp, they charging with fixed bayonets, without a moment's warning and capture half of our regiment, rushing the boys rapidly through the dark woods into the rebel works." The commotion had Cyrus Hall call out his Federal brigade as he believed they

[10] "Newspaper Clippings of Famous Vicksburg Fight," *Atlanta Constitution,* February 18, 1900, 6.

were under a general attack. This order gave the saboteurs the time they needed to escort their prisoners through the darkness and back to the Rebel lines. After marching their prisoners into Vicksburg, "Our men and the Georgians got crackers, clothing, hats, and many other treats," wrote Lieutenant Calvin Smith of the 31st Tennessee. Anticipating a counterattack, Stevenson reinforced Barton by ordering the 43rd Tennessee into the trenches to the left and right of South Fort. In the weeks that followed, Chief Engineer Lockett had fatigue parties from Stevenson's Division build rifle pits along the bluff line, stretching from South Fort to the Marine Hospital Battery.[11]

Both sides remained on edge as any Rebel breakout, it was believed, would be along the river near South Fort. Confederate messengers regularly used this portion of the line to slip through gaps in the stretched out Federal line. According to Illinois Private Reynolds, "The rebel Gen. Johnston with a strong force is constantly threatening our rear, making it necessary for us to strongly fortify and picket both front and rear, and we are sometimes on these lines for several days and nights without relief." On May 30, scouts reported that the Federals to the right of the corduroyed Warrenton Road had withdrawn. Colonel Robert Henderson was instructed to send out a small reconnaissance force to determine the truth of the report. Captain Lowndes Calhoun of the Calhoun Guards (K) was given command of the detachment. "I moved before day," recalled Calhoun, "and was not long in ascertaining the fact that they were in force, and with much difficulty again reached the fortifications, followed by the enemy, who were, however, soon repulsed by our artillery and riflemen in the trenches."[12]

June 1863, Siege of Vicksburg

Conditions in the humid Southern trenches deteriorated quickly under a burning June sun, supplemented by drenching rains, damp fogs, and heavy dews. Not having sufficient numbers to rotate his troops off the front line, General Pemberton forced his men to live day and night in the squalid rifle pits. With an unlimited supply of ammunition, the emplaced Federal batteries, some 220 guns, continuously fired shells into the Rebel breastworks each

[11] Reynolds diary, 18; Bearss, *The Campaign for Vicksburg*, 3:887; Carnes, "'We Can Hold Our Ground,'" 31; W. D. Taylor, *"A fit representation of Pandemonium,"* 350.

[12] Reynolds diary, 18; Wildman, *Iowa's Martyr Regiment*, 106; Calhoun, "Siege of Vicksburg; Historical Address Delivered by Judge W. L. Calhoun," in Bogle and Calhoun, *Historical Sketches*, 34.

day. At night, the mortars fired from the river would rain down on their positions, making sleep extremely difficult. Vicksburg historian Edwin Bearss points out that even worse for Southern soldiers was the damage done by Federal sharpshooters. "Posted in the hollows," Bearss writes, "their rifle-muskets steadied and sighted toward the Rebel defenses," they waited for exposed heads and shoulders to be silhouetted against the skyline. Lieutenant Underhill, on Stephen Lee's staff, wrote from the Rebel trenches, "It is astonishing to compare the ammunition they waste with the lives they destroy. All along our lines there is one unceasing rattle of musketry from the enemy's parallels." On May 24, Calvin Smith wrote in his diary, "The Yankies would make ditches after dark and would sharp shoot our men from light to dark as regular as clock work." Confederate casualties during the siege included 875 killed and 2,169 wounded.[13]

Federal trenches and underground mines, mostly in areas that had been previously attacked, steadily moved closer to the Rebel rifle pits along numbered approaches. In response, Confederates made several night sorties to slow Grant's progress. On June 22, Georgians in Cumming's Brigade made a gallant strike on the Hall's Ferry Road, managing to capture a colonel and twelve men. The following night, in retaliation, the Federals attacked one of Cumming's picket stations. By late June as little as thirty yards separated the two armies in some places. Johnny Reb and Billy Yank were soon talking back and forth and, during short truces, actively trading coffee for tobacco. Fortunately for Barton's men, their trench lines were never seriously threatened until the final week before the July 4 surrender.[14]

Above all, the Confederate soldiers in Vicksburg were hungry. They started on half-rations, and food supplies only diminished with each passing day. Each night, small details cooked the food behind the lines and brought the rations to the troops in the rifle pits. Water was drawn from the Mississippi River and carried to the trenches in barrels. The flour supply was exhausted by June, forcing the men to bake their bread using ground peas and cornmeal.[15]

Captain Lowndes Calhoun remembered dividing one small loaf of pea bread with Lieutenant Benjamin Franklin Walker, both of the Calhoun

---

[13] Bearss, *The Campaign for Vicksburg*, 3:955; Carnes, "'We Can Hold Our Ground,'" 30; Stephen E. M. Underhill to his mother, June 28, 1863, in "Lieutenant S. E. M. Underwood...Account of Vicksburg Campaign."

[14] Ambrose, "Struggle for Vicksburg," 60.

[15] W. D. Taylor, "*A fit representation of pandemonium*," 33.

Guards, "with mathematical precision." On June 1, Captain L. P. Thomas of the Gwinnett Beauregards (A) wrote satirically in his diary from the "Hill Trench Hotell" that the bill of fare for the officers in his Gwinnett company consisted of cold pea bread, accented with brown sugar and peanuts. All was digested with the help of murky "Chalk Water—from branch." Soldiers from both sides complained throughout the siege about the poor quality of the water. Hordes of disease-carrying insects only added to the misery of those living in the stench-filled trenches.[16]

On June 18, Billington A. Wilson of the Walton Blues (G) wrote his family a thoughtful letter. He was not ready to give up the struggle, and, like Robert E. Lee, believed that duty supplanted all other desires. "I love you and my children as dear as any one heart," he wrote his wife, "but if I never get home until I desert, you will never see me. I had rather die than bear the name of a deserter." Deeply religious, he was at peace with death. B. A. was proud of his young sons, Johnny and Buddy, for planting the family crop, and he beseeched the boys to look after their mother and younger sisters, and "if your country calls for your assistance to defend the rights guaranteed to us by our forefathers, step to defend our rights."[17]

On June 15, the twenty-ninth day of the siege, Barton's men were confronted with a new danger: the arrival of General Francis Jay Herron and his division of 5,000 men from the Department of Missouri. Made up of two Midwestern brigades with sixteen guns in its light artillery, these fresh troops began crossing the river at the ruined remains of Warrenton, just four miles from the southern-most point of Pemberton's line. It was an exhausting four miles, however, across ridges and then down into heavily wooded valleys that left the Midwesterners struggling with the humidity as they drew closer to the trenches belonging to Seth Barton's Georgians.[18]

Upon arriving at the siege lines, the five infantry regiments in General William Vandever's Brigade spread out to hold the Warrenton Road and stabilize Herron's mile and a half long left wing, while General William W. Orme's Brigade moved to the right and covered the line to Hall Ferry Road. The next morning Herron's men drove in the pickets from the 40th Georgia, who had been guarding the Warrenton Road. That night, Federal fatigue parties, escorted by strong patrols, fanned out to begin constructing rifle pits

---

[16] Thomas, "Vicksburg Menu," vol. 1, pg. 107, Thomas collection, BRC.

[17] Billington A. Wilson to home, June 18, 1863, UDC collection, 6:365. B. A. Wilson was killed at Bentonville in the last battle of the war for the 42nd Georgia.

[18] Wildman, *Iowa's Martyr Regiment*, 106.

on both the east and west sides of Stouts Bayou. Work was slowed by stubborn fire from skirmishers in the 40th Georgia. Twenty-four-year-old Bendick Bendickson, a Norwegian corporal in the 38th Iowa, was shot in the head and killed by a Rebel sharpshooter.[19]

The vegetation and terrain also did not favor the Federal advance. "In low places between hills the canebrake grows so rank and thick that it almost impossible to get through," wrote Edward Reynolds in his diary, "and we cut openings carrying and piling it in open places. After a few days in the sun, this wilting cane lets off a stink that would make a dog sick." On the morning of June 17, General Barton reported to General Stevenson that his men were unable to retake the picket posts in front of the 40th Georgia because "the enemy was found in so great force that the loss which must have ensued in forcing the point would have been out of proportion to the value gained."[20]

Between this newly forming Federal line and Barton's curious Georgians was an 800-yard plateau that was cut up by ravines. Well-placed Confederate artillery had a commanding view of the plateau, and any direct assault by Herron's regiments would have to cross this no man's land and pay a frightful cost. Five hundred yards in advance of the Confederate line on a ridge east of Stouts Bayou was a small listening post soon to be dubbed by both sides the "Cane Pit." Herron's first victory of sorts was the capture of the Pit and with it four embarrassed privates of the 42nd Georgia—one being Thomas Wilkerson of the Harper Guards (E). The Cane Pit quickly became a source of contention and changed hands on more than one occasion. Captain Lowndes Calhoun recounted that Private David Garrison of Atlanta was killed in the Cane Pit on June 22, resisting "so bravely an attack on his post, killing the officer commanding the assault, and fighting until riddled with bullets." Company B of the 20th Wisconsin claimed that, on that evening, they had rushed a forward rifle pit and captured thirteen soldiers belonging to the 42nd Georgia. Calhoun maintained that the enemy buried Garrison and marked his grave with a board that read "Bravest of the brave." As a member of the 20th Wisconsin wrote home on June 27, "We take, almost every day, from four to twenty prisoners, and the 41st and 42nd Georgia must be pretty well thinned out by this time."[21]

---

[19] Wildman, *Iowa's Martyr Regiment*, 110.

[20] Reynolds diary, 18; *OR*, field report from Barton, June 17, 1863, *OR*, ser. 1, vol. 24, part 2, 342.

[21] Bearss, *The Campaign for Vicksburg*, 3:951; record of events, Company B, 20th Wisconsin Infantry, in Hewitt, et al., eds., *Supplement to the Official Records*, part 2, vol. 76, ser.

Most disconcerting to Carter Stevenson and his Confederates was the deployment of three Federal artillery batteries. The 1st Missouri Light Artillery deployed on the Warrenton Road and aimed four 10-pound Parrotts and two 3-inch Armstrongs at Barton's rifle pits. An equal complement of guns was unlimbered east of Stout's Bayou. On June 24, Herron's Federals, under the engineering direction of Captain Arnold Hoeppner, began digging parallel trenches within 400 yards of Barton's line and had worked their sharpshooters to within 150 yards of the Georgian's position. In late June, two large rifled 42-pounder guns were taken from the warship *Benton* and hauled by large wagon teams to the "Lone Tree." Located some 400 yards from South Fort, this oak tree next to the Warrenton Road was the only point of cover in the plateau facing the Rebel fort.[22]

On the morning of July 1, the "Benton Battery," as Herron designated it, opened up at near point-blank range on South Fort. Everything seemed to shake as the large ship guns battered the earthen fort. Gunboats anchored in the Mississippi added their firepower to the onslaught. Fortunately for the men of the 40th Georgia inside the fort, Captain W. C. Capers and his experienced Louisiana artillerymen were ready to respond with their 10-inch mortar gun. By noon, Capers had found the range and dropped a mortar shell directly between the two guns, killing two and wounding four, all from the 34th Iowa. The guns were not damaged, but their platforms would have to be rebuilt. They were not ready to resume the bombardment until July 3. Carter Stevenson would write a month after the surrender that the favorable ground in front of Barton's and Reynold's brigades prevented the enemy from getting close to his lines. When the Yankees did approach, "they were invariably disposed by the daring sallies of the garrison."[23]

88, 213; Calhoun, "Siege of Vicksburg; Historical Address Delivered by Judge W. L. Calhoun," in Bogle and Calhoun, *Historical Sketches*, 34; "From the Twentieth Wisconsin Regiment," *Milwaukee Daily Sentinel*, July 7, 1863. The sixteen men from the 42nd Georgia captured in the Cane Pit were returned to their regiment at the final surrender of Vicksburg. They were paroled with the rest of the regiment on July 7, 1863.

[22] Engineer report of Captains Frederick Prime and Cyrus Comstock, November 29, 1863, *OR*, ser. 1, vol. 24, part 2, 174; report of Major General F. J. Herron, June 24, 1863, *OR*, ser. 1, vol. 24, part 2, 317.

[23] Wildman, *Iowa's Martyr Regiment*, 124; report of Major General F. J. Herron, June 24, 1863, *OR*, ser. 1, vol. 24, part 2, 318; report of General Carter Stevenson, July 29, 1863, *OR*, ser. 1, vol. 24, part 2, 345.

Surrender of the Hill City

Throughout June, the Southern defenders of Vicksburg strained their ears to hear cannon fire in the distance. All held the false hope that Joseph Johnston's army would soon be arriving to break the siege. "Oh! For Genl Johnson & plenty of Pea bred," wrote L. P. Thomas on June 21, the thirty-fourth day of the siege. It was July 1, however, before the Relief Army of 32,000 troops with seventy-eight cannons began marching for the Big Black. In the meantime, Lincoln's government had spent the month of June pouring in reinforcements to Grant's army, and there were now 77,000 officers and men in the Army of the Tennessee. Johnston had little prospect for success against such a large well-supplied force, and, as a cautious general by nature, he restricted his ambitions to creating enough of a distraction that Pemberton might find a way to break out of Vicksburg.[24]

Lieutenant Joseph Danforth Barnes of the 20th Iowa, like his superiors, was convinced that any Southern escape from the doomed city would be over and through his regiment. The farthest left of any of Herron's regiments, the 20th Iowa sat against the Mississippi River and across from the 40th Georgia. "You were not allowed to remove any of your clothing at night," Barnes wrote, "and you must sleep with your trusty rifle by your side, and your cartridge box buckled around you with forty rounds on the inside."[25]

Pemberton, in turn, asked his senior generals if their weakened men had enough energy left to fight their way out of Vicksburg and then conduct the long marches needed to escape Grant's army. Most of the generals believed the task to be impossible, as the men in the foul trenches were feeble at best. Nearly every home in Vicksburg had, by then, become a military hospital. The Confederate medical director counted 5,878 officers and men who were under his department's care, including 152 officers and 1,694 enlisted men from Stevenson's Division. Seth Barton wrote Carter Stevenson that only half his men were fit to take the field. Rations, at this point, were down to paltry amounts. On July 1, a debilitated L. P. Thomas noted in his diary that breakfast was "boiled pease & brown sugar" and dinner had improved to "mule beefe Pease & brown sugar—no bread." Alfred Cummings and A. W. Reynolds said the same about the condition of the men in their brigades.[26]

---

[24] L. P. Thomas, "Gen'l Johnston and Pea Bread," vol. 1, pg. 107, Thomas Collection, BRC.

[25] Barnes, *What I Saw You Do*, 27.

[26] Bearss, *The Campaign for Vicksburg*, 3:1279; L. P. Thomas, "Gen'l Johnston and Pea

Only Stephen D. Lee, among Stevenson's senior officers, thought an escape should be attempted, "taking in view its importance and the interest of our Confederacy." In a July 2 letter to his commander, Stevenson wrote, "my men are very cheerful, but from long confinement in the trenches and short rations are necessarily much enfeebled, and a considerable number would be unable to make the march and undergo the fatigues which would probably be necessary in a successful evacuation of this city." Acknowledging the obvious, Pemberton began surrender negotiations with Grant at the same time that the two Eastern armies began fighting a decisive three-day battle at Gettysburg, Pennsylvania. When Grant finally agreed to the parole and exchange of the Vicksburg defenders, rather than transporting them to prison camps in the North, a deal was made.[27]

On the bright and pleasant morning of July 4, 1864, soldiers on both sides bantered back and forth. They had been doing so all through the previous day after an armistice had been declared between the two armies. Captain William Perkins Black of the 37th Illinois, already a recipient of the Medal of Honor at Pea Ridge, wrote his mother that he had a short talk with Captain Thomas Hill of the Fulton Blues (I), after which they, "parted again, he to his fort, I to my rifle pit and so ended the courtesies of war." When Captain Lowndes Calhoun of the Calhoun Guards (K) was told of the impending surrender, he held to the opinion that, "if left to him, he would make a general assault on the enemy and sacrifice every man." After forty-seven days under the direct sun, the twenty-five-year-old Calhoun, who was by now, "as dark as an Indian," was not in a position to give such commands.[28]

At 10:00 A.M. on July 4, Rebel soldiers raised white flags above their works. Each Confederate column marched out of their entrenchments and formally surrendered in the fields between the two antagonists. The 495 members of the 42nd Georgia dressed ranks, stacked their rifles, and laid their flag poles and colors across the tee-pee shaped stacks of guns. Next, the men placed their knapsacks, belts, cartridge-boxes, and cap pouches on the ground next to each circle of stacked rifles. Captain Calhoun called the surrender, "a spectacle that has seldom been witnessed in the world's history." According

---

Bread," vol. 1, pg. 107, Thomas Collection, BRC.

[27] General S. D. Lee to General Carter Stevenson, July 1, 1863, *OR*, ser. 1, vol. 24, part 2, 348; operations report of General Carter Stevenson, July 29, 1863, *OR*, ser. 1, vol. 24, part 2, 346.

[28] Mullins, *The Fremont Rifles*, 211; "W. L. Calhoun Biography," James G. Kenan Research Manuscript Collection, 2.

to Edward Bearss, there was a subdued tone to the whole affair, with barely a sound except for a few orders of command. For the Yankee soldiers standing nearby, however, a spontaneous cheer erupted. The siege had finally ended. "To me it was a sad day," thought Calhoun, "and I felt that a soldier's death was preferable." With heads downcast, the Georgians marched back to their original trenches and waited for rosters to be confirmed and signed parole forms issued. Federal records show that 30,638 men had made up the Vicksburg garrison. Of that, Barton's 1st Brigade counted 1,952 enlisted men and 150 officers.[29]

Only Blackjack Logan's Division and William Vandever's Brigade of Herron's Division, which had been posted in front of the 42nd Georgia, were sent through the Confederate lines and into the streets of Vicksburg to serve as a provost guard. After passing through the camps of Barton's men, Lorenzo Bayes of the 38th Iowa wrote in his diary that defeated Rebels were "a hard looking nation and nearly starved to death." General Herron, upon seizing Barton's works in his immediate front, reported that "Several of these were well built, and from their strength could not have been carried by assault without heavy loss."[30]

A controversy soon emerged over whether Rebel officers would be allowed to take their black servants with them as part of the parole. This understanding was more complicated than it seemed because many of the slaves had become integral members of their Confederate regiments. "Henry," for example, attended to the needs of Dr. John Wilson, surgeon of the 42nd Georgia, and "won the esteem of the entire command by his willingness to do any favor for any member of the regiment." Grant was willing to allow a master to take one slave if that was the declared desire of the servant, but General James McPherson, speaking for the abolitionist wing of the Union army, filed a strong protest and orders were countermanded, and the passes were revoked on the afternoon of July 7.[31]

---

[29] Wildman, *Iowa's Martyr Regiment*, 126; Calhoun, "Siege of Vicksburg; Historical Address Delivered by Judge W. L. Calhoun," in Bogle and Calhoun, *Historical Sketches*, 35; Bearss, *The Campaign for Vicksburg*, 3:1292; *OR*, ser. 1, vol. 24, part 2, 324. Barton's Brigade according to surrender documents included one general, five colonels, three majors, thirty-nine captains, ninety-one lieutenants, and two chaplains ("The Vicksburg Captures—Officers and Men," *Southern Watchman*, September 2, 1863, 1).

[30] Wildman, *Iowa's Martyr Regiment*, 127; field report of Major-General F. J. Herron, July 6, 1863, *OR*, ser. 1, vol. 24, part 2, 319.

[31] "The Inside of Atlanta: Reunion of Veterans of the 42d Georgia Regiment," *Macon Telegraph*, July 23, 1886, 4.

On Friday, July 10, Barton's regiments were marched into Vicksburg, where each man gave his parole of honor and signed by name or mark. The next day, the first of the unarmed Confederate columns left Vicksburg after going through Federal inspection teams. The rank and file could carry their clothing and what rations they could carry out of the Confederate commissary. Officers kept their side-arms and were allowed one horse. Slaves attempting to accompany their masters were turned back and forced to return to Vicksburg, where they were assigned menial tasks by Federal officials. Freedom would be months away for these black men. Carter Stevenson's Division was the last to depart on July 12.[32]

Left behind in Confederate hospitals were 3,600 soldiers too weak to depart Vicksburg. Most of these soldiers were loaded onto steamboats and taken down the Mississippi River for a formal exchange in Mobile Bay, Alabama. Fifty-six men from the 42nd Georgia boarded the U.S. Steamer *St. Maurice* on July 30, 1863. They stopped in Federal-held New Orleans where the sickest of the patients were left. This explains why a handful of men from the 42nd Georgia are buried at Cypress Grove Cemetery in the Crescent City. The *St. Maurice* continued to Mobile, where the Confederates from the Vicksburg hospitals were formally exchanged off Fort Morgan on a hot August 5, 1863.

Returning Home

In the days between surrender and parole, there was much visiting between Yankee and Rebel camps. Three days' worth of Federal rations were issued to the 30,000 near-starving Confederates. Several Federal soldiers sought out Captain Lowndes Calhoun of the Calhoun Guards (K). In 1862, they had been his prisoners at the Cotton Mill Prison in Madison, Georgia. Many walked several miles to confront him in a reversal of roles. Some came to praise the twenty-six-year-old son of the Atlanta mayor, but others cursed him. Captain Ira Ford of the 18th Wisconsin invited Calhoun and nine of his former guards to his tent where he offered them coffee. "They had not had a cup of coffee for four months," recalled Ford. "I told them that I knew just how they felt."[33]

Lieutenant Seth Maxwell of the Milton Tigers (C) wrote that on July 12 they were "permitted to march out with side arms, private property, and one

[32] Bearss, *The Campaign for Vicksburg*, 3:1302.

[33] Erquitt, "Cotton Mill Prison Canteen Revisited," 30.

four horse wagon to the Regt." Their destination was Demopolis, Alabama, where Pemberton hoped to hold his surrendered army together until an exchange could be arranged. The Georgians reached the Big Black River on the first night of the march and camped. Over three days, according to Maxwell, they marched east over dusty roads sixty-four miles to the tiny community of Cato, Mississippi. Many of the weak, including Captain Paris and Lieutenant Maxwell of the Milton Tigers, paused here to rest for a few days. The main body of the regiment pushed on another seventy-five miles to Enterprise, Mississippi. By this time, General Pemberton's hope of holding his army together had vanished, as officers and soldiers were determined to go home first. Jefferson Davis reluctantly allowed Pemberton to grant furloughs of up to thirty days. The Georgians were soon boarding boxcars on the Mobile and Ohio Railroad.[34]

On the evening of July 22, Maxwell's train pulled into Mobile, Alabama. Another train took them to Montgomery, Alabama, where they camped the night of July 24. The next evening, they reached Atlanta. The regiment was officially disbanded, and the grateful men went home, welcoming an opportunity "to spend a short time with dear friends and loved ones," as well as regain their physical strength.[35]

The Georgia men who arrived home in late July were physical wrecks, not much better than skeletons. In the six months in Mississippi, they had been brutally tested by weather, starvation, battle, disease, and siege warfare. An astonishingly high number of soldiers from the 42nd Georgia died in the three months after the Vicksburg surrender. According to incomplete Confederate service records, some two dozen men, who had been left in the Vicksburg hospitals after the surrender, perished in the weeks that followed. Among the dead was Captain Thomas Davis, who had led the Walton Blues (G) since William Malcom had been mortally wounded at Chickasaw Bayou. Another twenty men died while trying to return to their native Georgia. About twenty more managed to reach home, only to die in their beds in the coming weeks. In short, more than sixty men in the 42nd Georgia died from the after-effects of the Vicksburg Siege.

Return to Service

By their condition of "parole," the surrendered Confederate regiments from

---

[34] Maxwell journal, 8; Bearss, *The Campaign for Vicksburg*, 3:1307.

[35] Maxwell journal, 9.

Vicksburg could not return to service until they were declared formally exchanged, which happened on September 12, 1863. Initial efforts were made in August to have the men report to Atlanta before going into "parole camps." Few men reported for duty on August 15, so those who did muster were sent home. Three days later the officers had more success, and a parole camp was established in Decatur, Georgia. The men drew arms and drilled until October 1, 1863.[36]

On September 30, Colonel Robert Henderson requisitioned two tents, as he was "destitute of the above." Not all the men reported back to service. Many were not physically able to return, and others simply did not want to continue the fight. Details were sent out to gather the stragglers. On October 17, thirty-year-old James A. Sharpton of the Walton Blues (G), was killed by Corporal Thomas A. Boyd. Jim Sharpton had deserted his company in December 1862 after the regiment returned from Kentucky. Boyd was attempting to arrest Sharpton at his home in Gwinnett County.[37]

Much of the ranks of the commissioned officers in the 42nd Georgia had been decimated from the campaigns of 1862 and 1863. Of the ten captains who left the training fields of Camp McDonald with their respective companies, only five of the captains were present in Decatur for the reorganization of the 42nd Georgia. Some companies were harder hit than others. The Fowler Guards (D) of DeKalb County, for example, lost all four of their original officers. Captain Nathan Clay resigned in bad health in Vicksburg in February 1863. First Lieutenant William White had resigned with rheumatism at Bean Station; 2nd Lieutenant Marshall Wright had died of measles in Knoxville, and 2nd Lieutenant James Young had been killed at the Battle of Chickasaw Bayou. Replacements were elected from the ranks. Private Elijah Clarke became the captain of the company, and privates Henry Burgess, Richard Morris, and James Moore gained the rank of lieutenant.

In late September and early October, friends of Robert Henderson of the 42nd Georgia began a letter-writing campaign to gain the promotion of their colonel to brigadier general. The campaign was led by the Newton County politician William W. Clark, who, at the time, was a member of the Confederate Congress in Richmond. President Jefferson Davis and Secretary of War James A. Seddon received endorsements on Henderson's behalf from

---

[36] Ibid., 10.

[37] September 30, 1863 Requisition Request, Service Records of Colonel Robert Henderson, NARA, M226, roll 28; Confederate service records of James A. Sharpton, NARA, M226, roll 54.

generals Howell Cobb, Carter Stevenson, and Seth Barton. Clark summarized the appeal by writing on October 8, "I am sure that Col. Henderson is deserving this advance—from his education—the manner he has conducted himself since he has been in the service and his high character as a man of good habits, correct and high toned deportment."[38]

Henderson's supporters were playing catch-up, however, as the friends of Abda Johnson had been campaigning on behalf of their colonel since the spring. On April 11, 1863, twenty-two senators and sixty representatives of the Georgia Assembly had signed a petition in Milledgeville urging the Confederate secretary of war to promote Colonel Johnson, of the 40th Georgia, to brigadier general. The legislators were under the incorrect impression that General Seth Barton had been promoted and therefore there was a vacancy at the top of the Georgia Brigade, then at Vicksburg. "Col. Johnson is an educated gentleman," they wrote in their petition, "has been in the army for more than a year, has often commanded temporarily the brigade, has high moral character, and unsurpassed integrity and the Department will not regret it if he is promoted." The principal backer of Colonel Johnson was Augustus Wright, who represented the 6th Georgia District in the Confederate House of Representatives. Wright wrote Jefferson Davis in April 1863 that Abda Johnson was a lawyer of distinction, and "the son of Col. Lindsay Johnson, one of our wealthiest and most influential citizens of my district." On October 1, 1863, General John Pemberton responded to an inquiry made by General Samuel Cooper, an adviser to President Davis and the highest ranking general in Confederate service. The former commander of the Vicksburg citadel called Johnson "an excellent officer and if his health will admit, is I think fully competent to take command." Pemberton then made the point that Abda Johnson had, for many months, been physically unfit for duty. His choice to lead the brigade was Robert Henderson, the second most senior colonel. He wrote Cooper, "I am also informed, He [Henderson] has already collected his full regiment for duty—I believe him entirely capable of exercising the command of a brigade should it devolve upon him."[39]

---

[38] W. W. Clark to President Davis, October 8, 1863, Rome, GA, NARA, M226, roll 28.

[39] Georgia legislators in Milledgeville to Confederate secretary of war in Richmond, April 11, 1863; Augustus Wright to President Jefferson Davis, April 11, 1863; Johnson's Confederate Service Records, NARA, M226, roll 32. Two letters endorsing Abda Johnson to Wright are in Johnson's file from Nathan Land (written in Manassas) and Mark Johnson (written in Cartersville); John C. Pemberton to Samuel Cooper, October 1, 1863, Richmond, Virginia.

On October 2, 1863, the newly reconstituted 42nd Georgia went by rail from Decatur to Big Shanty and camped there for the night. The next day they reached the small mountain community of Chickamauga, scene of a horrific battle fought there only fourteen days earlier. Reinforced with two divisions under General James Longstreet from the Army of Northern Virginia, Braxton Bragg's Army of Tennessee had driven the Federal Army of the Cumberland, under William Rosecrans, out of northwest Georgia. The two-day Battle of Chickamauga produced more than 32,000 casualties and would stand as the largest Confederate victory in the western theater. The 42nd camped on the battlefield until October 18. With the beleaguered Federals nearly surrounded in Chattanooga, Bragg had an opportunity to tighten the noose.

Carter Stevenson's Division was ordered into the long valley between Chattanooga and Knoxville with the intent of keeping the timid Ambrose Burnside, and his Army of the Ohio locked up in Knoxville. They were a logical choice for Bragg as Stevenson's men had operated in eastern Tennessee for the better part of 1862 and knew the lay of the land. Traveling twenty-eight miles by train, the 42nd reached Charleston, Tennessee, on October 19, 1863. Now on foot, they gradually worked themselves closer to Knoxville, reaching Sweetwater Station on October 28 and Loudon, Tennessee, on October 31.[40]

It was during fall 1863 that turmoil in the Confederate high command led to one of the worst strategic mistakes the Southern military would make in the war. Bad blood between Braxton Bragg and many of his generals grew to a boil after the Battle of Chickamauga. No relationship was worse than that between Bragg and Lee's "Old War Horse," General James Longstreet. The Georgian had brought two divisions of his corps down by rail from the Army of Northern Virginia, and they had contributed immensely to the Rebel victory at Chickamauga. At the suggestion of Jefferson Davis, Longstreet's men were ordered out of the Confederate lines around Chattanooga and into eastern Tennessee. Longstreet's forces were assigned the task of driving Burnside out of Knoxville and then, in the spring, returning to Virginia and Robert E. Lee's army. Problems abounded with this plan.

Bragg was dividing his army at a critical moment, sending some of his best troops on a mission that had little prospects of success. Further, the prin-

---

[40] Cozzens, *The Shipwreck of Their Hopes*, 103; Nash, *Civil War Diary of Francis H. Nash*, 18.

cipal commander in the Western Theater would now be heavily outnumbered in the coming battles around Chattanooga. Longstreet was no better off as he would not have enough troops to capture Knoxville. Knowing this, the departing general requested that Stevenson's Division that was already near Knoxville be assigned to him. Bragg refused and ordered Stevenson back to his positions around Chattanooga in early November. By November 13, the 42nd was holding the valley between Lookout Mountain and Missionary Ridge. Had Stevenson been assigned to Longstreet, in the many what if's of history, the Battle of Knoxville may have turned out differently, and the men of the 42nd would have found themselves new members of Robert E. Lee's Army of Northern Virginia.[41]

The Confederate Army of Tennessee was reorganized on November 12, 1863, under Special Order #294. Under this directive, Corps commander Daniel Harvey Hill was replaced by John C. Breckinridge. Several brigades were shuffled about by General Braxton Bragg in the reorganization, including Barton's Brigade, which moved from Stevenson's to Stewart's Division. Bragg was, in part, punishing generals who had signed a petition calling for his dismissal as commander of the Army of Tennessee. General Seth Barton was replaced by forty-five-year-old Marcellus P. Stovall, an Augusta native who had led the 3rd Georgia Battalion and had fought well at Stones River. Stovall's appointment meant that neither Abda Johnson nor Robert Henderson would gain the appointment to brigadier general.[42]

Marcellus Stovall, the new commander of the Georgia Brigade, was, unfortunately, something of a broken man. His only son, twenty-year-old Henry Stovall, had died of typhoid fever in a Confederate hospital in Rome, Georgia, on March 31, 1863. Henry had been the sergeant major of the 3rd Georgia Battalion and "an estimable young man, who was fast growing into favor with his military associates as a model soldier and capable officer." His death took a toll on his father, and General Marcellus Stovall would, in the coming campaigns, often be absent from his troops as he tried to regain his

---

[41] Cozzens, *The Shipwreck of Their Hopes*, 105; The Battle of Knoxville had personal meaning to L. P. Thomas, as his uncle, Colonel Henry Philip Thomas of the 16th Georgia, was killed in the unsuccessful frontal assault on Fort Sanders on November 19, 1863. Henry's possessions were brought home by his servant, Ben Thomas. Henry was carrying the Revolutionary War sword of his grandfather at the time of his death. His overcoat had sixteen bullet holes in it.

[42] Special Orders 294, issued by General Braxton Bragg, November 12, 1863, *OR*, ser. 1, vol. 31, part 3, 685.

own health. When present, though, Stovall was a capable brigade commander. On March 25, 1864, an officer for the 52nd Georgia, writing under the pen name CHOICE for the *Southern Watchman*, noted that the entire brigade had confidence in the general, noting that "His mild, pleasant manner and manly deportment will at once make a favorable impression on all who know him."[43]

The Augusta native had enrolled at West Point in 1836 as a teenager but had not graduated. It is believed that Stovall's cadet roommate was William Tecumseh Sherman of Ohio. According to a Bill Arp story, Marcellus was visited at West Point by his sister Cecilia, "a great belle and beauty in her day." Sherman was smitten by her charms, but his advances were turned down. According to Arp, when Sherman was advancing his Union armies across the Etowah River in May 1864, he came across "Etowah Heights," the magnificent home of his unrequited love, Cecilia Stovall Shelman. The mistress of the home had fled before his arrival, but the Federal commander placed guards around her home for its protection. General Sherman also wrote her a note, saying "You once said that you pitied the man who ever became my foe. My answer was that I would ever shield and protect you. That I have done. Forgive all else; I am but a soldier."[44]

General Alexander Peter Stewart, their new division commander, was a Tennessean who was well regarded in the Army of Tennessee, having fought particularly well at the Battle of Chickamauga. The slender, red-headed Stewart graduated from West Point in 1842 in the same class as James Longstreet, William Rosecrans, and D. H. Hill. Before the war, Stewart was a professor of mathematics at Cumberland University in Lebanon, Tennessee. Stewart was assigned five brigades for his division: his old Tennessee brigade now under Otho French Strahl, two Alabama brigades under Daniel Adams and Henry Clayton, a Louisiana brigade under Randall Gibson, and finally Stovall's Georgia Brigade. Stewart was keenly aware that Stovall's men were exchanged Vicksburg prisoners and therefore might be militarily questionable. It is a commonly voiced belief among historians that the Confederate regiments exchanged after the Vicksburg surrender never managed to regain their intensity and will to fight. But, if there were an exception to this rule, it was the 42nd Georgia.[45]

---

[43] Newspaper clipping found in the notebook of Anna M. Stovall; "Camp 52nd Georgia Regiment, 1864 March 30."

[44] Cunyus, *History of Bartow County, Georgia*, 250–53.

[45] Elliott, *Soldier of Tennessee*, 141.

Missionary Ridge

While the Confederate high command bickered with one another, Ulysses S. Grant was restoring order to the Federal Army in Chattanooga. George Thomas, now the "Rock of Chickamauga," was put in charge of the Army of the Cumberland. Major General William Sherman marched elements of two corps, some 20,000 men, across north Alabama to relieve Chattanooga. Joining Sherman and Thomas, General Joseph Hooker, the former commander of the Army of the Potomac, brought another two corps, the XI and XII Corps, roughly 15,000 troops, from the eastern theater to Chattanooga. By November 24, Grant had 56,000 troops to Bragg's 44,000 men, and he was ready to go on the offensive. That evening Hooker charged up and captured Lookout Mountain in the noted "Battle Above the Clouds." Hooker now threatened Bragg's left flank on the south end of Missionary Ridge. Bragg ignored this danger and placed most of his forces on the north end to contest the anticipated flanking attack from Sherman's divisions. This decision meant that Breckinridge's Corps would, by itself, have to hold the southern two-thirds of Missionary Ridge. Stewart's Division was given the important responsibility of holding the extreme southern flank of the Confederate army.[46]

On November 24, 1863, John Breckinridge ordered Stewart to send two of his regiments and an artillery battery to guard Rossville Gap. Located at the bottom of Missionary Ridge just inside the Georgia state line, the gap was a break in the ridge that provided a clear path into the Confederate rear. Back in September, Rosecrans's Army of the Cumberland had escaped through this pass using the Ringgold Road which ran from Chickamauga to Chattanooga. The hamlet of Rossville at the eastern end of the gap also held substantial stores for the Rebel army. Stewart sent Stovall's two largest regiments, the 42nd and 43rd Georgia, as well as Dawson's Georgia Battery under Lieutenant Ruel Wooten Anderson, to Rossville Gap. General Stovall would remain at the top of Missionary Ridge with his three remaining regiments. Command of this valley detachment would fall to Colonel Robert Henderson, who was senior to Colonel Henry Kellogg of the 43rd Georgia. Henderson was short one company, the Milton Tigers (C), who were detailed to guard the brigade commissary supply wagons fifteen miles to the rear near Bird's Mill.[47]

---

[46] Ibid., 139–45.

[47] Cozzens, *The Shipwreck of Their Hopes*, 11; Elliott, *Soldier of Tennessee*, 146; Maxwell

At daybreak on November 25, the leader of the 42nd selected a position on the crest of a small eminence some 300 yards to the front of the gap under a narrow belt of timber and placed his battery in the interval between the two Georgia regiments. Company A, the Gwinnett Beauregards, was pushed ahead another 150 yards to serve as skirmishers. All was quiet until 11:00 A.M. when a strong line of Yankee skirmishers appeared 300 yards distant. These were members of the 27th Missouri Infantry from Peter Osterhaus's First Division of the XV Corps. Behind them was Hooker's army coming down from Lookout Mountain. For the last several hours the Federal columns had been slowed down as timber bridges were laid across Chattahoochee Creek. Henderson, with only two regiments, was in no position to stop such a large force. "We engaged him at once," reported Colonel Thomas Curly of the 27th Missouri, "and in such a manner as to cause him to open upon us with his artillery, but with little injury to us, his sharpshooters doing all the injury we received."[48]

For the next two hours, the Georgians held the Federals at bay. Henderson was impressed by Captain Anderson and his battery, whose guns "were managed with skill, their missiles almost invariably exploding in the right place, causing the enemy to scatter and retire in confusion." The German-born Osterhaus sent his 1st Brigade, under General Charles Woods, eight infantry regiments in all, up along the crest of the ridge along the right side of the valley. Anderson punished them with artillery fire as they climbed the ridge. Four Iowa regiments of the 2nd Brigade under Colonel James Williamson ascended the steeper Missionary Ridge on the left of the gap. Down in the valley, Henderson's 600 or so men were in danger of being flanked on both sides or even being surrounded and captured. The Rebel colonel ordered a retreat, "the infantry falling back in good order, followed by the artillery, the skirmishers retiring slowly, contesting every inch of ground."[49]

L. P. Thomas and his Company A bought precious time for the artillery to escape the gap and the infantry to establish a position on the last ridge in

---

journal, 11.

[48] Report of Colonel Robert J. Henderson on the Battle of Missionary Ridge, December 12, 1863, in Hewitt, et al., eds., *Supplement to the Official Records,* part 2, vol. 31, ser. 54–55, 170–73; report of Colonel Thomas Curly, 27th Missouri Infantry, December 13, 1863, *OR,* part 2, vol. 31, ser. 55, 610–11.

[49] Report of Colonel Robert J. Henderson on the Battle of Missionary Ridge, December 12, 1863, in Hewitt, et al., eds., *Supplement to the Official Records,* part 2, vol. 31, ser. 54–55, 170–73; report of General Peter J. Osterhaus, November 23–27, 1863, *OR,* ser. 1, vol. 31, part 2, 598.

the valley. Henderson would, in the subsequent report of the fight, praise "the coolness and gallantry" of Captain Thomas and his company of skirmishers. The infantry, in their new position, stood for another hour firing volleys into the advancing Bluecoats. With time, however, Osterhaus's men gained the heights on the Confederate right and "commenced enfilading my line from right to left." There was nothing more the badly outnumbered Georgians could do, so they retreated from Rossville Gap and headed east on the Ringgold Road. Federal colonel Curly reported that his regiment, fighting at Rossville Gap, had killed seven Confederates and wounded thirty-six more. They had also captured, "large quantities of arms, fixed ammunition for artillery and infantry, and a large supply of commissary and quartermaster's supplies."[50]

This small fight at Rossville Gap was a prelude to one of the most dramatic events in the war, the Federal capture later that day of Missionary Ridge. General Sherman's Army of the Tennessee started that morning with an attack on the Confederate right flank. They met stubborn resistance from the veteran divisions of Patrick Cleburne, William Walker, and Carter Stevenson. To take some of the pressure off Sherman's faltering lines, Ulysses S. Grant ordered George Thomas and his Army of the Cumberland to seize the rifle pits at the base of Missionary Ridge. To the astonishment of almost everyone on both sides, the Federal soldiers did not stop at the rifle pits but began to scale the imposing mountain. The Confederates, it turns out, were in no position to stop them. They had dangerously extended their lines on the southern portion of the ridge, had laid out their defensive line on the natural crest as opposed to the military crest, and, if that was not enough, General Bragg's confusing orders had created gaps in their already thin defensive lines. As the Yankees crested the hill, the Rebels fled down the other side of the mountain in chaos. Missionary Ridge was the most humiliating defeat suffered by the Army of Tennessee in the entire war.

What is less known about this famous battle is the role of Joseph Hooker's army, which came up the southern side of Missionary Ridge and rolled up Stewart's overextended defenders. John Breckinridge had tried to protect the Confederate left flank by personally marching Clayton's tired brigade, under the temporary command of Colonel James Thadeus Holtzclaw,

---

[50] Report of Colonel Robert J. Henderson on the Battle of Missionary Ridge, December 12, 1863, in Hewitt, et al., eds., *Supplement to the Official Records*, part 2, vol. 31, ser. 54–55, 170–73; report of Colonel Thomas Curly, 27th Missouri Infantry, December 13, 1863, *OR*, part 2, vol. 31, ser. 55, 610–11.

down to the southern base of Missionary Ridge. The Alabamians had been driven off Lookout Mountain the night before. Instead of finding Henderson's two regiments in control of Rossville Gap, they encountered three divisions of Hooker's Corps beginning to work their way up the southern end of Missionary Ridge. It was about 4:00 P.M. when Charles Cruft's 1st Division of IV Corps moved directly up the crest of the ridge, with John Geary's Division sweeping along the western side and Osterhaus's Division moving north along the eastern slopes of Missionary Ridge. Holtzclaw and his Alabama Brigade never had a chance and were soon surrounded on three sides. Almost half the brigade, 660 of these Confederates, were captured. General Osterhaus claimed that his Federal division alone captured 2,000 of Bragg's 4,146 men, who became prisoners of war that afternoon. Stovall's Brigade lost 69 men out of 1,291 engaged in the day's action. Also lost were the brigade's large store of supplies. Commissary Captain Joseph Simmons had not been able to find transportation for his goods in the sudden Rebel collapse.[51]

Retreating from Rossville Gap, the 42nd, and 43rd Georgia under Colonel Robert Henderson reached a point where the Red House Ford Road and Lafayette Road fork. "Having no knowledge of the country," wrote Henderson, "I was greatly confused as to the direction I should take." One road went north parallel to the rear of Missionary Ridge. The other road headed south toward Ringgold. Henderson marched his men southward until 5:00 P.M. when they came to a bridge over a country stream known as West Chickamauga Creek. Osterhaus and his Federals were not in pursuit as they had headed north at the fork attempting to cut off as much of the Confederate retreat from Missionary Ridge as they could.[52]

Colonel Henderson chose to defend the bridge, which he did until midnight when a written dispatch reached him from Braxton Bragg. The overall Confederate commander ordered Henderson to burn the bridge and to then to "defend the passage of the ford against the enemy to the last extremity." This directive put Henderson in a difficult position as he had learned from

---

[51] Elliott, *Soldier of Tennessee*, 152–54. The Milton Tigers (C) under Captain Henry Paris were detailed to protect the supplies but had no animals to move the 267 pounds of beef, 607 pounds of bacon, 979 pounds of flour, 446 pounds of corn meal, 122 pounds of peas, 30 pounds of rice, 30 pounds of candles, and 47 pounds of soap. The supplies were abandoned, according to the statement made by Henry Paris in Dalton, GA, on November 30, 1863 (Confederate Service Records of Henry Paris, NARA, M226, roll 46).

[52] Report of Colonel Robert J. Henderson on the Battle of Missionary Ridge, December 12, 1863, in Hewitt, et al., eds., *Supplement to the Official Records*, part 2, vol. 31, ser. 54–55, 170–73.

reliable local citizens that there were, in fact, three possible places to ford the creek. He lacked the troops to defend all three crossing points. At 9:00 A.M. on November 26, without orders, he directed his two regiments to fall back toward Ringgold, "for the safety of my command and the better protection of the main army." Outside of Ringgold, he placed his two regiments in a line of battle and held the position until 2:00 P.M. Seeing no Federals, the Georgians fell back to Ringgold where they found the rest of Stovall's Brigade. It was Thanksgiving Day 1863.[53]

---

[53] Report of Colonel Robert J. Henderson on the Battle of Missionary Ridge, December 12, 1863, in Hewitt, et al., eds., *Supplement to the Official Records*, part 2, vol. 31, ser. 54–55, 170–73.

Chapter 5

The Atlanta Campaign of 1864

"I was sitting on Kennesaw, my horse, under a heavy cannonade of solid shot. I saw one coming, a solid ball, very rapidly, which passed a little to my right. The first one I ever saw, which seemed to be coming on an air-line right straight to me."
—L. P. Thomas at the Battle of Kolb's Farm, June 22, 1864

"Water very good, Yankees plentiful"
—Private Zachariah Armistead, Walton Tigers, at New Hope Church, May 29, 1864

Dalton and the Atlanta Campaign

After the debacle on Missionary Ridge, Braxton Bragg resigned his commission as commander of the Army of Tennessee and traveled to Richmond to become Jefferson Davis's military adviser. The shattered western army of the Confederacy that Bragg left behind attempted to pull itself together near the Georgia town of Dalton, on the Western and Atlantic Railroad. On December 10, 1863, the star-crossed Army of Tennessee was under the temporary command of General William Hardee. With Marcellus Stovall absent, his Georgia Brigade was temporarily under the command of Colonel Robert Henderson. This meant that the men of the 42nd took their orders from twenty-six-year-old Major William Henry Hulsey. The young officer's boyish face seemed to belie the powerful image he portrayed while riding his fiery black stallion. "No officer in the command," remembered Captain Andrew West, "attracted more attention or drew around him more enthusiasm." Stovall's Brigade was one of four brigades in the infantry division of Major General Alexander Stewart. On December 14, the 42nd Georgia reported 444 men present on its muster roll, of whom 412 were in fighting condition. On paper, the 42nd had 822 names, but half those names were of men who were dead, disabled, or gone never to return. With 444 men present, it was still, by far, the largest regiment in the brigade. The 40th had 223 present;

41st, 269 present; 43rd, 283 present; and the 52nd, 279.[1]

President Jefferson Davis had expected "Old Reliable" William Hardee to assume the command of the Army of Tennessee on a permanent basis, but the Georgian surprised the Confederate president and turned down the position. The task of replacing Braxton Bragg was then given to the dapper Joseph Eggleston Johnston. Private Sam Watkins, in his famous 1882 book *Co. Aytch*, described Johnston as a "perfect dandy." Watkins wrote, "He ever wore the very finest cloths that could be obtained, carrying out in every point the dress and paraphernalia of the soldier, as adopted by the war department at Richmond, never omitting anything, even to the trappings of his horse, bridle and saddle." A professional soldier since his days at West Point, the fifty-seven-year-old Johnston had been appointed as one of the Confederacy's five full generals. The small-in-stature Johnston was both a Virginian and a Southern aristocrat. He and Jefferson Davis despised one another, as these extremely sensitive men found the other to be intolerable. Communications between Richmond and the commander of the Army of Tennessee in the coming campaign would be strained at best, with Davis too often listening to his Richmond cabinet and the discredited Braxton Bragg while discounting the cryptic reports of his field commander.[2]

Johnston was, above other things, a competent commander, and he did much during the Dalton winter to bring order to the Army of Tennessee. The soldiers found their rations improved, new clothing and shoes issued, and daily military drills a part of the camp routine. The men soon had a great deal of respect and affection for their "Old Joe." A quartermaster report for the 42nd showed the extent of the refitting. In spring 1864, the regiment received 100 haversacks, 89 cartridge boxes, 89 cap pooches, 89 waist belts, 59 canteen straps, 103 bayonet scabbards, and 88 bayonets. Between the different companies, two different rifles were used in the 42nd. Major Hulsey signed for 66 British Enfield 57' caliber rifles, along with 21,110 minie ball cartridges. He also accepted 24 54' caliber Austrian rifles with 4,444 cartridges of that caliber and 3,490 percussion caps.[3]

---

[1] The December 13, 1863, muster roll was the last full accounting of the 42nd Georgia that would survive the war (Andrew J. West, "Anniversary of the Battle of Atlanta; Thrilling Story of the Enormous Encounter," *Atlanta Georgian & News*, July 22, 1904; "Organization of the Army of Tennessee, with regimental strength," December 14, 1863, *OR*, ser. 1, vol. 31, part 3, 822–28.

[2] Watkins, *"Co. Aytch": A Side Show of the Big Show*, 131.

[3] Ordinance report for the 42nd Georgia for the quarter ending June 30, 1864, Confederate service records of Col. William Hulsey, NARA, M226, roll 31.

The Atlanta Campaign began in earnest on May 7 in the mountain passes near Dalton. Two weeks earlier, on April 19, General Joseph Johnston held a grand review of the reconstituted Army of Tennessee on an open plain at Taylor's Ridge. With Major Hulsey on leave, Captain L. P. Thomas was the officer of the day and at the head of the 42nd Regiment during the review. He was in especially good spirits about his own Company A. "We are in fine health," he wrote his wife Jennie. "I have about 65 (men) present, and all for duty. I never saw the health of the men better." Company A, however, would be without 1st Lieutenant Darling Peeples McDaniel, who died in the Dalton winter camp on January 21, 1864, when, straining the fortunes of war, a large tree fell on him. Lieutenant John Medlock was now second in command of the company.[4]

"Uncle Joe," it seems, had done wonders that spring to restore morale and discipline to the 36,000 infantry soldiers present that day. The Army of Tennessee was now divided into two corps. The first, under Lieutenant General William Hardee, passed in double column. Hardee had four divisions under generals Benjamin Cheatham, Patrick Cleburne, W. H. T. Walker, and William Bate. The second corps, now under Lieutenant General John Bell Hood, came next in the mile-long procession. His three divisions were led by generals Thomas Hindman, Carter Stevenson, and Alexander Stewart. Most of Hindman's regiments were from Alabama or Mississippi, brought up from coastal duty on the Gulf after Shiloh, or having been organized in spring 1862. Stevenson's brigades were filled with mountain men from east Tennessee, north Georgia, and north Alabama. Stewart had at his disposal one Louisiana brigade, two from Alabama, and Stovall's Georgia Brigade. With flags waving in the spring breeze, it must have been an uplifting day for Captain Thomas and his men of Company A, 42nd Georgia Infantry, Stovall's Brigade, Stewart's Division, Hood's Corps, Army of Tennessee.[5]

The pageantry of the grand review, however, could not belie the fact that the Southerners faced an almost impossible task. Marching toward them from Tennessee was General William Tecumseh Sherman and three Yankee armies; the Army of the Cumberland, under General Thomas, with 60,773 troops; the Army of the Tennessee, under General McPherson, with 24,465 men; and the Army of the Ohio, under General Schofield, with 13,553 soldiers. With the addition of several thousand cavalrymen, Sherman would

[4] Confederate service records of Darling P. McDaniel, NARA, M226, roll 40; L. P. Thomas to Jennie, April 17, 1864, in Huff, *Four Families*, 84.

[5] Daniel, *Soldiering in the Army of Tennessee*, 4–6.

have more than 100,000 men at the inception of the Atlanta campaign. Johnston and his Confederate army would be reinforced in the weeks ahead by Bishop Leonidas Polk's Mississippi Corps of some 12,000 veteran troops, but even with the arrival of this third corps, Johnston would be outnumbered two to one in the coming campaign.[6]

Quartermaster Sergeant Charles Alden Rowland

In late December 1863, a new quartermaster was needed for Stovall's Georgia Brigade. Captain J. P. C. Whitehead, the twenty-five-year-old adjutant general of the brigade, received letters from at least two regimental commanders recommending Henry A. Newman of the 43rd Georgia for the position. Captain Newman had been the acting quartermaster for the brigade through the Vicksburg campaign, and, according to Colonel Henry Kellogg of the 43rd Georgia, Newman was the only quartermaster "who brought any money out of Vicksburg." It was from these smuggled funds that the brigade had been able to reorganize in parole camp, "and get up its transportation, clothing, and pay." General Stovall and Captain Whitehead considered Newman for the promotion, but instead, chose not to formally fill the position. Instead, they recruited Whitehead's brother-in-law, Charles Rowland, to serve as a civilian on the brigade staff and handle the duties of a quartermaster.[7]

The thirty-one-year-old businessman from Augusta arrived in camp in January 1864. As a relatively young man, Charlie Rowland had done much in Augusta as a business partner in Fleming & Rowland, a director in the Mechanics' Bank, and an elder of the First Presbyterian Church. He had spent the first two years of the war as a commissioner for the Confederate Government in Florida. But perhaps his greatest contribution to this telling of the story of the 42nd Georgia would come as a devoted letter writer. Fortunately, many of Rowland's witty, opinionated, and observant letters to his

---

[6] McDonough, *William Tecumseh Sherman*, 471.

[7] John Philpot Curren Whitehead (1838–1906) grew up on his family plantation outside of Augusta, known as Ivanhoe. An 1860 University of Georgia graduate, Whitehead was a young attorney in Augusta at the start of the war. Captain Whitehead was among the severely wounded during the July 22, 1864, Battle of Atlanta. Two days later, Hugh Wylie of the 42nd Georgia assumed his position as adjutant of Stovall's Brigade. After the war, Colonel Whitehead moved to Texas where he was a successful lawyer and politician; Henry Newman finished the war at the Greensboro surrender as a captain and assistant quartermaster of Hill's Division, Office of the Governor (Henry C. Kellogg, 43rd Georgia, Cassville, Georgia, to J. P. C. Whitehead, Stovall's Brigade, December 20, 1863, Missouri State Archives).

wife, Katie Whitehead Rowland, have survived.[8]

According to his first letter, Charlie Rowland arrived in Dalton while Marcellus Stovall was on leave in Augusta, so the brigade at that time was led by Colonel Abda Johnson of the 40th Georgia, "a very pleasant man and good officer." Rowland's initial impressions of the common soldiers in the brigade were not in the least positive. "The great majority of men are the ignorant of the upper counties and are poor specimens of Confederate soldiers," he observed. "It would be awful to be associated with them in the ranks." The new civilian quartermaster was thankful to be sleeping in a cabin attached to the brigade headquarters and not spending his nights as a camp sentinel, "pacing up and down before our door having to stand the midnight frost." He soon learned that army life never ceased. "I cant bear the idea of working on Sunday and yet I was engaged all the morning in writing orders for the different regiments," he wrote. Rowland had brought several of his slaves with him and would frequently mention Jno and Willie, whom he assigned to a wagon that transported supplies to the brigade. The Augusta native usually wore his old Home Guard militia suit and thereby looked the part of a common soldier. "When I wear my good jacket," he explained, "the men call me Captain, but seem to be in doubt when I have the other on."[9]

General Joseph Johnston spent the winter and spring improving the morale and discipline of his army. By liberally granting leave to the men on a rotating basis, he both preserved camp supplies and refreshed the men's health and spirits. For many of the men, it was their first visit home in two years. For those remaining in camp, winter 1863–1864 was filled with refresher drills and training at a company, brigade, and division level. It turned out to be a frigid winter, but when weather permitted, drills, target practice, or marches were undertaken both in the morning and the afternoon. Several snowball fights brought some athletic exuberance to the camp. Lieutenant James Gresham of the Walton Tigers wrote on December 28, "Major Hulsey sent word to bring my company out, as he wanted to storm Major Palmer's Battalion." Two hours later they were fighting another battalion. Gresham reported, "Everything was taken in good fun, but it was tough play. The ground was speckled with blood from bruised noses. I was taken prisoner while dashing after their flags, but my company each time fought their way in and rescued me after a good deal of rough usage."[10]

---

[8] Rowland, preface to *Charles Alden Rowland*, n.p.

[9] Rowland to Katie Rowland, January 22, 1864, ibid., 17–18.

[10] Daniel, *Soldiering in the Army of Tennessee*, 24; Goodson, *Letters and Diaries*, 66.

A robust religious outpouring swept over many of the Southern men in that cold Dalton encampment. As Reverend S. M. Cherry, the Presbyterian chaplain of the 37th Georgia, reported, "Protracted meetings were being held in almost every brigade; thousands of our soldiers were thronging our crude camp altars, hundreds were giving their hearts to God, and scores were nightly asking for certificates of Church membership." This Christian revival bonded the men and extolled the virtues of sacrifice, commitment, and bearing hardships with patience. Charlie Rowland wrote on February 28 of a rare communion performed by Chaplain Parks. "It was most impressive to see those hardy careworn soldiers come round the table," he wrote home. "There were 113 and many of them have had their last communion on earth." The men of the 42nd Georgia were diligent participants. Lovick Thomas wrote to Jennie near the end of winter camp that a Christian Society had been organized in the regiment. "Tis a mournfull working of the spirits all through our army,—a good many were baptized today—and from accounts tis a daily occurrence—with an army of Christian soldiers—what we may not accomplish." The parson had offered Lovick the presidency of the society, but he had declined, saying "A humble member is as much as I desire to be."[11]

There were also, however, grim realities to deal with in camp. "We have two men, both of Co. E of this Regt. condemned to be shot to death with musketry," he wrote Jennie. The two men had been caught trying to desert to the enemy. "I had hoped that such a thing would never occur in our Regt.," lamented Thomas. He visited the doomed men with Major Hulsey and believed "the Poore fellows" were not prepared for such a severe decision from the military court. Fifteen deserters in all were condemned to die in front of a firing squad. On the day of the execution, Colonel Robert Henderson approached General A. P. Stewart about gaining the reprieve of one of the young Georgians. According to Bromfield Ridley of Stewart's staff, "Colonel Henderson impressed General Stewart that the soldier was a half-witted fellow, partially devoid of reason, and almost bordering on idiocy." Stewart, Henderson, and the father of the condemned soldier rode to the headquarters of General Johnston. They returned with minutes to spare with a stay of execution for the young soldier. The other fourteen condemned men were brought in front of Stewart's Division, "and to the tune of the 'Dead March'

---

[11] Rev. S. M. Cherry to Rev. Robert J. Harp, May 31, 1864, in Mathisen, *The Routledge Sourcebook*, 351; Daniel, *Soldiering in the Army of Tennessee*, 123; Rowland to home, February 28, 1864, in *Charles Alden Rowland*, 23; L. P. Thomas to Jennie P. Thomas, May 1, 1864, Thomas papers, Santa Fe.

were conducted around the square, an ambulance following with their coffins." They were blindfolded and executed by a contingent from the provost guard.[12]

Buzzard Roost

It had not been an entirely quiet winter. General Sherman probed the Confederate lines in late February 1864, assessing Rebel strength and searching for a vulnerable spot. He found that Joseph Johnston had placed his 40,000 Confederates in a strong defensive position, about five miles west of Dalton on an imposing summit known as Rocky Face. Sherman's reconnaissance in force was led by the 2nd Division of the XIV Corps under General Jefferson C. Davis. They were involved in heavy skirmishing that took place from February 22 until February 26. From these engagements, Sherman determined that Johnston was entrenched in a formidable position, one that he would not try to assail again until May 1864.

The 42nd found themselves in the thick of the fighting that took place in late February 1864. The regiment constituted the picket line for the Georgia Brigade that was holding an advanced position near the base of Rocky Face Ridge. The 42nd was defending Mill Creek Gap, which was better known to Georgia locals as Buzzard Roost Gap. The tracks of the Western & Atlantic Railroad ran through this strategic pass. Two companies—the Walton Blues (G) and the Walton Tigers (H)—were placed along a low foothill near the gap. Another two companies—the Milton Tigers (C) and the Fulton Blues (I)—were held in reserve behind them, with the rest of regiment deployed above them, along the crest of the ridge.

Commanding Companies G and H was Captain Joshua Marion Mitchell, a thirty-eight-year-old physician and farmer from Monroe, Georgia. He had led sixty men from Walton County since summer 1862 and had earned their respect as a brave and just company captain. On the afternoon of February 24, Mitchell deployed his men twelve feet apart with instructions for each man to build a rock pile that would be breast high and three feet wide at the top. By nightfall, the various rock piles were finished. At sunrise the

[12] Ridley, *Battles and Sketches of the Army of Tennessee*, 285. Three men from Company E—Drury Brewer, Madison Smith, Fenton Taylor—were court-martialed in spring 1864; the men in the firing squad did not load their own guns for the execution. Half of the weapons were loaded with ball and powder and half with only powder. Those firing the guns did not know whether their gun was loaded or not.

next morning, the foresight of Mitchell's preparations became apparent as 750 members of the 60th Illinois and the 10th Michigan charged their position in what Federal General Jefferson Davis called, "one of the severest little fights of the war."[13]

The Michigan men held winter furloughs in their pockets but had agreed to fight in one more engagement before going home. The Midwesterners stopped at a sharp ridge before crossing a small valley of spurs and ravines that separated the Yankees from the base of Rocky Face. A spirited skirmish fight in this 250-yard broad valley consumed most of February 25 before the 60th Illinois increased the stakes by charging headlong into the valley at about 3:00 P.M. They were soon followed by the 10th Michigan. Colonel Christopher John Dickerson commanded the 10th Michigan that day. He reported that when his men descended the ridge "the enemy opened a most galling infantry fire upon us from the top of Rocky Face and in our front, while two or more batteries of artillery on our right and in front throw shot and shell, raking the ravines and sweeping the tops of the spurs." Dickerson led his men across the valley to the base of Rocky Face, where they planted their colors in front of the 42nd Georgia. For the next thirty minutes, the men of the 10th Michigan lay on the ground awaiting a general assault from the XIV Corps to follow behind them. Realizing that no such help was coming, Colonel Dickerson ordered a general retreat.[14]

From the perspective of the Confederates, Dr. Mitchell's men, who had been firing from behind their individual rock islands, were now "slowly falling back, but stubbornly contesting every foot of the ground." Commanding the reserves was 1st Lieutenant John Smith Stubbs, a twenty-three-year-old school teacher from Fayetteville, Georgia. Colonel Henderson had left him with a stern order: "I charge you, whatever you do, do not let Mitchell be driven in." Stubbs led companies C and I rapidly down the steep ridge to the relief of Mitchell's two companies. Stubbs found Captain Mitchell in the smoke of battle and bleeding profusely from an arm wound. Mitchell placed Stubbs in command, an action the young lieutenant protested, as Lieutenant Henry Britton Mikel of the Walton Blues was the next most senior officer on the field. In the din of musket and artillery fire, however, it was not clear where exactly Mikel of Company G was in the battle line. Stubbs would have

---

[13] Nash, *Civil War Diary of Francis H. Nash,* 23; Dr. R. J. Massey, "A Brave Georgia Captain Who Captured Big Force," *Atlanta Constitution*, April 9, 1911, 3; Robertson, *Michigan in the War,* 307.

[14] Robertson, *Michigan in the War,* 306 (report of Colonel Dickerson).

to lead the counterattack. His two companies were deployed between the two Walton companies, and the entire line was ordered to advance. "With the old Rebel Yell," wrote James Monroe Gresham, the 2nd Lieutenant in Mitchell's company, "we made a daring charge and swept them before us."[15]

Dickerson's Michigan men had, to this point, been retreating in an orderly fashion. In a large gully, halfway across the valley he had stopped the men and had them "dress on the center." The advancing Confederates found the Yankees in the vulnerable depression and fired a volley into their ranks, thus provoking a route of the Bluecoats. Colonel Dickerson reported that "As they left the ravine I jumped across it, about which time a shell exploding a short distance from me threw a quantity of earth into my back with such force that it knocked me down upon my face, and I found myself gasping for breath." As soon as he could recover, Dickerson was back on his feet, only to be shot in the right heel with a musket ball. As the Georgians overran the position, Gresham single-handedly captured the wounded Federal colonel and five of his men. The Michigan regiment reported twenty-two killed and fifty-six wounded; the 60th Illinois reported losing another forty-four men. Captain Mitchell reported the capturing of Dickerson and twenty-five of his men. Brave young lieutenant James Williams of the Walton Tigers was killed in the battle as was James Darby of the Milton Tigers. General Stovall would mistakenly report that the 42nd had casualties of one man killed and eight wounded in the affair. "The Tigers," wrote Lieutenant Gresham, "never experienced a harder fought battle than the little fight at Rocky Face." Lieutenant Stubbs would carry Dickerson's captured sword until the close of the war.[16]

Rocky Face

Lovick Thomas was probably not present at Buzzard Roost in the heavy skirmish of February 25. He was sick much of January and made no reference to the battle in later accounts of the war. Jennie Thomas wrote in her journal that Lovick had reported to a hospital in Atlanta on January 30 with "chills." As a result, he stayed home in Lawrenceville until March 28. On April 17,

---

[15] Goodson, *Letters and Diaries*, 64 (account of Lieutenant James M. Gresham).

[16] Robertson, *Michigan in the War*, 306 (report of Colonel Dickerson); Hughes and Whitney, *Jefferson Davis in Blue*, 223; Dr. R. J. Massey, "A Brave Georgia Captain Who Captured Big Force," *Atlanta Constitution*, April 9, 1911, 3; report of General Marcellus A. Stovall, February 28, 1864, *OR*, ser. 1, vol. 32, part 1, 479–80; Gresham, "A Partial History of the Walton Tigers," 662.

the clean-shaven Captain Thomas was writing home from Dalton explaining the rigors of picket duty: "I had to be up all last night visiting and instructing the men on duty. I suppose that I did not sleep more than an hour." To add to the discomfort of their forward position, it was still raining and sleeting much of the time. Thomas also had to contend with prostitutes trying to slip into the camps. When he stopped one from this "accommodating tribe," she tried "the effects of her smiles and conversation combined," but to no avail. Thomas was accused of being "a hard knot to get by."[17]

As the time for major fighting grew near, Lovick wrote Jennie to thank her for shipping to him assorted foods. He described the "bill of fare" for the special camp breakfast: "Corn Bread Warm, Fried Pease, Boiled Hocks-ham, cold light rolls, soft Ginger Cake, Tea Cake, Tomato Pickles, Coffee with sugar in it, and fresh fish" which completed the menu. "The first mentioned article Bread, was furnished by the (Confederate) Government, the last by one of my men to keep on the good side of me, so that he could get permission to go fishing again." Everything else came by way of Gwinnett County. Lovick wrote Jennie that the officers had finished her lettuce and onions the previous night. The tall, lanky captain especially missed his wife on the warm, pleasant spring evenings. He wrote that if he were home with Jennie, they would, no doubt, "enjoy each other's society" and "be very apt to take a nap together, would we not?"[18]

Most Southerners were still defiant at the beginning of 1864. Despite the loss of most of Tennessee, Arkansas, and Louisiana, and the serious defeats at Gettysburg and Vicksburg in summer 1863, there remained faith that independence would be achieved. Jennie wrote "My dear Lovick" on April 29, 1864. "I feel very hopeful in regard to our prospects now, nationally," she penned. "I scarcely doubt our success in the coming struggle, but our dependence should be upon God, and not in our puny arm for Victory, and we should ascribe all the honor to him." Southerners had convinced themselves that Abraham Lincoln would lose his re-election in fall 1864 if things did not go well for the Union armies that year. L. P. Thomas was particularly pleased that a winter of rest and recruitment had brought his 42nd Georgia regiment up to 700 men, all "in excellent fighting trim."[19]

---

[17] Jennie P. Thomas, diary, March 26, 1864, in Huff, *Four Families,* 82; L. P. Thomas to Jennie, April 17, 1864, ibid., 83.

[18] L. P. Thomas to Jennie P. Thomas, May 1, 1864, Thomas papers, Santa Fe.

[19] Jennie P. Thomas to L. P. Thomas, April 29, 1864, in Huff, *Four Families*, 84; L. P.

General Sherman launched his invasion of Georgia in early May 1864. His destination was Atlanta, the industrial and administrative capital of the western Confederacy. The stakes were high for both nations. Four railroads ran through Atlanta, making it the hub of travel and commerce from all points of the compass. Seven newspapers operated from the "Gate City." By this point in the war, with an 1864 population of 20,000, only Richmond was a bigger Rebel-held city than Atlanta. Behind Rocky Face Ridge held by the Confederate Army of Tennessee, the land began to flatten out and to the southeast, only 100 miles away, lay Atlanta. It would be critical for Johnston to check the Federals in the mountains of North Georgia, where strong defensive positions could be established.

The ninety-day Atlanta Campaign was a logistical masterpiece under the direction of William Tecumseh Sherman. To keep his large army fed and equipped, the Federal field commander had to stay close to the Western and Atlantic Railroad, his sole supply line. Sherman calculated that on every day of the campaign, he would need 145 boxcars of food and supplies. The Federal armies would have to cross three rivers—the Oostanaula, the Etowah, and the Chattahoochee—before reaching Atlanta. Sherman's first challenge would be to dislodge Johnston from his fortified positions along Rocky Face. Rising 700 feet above the valley floor, the steep Rocky Face ran north-south for roughly ten miles. There were only three passages through the ridge—Mill Creek Gap, Dug Gap, and Snake Creek Gap. Stewart's Division was entrusted with holding Mill Creek Gap. Stovall's Brigade, which included the 42nd, was placed on a ridge below Rocky Face to the right of Mill Creek and the Western & Atlantic Railroad. Today cars and trucks on Interstate 75 zoom through this portion of the gap, but in May 1864 every yard would be stoutly contested.[20]

On May 1, Lovick wrote Jennie that Alexander Stewart had called his regimental officers together. The Tennessee general talked of "character—and honors already gained—and how much was at stake—and that the position must never be deserted—to fight until the last round of ammunition is gone. And if necessary to club their muskets—and when they are broken and useless to use rocks." His words were timely and well-heeded as the Federals were, even as he spoke, approaching with the tell-tale boom of distant cannonading announcing their progressions. The Rebel supply wagons were

Thomas, "The Battle of Resaca: The Part the Forty-Second Georgia Took in It," *Atlanta Constitution*, February 2, 1890, 12.

[20] McDonough, *William Tecumseh Sherman*, 470.

brought to the safety of the rear in preparation for the coming onslaught. Brigade leader Marcellus Stovall, his health regained in Augusta, was now back with the troops. Colonel Robert Henderson returned to his accustomed role as leader of the 42nd Georgia. Only Colonel Robert Maddox, who had been second in command to Henderson during the Kentucky and Vicksburg campaigns, would not be back with the 42nd. Maddox had accepted command of Governor Joseph Brown's second brigade of state reserve militia troops. Thomas wrote of his Georgia regiment that we "now moved out on our lines—as determined a set of men as ever I saw, eager for the fray."[21]

Joseph Johnston believed that Sherman would hold the Confederates defending Rocky Face with one force, and then flank the Rebels from the north with another. Hood's Corps was given the vital task of stopping this anticipated Federal attack. Alexander Stewart's Division, with its left at Mill Creek Gap, ran its lines north along the thin crest of Rocky Face Ridge. Stovall's Georgians anchored the far left, with responsibility for holding Mill Creek Gap. Hood's other two divisions, those of Stephenson and Hindman, were in trenches going eastward across Crow Valley, where Johnston anticipated the main attack would fall. Hardee's Corps, the other half of Johnston's Army of Tennessee, would cover Rocky Face Ridge south of Mill Creek Gap, including the critical Dug Creek Gap, as well as hold two divisions in reserve in the town of Dalton.[22]

"About the 5th instant the soldiers were called from their camps to meet the enemy in the vicinity of Dalton—," wrote Georgia chaplain S. M. Cherry, "they literally went from the altar to their entrenchments—from their knees to the battle with their foes—still singing the songs of Zion and supplicating the throne of grace as they surrounded the fires of bivouac, or waited to receive the fire of the foe." On May 7, General George H. Thomas's Army of the Cumberland directly assaulted Mill Creek Gap and Dug Gap along Rocky Face Ridge. At the same time, Major General John M. Schofield brought his Army of the Ohio down from Knoxville and was approaching Crow Valley from the north along the East Tennessee and Georgia Railroad.[23]

For three days, the Georgia men at Mill Creek Gap brushed back assaults made by General James Dada Morgan's Brigade from the XIV Corps.

---

[21] L. P. Thomas to Jennie P. Thomas, May 1, 1864, Thomas papers, Santa Fe.

[22] McMurry, *Atlanta 1864*, 60–61.

[23] Rev. S. M. Cherry to Rev. Robert J. Harp, May 31, 1864 **[in? or are these separate sources?]** Mathisen, *The Routledge Sourcebook*, 351. "Instant" was a term which meant in the same calendar month as the date the letter was written.

The Secessionists were aided in their defense by an artificial lake that ran across the gap, which their pioneer detachments had constructed by impounding the waters of Mill Creek at the two railroad culverts. General Jefferson C. Davis, who commanded the 2nd Division in the XIV Corps, wrote that the fighting consisted of heavy skirmishing, but "when attempts were made to feel of the enemy's works, it became almost a general engagement in severity." Joseph Bogle of the 40th Georgia wrote that, "the lines of the 40th and 41st ran about due North and South, the line of the 42nd made a sharp angle diagonally up the mountain to the Northwest, and when the Yankees tried to charge through the gap the 40th and 41st were immediately in their path, but before they reached us the 42nd made a counter charge down the mountain on their flank and routed them before we could open fire." He enthusiastically recalled, "How we did wave our hats and cheer the 42nd with the old Rebel Yell as they made the charge down the mountain, in splendid order, in full view of the rest of us!" According to a correspondent's account published in the *Daily South Carolinian*, the Federals made four assaults on May 10 but "were handsomely repulsed by Stovall's Brigade, leaving several hundred of their dead upon the field."[24]

The wild card in this unfolding confrontation was Sherman's third wing, the Army of the Tennessee under General James Birdseye McPherson. Johnston knew little of what was happening to his northwest as his cavalry under Wheeler was massed on his right preparing for a great mounted battle. If McPherson moved his army into Georgia from Chattanooga, Johnston believed it would be to capture Rome, Georgia, well to his southwest. General Leonidas Polk, commanding Rebel forces in Alabama and Mississippi would have to contend with McPherson and defend Rome. Joseph Johnston's attention was directed to his defenses around Dalton, and he was blind to other possibilities. McPherson's Army of the Tennessee marched fourteen miles southward along the western side of Rocky Face and past Hardee's positions. Instead of continuing toward Rome, however, McPherson completely surprised the Rebel high command by turning east and entering Snake Creek Gap. Historian Richard McMurray called it "one of the great strategic

[24] Report of Operations of General Jefferson C. Davis, May 1–August 22, 1864, *OR*, ser. 1, vol. 38, part 1, 625–26; Mathisen, *The Routledge Sourcebook*, 351; McMurry, *Atlanta 1864*, 63; Bogle, "Some Recollections of the 40th Regiment, C.S.A., in the Civil War," in Bogle and Calhoun, *Historical Sketches*, 18; "The Dalton Front," *Daily South Carolinian*, May 13, 1864. Colonel William Curtis of the 41st Georgia was killed that day, standing on the breastworks, encouraging the 42nd Georgia forward in their advance.

marches of the war" as the Federals found Snake Creek Gap "unoccupied, unguarded, unobstructed, and unobserved." An aggressive McPherson could have pushed through the narrow three-mile passage, stormed across Sugar Valley, and captured the railroad village of Resaca. This action would have placed his 25,000 men between Atlanta and the Army of Tennessee. Fortunately for the Rebels, McPherson hesitated, giving time for Bishop Polk and his Army of the Mississippi to reach Resaca and set up a defensive position.[25]

From a strategic perspective, however, Johnston now had a sizable Union force between his position and Atlanta. His strong defensive position near Dalton would have to be abandoned. The disappointed Rebels managed to orchestrate a smooth retreat from Dalton down to the endangered Resaca. Stewart's men were the last to withdraw from their Rocky Face position. Late in the evening of May 12, the 42nd marched through the streets of Dalton. Only Wheeler's cavalry stood between them and the advancing Yankees. On the evening of the thirteenth, Stovall's Georgia Brigade began filing into the trenches at Resaca.

Battle of Resaca

Forced to retreat from their defensive stronghold around Dalton, the Rebels fell back twenty-one miles, staying close to the tracks of the Western & Atlantic Railroad. They were soon in open territory, a rural land of rolling hills, farm fields, and forest. Just north of the railroad village of Resaca, Johnston ordered his army to dig-in. By May 14, 1864, the Confederates were behind a four-mile protracted line of freshly dug trenches shaped like a horseshoe that stretched from the village of Resaca in the south, went north-south along a ridge on the eastern side of Camp Creek, and then turned east-west on a line that went all the way to the Conasauga River. Polk's men defended Resaca, Hardee held the middle, and Hood's troops faced northward. Behind the Confederates were the tracks of the Western & Atlantic, and just beyond the rail line was the Oostanaula River, flowing west to east. Stovall's Brigade held the far-right flank of the Confederate position. Only Baker's Alabama Brigade was farther east on the Confederate front. General William Tecumseh Sherman was pleased to have pushed the Southern army out of Dalton without a major battle. Now, he had his Rebel counterpart in a tight spot with a river directly behind the Confederates' position. The Federal commander ordered his three armies forward with the expectation that they might

[25] McMurry, *Atlanta 1864*, 61–63.

administer a decisive blow at an early stage in the Atlanta campaign. What followed was a bloody two-day battle that was fought aggressively by both sides. Combined casualties exceeded 6,000 soldiers, which was far greater than the more famous Battle of Kennesaw Mountain, which would be fought by the same armies a month later.[26]

May 14, the first day of the Battle of Resaca, began with a series of unsuccessful and disjointed Federal attacks across the Camp Creek Valley and up open ground to the Rebel trench line. By late afternoon, a more confident Johnston saw a small opportunity to outflank and counterattack the aggressive Yankee invaders. Seeing that the Union left did not reach all the way to the Conasauga River, the Confederate commander ordered General Hood to move two divisions forward and attack the exposed Union left flank. The 32-year-old John Bell Hood had the ideal temperament for this assignment. His frontal assaults at Gaines Mill, Antietam, and Chickamauga had produced tremendous casualties in his ranks, but they had also swung the tide of battle in favor of the Confederate army. Unfortunately for the Southerners, the 6'2" blonde-haired Hood was no longer able to lead the charge himself. His left arm was useless, having been mangled by shrapnel at the Battle of Gettysburg. His right leg was now a 4½-inch stump; the rest was amputated after a thigh wound in the charge at Chickamauga. The invalid Hood would now go into battle, behind his men, with his crutches strapped firmly to his saddle.[27]

The Confederate divisions of Carter Stevenson and A. P. Stewart were assigned the task of making the flanking attack. At 5 P.M. Stevenson's men, who were on the crest of the ridge near the Resaca-Dalton Road, began to pivot, starting a great wheeling movement towards the Federal positions. Stewart's men, who had the most ground to cover, struggled to keep up in the difficult terrain and soon lost contact with Stevenson on their left. Stevenson's two brigades, which were much closer to the Yankees, overwhelmed a brigade in the IV Corps under General David Stanley. After occupying the Federal trenches, Stevenson's men pushed on into an open field where they took heavy losses as they attempted to capture the six cannons of Simonson's 5th Indiana Battery. A member of the 150th New York watched the attack and wrote, "With the artillerymen it was a question of how fast, with their decimated ranks, they could load and fire, the grape and canister tearing great gaps through the enemy's lines at each discharge. With the Confederate troops, it was a question of whether they could bear this destruction until

---

[26] S. Davis, "Sherman in North Georgia: The Battle of Resaca," 50.

[27] Castel, *Decision in The West*, 58–62.

they should reach the guns, and either kill or capture the gunners." Only the timely arrival and brave bayonet charge of James Robinson's 3rd Brigade from Alpheus Williams's 1st Division of Hooker's XX Corps saved the Yankee artillery from capture.

Stewart's four brigades provided no help to Stevenson's valiant attackers as they were probing far off to the right until darkness fell. Sherman was fortunate that the Confederate attack on the evening of May 14 was poorly coordinated. General Stovall reported that his brigade had started the day holding the far right of the Confederate army with their backs to the Conasauga River. He wrote,

> I moved up the banks of the river by the right flank until I could get into the designated position, when I changed the direction from right to left and advanced in line of battle. During this charge, notwithstanding the fire of grape and canister to which my command was subjected was terrific, still none faltered and halted only in obedience to orders, not, however, until they had come up with the first line to which I was acting as a support.

At 9 P.M. they gave up the advance and returned to the Confederate entrenchments. For Hood and the Confederates, it was an opportunity lost.[28]

May 15 saw heavy fighting centered on Joseph Hooker's 1:30 P.M. attack in columns, by two of his divisions, some 12,000 men, directly down the Dalton-Resaca Road. Concentrated Confederate fire decimated "Hooker's Ironclads" and halted their advance at the foot of the Rebel entrenchments. As the firing died down, Joseph Johnston believed he saw a second opportunity to crush Sherman's left flank. Around 3:00 P.M., Stewart and Stevenson received orders to resume the flanking attack of the previous day. By this point in the battle, however, Stevenson had active Federal infantry directly in his front and he viewed this as reason enough to stay in place. A. P. Stewart, on the other hand, was far enough on the right flank of the Rebel line to advance into a small unoccupied piece of forested land. His Butternuts moved forward with the tracks of the Western & Atlantic on their left and the Conasauga River on their right. Stewart's plan was to march north a half-mile, putting his division past the Federal line, and then wheel left into a battle line facing west. From here, his soldiers could cross the railroad tracks and sweep into the exposed Federal flank. Stewart formed his division into

---

[28] Benton, *As Seen from the Ranks*, 136; Secrist, *The Battle of Resaca*, 36; field report of General M. A. Stovall, June 2, 1864, *OR*, ser.1, vol. 38, part 3, 822–25.

two battle lines, with Clayton's Alabama Brigade composing the left side of the front line and Stovall's Georgians lining up on the right. Gibson's Louisiana men formed a second line behind Clayton and Baker's Alabama men did the same behind Stovall. Maney's Brigade of Tennesseans, which had been detached from Cheatham's Division, constituted Stewart's reserve.[29]

By 4:00 P.M., Stewart had reached his push-off point, but he was now aware of "heavy movement of the enemy" in his front. Two of Alpheus Williams's four brigades had done their own wheel movement to meet the anticipated charge of Rebel infantry. Thomas Ruger's 2nd Brigade, made up of six Federal regiments, had arrived first and anchored their position on the top of a knoll where the dwelling-house and outbuildings of John A. Scales were located. Ruger's soldiers immediately began tearing apart the buildings and fences, using the wood to construct rudimentary breastworks that soon enclosed the entire knoll. A second Yankee brigade, under James Robinson, soon arrived and further extended the Federal line. "I moved my brigade to the crest of the hill in front of the line then occupied," wrote Colonel Robinson, "and directed it in position on the left of General Ruger's brigade, and upon the left flank of the division, and in fact the left flank of the army."[30]

It was soon clear to General Stewart that his Confederate division would not be flanking an unprepared Federal line that was facing south; instead, his advancing troops were about to charge across an open farm field at a Federal force of comparable size that was dug in and facing east. The two Federal brigades holding the knoll were veteran troops, men who had fought with distinction at Gettysburg in 1863. Stewart alerted his Corps commander John Bell Hood of his predicament, but, receiving no response, followed the order and began his assault. A tardy Hood would send his inspector general to Stewart to suspend the advance, but Colonel Cunningham arrived minutes after the attack had commenced.

Stewart's Division stepped off promptly at 4:00 P.M., crossing the tracks of the Western & Atlantic and into a wood line that shielded them from Federal shelling. Because the Confederates were attempting a half wheel to the left as they advanced, Clayton's Alabama Brigade was the first to emerge

---

[29] Rebels referred to the men of Hooker's Corps as "Hooker's Ironclads."

[30] Robinson initially deployed the 101st Illinois and 143rd New York in his first line and behind them formed a second line made up of the 61st Ohio, 82nd Ohio, 45th New York, and 82nd Illinois; field report of General Thomas H. Ruger, September 11, 1864, *OR*, ser.1, vol. 38, part 2, 56–62; field report of Colonel James Robinson, July 24, 1864, *OR*, ser. 1, vol. 38, part 2, 84–85.

from the woods. Stephen Cook, an assistant surgeon in the 150th New York, wrote home that, "We had but nicely completed our breastworks when a rattling fire in the woods in our front told us that their advance had met our skirmishers and soon out of the woods came the skirmishers literally running for their lives." Cook believed the Federal skirmishers were in imminent danger of being shot by their own men, "as the rebels were pushing right on after them, emerging from the woods very close to the last of them." The Federal surgeon was mesmerized by the advancing Southerners. "Oh! It was a grand sight to see them pour out of the woods, form in double columns and advance at a quickstep towards our unsupported left."[31]

Clayton's Alabamians were separated from the two lines of Yankee infantry by 150 yards of farm field. According to Charles Benton of the 150th New York, the long gray lines approached, "with good alignment and steady front." The order to fire did not come until the Rebels were halfway across the field. The Federal rifle volley tore into the advancing Southerners at close range. An anxious Stephen Cook reported that Clayton's gray line "first faltered, fell back a few steps, then rallied and poured at us an unmerciful fire from guns that outnumbered us four or five to one." Benton recalled "the bullets splintering the rails, and passing us with that peculiar, zipping sound so familiar to veterans." Cook admitted that he gave serious thought to running for the rear, but "Pride would not let me, for there was the eyes of all the boys in the regiment to see me if I did run, and what would they say to me afterwards?"[32]

At a distance of seventy-five yards, the two sides fired at one another for forty-five minutes. The exposed Confederates took more than 400 casualties before falling back. Benton recalled, "They were retreating in disorder now, but scattered over the field were hundreds of their dead and wounded who could not retreat. The fire of our line had been very deadly and effective." Colonel John Fesler of the 27th Indiana reported that "the Thirty-eighth Alabama regiment made a charge on this regiment and was handsomely repulsed, with a loss of their colors, 35 prisoners, including Colonel (Augustus) Lankford, commanding the regiment, and 33 killed."[33]

---

[31] Stephen G. Cook to home, May 22, 1864, written in Cassville, GA, and published under the pen name "Fred Fulton" in the *Poughkeepsie Daily Press*, June 21, 1864.

[32] Benton, *As Seen from the Ranks*, 136; Cook to home, May 22, 1864.

[33] Ruger's 2nd Brigade consisted of the 27th Indiana, 2nd Massachusetts, 13th New Jersey, 107th, 150th New York, and the 3rd Wisconsin; Benton, *As Seen from the Ranks*, 140–41; field report of Colonel John Fesler, 27th Indiana, September 6, 1864, *OR*, ser. 1, vol. 38,

Stovall's Brigade had an impenetrable thicket of dense undergrowth to first navigate and did not march out from the forest until after Clayton's charge had faltered. The 42nd Georgia anchored the far left of Stovall's battle line, which meant they would be charging into the 13th New Jersey, which held the center of the Federal line. First to appear was the skirmish line, which was Company F of the 42nd under 34-year-old Captain James Madison Summers. Fifteen men in the Newton Rifles became casualties that day, but it could have been worse. A lighted cannon ball landed in the middle of the company. In the ensuing pandemonium, it was a nimble Summers who picked up the bomb and threw it in a ravine, where it instantly exploded.[34]

When Stovall's Georgians emerged from the woods, their "unearthly yell" was met with a hail of canister and bullets. According to Summers's account, the men of the 40th Georgia immediately fled or dropped to the ground. The 41st, 42nd, 43rd, and 52nd continued forward into the maelstrom. Baker's four Alabama regiments passed over the prone Georgians of the 40th in their charge, but there were not enough men to make a difference. General Ruger called this second Confederate attack on his improvised farmhouse redoubt to be "fought more obstinately than his first," but the results were the same. Private Benton of the 150th New York agreed, reporting that the second Rebel attack was "more persistent and got nearer to our line, though they were finally driven back before the storm of lead, and the dead and wounded in our front were thicker than ever." Colonel James Robinson, who commanded the 3rd Brigade, thought the second wave of Confederates "came forward with a reckless desperation, which indicated a determination to break our line at every hazard. But his rash purposes were doomed to the same signal failure as before." Confederate division commander Alexander Stewart, who had three horses shot out from under him that afternoon, finally saw the futility of the attack. He called back the brigades of Gibson and Maury before they became seriously engaged.[35]

A victorious Federal General Thomas Ruger reported that his brigade casualties at Resaca were 15 killed and 135 wounded. Confederate casualties

---

part 2, 63.

[34] The 43rd Georgia held the far right position in Stovall's line of battle; field report of General M. A. Stovall, June 2, 1864, *OR*, ser.1, vol. 38, part 3, 822–25; "Captain J. M. Summers, Hero of Gray, is Dead," *Atlanta Constitution*, February 28, 1915, 10; James M. Summers, "Battle of Resaca," *Semi-Weekly Atlanta Journal*, June 28, 1901, 8.

[35] Benton, *As Seen from the Ranks*, 141; field report of General Thomas H. Ruger, September 11, 1864, *OR*, ser.1, vol. 38, part 2, 56–62 field report of Colonel James Robinson, July 24, 1864, *OR*, ser. 1, vol. 38, part 2, 84–85.

were much worse. General Marcellus Stovall grimly reported the tally for his Georgia Brigade on June 2: "The casualties in this engagement alone, amounted to 270 in killed and wounded. Since the 7th of May, they comprise an aggregate of 440, among whom are 6 out of the 8 field officers with whom I commenced the campaign." Baker's Alabama Brigade took 250 casualties in the ill-fated charge on May 15.[36]

The 42nd Georgia Regiment had followed their gallant Colonel Robert Henderson into the open field that afternoon. The colonel was "riding on his dashing charger *Old Roan*" and headed for "the thickest of the fray." L. P. Thomas remembered Henderson's drawn saber flashing in the sunlight until it was "carried away by a cannon ball." The 42nd took 117 casualties in their doomed charge at Resaca, by far the most of any regiment engaged, making May 15, 1864, the bloodiest single day in the regiment's distinguished history. Scores of the best men in the regiment fell that day. Chief among them was Colonel Henderson, who attempted to rally his men in the middle of the open field after the failed attack. The men, according to Thomas, "rushed around him, grape and canister flying and balls whistling." It was here that a minie ball tore a serious cut along the colonel's temple. Captain Summers recalled seeing Henderson with "blood running from a wound in the side of his head and dripping off his feet." Wiley Jackson Baggett of the Calhoun Guards (K) wrote that the wounded Henderson was "crying out that he would rather die than see his regiment give way."[37]

Command of the regiment now fell to Major Hulsey, but the second-in-command, who had been leading the left side of the battle line had already been severely wounded in the same charge. The same was true of Captain Lowndes Calhoun, who had been at the head of the Calhoun Guards when he fell, shot in the hip. The soldiers who carried the regimental flags were always conspicuous targets, and Resaca was no exception. Company E from Newton County always served as the Color Company of the 42nd. Color Sergeant Seth Huff from this company was killed while crossing a fence thirty yards from the Federal line, the regimental flag falling on the rail fences. With the Rebels now in retreat, John Burton Graham of the Newton Rifles and

---

[36] Castel, *Decision in The West*, 176–78; field report of General M. A. Stovall, June 2, 1864, *OR*, ser.1, vol. 38, part 3, 822–25; field report of General Thomas H. Ruger, September 11, 1864, *OR*, ser.1, vol. 38, part 2, 56–62.

[37] Lovick Thomas, "Reunion Speech Given at Social Circle, Georgia, July 22, 1899," vol. 1, pg. 14, Thomas collection, BRC; James M. Summers, "Battle of Resaca," *Semi-Weekly Atlanta Journal*, June 28, 1901, 8; Wiley Baggett letter in Goodson, *Letters and Diaries*, 68.

James Boyd of the Harper Guards raced back under fire to retrieve the colors.[38]

Twenty-nine-year-old Lovick Thomas had begun the attack as one of ten captains of the 42nd Georgia. Minutes later he would find himself in charge of the entire regiment, as the 600 or so men struggled to reform behind the railroad tracks. Captain Summers wrote that Thomas "aided in rallying the men," and conducted an ordered retreat to the Southern trench lines. The men of the 42nd soon reestablished their defensive position and spent the night sleeping on arms. The next morning, Lovick, as he counted eight new holes scattered in his clothing and in an old army blanket that he had worn across his shoulders, realized he was lucky to be alive. His eventual promotion to major would be dated back to May 15, 1864, and the charge at Resaca.

As shattered as the 42nd was from their ill-fated charge, their job was not done. By the evening of March 15, Johnston's Army of Tennessee found themselves in a precarious situation. Federal infantry had drawn within 500 yards of the railroad bridge across the Oostanaula River and Federal artillery could place interdicting fire on Johnston's two river crossings. The Confederate commander had no option but to order a nighttime retreat over the two river bridges. Stewart's battered division, being farthest from the Oostanaula River, would cover this withdrawal. Captain Summers wrote that he was among the 150 men who held the trenches under Federal artillery fire until 1:00 A.M., when they made a perilous escape to the last bridge, minutes before it went up in flames. Wiley Baggett wrote in his memoir of the sad spectacle of the wounded laid out at the Resaca train depot under the watchful eye of Stewart's rear guard. At midnight, two trains were dispatched to pick up these Rebel casualties. Unfortunately, the yells from Federal infantry seizing the empty works panicked the last of the Confederates. According to Baggett, "At that point the hands and one train ran off, and the other train and the wounded was captured by the Yankees!" It was during the subsequent retreat that Marcellus Stovall, who had been fighting a severe cold complicated by bronchitis, relinquished the command of his brigade to Colonel Abda Johnson of the 40th Georgia. No longer able to ride a horse, Stovall left in an

---

[38] The first flagbearer for the 42nd was Thomas D. Goodson, who died at Bean Station, Tennessee. John Ingram, who died at Vicksburg, succeeded him. Seth Huff followed Ingram. The longest-serving and final flagbearer was William F. Edwards of Newton County (Lowndes Calhoun, "Flag That Was Followed by Bloody Forty-Second," *Atlanta Journal*, April 27, 1904, reprint of an address given on Confederate Memorial Day at the Georgia Capitol).

ambulance. He spent two weeks recuperating in Augusta with his family and returned to his troops June 1.[39]

General Randall Gibson wrote in his diary that his Louisiana Brigade and Stovall's Georgia Brigade, both under his overall command, "constituted the rear guard, and were posted on the railroad equidistant from the two heels of the horseshoe, which had been the formation of our army, its flanks resting on the Etowah River." Abda Johnson reported that the retreat on the night of the fifteenth had bordered on chaos, writing, "This arose from the fact that the command had passed over ground unfavorable for marching, and could not be properly closed up." He reflected, "that we were double-quicking to get into position; that the road across which we had to move out from the line of battle was covered with troops, and that when a few companies at the head of the column had crossed the railroad a train went by, cutting the column." Gibson wrote, "It was daylight when we reached the pontoon bridge over the Etowah, where I found General Hood, in person. This was the most anxious night I ever had in the army."[40]

Battle of New Hope Church

General Sherman hoped at first that he could overtake and defeat Johnston's army somewhere in the thirty-five miles between the Oostanaula and Etowah Rivers, a relatively open and level area of north Georgia. The terrain would be ideal for the Union forces to envelop and crush the outnumbered Secessionists. Johnston's Confederates, though, wisely fell back quickly, first to Calhoun, then Adairsville, and finally Cassville. The fertile farmland they passed through that spring was clothed in fields of wheat, rye, and corn. Brigade quartermaster Rowland called the country "an immense garden spot." Studying his topographical maps, Sherman predicted that Johnston would make his next defensive stand on the hilly terrain south of the Etowah River. To follow as closely as possible with such a large Federal army, Sherman sent his armies down several different roads heading southward. Johnston, however, chose not to cross the Etowah River. On May 17, Charlie Rowland was writing from under a tree in Kinston "as the wagon trains of the Rebel army extend for miles from Calhoun to Cassville." He was cautiously optimistic

---

[39] Elliott, "I Regard Maj. Gen. Stewart as the Best Qualified…," 144; Goodson, *Letters and Diaries*, 68; C. L. Bragg, *Distinction in Every Service*, 95.

[40] "Eloquent Tributes," *New Orleans Daily Picayune*, March 26, 1891, 6; field report of Colonel Abda Johnson, June 2, 1864, *OR*, ser. 1, vol 38, part 3, 825–26.

that Johnston was about to engage Sherman in a large battle on favorable terrain. "The army is yet confident," he explained, "and relies on General Johnston's extreme caution and prudence."[41]

On May 19, a hot, muggy day, the dapper Virginian drew his army up behind Two Run Creek in front of Cassville. Johnston had a bit fewer than 74,000 soldiers—which, according to historian Albert Castel, was "the largest force ever assembled by the Confederacy in the West." Johnston understood that Sherman was in a vulnerable position with his three armies spread out on several different country roads. The moment for the Napoleonic counterstroke had arrived. The Confederate commander believed that his army could fall on one Federal column and destroy it before the other columns could join the battle. Johnston issued a General Order to his troops: It was time "to turn and meet his advancing columns." The Rebel counterattack had been ordered. Johnston was blunt, writing, "I will lead you to battle."[42]

The anticipated great battle on the plains near Cassville never took place. The chess board was set; the game never played. Hood followed Johnston's battle plan, advancing and placing his corps in a position to pounce on Hooker's XX Corps, the Federal left flank, as it marched up to confront Polk's Corps. Unfortunately for Johnston's well-laid plan, Hood soon had second thoughts about the surprise attack after his scouts spotted the lost Federal cavalry divisions of McCook and Stoneman coming down a road behind Hood's Corps. An attack on the XX Corps by Hood would have left his rear and flank vulnerable to a Federal cavalry counterattack. Hood, demonstrating the better part of valor, retreated without orders and soon the entire Confederate army was retreating across the Etowah River. "I thought a fight was before us as we threw up breastworks, planted cannon, etc. all night," wrote Sergeant Richard Van Wyck of the 150th New York Volunteer Infantry, "but when morning came nothing was seen of the enemy, although they were erecting like fortifications till nearly morning."[43]

The 42nd covered the retreat, holding the imposing Etowah railroad bridge until the night of May 20, when it was burned by Confederate engineers. Only the stone pillars were left standing. It was a surprised, indignant, and disappointed Rebel army that fell back; they had expected to be in a battle on the scale of a Chancellorsville, but it was simply not in Johnston's nature to initiate such a grand roll of the dice. Better, thought Johnston, to wait for

---

[41] Rowland to home, May 17, 1864, in *Charles Alden Rowland*, 26.

[42] Castel, *Decision in the West*, 198; Cunyus, History of Bartow County, Georgia, 225.

[43] Van Wyck, *A War to Petrify the Heart*, 205.

the perfect opportunity. The morale of the Army of Tennessee never fully recovered from this episode. "There is a great many deserting into our lines," wrote William Shrimp of the 46th Pennsylvania on May 22. "They all say the Southern Confederacy is gone up and I do tell you I am glad to hear it." At the same time, it must be noted, General Robert E. Lee and his army of 60,000 was aggressively attacking a Federal army in Virginia twice their size in the Battle of the Wilderness. The retreating western Confederates could not help but feel the sting of comparison between the two Southern armies.[44]

After weeks of marching, both opposing armies, separated by the Etowah River, finally rested for three days. Charlie Rowland wrote that the Georgia Brigade had, for weeks, been "fighting by day and marching by night." He wrote his wife that, "I have stood it very well and no bed has ever been so grateful as old Mother earth after riding all night in the saddle." He reflected, "Our retreat was made during the beautiful moonlight nights, and it made me sick at heart to see the splendid country fall into hands of the enemy. The fields were green with wheat, oats, rye, and clover, and was beautiful in the extreme." The 42nd Georgia camped on a treeless ridge near the river. In the scorching sun, they could see volumes of black smoke defiling the otherwise clear sky as Federal engineers destroyed the Etowah Iron Works.[45]

The New Yorker Van Wyck, on the other side of the Etowah, considered little Cassville to be, "the finest [village] yet seen in Ga., in a rich cotton district, fine dwellings, public institutions, streets well laid out, and among many of the houses good libraries of excellent works evince the character of the place." Unfortunately, in a precursor to the famous March to the Sea, "the enemy had not vacated the place more than a quarter of an hour before a perfect sack was instituted by our troops." According to Sergeant Henry Morhous of the 123rd New York, many of the Federal soldiers took this opportunity in Cassville to minimize what they would carry in the upcoming campaign. "The weather was extremely hot," he explained, "and many of the boys in the Regiment threw away blankets, tents, knapsacks, and all save haversack and accoutrements."[46]

On May 23, 1864, Sherman sent his 85,000 soldiers across the river at

---

[44] Nash, *Civil War Diary of Francis H. Nash*, 29; William T. Shrimp to wife, May 22, 1864 in Civil War Times Illustrated Collection of Soldiers' Correspondence and Diaries.

[45] Rowland to home, May 21 and 23, 1864, in *Charles Alden Rowland*, 28, 31.

[46] Van Wyck, *A War to Petrify the Heart*, 206; Morhous, *Reminiscences of the 123rd New York*, 93.

three points. He had no interest in directly assaulting Johnston, whose Confederate army defended the Western & Atlantic Railroad along the long, narrow eighty-foot deep gorge known as Allatoona Pass. Instead, Sherman strayed from his previous habits and directed his collective army away from his railroad supply line, moving southwest toward Dallas, a crossroads village about fifteen miles from the Etowah. His command was leaving Appalachian Georgia for a flatter terrain covered by dense vegetation. Primitive dirt trails connected the farms and hamlets, but it was mostly a vast, unending "Jack Oak" forest. Too far from the W & A, the Union army would be sustained by their 5,150 supply wagons that were pulled by thousands of mules and horses. From Dallas, Sherman would be just sixteen miles west of Marietta. He expected that Johnston would give him a fight somewhere near Marietta, a noted railroad town sitting only eight miles north of the Chattahoochee River. This time, however, Johnston had anticipated the Federal flanking movement and had his Butternut army in motion. Early on the morning of May 25, Hood's Corps reached New Hope Church, a point on the road between Dallas and Marietta. Hardee's Corps was occupying the town of Dallas, and Polk was moving to fill the gap in the center.[47]

Stewart's Division occupied Hood's center anchored on a low ridge by the little country log church where three dirt roads came together. Stevenson's Division took a position on Stewart's right, with Hindman on his left. A morning reconnaissance alerted Stewart to the fact that Joseph Hooker's entire XX Corps from the Army of the Cumberland was advancing directly on his front. Hooker's forward division, under the 6'6" John Geary, was as confused as the Confederate skirmishers trying to slow their progress. The topography for an advancing army was difficult. Historian Russell Blount described the woods as "thick, nearly impenetrable, and in a wild primordial state." From captured prisoners, Geary learned that Hood's entire corps was somewhere hidden in the woods ahead. It was late in the afternoon before Hooker had all three of his divisions, some 16,000 veterans of the Army of the Potomac, in place for a concentrated assault. Stewart waited with the brigades of Stovall and Clayton and parts of Gibson's and Baker's brigades. These 4,000 men were backed by the sixteen field cannons of Eldridge's Artillery Battalion. Clayton's men had piled up logs to fight behind, but Stovall's soldiers were less secure, in a line of battle by the little Methodist church, with only trees and tombstones to crouch behind.[48]

---

[47] Blount, *The Battles of New Hope Church*, 22.
[48] Ibid., 10.

Joseph Hooker deployed his corps in columns of brigades, one after the other, a formation designed for maximum penetrating power. The Federal division of Alpheus "Pop" Williams, the very men who had decimated Stewart's ranks at Resaca, led the assault. His division was in three battle lines of a brigade each. From the Confederate perspective, General Stewart realized that he would be able to concentrate his artillery and infantry fire to an area the width of an advancing brigade. After pushing through the underbrush for a mile, the Yankees broke into a trot as they descended into a small creek valley. "We went part of the time on double quick," wrote John Gourlie of 123rd New York, "and the first thing we knew had run right on to the Reb's breast works where they had their artillery masked." Upon descending into the ravine, the Bluecoats were suddenly met with a "hurricane of bullets, shrapnel, and canister." The Georgians fired into the dense trees and blackjack bushes as soon as they made out the lines of approaching Federal infantry. The attack faltered about forty yards from Stewart's lines. Taking shelter behind trees and boulders, or lying prostrate on the ground, the Yankees did their best to return the fire at the Rebels above them. "We lay down & by doing so saved ourselves a great deal from the effect of their fire," wrote Gourlie. "The grape and canister just fairly whistled over us cutting off the leaves in showers." Because of the dense landscape, the Federals had not been able to bring their artillery into position for the attack. The sixteen Confederate guns, however, pinned down the Yankee lines. Before the day was over, these Rebel cannoneers had expended some 1,600 rounds of ammunition. Geary reported that "the discharges of canister and shell from the enemy were heavier than in any other battle of the campaign."[49]

The fight came to an end in the growing darkness and the crackling of thunder. General John Geary described the night as "intensely dark, and a very severe thunder-storm, with cold, pelting rain, added to the gloom." Sergeant Rice Bull of the 123rd was even more descriptive, writing, "The rain came down in torrents, the lightning was blinding; then the darkness so black it could almost be felt." The victory brought relief and Rebel yells from the Southern lines. Captain L. P. Thomas, in his first battle as a regimental commander, was proud of his men. Holding the cemetery at New Hope Church,

---

[49] McDonough, *William Tecumseh Sherman*, 489; Castel, *Decision in the West*, 222–226; Corporal John C. Gourlie to his brother, May 29, 1864, in Civil War Times Illustrated Collection of Soldiers' Correspondence and Diaries; Corporal John C. Gourlie to his father, June 8, 1864; Blount, *The Battles of New Hope Church*, 61; report of General John W. Geary, Second Division, September 15, 1864, *OR*, ser. 1, vol. 38, part 2, 122–25.

they had been at the center of the Confederate defenses. He wrote, "This fight lasted over two hours, in an open woods, without the slightest protection of breastworks of any kind, and although confronted by overwhelming numbers, not a man was seen to waver." This was also true of division commander Alexander Stewart, who rode up and down his battle line on his big roan. When asked by one of Joseph Johnston's couriers if he needed reinforcements, Stewart, who was outnumbered three to one, curtly replied, "My own troops will hold the position." So it was, as corps commander John Bell Hood economically wrote, that "Hooker's entire corps was driven back by three brigades of Stewart's division." Bromfield Ridley called New Hope Church, "an epoch in the history of Stewart's division that is pointed to as a memorial of heroic valor just as Cleburne's men point to Ringgold and Cheatham's to near Kennesaw."[50]

Williams's Federal Division took 745 casualties, and Geary's Division reported another 509 casualties. Lovick Thomas reported his losses in the 42nd as "not so great, amounting to only 2 killed, 20 wounded, and 3 missing." Service records show that Philo Brooks (A), Silas Kirby (A), Allen Jenkins (B), John Thomas Barron (B), Thomas Webb (C), and David H. Malcolm (H) all lost their lives in that fierce contest of May 25, 1864. The Battle of New Hope Church ranks with Chickasaw Bayou, Resaca, Atlanta, and Bentonville as one of the bloodiest days in the regiment's history. Both Lovick Thomas and Colonel Abda Johnson of the 40th Georgia praised the actions of Captain James Summers and his Newton Rifles (F) in their after-battle reports. Summers's company served as skirmishers at New Hope Church with orders to "hold the ridge in front against any force less than a line of battle." The men from Newton County did their job well, "contesting every inch" and then moving to the left and then rendering "important service in protecting the left flank of the brigade."[51]

Both sides entrenched where they knelt on the night of May 25, with lightning flashes providing brief glimpses of their work. Parapets were dug

---

[50] Report of General John W. Geary, Second Division, September 15, 1864, *OR*, ser. 1, vol. 38, part 2, 122–25; Bull, *Soldering*, 117; report of Capt. L. P. Thomas, 42nd Georgia, of operations May 14–25, 1864, *OR*, ser. 1, vol. 38, part 3, 827–28; Blount, *The Battles of New Hope Church*, 59; report of Lt. Gen. J. B. Hood, undated 1864, *OR*, ser. 1, vol. 38, part 3, 761–62; Ridley, *Battles and Sketches of the Army of Tennessee*, 303.

[51] Report of General Alpheus S. Williams, May 1–July 28, 1864, *OR*, ser. 1, vol. 38, part 2, 26–36; report of Capt. L. P. Thomas, 42nd Georgia, of operations May 14–25, 1864, *OR*, ser. 1, vol. 38, part 3, 825–26; report of Colonel Abda Johnson, 40th Georgia, June 2, 1864, *OR*, ser. 1, vol. 38, part 3, 825–26.

about three feet deep with the shoveled dirt being thrown over logs that had been piled in front of the ditches. Skid poles were then planted into the ground to support a head log. The now dirt-caked soldiers would fire their rifles in the space between the parapet and the head log. Adding to the zigzag of trenches would be an abatis, which was a row of sharpened stakes in front of their parapets. For the next five days, there was only a ravine separating the two armies with troops on both sides constantly under fire from sharpshooters. The tiny valley earned the name "Hell Hole" from Yankee veterans. It could be seen as a precursor of World War I trench warfare. Alpheus Williams reported, "We have these affairs almost every night and there is not a minute of the twenty-four hours that popping is not going on." It also rained incessantly, and the trenches filled with water. Wood fires drew unwanted attention, which meant that cold hardtack was the meal of the day for the men in the rifle pits.[52]

On May 27, Charlie Rowland wrote, "We have lost in wounded 5 field officers and our Brigade is growing small." He mentioned the wounding of Colonel Robert Young, from the 40th Georgia, and Major Mark Nall of the 41st Georgia. "Having no commission, sword or horse," he explained, "I have remained with the ordinance train about a mile off during the day and go up at night with the ambulance to fix up some comforts for the staff to eat." Rowland would stay awake all night so that the rest of the field staff could gain some sleep. The brigade quartermaster decided one night to bring his slave Jonas with him, "as this was the occasion when a man wanted his servant." Unfortunately, Jonas was terrified of the minie balls whistling about and disappeared into a hole under the ambulance. "I will try Edward now and see if he can do any better," wrote Charlie.[53]

Johnston's fortified line now ran for several miles from Dallas eastward to Pickett Mill. Sherman attempted to break the stalemate by having General Oliver O. Howard march two of his divisions through rugged terrain on a flanking mission around the Confederate right flank. On the late afternoon of May 27, Howard sent the divisions of Woods and Hazen against the hastily thrown up entrenchments of Granbury's Texas Brigade. Much like New Hope Church, the Yankees at Pickett's Mill were attacking without the support of artillery, across heavily wooded ridges and ravines. The Confederates

[52] Blount, *The Battles of New Hope Church*, 75; report of General Alpheus S. Williams, May 1–July 28, 1864, *OR*, ser. 1, vol. 38, part 2, 26–36.

[53] Rowland to home, May 27, 1864, in *Charles Alden Rowland*, 34.

were under the overall direction of their best division commander, the Irishman Patrick Cleburne. Attacking in a piecemeal fashion, the Bluecoats were slaughtered, taking 1,580 casualties. They were beaten back at all points. Today, Pickett's Mill is one of the better-preserved Civil War battlefields in Georgia, maintaining much of its original terrain.

After the Federal debacle at Pickett's Mill, John Bell Hood immediately saw an opportunity to launch an aggressive counterattack. After gaining permission from General Johnston, Hood pulled his entire corps out of the battle line around midnight. Thousands of men, including the 42nd Georgia, assembled into columns and began a long night march to the northeast through the tangled woods. The seven-mile journey took the entire night. At dawn on May 28, Hood, emerging from the woods, believing he would find, after the chaos of the previous day's battle, Sherman's left flank unprotected, or, in military parlance, "in the air." He hoped, as Jackson had done at Chancellorsville in May 1863, to send his screaming men forward and begin rolling up the Federal line. Once underway, his surprise attack would be augmented by frontal assaults by Polk and Hardee's Corps. If it all worked, Sherman would be finished in Georgia. Unfortunately, the physically disabled general found the Yankees waiting for his men behind prepared breastworks. Richard Johnson's division of the XIV Corps had spent the night entrenching at right angles to the Union line. It was not a position even the aggressive Kentuckian believed should be attacked. His corps spent the day marching back to their original spot in the Rebel trench line. There they collapsed into the trenches exhausted from their twenty-four-hour excursion for naught.

On May 27, Stovall's Georgia Brigade was reinforced by the 1st Regiment Georgia State Line, a militia regiment of 550 men under Colonel Edward Galt. Originally recruited as railroad guards, the new men had spent the last fourteen months guarding the W & A Railroad from their posts at Cartersville and Etowah. Galt found Stovall's men, "in fine spirits & confident." The militiamen would fight alongside the 42nd Georgia through summer 1864. They were composed of men formerly exempted from military service, boys younger than seventeen and men older than fifty. Despite their ages, these men would prove to be courageous and effective fighters who were spurred by the reality that they were defending hearth and home. The 700 men of the 2nd Regiment Georgia State Line were assigned to Cumming's Georgia Brigade in Stevenson's Division. Since both militia regiments ultimately reported to the Georgia governor, they were goaded by the hardened Rebel veterans as "Joe Brown's Pets." Most of the men from the 1st Georgia came from counties in north Georgia. Colonel Galt was not pleased on May 31 when whiskey rations were issued to his men, and his Colonel James Bryan became drunk when "an engagement was most probable."[54]

On May 31, twenty-five-year-old Harvey Rainey of the Milton Tigers (C) wrote a letter home from the trenches. He had just returned to his command, having missed the Battle of New Hope Church. Rainey informed his wife that Thomas Barron had been killed instantly after being shot in the left breast. Skirmishing was now constant, he expounded, with five men in the 42nd wounded that morning, including Thomas Mann, who was "shot in the right cheek." The wound was not serious, he believed, though, "he [Mann] bled considerably." H. A. Rainey was bone tired from spending the previous night in the important work of repairing the breastworks, with the parapets now some nine feet in height. With the men mired in the trenches, Rainey viewed the health of the army as being "on the decline," but was pleased that, "We got Plenty Meat & bread to eat."[55]

Sixteen-year-old Corporal Zachariah Jabez Armistead of the Walton Tigers (H) was another soldier returning to the front. On Sunday, May 29, he managed to find the cooks of the Stovall Brigade. Armistead had been on a furlough to Newton County, probably assigned to escort his twenty-year-old brother Thomas home safely. Thomas Armistead had been badly wounded in the hip during the ill-fated charge on the Resaca battlefield. Along with a third Armistead brother, James, they were all members of the Walton Tigers. Zachariah reported, in a letter home to his brothers, written on the thirty-first, that he found the boys in the company "in fine spirits tho very badly worn out." He was relieved that, "Our Regt. Was not half as bad cut up as we heard it was."[56]

Still, only forty-four of the seventy-five men in the company were present for duty that day. Zachariah marveled that his Captain Joshua Mitchell was still in good health and "hasn't forgot how to sing yet." He told of heavy skirmishing up and down the line and wrote that 700 "yanks" had been buried the previous day, with another 300 still lying in the field. Zachariah knew that his brothers would want to know that "our 16th Ohio, who always fought our Regt. Was cut up very bad Sunday Knight." The 42nd was now "held back as a reserve about fifty yards from the ditches." Ironically, since the Yankees were firing uphill and generally too high, being farther back

[54] W. H. Bragg, *Joe Brown's Army*, ix, 85.

[55] Harvey A. Rainey to his wife, May 31, 1863, Rainey letters.

[56] Zachariah Armistead to John Armistead, May 31, 1864, UDC collection, 4:354–59.

made it a more "dangerous a place than in the breastworks." Zachariah longed to be home and to be with a girl in Covington who had given him a bouquet of flowers. He asked his family not to worry for him, as "I don't want you all to be, for if I am killed it won't be the first boy who has fell in the defense of his country. Water is very good. Yankees plentiful."[57]

Kennesaw Mountain

Conceding that his Dallas flanking movement had been a failure, Sherman, on June 1, began moving his battle lines to the left to disengage from the Confederates and regain contact with the railroad and his supply lines. That evening his cavalry seized the now unguarded Allatoona Pass and, in the coming days, Northern engineers would rebuild the bridge over the Etowah River. Soon the Federal foot soldiers could hear the welcome whistle of approaching supply trains coming south from Chattanooga. By June 4, the Federal army had reached the small village of Acworth on the Western & Atlantic Railroad. More good news for Sherman came on June 8 with the arrival of the XVII Corps under Francis Blair. These reinforcements, 9,000 veteran troops in all, roughly equaled the Federal losses sustained since the Atlanta Campaign began at the beginning of May. Johnston, on the other hand, could expect no reinforcements in this campaign of attrition. His Southern army continued to shrink in the face of both battlefield casualties and an increasing rate of desertion. On Sunday, May 29, Charlie Rowland found himself in the public square of Marietta writing a letter home. "Col Kellogg [43rd Georgia] was wounded Thursday and this makes 6 out of 8 field officers in the Brigade," he informed Katie. "We have suffered so much I hope the Division will have an easy place in the general fight."[58]

Two days later, on June 1, 1864, Lovick wrote to Jennie that five of his men had been wounded in the previous day's skirmishing along the "Hell Hole." The tired Confederate captain summarized the Georgia campaign to date, noting that his regiment had sustained nearly 150 casualties. "Col. [Abda] Johnson now commands the Brigade," he reported. The Georgians, he elaborated, had "brought up the reare in all the marches and been in all the fights losing more I have no doubt than any other in killed and wounded." Stovall's Brigade had been complimented in Napoleonic style by General Joseph Johnston, "the other day by his calling it the *Old Guard*." Morale had

[57] Ibid., 354–59.

[58] Rowland to home, May 29, 1864, in *Charles Alden Rowland*, 35.

"not cooled. We are still confident, and I might say eager for the contest to be settled." On June 4, Johnston ordered the first of what would be four retreats from the Dallas-New Hope Church Line to the Kennesaw Mountain area. His new ten-mile "Mountain Line" ran from Lost Mountain to Brush Mountain. His troops blocked the Western & Atlantic a few miles south of Big Shanty. Polk's Corps held the center, including Pine Mountain, which was about a mile in front of his breastworks. Hood's men would hold the right flank.[59]

On June 5, Lovick wrote Jennie from their new defenses near Lost Mountain. The 42nd had covered the retreat of the Stovall Brigade from New Hope Church by skirmishing "day and night." Upon reaching the base of Lost Mountain, the tired 42nd was sent out at dawn to push the Yankee skirmishers back to their main line of entrenchments. Lovick wrote Jennie that Silas Kirby of Company A had been mortally wounded in the day's skirmish and that James Tallent of Company B and "a citizen of Lawrenceville" had been wounded in the head. Lovick predicted another battle soon. Francis Nash reported that the 42nd crossed the railroad on the ninth and set up a position on Brush Mountain. They would maintain this position until the eighteenth. Lovick also picked up his pen on that day to report that he finally had the opportunity to put on some clean clothes "and feel more like a white man than I have in some time." Lovick and Sergeant Major Jim Griffin "sent Boy Tan to Oxford for some vegetables, which he brought up with other nice things & which we enjoyed very much." Lovick reminded Jennie of the time at the beginning of the war when she had visited him at Camp McDonald. She had boarded with a Mr. Roberts in Big Shanty. His home had a fine view of the majestic Kennesaw Mountain. Lovick lamented how "the Yanks must be all around him [Mr. Roberts] now." He asked Jennie to remember "what a fine view and beautiful scenery it presented, as we promenaded the piazza of his house."[60]

On June 14, 1864, generals Johnston, Hardee, and Polk went to the top of Pine Mountain to observe enemy positions. During this reconnaissance, Leonidas Polk was hit in the chest by a well-directed round from the 5th Indiana Battery, killing the corps commander instantly. The loss of such a high-ranking general, who was also revered as the Episcopal bishop of Louisiana, came as a shock to the foot soldiers of the Army of Tennessee. The

---

[59] L. P. Thomas to Jennie P. Thomas, June 1, 1864, Thomas papers, Santa Fe.

[60] L. P. Thomas to Jennie P. Thomas, June 5 and June 9, 1864, in Huff, *Four Families*, 85–86; Nash, *Civil War Diary of Francis H. Nash*, 30.

Rebels retreated from Pine Mountain that evening. Johnston was concerned that his lines were overextended and made frequent adjustments first to the Gilgal Church Line, then the Mud Creek Line, and finally the Kennesaw Line. Meanwhile, drenching rain storms soaked both armies for thirteen straight days. On June 16, sixteen-year-old Zachariah Armistead wrote his brother Thomas from the top of Kennesaw Mountain. He was on a detail guarding a signal station and found it "a beautiful place to see everything." From his position, he was safe from sharpshooters, but thought "a fellow don't feel right up here and see fighting going on." The next day Zachariah was back with the Walton Tigers and "very apt to be in all the fighting from know on." He hoped that brother Thomas's wound "would not prove fatal and that some future day you may be able to return to the Command and get revenge out of that yank for shooting you."[61]

Sherman was confronted with a Confederate army secure behind formidable positions, with the center anchored on Kennesaw Mountain. It was rugged, hilly land that was, once again, not conducive for mass troop movements, thereby reducing Sherman's overwhelming advantage in infantry numbers. The constant rain only made conditions worse. On June 20, Schofield's Army of the Tennessee gained some success trying to turn Johnston's left flank. The Confederate commander reacted immediately by pulling Hood's Corps away from his right flank and marching these Rebel divisions behind his lines, and then placing these 8,000 or so troops on his left flank. On June 21, the 42nd Georgia marched through the town of Marietta and camped for the night a few miles down the Powder Springs Road. Their orders were to link up with Hardee's Corps on their right and extend the Confederate line to Olley's Creek, with an emphasis on holding the Powder Springs Road.

By early afternoon, the First Division of Hooker's XX Corps began pushing up the now protected Powder Springs Road toward Marietta, which was four miles distant. The Federals reached a farm belonging to the Kolb family, where the 123rd New York and the 14th Kentucky were sent ahead as a reconnaissance in force. The Bluecoats soon encountered a strong Confederate battle line only 800 yards ahead. Hood was as surprised as Hooker to find a large force directly in their front. The rash Hood chose this moment to attack, even though he did not know the terrain ahead nor how Hooker had positioned his forces. Hood was gambling that he would catch Hooker's

[61] Z. J. Armistead to his brothers, June 16, 17, 1864, UDC collection.

Federals in a column of march and thereby be able to deliver a crushing flank attack. Hood did not inform General Johnston of his intentions. It was a bold decision fraught with danger.[62]

The Battle of Kolb's Farm commenced at 5:00 P.M. with Carter Stevenson's Division moving southwest, down both sides of the Powder Springs Road. Hindman's Division advanced on their right, hoping to eventually wheel into the flank of the Federal column. Hood held Stewart's Division, with Stovall's Brigade, in reserve. The Rebel attack was a disaster from the beginning. Stevenson's men had trouble pushing the 14th Kentucky aside, and their lines were wrecked by shot and canister from Federal artillery. Hindman's formation made little progress on the right. Both divisions barely made it to the main defensive line of Alpheus Williams, where a few volleys sent them into a full retreat. It was more of a one-sided slaughter than a battle. In less than two hours, Hood lost about 1,000 men, with Stevenson suffering 807 killed and wounded. Oddly, neither Sherman nor Johnston knew of the fight until late in the evening.[63]

The men of the 42nd stood in order of battle that late afternoon waiting to be called into the fight if an opportunity presented itself. None did. They were torn between wanting to get into the struggle for Kolb's Farm and equally relieved that they were not in the fight. "This is not a pleasant situation," wrote Abraham Moses Colton, a forty-one-year-old private in the Walton Tigers (H), to his wife. "The minie balls did not reach us, but the cannon balls and shells fell thick around us. But only one man was killed and two wounded in our regiment."[64] In a letter to his wife, Captain Thomas identified the dead soldier as James Arnold, a man he had enlisted in the Gwinnett Beauregards (A). "Little Jim" was killed by a solid shot. Charlie Rowland was with the ambulances and participated in the burial of Captain Jonathan Bowen of the 41st Georgia. "Although he only lived 17 miles from here," wrote the civilian quartermaster, "as we expect to fall back, he was buried just as he came off the field; such are the horrors of war as daily witnessed by someone in the army."[65]

The next morning, Stovall's Brigade moved a half mile to the rear and had a day of rest. Captain Julius Barclay of the 52nd Georgia, in a letter to

---

[62] Castel, I, 294.

[63] Hess, *Kennesaw Mountain*, 33, 35–41.

[64] Abraham Moses Colton to home, June 23, 1864, UDC collection, 2:41–42.

[65] L. P. Thomas to Jennie P. Thomas, June 1, 1864, in Huff, *Four Families*, 87; Rowland to home, June 19, 1864, in *Charles Alden Rowland*, 40.

Colonel Weir Boyd in Dahlonega, wrote, "It is the first day for a long time that our ears have not heard the…whir of the shell…and can enjoy the small ration of real genuine coffee." Private Colton reported that "the army is putting up breastworks to-day and I learn that General Johnston intends to establish a line from Kennesaw Mountain to the Chattahoochee River, a distance of sixteen miles." On June 24, Colton reported that the men of the 42nd were "digging on ditches." Of Captain Mitchell, he reported, "My old cap is a poor thing for hot sunshine and rain." Colton lamented, "Poor soldiers, indeed, more like hogs, so dirty, a lousy set of fellows." The rain never seemed to stop. "If we don't get rheumatism when the war is over," wrote Charlie Rowland, "it wont be because sleeping on the wet ground can produce it."[66]

General Johnston did not have the men to present a sixteen-mile front. His line was now in a six-mile horseshoe protecting Marietta. Kennesaw Mountain held the top of the horseshoe and Hood's Corps, now feverishly building breastworks, held the far left side of the Rebel line. It was enough to effectively block "Old Billy's" army for the time being. Major William Hulsey returned to duty with the 42nd Georgia on June 23, after recovering from his wound received in the charge at Resaca. He found his command "erecting a line of entrenchments in an open field on Appling's Farm." The men held this position, skirmishing with the Federals in front of them, until the night of July 3.[67]

With flanking movements "played out," Sherman ordered a general frontal assault on Kennesaw Mountain for June 27. After viewing the embrasures for three batteries Hood's men had built across Powder Springs Road, Sherman, on June 25, exempted Schofield's Corps from a frontal attack on the Confederate left flank. Instead, the Army of the Cumberland would continue to push south on the Sandtown Road and attempt to find a passage around the Confederate army on the other side of Olley's Creek. June 26 was Pentecost Sunday and a quiet day all around.[68]

On the morning of June 27, eight of Sherman's fifty-four infantry brigades, some 15,000 troops, charged up the hills into the Confederate lines

[66] Kinsland, *"A Band of Brothers,"* 23–24; Colton to home, June 23, 1864, UDC collection, 2:41–42; Rowland to home, June 1, 1864, in *Charles Alden Rowland*, 40.

[67] Hulsey, report to General Clayton on the Operations of the 42nd Georgia, August 9, 1864.

[68] Castel, *Decision in the West*, 304.

above them. The charge was a colossal failure with no breakthroughs anywhere on the Rebel's fortified front. The hardest fighting was at Cheatham's Hill against Maney's Brigade of veteran Tennesseans. The accounts of the doomed assault caught the attention of the nation, especially with the death of two young brigadier generals from Ohio, Charles Harker and Dan McCook. Sherman reported nearly 3,000 casualties in the day's action, while Johnston's loss was close to 700 men.

Sherman got better news from Schofield's XXIII Corps that evening. They had pushed down the Sandtown Road and occupied Nickajack Valley, which ran parallel to the Chattahoochee River. The Bluecoats were beyond Hood's left and closer to the Chattahoochee than were the Confederates. Only Jackson's cavalry and the Georgia State Militia, under Gustavus Smith, stood in front of Schofield's small corps.[69]

Johnston had been flanked again. For Sherman to take advantage of this, however, he would have to be willing to abandon the railroad a second time and swing McPherson's Army of the Tennessee all the way from Kennesaw, on the Federal left flank, around to the right flank held by Schofield's men. There was no better alternative. On June 30, McPherson began the long march around George Thomas and his Army of the Cumberland, which was holding Johnston's army in place. A bolder general than Joseph Johnston would have seen how extended, and thin Schofield's line was and attacked it with a strike force. Instead, on July 1, the Confederate commander ordered another retreat. "On the night of the third," reported Major Hulsey, "I received orders to go on picket with my command, with orders to withdraw as quietly as possible." The 42nd held the trenches until 1:00 A.M. before bringing up the rear of the Confederate retreat.[70]

---

[69] Ibid., 321.

[70] Hulsey, report to General Clayton on the Operations of the 42nd Georgia, August 9, 1864.

Chapter 6

The Battle for Atlanta

"This I do know, he will never lose his army and he will not likely have it defeated if any movement on his part can prevent it."
—Captain Lovick Thomas on the military strategy of General Joseph E. Johnston, July 5, 1864

Jennie's Dilemma

In 1860, 75 percent of Georgia's adult white male populace and 90 percent of its slaves were directly engaged in agricultural pursuits. Apart from company K, composed principally from the business ranks of Atlanta, almost every soldier in the 42nd Georgia came from a farming background. Even within the officer class, many had their fortunes, if not their professions, tied to the land. Dixieland was agrarian, and the men of the 42nd Georgia were truly Sons of the South. Thomas Jefferson had spoken for all Southerners when he declared, "Those who labour in the earth are the chosen people of God, if ever he had a chosen people." These Georgia farmers had left their homes in spring 1862 with at least one comforting thought; that even in times of war the land would produce its bounty and their families would not suffer in want during their extended absence. Captain Lovick Thomas left home in a different situation. He was a merchant, co-owner of a Lawrenceville dry goods store. To provide for Jennie and the children, Lovick would have to be creative while he was serving his new nation.[1]

In December 1862, while on sick leave from Tennessee, Captain Thomas had put in motion a plan to provide for his family in a war he had already concluded would be long and drawn out. As a first step, he shrewdly chose to sell his interest in the general store in Lawrenceville. Wars throughout history have meant inflation and subsequent currency devaluations as governments pay for their armies with paper and promises. Lovick believed that the speculator with goods to barter held a better position than an owner of suspect specie. With the money generated from the sale of his mercantile

[1] Kibler, *The Classical Origins of Southern Literature*, 74.

ownership stake, L. P. immediately invested in basic staples that he believed would hold their value in the coming economic crisis. While on leave, he purchased ten boxes of sun-cured tobacco leaf and a town lot in Atlanta, which subsequently served as the first post-war home of the Thomas family.

One of the boxes of tobacco was sent to Jennie in Lawrenceville and the others entrusted to his brother-in-law, Joseph Porter, for storage in an Atlanta warehouse. When peace returned, Lovick hoped that his tobacco leaves, safe in storage, would give him and Jennie the means to establish a new beginning in Atlanta. By spring 1864, the ever-resourceful Jennie found herself using the contents of the tobacco box to barter for the daily supplies of life. To the same end, she traded from a fifty-pound barrel of sugar that Lovick had wisely sent to her from Vicksburg. The couple also speculated in coffee, the third leg in their list of necessities of life. Jennie's diary entries and letters to Lovick are filled with price lists, which bore testimony to their prediction of ever-rising inflation. On April 14, 1863, she wrote in her journal, "I only buy what I *need*, no finery."[2]

In late spring 1864, along with her daily concern for her husband at war, the safety of whom she entrusted to God, and the pressure of providing for the welfare of her young children, which she entrusted to her tobacco, Jennie watched as Sherman's three Federal armies advanced through Georgia. Newspaper accounts no longer told of battles in faraway Virginia or near the Mississippi River. In May and June, she keenly followed Sherman's progress down the Western & Atlantic Railroad through familiar Georgia towns: Ringgold, Dalton, Resaca, Calhoun, Adairsville, and Kinston. Atlanta was the grand prize, and, with each day, his vandals were ever closer.

Should she remain in Lawrenceville with her small family or attempt to flee ahead of their advance? If she abandoned their Gwinnett home, where should they go? On June 9, Lovick responded to her questions and concerns by writing, "I can't advise you my darling what to do, only to keep out of the enemy's way & if there should be at any time a probability of them coming along, get off." On June 25, two days before the Battle of Kennesaw Mountain, Lovick put her retreat in military terms, writing, "If you do have to fall back, I would advise you to do so in good order, not hastily. Nor do not become demoralized from a retrograde movement and get frightened." He did not see a danger to Lawrenceville unless the Confederates retreated across

[2] Jennie P. Thomas, diary, April 14, 1863, in Huff, *Four Families*, 78.

the Chattahoochee River. "They seem to be still moving to our left," he reported, "which is right off in the opposite direction from Lawrenceville." The military analogies, which showed the degree to which the war now consumed his thought processes, continued. "You could either go to Oxford or some other place nearby first," he wrote, "and if the enemy still move on, and you think you can't risk the position, fall back a little and so on." He then added, "Take care of your little army, and the baggage train. You must, my darling Jennie, do most of the maneuvering yourself. Display Generalship if the emergencies arise." Financially, Lovick had nothing to offer should it be necessary, as he had only been paid for one month of service since leaving Dalton. He hoped "to draw before you are pushed for it."[3]

Chattahoochee River

On the evening of July 2, Johnston, having been flanked again, abandoned his Kennesaw line and the town of Marietta and pulled back to the Smyrna Line, a six-mile impregnable bridgehead on the northern banks of the Chattahoochee River. Teams of slaves had recently built a double-line of deep ditches and impressive redoubts, all lined with rows of sharpened abatis logs. Though the Rebel soldiers had the Chattahoochee River at their backs, they were confident that they could defend the railroad bridge and the various pontoon bridges that crossed the river. "How long we will remain here, I can't say," wrote Lovick on July 4, 1864. We have good works, at the River, and are now fortifying here," he added. According to Francis Nash of the Milton Tigers (C), the 42nd Georgia was in a line of battle on the heights of the riverbank just south of the railroad bridge. They were the extreme left of the Confederate army, in sight of the village of Vinings Station that was across the river and only eight miles from Atlanta. "We have a good position and can't be whipped out of it," Captain Thomas wrote. Union commander William Sherman agreed with Thomas, writing William Halleck in Washington that the Confederate line extended for five miles, "having in its whole extent finished abatis and parapet, with glacis obstructed with chevaux-de-frise, and all manner of impediments." Heavy skirmishing took place here on the afternoon of July 4, 1864, as the Federals probed the strong Confederate lines.[4]

---

[3] L. P. Thomas to home, June 9 and 25, 1864, ibid., 86, 87.

[4] McDonough, *William Tecumseh Sherman*, 514; L. P. Thomas to Jennie P. Thomas, July 5,1864, in Huff, *Four Families*, 89; Nash, *Civil War Diary of Francis H. Nash*, 32; report of W. T. Sherman to H. W. Halleck, July 11, 1864, *OR*, ser. 1, vol. 38, part 5, 114.

Major William Hulsey returned to duty in the first week of July and assumed command of the 42nd Georgia. Hulsey was not fully recovered from his wounds received in the charge at Resaca, and, thus, Lovick Thomas continued to direct the hands-on movement of the regiment. "I will still act in the capacity of major," wrote Thomas. "I will not regret the purchase of my horse," he added, "as I think it a good investment, and I am getting the benefit thereof every day." A box of cakes arrived from Jennie on July 3. Lovick had the 42nd out skirmishing with the Army of the Cumberland at the time. Major Hulsey, who was in camp and not in the field with his regiment, opened the box and put the cakes in L. P.'s haversack. As it turned out, there had been no rations that day, and when Lovick returned, he found only one surviving cake left in the haversack. Evidently, foraging was not restricted to the Georgia countryside.[5]

Camp life, however, especially for men who had survived the Vicksburg Siege, was generally tolerable. Close to their supply base in Atlanta, the men were usually receiving full rations and sanctioned trips to home brought supplements to their diet. Jim Griffin's servant, "Boy Tan," was still making trips to Oxford for vegetables. The officers had a little flour to cook with, and creativity was the order of the day. Mess groups would often trade amongst themselves to put a meal together. Friendships and favors often brought rewards to those who nurtured them. Lovick had befriended J. B. Groves, a sergeant in the 1st Georgia State Line, who oversaw the brigade's tool wagon. Returning the courtesy, Groves agreed to carry a change of clothes for Captain Thomas in his wagon.[6]

Lovick Thomas's religious ardor had not faded since the Dalton revivals. "I had a present of a bible from Chaplain Parke of the 41 Reg.," Lovick informed Jennie on July 5. His fellow officers noticed L. P.'s preference for reading the Good Book each day. "I find my example is doing good," he wrote, "as my Bible is found often laying about convenient to mess mates, [William] Hulsey, [Hugh] Wiley, [James] Griffin, [William] Durham, the later the adjt for Capt W [Weaver], now and then." This was Lovick's first mention of eighteen-year-old William Durham, who would finish the war as adjutant for the 42nd Georgia. "I have got them all to promise a more regular attendance on divine service," remarked Thomas, "& the abstinence of all swearing in our Mess." On reflection, he added, "I may yet do some goods

[5] Hulsey, report to General Clayton on the Operations of the 42nd Georgia, August 9, 1864; L. P. Thomas to home, July 5, 1864, in Huff, *Four Families*, 89.

[6] L. P. Thomas to home, July 4, 1864, in Huff, *Four Families*, 88.

by precept and example, even in the Army, and with God's assistance I hope I may be enabled to do much more when this wicked war is over."[7]

Changes in command on July 7 directly affected the 42nd Georgia. Their respected division commander Alexander P. Stewart was promoted to lieutenant general and given command of Polk's former corps, now called the Army of Mississippi. Stewart's old division was not given to Marcellus Stovall, as Lovick had hoped, but to thirty-seven-year-old Henry De Lamar Clayton of Alabama. Clayton was a principled Alabama politician and a competent military commander. He had been a Southern Rights advocate since his days as a student at Emory and Henry College and then as a successful lawyer and prosperous planter near Eufaula, Alabama. Clayton had been severely wounded at Stones River while leading the 39th Alabama, but he had returned in time to command a brigade at Chickamauga. General Clayton would prove to be an excellent division commander.[8]

With the Chattahoochee River at their backs, the Confederates knew that it was only a matter of time before Sherman found a way to flank their impressive position. "The enemy may cross & can cause the same on our part," thought Thomas. The captain from Gwinnett was certain of one thing. The invaders "must fight before getting to Atlanta," he wrote Jennie. On the night of July 9, the Rebels retreated across the Chattahoochee River and burned the railroad bridge behind them. Their new defensive position on the opposite bank was called Johnston's "River Line."[9]

The new river defenses that the Army of Tennessee retreated to were unlike anything seen by either army. They were the brainchild of General Francis Asbury Shoup, chief of artillery for Johnston's army. He had gained permission from Old Joe to construct these five miles of fortifications only two weeks earlier. Using gangs of slave labor, Shoup built a series of aboveground forts on the Atlanta side of the Chattahoochee. Each "shoupade" was shaped like an arrow pointing toward the river, with a double row of logs, which were backfilled with dirt. These shellproof redoubts were placed to have interlocking fields of fire. Shoup believed his fortifications gave Joseph Johnston a strategic advantage. In theory, Johnston could now hold the bulk of Sherman's large army with a small force distributed among the impregnable shoupades. General Johnston would then be free to move the majority of his infantry in either direction and aggressively respond to any move made

---

[7] L. P. Thomas to home, July 5, 1864, ibid., 89.

[8] H. J. Walker, "Henry De Lamar Clayton," 138–140.

[9] L. P. Thomas to home, July 5, 1864, in Huff, *Four Families*, 89.

by Sherman or even surprise the Federals with a flank attack of his design.

Unfortunately for Confederate grand strategy, Johnston was tepid at best about the whole idea. While Shoup saw his line as "the savior of Atlanta," Johnston saw it only as another temporary line of defenses. Within days of occupying the River Line, Schofield's Bluecoats found a submerged stone fish dam near Roswell and began crossing the Chattahoochee River miles to the east. Like the Maginot Line in World War II, Shoup's River Line was never to be challenged. Johnston saw peril where others might have seen opportunity. He did not contest the Union crossing near Roswell but merely conceded that he had been flanked again. The Rebel commander abandoned the River Line without a fight and withdrew the Southern army to a ring of trenches, three miles from the center of Atlanta.[10]

Change of Command & Desertions

Division commander Henry Clayton reported that, on the night of July 9, three of his brigades crossed the Chattahoochee and headed down the Turner Ferry Road toward Atlanta. Holtzclaw's Alabama Brigade held the crossing through the night before joining the retreat. Stovall's Brigade would skirmish in the woods by the river for the next week, with William Hulsey reporting the loss of two men from the 42nd Georgia. The young major would describe one of the fatalities, Sergeant Henry House (I), "as the best soldier I ever knew." On July 12, General Leggett of the XVII Corps reported that "Stewart's division, of Hood's Corps, is opposed to us in the region of Sandtown Ferry. A band of guerrillas, consisting of Georgia State Troops, was on this side of the Chattahoochee, just beyond the Sweet Water, last night."[11]

Henry Campbell of the 18th Indiana Battery considered the Chattahoochee a difficult river to cross even on a pontoon bridge. He wrote in his diary, "It is not a very wide stream but is swift, muddy, with great large limestone rocks scattered all through it. The banks are steep, descending into deep water very abruptly, and covered with a thick growth of cane." By July 12, 1864, Sherman had crossed the last natural barrier between his three armies and his ultimate objective. Atlanta lay at his feet and before his guns. "Cump" Sherman gave his infantry, who were camping along the Chattahoochee River,

---

[10] Shoup, "Works at the Chattahoochee River," 262–65.

[11] Clayton, unpublished account of military service, military papers of Henry De Lamar Clayton, 7; Hulsey, report to General Clayton on the Operations of the 42nd Georgia, August 9, 1864; report of General M. D. Leggett, July 12, 1864, *OR*, ser. 1, vol. 38, part 4, 130.

five badly needed days to rest and refit. His cavalry, under General George Stoneman, was sent on a raiding mission south of Atlanta.[12]

The historian Albert Castel noted that in the thirty days since leaving Acworth, Sherman had only advanced fifteen miles and had "lost about 9,000 men in combat and approximately the same number from disease, illness and fatigue." Both armies had fought and marched through a humid Georgia summer that alternated between rain and mud—or heat and dust. Far removed from the parades and "send-offs" of 1862, the uniforms of both armies were now in tatters, and biting and stinging insects made life a marching production of Victor Hugo's popular book *Les Misérables*. The journal of Captain Andrew West, adjutant of the 41st Georgia, reported that it had rained for forty-eight consecutive days. Johnston's ragged Confederates had also taken heavy casualties, and they were physically beaten down by the long, contested retreat. Even with the addition of the Georgia militia, the Atlanta defenders were still heavily outnumbered.[13]

On July 13, Abraham Colton of the Walton Tigers (G) wrote his wife that he was "worn out and if the army has much marching I will not be able to go with them. I am growing weaker." Stovall's Georgia Brigade had just returned from picket duty that morning. Colton told his wife that since the Georgia campaign began, his company had suffered three killed and twelve wounded. Close to the supply warehouses in Atlanta, rations had become consistent but highly repetitive, as each meal's menu was often derided as a choice between bread and meat or meat and bread. Slaughterhouses and meatpacking plants in Atlanta supplied the troops with Florida beef, pickling the cuts in brine before transporting them to the army. The soldiers found the processed meat to be unpalatable unless boiled over the fire for an extended period. Patrick Stovall of the 40th Georgia wrote, "The beef we draw now is as poor as you ever saw. And when we boil it, it ain't got no more grease on it than a piece of wood." Additions to the diet were always welcome. "Cap't Mitchell has written his Bro. Tom," wrote the forty-three-year-old Private Colton, "to get up some vegetables and bring to the company."[14]

With both armies now on the outskirts of the Gate City, important stra-

---

[12] Campbell, *A Hoosier's Civil War Diary*, July 19, 1864.

[13] Castel, *Decision in the West*, 343; Alexander J. West, "The Private Soldier," 1899 Speech at 42nd Georgia Reunion, Thomas collection, BRC.

[14] S. Davis, *What the Yankees Did to Us*, 45; Whatley, *The Typical Georgia Confederate*, 33; Abraham Moses Colton to home, July 13, 1864, UDC collection, 2:41–42.

tegic decisions had to be made. With 106,070 men, Sherman's plan for capturing Atlanta would be to pressure Johnston's 54,000 men from the north with the armies of Schofield and Thomas while swinging McPherson's Army of the Tennessee well to the east of Atlanta toward Stone Mountain, where McPherson could flank the thinning gray lines. Jefferson Davis, back in Richmond, had by now lost faith in his western commander and Old Joe's "Fabian" strategy of avoiding "major pitched battles by maneuvering to trade space for time." Joseph Johnston gave no indication to the Confederate president that he was committed to a significant battle in defense of Atlanta. The cabinet in Richmond was advising their president that the Confederacy would be close to collapsing with the surrender of Atlanta. The Gate City of the South had to be held.[15]

Captain Lovick Thomas, on the other hand, was an unequivocal supporter of Joseph Johnston. In the captain's mind, the preservation of the Army of Tennessee was paramount. "Atlanta could be rebuilt," he wrote Jennie. "Our army could not." L. P. thought it better "to lose half a doz[en] such places than an army." What Thomas could not explain was how to supply a Southern army without this strategic supply center. The Atlanta Rolling Mill and the Tredegar Iron Works of Richmond were the only producers of iron rail lines and naval gunboat plating operating in the South. The Atlanta Arsenal was the city's largest employer with 5,500 workers, operated from nine buildings on fifty acres at the "Old Race Track" across from Lovick's yet unseen town lot. In early July, the arsenal was turning out 30,000 rounds of rifle cartridges and 400 rounds of 20-pounder Parrott artillery ammunition each day. The second-largest employer in the city was the Confederate Quartermaster Department, responsible for supplying Johnston's army with uniforms, shoes, blankets, medicines, and camp equipment. The Commissary Department used its meat-packing plants and warehouses to feed the Army of Tennessee.[16]

The ring of earthworks and forts that surrounded Atlanta testified to the vital importance of its munition factories, army warehouses, military hospitals, and railroad facilities. As historian Stephen Davis wrote of the city, "probably only Richmond and Washington were more heavily protected by entrenchments and embrasures." On July 5, as his army retreated to the

---

[15] S. Davis, *What the Yankees Did to Us*, 87–88; Hood, *The Lost Papers of Confederate General John Bell Hood*, 53.

[16] L. P. Thomas to Jennie P. Thomas, July 5, 1864, in Huff, *Four Families*, 89; S. Davis, *What the Yankees Did to Us*, 43,77.

Smyrna Line on the north bank of the Chattahoochee River, Confederate General Joseph Johnston issued orders that the hospitals and munition works in Atlanta be evacuated. The abandonment of Atlanta by the Confederate government had officially begun.[17]

On Sunday, July 17, President Davis replaced Johnston with one of his three corps commanders, John Bell Hood. In so doing, Davis passed over the more senior and more respected William Hardee of Georgia. Because Hardee had declined the appointment six months earlier, Davis did not ask him again. The change in command, at this crucial moment in the campaign, broke the morale of many of the Southern soldiers. Shock in the Rebel ranks soon gave way to indignation and then to outrage and despair. Lieutenant James Monroe Gresham of the Walton Tigers (H) wrote that the removal of Johnston in favor of Hood, "came near creating mutiny in the western army." Historians and Civil War enthusiasts have and will forever question the wisdom of Jefferson Davis's decision, with Sherman at the gates of Atlanta, to replace Johnston with Hood. Perhaps most intriguing is the speculation of what Johnston might have done differently in the critical months that followed.[18]

Desertions had been a problem for the Army of Tennessee since the disappointing retreat from Cassville. Filled with a sense of futility and despair and unwilling to endure the hardships and perils of war any longer, Southern soldiers were slipping over to the Union lines with each passing day. Charles Houghton of the 141st New York wrote from Cassville, "We took 30 or 40 prisoners yesterday. They all say that they are caesfide [satisfied] that they are whipped and will not fight any longer." Like Washington at Valley Forge, General Johnston found it necessary to order the execution of sixteen would-be deserters as a deterrent. Confederate companies from counties in north Georgia were particularly hard hit by men disappearing into the night. With the war at their doorstep, these deserters felt a greater responsibility to themselves and their threatened families than to a dying Confederacy. All ten companies in the 42nd Georgia, however, were not from the mountains of north Georgia. They had been drawn from counties in the Piedmont region of the state, principally around Atlanta and east of the city. The regiment had held together through the tough fights at New Hope Church and Kennesaw

---

[17] S. Davis, *What the Yankees Did to Us*, 63, 77.

[18] Gresham, "A Partial History of the Walton Tigers," 662.

Mountain.[19]

With the retreat past the Chattahoochee River, however, the 42nd began to lose men to the siren's call of desertion. Two companies, the Gwinnett Beauregards (A) and the Milton Tigers (C), were especially hard hit. According to military service records, the Gwinnett and Milton companies each lost seven of their men in July, compared to five desertions in total for the other eight companies of the 42nd Georgia. William Morris, Harvey Rainey, and Joseph Thomason were all deserters from Company C who were rounded up by Federal horsemen in Alpharetta around July 21. In accordance with Union military protocol, they were shipped to Louisville, Kentucky, where, on July 27, they took an Oath of Allegiance to the United States government and agreed to remain north of the Ohio River until the war ended. Similarly, Yankee cavalry rounded up Daniel Brogdan, Daniel Keheley, John C. King, and William Worsham, all deserters from Company A, in Lawrenceville in July. They accepted the same terms in Louisville in lieu of transfer to a Yankee prison.[20]

In all, a few thousand former Confederates were given their freedom as long as they remained in Indiana or Ohio. One example was Private John Henry Smith from the Calhoun Guards (K). After giving his Oath of Allegiance in Louisville, Smith worked at the Ohio Falls Car Company in New Albany, Indiana. In 1869, he married a local Hoosier girl named Sarah Doster. Smith would later return to Georgia with his Midwestern bride, finishing his days as a farmer in Jackson County.[21]

The Battle of Peach Tree Creek

Upon assuming command of the Army of Tennessee on July 19, General Hood found his fellow Southern soldiers holding a line north of Atlanta from the Chattahoochee railroad bridge eastward to just beyond the Peachtree Road. According to a captured Rebel document dated July 18, Hood was down to 44,400 effective infantrymen in his three corps, a far cry from the 77,000 men that had swelled the ranks of the Southern army at the end of May. Hardee's Corps of 18,400 effectives held the center, with Stewart's

[19] Charles A. Houghton to his wife, May 20, 1864, in Civil War Times Illustrated Collection of Soldiers' Correspondence and Diaries; Castel, *Decision in the West*, 261.

[20] Henderson, *Roster of the Confederate Soldiers of Georgia, 1861–1865*, 4:510–21, 532–39.

[21] Confederate pension application of John Henry Smith filed in Jackson County, 1917, record group 058-01-001, Georgia Archives, Morrow, GA.

12,400 men to his left, and Hood's former corps, some 13,600 men now under the temporary command of Benjamin Cheatham, on the right. Hood's eyes and ears were his 6,000 Rebel horsemen. Red Jackson's cavalrymen guarded the Chattahoochee crossings to the west and Joseph Wheeler's cavalry, along with 700 members of the Georgia Militia under Gustavus Smith, watched the eastern approaches to Atlanta. Sherman outnumbered Hood's army by more than two to one. The men of the 42nd Georgia were part of the 1,200 men in Stovall's Georgia Brigade. Stovall himself was again absent, and his men were under the direction of Colonel Abda Johnson of the 40th Georgia. Stovall's veterans were one of four brigades that made up the third division of Cheatham's (formerly Hood's) Corps under the leadership of General Henry Clayton of Alabama.[22]

Hood quickly adopted Johnston's plan of attacking George Thomas's Army of the Cumberland as it attempted to cross Peach Tree Creek, a wide stream running east to west between the village of Buckhead and the city of Atlanta. The largest of the three Union armies would be by itself, as Schofield's small Army of the Ohio was too far to the northeast to assist Thomas. Seeing this, Hood planned to attack Thomas with both Stewart's and Hardee's corps, hopefully when Thomas was most vulnerable—on the move and in the open. For this to work, Cheatham's Corps would have to hold Schofield's advancing force from the northeast as well as cover the eastern flank of the Confederate army. What Hood could not account for, until it was almost too late, were Sherman's favorite troops, McPherson's Army of the Tennessee. Sherman had, throughout the campaign, used these trusted Midwesterners as his flanking force. Around noon on July 20, word reached a horrified Hood that Sherman had successfully swung the 28,000 infantrymen under McPherson all the way to Decatur, a railroad town on the Georgia and Augusta Railroad only six miles from Atlanta. In short, both commanders entered July 20 committed to attacking their opponent's right flank, while oblivious to the danger posed to their own right flanks.

Hood stayed with his plan to go on the offensive with two-thirds of his army, but the Confederates did not execute the attack well. The change in command had pushed everything back, and there was no longer the careful planning and execution that Johnston had brought to each undertaking. Edward McMorries of the 1st Alabama Regiment noted the absence of Johnston's cautious reserve when he wrote, "The most perfect order and system in

[22] Jenkins, *The Battle of Peach Tree Creek*, 22–24; Ecelbarger, *The Day Dixie Died*, 214.

movements of the army [under Johnston] suddenly changed into utter confusion. Cavalry were hurrying in one direction; artillery flying in another; infantry, double-quicking in another; and everywhere confusion." Where Johnston had selected open terrain near Clear Creek, Hood's men would be going through heavily wooded ravines. Keeping order in the ranks would be almost impossible. The 11:00 A.M. attack on this sweltering day was delayed for several hours.[23]

With Schofield pushing toward Atlanta from the northeast, the new Confederate leader responded by moving Benjamin Cheatham's Corps, which included Stovall's Georgia Brigade, two miles south and east to cover the Rebel right flank as far as the Decatur Road. This left a large gap between Cheatham and Hardee. Valuable time was lost as Stewart and Hardee shifted their formations a half-mile to the east. When the Confederate attack came in the late afternoon, it was furious, desperate, and disjointed. It did not help the Secessionists that the ever-cautious commander of the Army of the Cumberland, George Thomas, had been given time to entrench some of his positions immediately upon reaching the high forested ground south of the creek. What might have been a brilliant tactical move on Hood's part at 11:00 A.M. would become a trap some five hours later. Several Federal lines were overrun in the Rebel attack, creating 1,900 Union casualties, but, in the end, the Confederates were thrown back at Peachtree Creek with heavy loses, some 2,400 killed, wounded, or missing. Particularly devastated were the men who had charged the deepest into the Federal lines, the brave Mississippi regiments under A. P. Stewart.

Removed from the main theater but fully engaged with the enemy, Cheatham's Corps spent the afternoon attempting to cover a five-mile-long front with a single line of soldiers. Stovall's five regiments were on the far right and deployed in the woods of what is today Atlanta's Morningside neighborhood near the campus of Emory University. They found themselves skirmishing with the advance troops of Schofield's XXIII Corps. Neither side had reserves to call upon, so it became a dangerous day of "Indian fighting" from tree to tree.

Privates George W. Cawley and William W. Turnbull of the Walton Tigers (H) were both killed in the sniping. The sad tally for the 52nd Georgia was ten men killed or wounded, including the death of Captain Julius Barclay of Union County. The 1st Georgia State Line troops lost their leader, Colonel

[23] McMorries, *History of the First Regiment*, 77.

Edward Galt, who was badly wounded in the left forearm and subsequently replaced by Colonel John Brown, the twenty-five-year-old brother of Governor Joseph E. Brown. General Clayton reported that "At night we fell back to the inner work around Atlanta and formed a line with the left resting on Peach tree street."[24]

The problem for Hood and his right flank was that Cheatham did not have enough men to reach the Decatur Road. On July 20, McPherson's Army of the Tennessee had only Joseph Wheeler's indefatigable cavalry between them and the capture of Atlanta. An aggressive McPherson could have pushed into the city and gained glory that day, but instead, he chose to move his forces cautiously down the Decatur Road, fearful that his long columns were vulnerable to a flank attack. His Federals pulled up only two and a half miles from the city center, where Captain Francis DeGress, of the 1st Illinois Light Artillery, unlimbered one of his 20-pound Parrott guns and fired the first artillery shells into the downtown of Sherman's long sought-after prize. The slow Federal advance gave Hood time to order his only reserves, Patrick Cleburne's division, to the east where, late in the day, they fortified the countryside south of the Georgia Railroad and interposed themselves between McPherson and Atlanta.

The Battle of Atlanta

Private Nash reported that on the evening of July 20, the 42nd moved a mile and a half south to the Georgia Railroad. They spent the night building breastworks, a constant activity on the long retreat. After a brief respite, the expected skirmishing started the next morning as "the enemy cannonaded us incessantly." The bigger fight on July 21 was to their right, where a fierce fight broke out between the three brigades of Patrick Cleburne and McPherson's advancing army. It was still early in the morning when Benjamin Leggett's best brigade under Giles Smith pushed Granbury's Texans off the top of Bald Hill, a prominent elevation where today Moreland Avenue crosses over I-20. Leggett planted his artillery on top of the summit, which was only a mile and a half from the center of Atlanta.[25]

John Bell Hood now focused on McPherson's Army of the Tennessee, which threatened his tenuous right flank. What the Confederate commander

[24] Jenkins, *The Battle of Peach Tree Creek*, 46; Clayton, unpublished account of military service, 10.

[25] Manigault, *A Carolinian Goes to War*, 225; Castel, *Decision in the West*, 383.

saw was an opportunity to turn the tables on his adversary. Sherman's three armies were separated from one another by several miles, and McPherson's left flank was "in the air." His cavalry escort under Kenner Garrard was not with him, as they were on a fool's errand wrecking the Georgia Railroad near Covington. This allowed Wheeler's Rebel horsemen free reign, and they were reporting to Hood that McPherson's large wagon train was parked in Decatur and open to capture. True to his reputation and in keeping with the wishes of his Richmond superiors, Hood composed an aggressive plan to attack McPherson's exposed left flank.[26]

As the sun rose over a stifling humid Atlanta on July 22, news had already spread down the line of Clayton's Division "that Hardee's Corps was off on a secret expedition." Indeed, they were. Historian Gary Ecelbarger called Hood's audacious plan of attack "unlike any seen in the Western theater of the Civil War to date." As darkness had fallen on July 21, William Hardee had begun pulling his entire corps from their line of trenches and started a long march south through the city of Atlanta. Stovall's Georgians could hear the long gray columns moving off behind their trench line. Hardee left downtown marching south four men abreast on the McDonough Road. The plan was to cover fifteen miles in the eight hours of summer darkness, making a wide circuit around the Federal lines. At dawn, Hardee and Wheeler's 17,000 men would then deploy and aggressively attack the flank and rear of McPherson's Army of the Tennessee. If everything went according to plan, it would be a devastating surprise attack with the potential to destroy one of Sherman's three armies. Hood's timeline, however, proved to be far too ambitious. Hardee was nowhere close to being in position at daylight, and it would be noon before his Confederates burst from the woods shouting their Rebel yell.[27]

When dawn came on the twenty-second, Cheatham's men were resting in their freshly built trenches, dug in on high ground on either side of the Georgia Railroad. They were one mile north of Bald Hill, crossing the railroad between present-day Haralson and DeGress avenues. Being directly under the Federal guns of Leggett's artillery, however, Cheatham ordered his corps to retreat a mile west to the safety of the inner defenses of Atlanta. These were the last and largest of the Confederate defenses that ringed the city. As

---

[26] Colonel Henderson's English neighbor and friend George Hewitt Daniel was caught up in Garrard's raid in Covington. He was at the train depot waiting for his daughter. Because he carried a pistol, the cavalrymen took Daniel to "Clark's Grove" where he was executed.

[27] C. I. Walker, *Rolls & Historical Sketches*, 113; Ecelbarger, *The Day Dixie Died*, 48.

the veterans of the Vicksburg Siege filed into these earthen trenches for the first time, they could not help but be impressed by the size and detail of the fortifications. Slaves under the direction of local engineer Lemuel Grant had spent months building these entrenchments, with each artillery position having plank flooring and wood boards to hold the interior walls in place. Huge wicker baskets filled with dirt buttressed the parapets which looked out onto dug-out moats and rows of sharp wooden palisades. It was not a position that any Federal general would want to assault directly.[28]

There was still work to be done on the new defensive line as cannons had to be carefully positioned. The men anticipated a possible Federal attack within hours. By 10:00 A.M., the Yankees of the 2nd Division of the XV Corps had moved forward and seized the former trenches of Cheatham's Corps and were busy reversing the lines to face west toward Atlanta. These Union veterans were members of one of three corps assigned to James Birdseye McPherson and his Army of the Tennessee. The Bluecoats cut a new trench line perpendicular to where the Georgia Railroad and the Decatur Road passed through a fifteen-foot-deep cut in the high ground. With both sides consumed with reinforcing their respective positions, it did not seem likely that there was going to be a fight on that hot, muggy day. But at 2:00 P.M., General Arthur Manigault, commander of a brigade in Cheatham's Corps, reported that all the men around him were "startled to hear heavy firing on our right, distant from two to three miles, which gradually drew nearer and increased in volume."[29]

The much-anticipated Battle of Atlanta was announced with the roar of cannon, rifle volleys, and the screams of the wounded. The Lincoln presidency and the fate of Southern Independence lay in the balance. The Battle of Atlanta on July 22 was the deadliest fight of the long Atlanta campaign, as some 10,000 Americans would be killed, wounded, or captured that day. As Gary Ecelbarger succinctly writes, "No single day of action in any theater for the remainder of the Civil War would match the carnage of this battle." The first two hours of this confrontation did not go well for the Confederates. A late morning repositioning of Thomas Sweeny's 2nd Division from the right flank to McPherson's left flank meant that there was now a sizeable Federal force directly in the path of Hardee's advancing Confederates, effectively blunting the potential for an immediate Southern breakthrough. Among the first Rebel casualties was Major General William H. T. Walker, a rail-thin

---

[28] S. Davis, *What the Yankees Did to Us*, 86.

[29] Manigault, *A Carolinian Goes to War*, 225.

Georgian known to all as a fierce fighter. His division's 4,000 tired but determined soldiers, the vast majority of whom were Georgians, charged fiercely but were rebuffed at every point.[30]

Joseph Wheeler's cavalry had ridden on to Decatur only to find the wagon train guarded by an infantry brigade under Colonel John Sprague. Three Federal regiments made a valiant stand in the town square, buying the supply train just enough time to slip away to the north. Despite facing long odds of reaching the wagons before they could escape, Wheeler launched a series of reckless charges into Sprague's stubborn lines, inflicting 242 casualties on the Federal brigade.[31]

It was not until 1:30 in the afternoon before Hood, whose headquarters was on a hill at the Atlanta City Cemetery, began receiving reports that were positive in nature. Patrick Cleburne launched his veteran Rebel division into the flank of Giles Smith's Midwesterners. Daniel Govan's Arkansas Brigade overran the breastworks of several Iowa regiments and began fighting their way north up Flat Shoals Road toward Bald Hill. James Smith's Texas Brigade, which came up on the right flank of Govan's Razorbacks, then swept into the half-mile gap between Frank Blair's XVII Corps and John Logan's XV Corps.

As the battle swirled in chaos, the Federal commander of the Army of the Tennessee, Major General James McPherson, attempted to ride from one beleaguered corps of his army to another. He found himself caught in the woods by a small squad of Confederates, who were part of Smith's Texas Brigade. Ordered to surrender by Captain Richard Beard of the 5th Confederate Regiment, the thirty-six-year-old Union commander spun his horse around and attempted to flee. A corporal from Mississippi named Robert Coleman shot him off his horse. The mortally wounded major general died within minutes. McPherson, whom some had previously speculated would be a future American president, was the highest-ranking Union officer to die in battle during the entire war.[32]

Bloody Diversion

As more of Hardee's Confederates entered the fight, Ben Leggett's nine regiments from Illinois, Wisconsin, and Ohio fought desperately to maintain

[30] Ecelbarger, *The Day Dixie Died*, 214.

[31] The Atlanta City Cemetery was renamed Oakland Cemetery in 1872.

[32] Ecelbarger, *The Day Dixie Died*, 110–17.

control of Bald Hill. The Rebel assaults came from the woods to the east forcing Leggett's men to turn and fight in the opposite direction of their trench lines. The gallant stand of Leggett's troops would leave their mark on Bald Hill, which from that time forward would be honored as "Leggett's Hill." As the Federal XVII Corps fought for its life, Hood realized the importance of keeping the XV Corps in place, isolated, and unable to come to the aid of its sister corps. To do this, Benjamin Franklin Cheatham and his three Rebel divisions would have to create a major diversion.[33]

Three days earlier, the Tennessee major general had been the senior division commander in Hardee's Corps. When John Bell Hood replaced Joseph Johnston as overall commander of the Confederate army, Cheatham, for reasons of seniority, had been given temporary command of Hood's former corps. Benjamin Cheatham was a wealthy forty-three-year-old Nashville planter and had earned a reputation in the war as both a hard fighter and a hard drinker. July 22 was undoubtedly a day for hard fighting. Shortly after 3:00 P.M., his infantry swung over the Atlanta breastworks and formed up in line of battle.

Cheatham chose not to make a diversionary push, but to instead make a full-scale general assault with virtually every single one of his regiments. His orders to his division commanders were to seize and hold the Federal trenches in their front. The attack would be made *en echelon* from south to north, with each division setting off fifteen minutes after the one to its right. With the church spires of Atlanta at their backs, the southernmost division set off first. These were 4,000 Rebels under Carter Stevenson moving west to east toward the base of the already hotly contested Bald Hill. Stevenson's attack, though, was across an open field and was little more than "a half-hearted assault," as it was beaten back in its entirety in less than half an hour. Like Resaca on May 14 and Kolb's Farm on June 22, Stevenson's Division had been unable to "take advantage of an exploitable Union position."[34]

Cheatham's second division, to the left of Stevenson, fought with more courage and had better results. This was Hindman's Division, now under the command of General John Calvin Brown. They would follow the Georgia Railroad with two brigades on either side of the tracks. North of Brown's position and last to step off would be the division of General Henry Clayton, with another 4,500 Confederates. Clayton placed Baker's Alabama Brigade

[33] While Leggett's Hill did not fall to the Confederates, it did fall in the 1960s to the bulldozer for construction of Interstate 20.

[34] Ecelbarger, *The Day Dixie Died,* 151, 154.

on the left and Stovall's Georgia Brigade on the right. Behind Baker would be Gibson's Cajuns and behind Stovall would march Holtzclaw's Alabama Brigade. One mile separated the Confederates and the jagged line of Federal earthworks.

Capture of the DeGress Battery

The Atlanta Cyclorama, generally conceded to be the world's largest painting at 371 feet in length, is now restored and in its new home at the Atlanta History Center. The centerpiece of this iconic 1886 work of art is the unfinished brick Troup Hurt House where disorganized Butternut soldiers, standing behind bales of cotton, are desperately defending their newly acquired position from a determined Federal counterattack. The earlier breaking of the Union line near this house and the capture of the guns of the DeGress Battery was one of the few high points for the Confederacy during the Battle of Atlanta. The story of how this event transpired is open to a debate that has far outlasted the lives of the men involved in the battle. At the center of this confusion were two companies of Illinois artillery that engaged the Rebel infantry that afternoon. One set of cannons was defending the railroad cut while the other artillery company was some 250 yards north of the rails, on the other side of the Hurt House. Different Southern units have claimed the honor of seizing these cannons, but historians have generally given the lion's share of credit for the Confederate success that afternoon to the South Carolina troops of Brigade Commander Arthur Middleton Manigault. Less attention has been paid to the actions and claims of the 42nd Georgia and their leader that day, Lovick Pierce Thomas.

The forty-year-old Manigault (his Huguenot family name is pronounced *Man-i-go*) was a Charleston native and Mexican War veteran. His brigade that afternoon consisted of the 10th and 19th South Carolina regiments and three Alabama regiments. The two South Carolina regiments were the only troops representing the Palmetto State in the Army of Tennessee. As part of Brown's 5,000-man division, they began the assault by marching parallel to the Georgia Railroad. Manigault's Brigade of 1,400 soldiers kept twenty paces north of the tracks while Colonel John Coltart's Alabama Brigade was instructed to keep twenty paces south of the iron rails. Jacob Sharp's Mississippi regiments were behind Manigault's men, and more Mississippi regiments under Colonel Samuel Benton backed Coltart's Alabama troops. As they set off from the Southern entrenchments, Brown's Division was confronted almost immediately by a strong Federal skirmish force. Supported by two Napoleon cannons of the 1st Illinois Artillery, these two Ohio infantry regiments under Colonel Wells Jones were soon swept back by overwhelming Confederate numbers. One hundred Yankee soldiers were taken prisoner, but the two Napoleons escaped capture, as they were quickly wheeled back to the safety of the Federal line by the Decatur Road.[35]

Under increasing artillery fire from their front and their flanks, the men of Brown's Brigade picked their way over hilly and forested terrain. Manigault's Brigade paused to reform their lines in a ravine that is now known as Springvale Park. Upon clearing the brow of the hill, it was a final 300-yard advance to reach the Federal breastworks. The steady progress of the Confederate infantry was shielded by a wooded forest that hid them for the most part until the last 100 yards. Seeing Manigault's lines emerge from the tree line, Federal officer George Bailey of Missouri recalled, "A hasty dressing of the lines without halting, a lowering of bayonets, a charge, and cheers that made the welkin ring! There was a conflict (for me) between fear and admiration."[36]

The 2,500 Federal troops opposing the oncoming Rebels were in the 2nd Division of Logan's XV Corps under General Morgan Smith. Many of the division officers were in unfamiliar positions that afternoon, having been promoted only hours earlier. This was the result of a chain reaction from the death of Army Commander James McPherson that morning. Major General John Logan now commanded the Army of the Tennessee, which meant Morgan Smith was not holding the trenches with his division that afternoon but was leading the XV Corps. As a result, command of the 2nd Division had fallen to General Joseph Lightburn, formerly a brigade commander. Lightburn had grown up in the western mountains of Virginia and had thrown his lot in with the loyalists.

William Tecumseh Sherman had commanded the VX Corps during the Vicksburg campaign, and these were special troops to the major general. Their corps badge, worn on the front of their hats, was a shield with a cartridge box in the middle with the corps motto "40 Rounds." Colonel Wells Jones of the 53rd Ohio, now in charge of Lightburn's 2nd Brigade, had one Indiana and five Ohio regiments in the trenches north of the Georgia Railroad. South of the railroad, however, it was a thin blue line of only three regiments, burdened with the responsibility of defending a half-mile front. Most of James Martin's Brigade in this portion of the line had been pulled

---

[35] This was near Delta Park. Arthur Manigault's grandfather, Peter Manigault, was considered the wealthiest man in North America in 1770.

[36] Bailey, *A Private Chapter of the War*, 4.

out two hours earlier, sent to help General Leggett defend Bald Hill.[37]

The artillery along this Federal front had also been robbed with two entire batteries sent elsewhere. What remained were four brass 12-pound smoothbore Napoleons (named for French Emperor Napoleon III) and two 10-pound Parrott guns of Battery A of the 1st Illinois Light Artillery positioned at the railroad, and four large rifled 20-pound Parrott guns from Battery H, of the same 1st Illinois Light Artillery, placed just north of the Troup Hurt House. This redbrick two-story house was the unfinished home of a Columbus plantation owner named George Troup Hurt. The house faced south in the direction of the railroad tracks, which were about 150 yards away. Battery A from Chicago guarded a fifteen-foot cut through a rock outcropping. Through this ravine ran the single-track railroad and twenty-five feet to its north and running parallel to the tracks was a dusty wagon road known as the Decatur Road. No trenches ran across the road, and it would be up to the artillery and a skirmish line of infantry to hold this pathway into the Federal rear. The four Napoleons were placed on the Decatur Road for this purpose, with the two 10-pound Parrott guns positioned south of the railroad behind the Federal entrenchments.[38]

Captain Francis DeGress protested that his four 20-pound Parrotts, positioned north of the Hurt House, were also vulnerable. The German artilleryman thought these heavy guns, with their greater range and caliber, were posted too far forward, "but I was assured that the Rebels would 'not charge.'" There was a shortage of light guns in the division and DeGress's siege guns were in a position "hardly fit for a howitzer." By cutting down "timber the greater part of the night" the twenty-four-year-old artillery captain "could only obtain a Range of 211 yards in my immediate front." His guns were in a dense thicket of blackjack oak trees, and the underbrush ran up to the rifle pits. DeGress was particularly concerned that only two companies of the 53rd Ohio covered his right flank.[39]

"Their flags fluttering lazily in the breeze indicating the length of line occupied by each regiment, and the numbers opposed to us," Manigault recalled. It was only a brief view because soon everything was hidden in artillery smoke, and "a deafening roar smote upon the ear, and a storm of bullets and

---

[37] Hess, *The Battle of Ezra Church*, 50; Ecelbarger, *The Day Dixie Died*, 157.

[38] Ecelbarger, *The Day Dixie Died*, 156–59. The cyclorama shows the Hurt House much closer to the tracks than it was in reality (Baxley, *Walk in the Light*, 108). Battery A was under the command of Lieutenant S. S. Smith.

[39] DeGress, report of Captain Francis DeGress to Major T. Maurice, 1.

canister tore through our ranks and around us." The 19th South Carolina lost three color-bearers in the space of ten paces. Seventy-five yards from the Union line, just north of the Decatur Road, the South Carolinians reached a white framed two-story eight-room house owned by Widow Pope. Members of the 10th South Carolina took control of the second-floor piazza and began shooting down upon the Yankee trenches. It had a telling effect and came at a critical moment as Manigault's line had begun to stagger and reel. "The line had lost its regularity," wrote Manigault, "warbling like the movements of a serpent, and things looked ugly."[40]

Coltart's Alabama Brigade had drifted south some 200 yards from the railroad tracks exposing Manigault's right flank. Meanwhile, DeGress was creating large gaps in the gray line with deadly rounds of double canister. Bailey on the Federal side saw the Confederates make a "slight recoil, a hasty reformation of two lines in one, hearty cheers, another charge, another bloody repulse." Samuel Benton's Mississippi Brigade had been brought up from behind Coltart and put in the growing gap between Manigault and Coltart. Benton's furious charge was a "grey torrent pouring over the field toward us," wrote Captain Henry Nourse of the 55th Illinois. Concentrated Federal volleys broke up the advancing Confederate lines, and the surviving Mississippians fell back to the tree line.[41]

Just as quickly, events changed dramatically as "a rejuvenated assault at and north of the railroad bed had begun to turn the tide of the contest." Bailey wrote, "A column of Confederates pressed through our lines through an unguarded cut in the railroad, and opened an inexplicable fire on our boys from the rear." With Palmetto sharpshooters above them firing from the Pope house, and now Confederates behind them, it was too much for most of the Yankees in the trenches, and they began a pell-mell retreat to the rear. Some Federals stayed to the end, defending their position in desperate hand-to-hand fighting for control of the breastworks.[42]

Manigault's success also had something to do with the arrival of fresh troops to their left. General Henry Clayton's Division of 4,500 Confederates was the northernmost of Cheatham's troops and on the left flank of Hindman's Division. They were the last troops to step off *en echelon* and join the general assault. Much like the Federals, officers in this Rebel division were in

---

[40] Manigault, *A Carolinian Goes to War*, 227; Ecelbarger, *The Day Dixie Died*, 227; Baxley, *Walk in the Light*, 108.

[41] Bailey, *A Private Chapter of the War*, 5; Nourse, "Atlanta Campaign," 338.

[42] Ecelbarger, *The Day Dixie Died*, 163; Bailey, *A Private Chapter of the War*, 5.

unaccustomed positions. Three of Clayton's four brigades were being led by officers in temporary command—Colonel John Higley directed Baker's Brigade, Colonel Bushrod Jones directed Holtzclaw's Brigade, and Colonel Abda Johnson of the 40th Georgia was directing Stovall's Brigade. Major William Hulsey, the nominal commander of the 42nd Georgia, was in a deep sleep from overnight duty as the "officer of the day," and was not awakened for the charge. Captain Lovick Thomas, with his "piercing eagle eye, raven locks and handsome beard," would lead the 42nd Georgia, as he had since Resaca, on their most significant charge of the war.[43]

The four brigades in Brown's (formerly Hindman's) Division had charged directly east by following the Georgia Railroad. Clayton's men, to their north, could not march in the same direction, as straight in front of them was a swampy creek valley some 250 yards across. The thick marsh also served to separate the 1st Division of the XV Corps, under General Charles Woods, from the besieged 2nd Division under Lightburn. Clayton could either veer left and attack the 2,000 troops in Woods's 1st Division or veer right and join Brown's assault on Lightburn's 2nd Division. The order to march "right oblique" was given, and the formation headed southeast in the direction of the Troup Hurt House. Lightburn's Division, already under attack by one Confederate division, would now have to contend with a second Rebel division.[44]

Clayton wrote in his battle report that, "I then ordered Colonel [Abda] Johnson to move forward and make the attack, forming a connection with Hindman's Division, upon his right, and that Colonel [Bushrod] Jones should follow at a given interval." The 42nd Georgia anchored the left side of Abda Johnson's formation. The 43rd Georgia, under Captain Joseph Storey, jumped out ahead as the brigade skirmishers. The First Georgia State Line troops under Colonel John Brown were positioned to the right of the 42nd Georgia. The 52nd, 41st, and 40th Georgia completed the line of battle. Colonel Abda Johnson did not like how it all commenced, later complaining that he, "knew nothing of the ground over which I was to move nor of the position of the enemy or how other troops were to move or [when they

---

[43] Andrew J. West, "Anniversary of the Battle of Atlanta: Thrilling Story of the Enormous Encounter," *Atlanta Georgian & News*, July 22, 1904; Ecelbarger, *The Day Dixie Died*, 179.

[44] Today Austin Avenue traverses through the lowland between Euclid and Highland avenues.

would] attack."[45]

When Clayton's Division set off, Federal infantry under Wood began following them, moving south parallel to the swamp. "Observing a considerable force of the enemy moving down his lines from my left to the right," reported Clayton, "I ordered Major Eldridge, commanding the artillery battalion attached to the Division, to move out and open fire. This was done, but with what effect I was unable to judge." Not completely satisfied that artillery alone was sufficient, Clayton ordered Colonel John Higley to take Baker's entire brigade and "change his front obliquely to the left and attack." General Clayton did not, under any circumstances, want Wood to be able to come to the assistance of Lightburn's Federal Division. Higley did the job "in good style, and together with the artillery, checked the movement from that direction."[46]

As Clayton's Division, now composed of only three brigades, continued toward the Hurt House, dispatches arrived ordering General Clayton to send support to Brown's Division, "which had now been repulsed." Henry Clayton ordered General Randall Gibson and his Louisiana troops to fall out of the advancing formation and march due south and form again behind Brown's Division. As it turned out, Gibson's men, some of the best fighters in Cheatham's Corps, would spend the day in a reserve role. Thus, the frontal attack that Clayton would make on the breastworks south of the Hurt House would be made without any artillery support and would be made by only two of his four brigades—those of Abda Johnson's Georgians and, behind them, four Alabama regiments under Bushrod Jones.

The first Confederates to breach the Federal line were from Manigault's Brigade in Brown's Division. General Manigault reported that as the battle lay in the balance, "our supports were coming up in capital style, not more than one hundred yards in rear." He was, no doubt, referring to Jacob Sharp's six Mississippi regiments. "The men saw them," wrote Manigault a year after the war, "and gathered confidence." The hesitation had ended and, "the brigade nearly as a whole, dashed forward and over the works, rifles, and artillery flashing in their faces." Sharps and Manigault's men were now one body pushing left and right and driving the Federals from their breastworks and into full retreat. Colonel James Pressley leading his 10th South Carolina into the trenches was shot in the shoulder in hand-to-hand fighting. Captain C.

[45] Losson, *Tennessee's Forgotten Warriors*, 181.

[46] Clayton, unpublished account of military service, military papers of Henry De Lamar Clayton, 21–22.

Irvine Walker, a Citadel graduate who assumed command of the 10th, wrote the next day that, "Col. P was desperately wounded, shot so close that the powder burnt his clothes."[47]

The 10th South Carolina stormed through the railroad cut. Charles Kimball of Company A of the 1st Illinois Artillery remembered that "the whole cut was enveloped in smoke, and they filed through unnoticed, and deploying in the rear of Battery, made a bayonet attack on it, simultaneous with a similar attack in front." The four brass guns now belonged to the men from the Palmetto State. Manigault's men together with the 41st Mississippi from Sharp's Brigade overran the breastworks north of the Hurt House and sent four Ohio regiments under Wells Jones into wholesale flight.[48]

Artillery Captain Francis DeGress, with his four large 20-pound Parrotts, had been firing canister, at a left oblique, into Manigault's advancing lines. As the Federal infantry around him gave way, DeGress reported that Colonel Theodore Jones of the 30th Ohio "ran to me and told me to save my Guns as everything on his left had given away and he had no support at all, still he held his front for several minutes while the enemy changed front and formed in rear of a large Brick House and in rear of our works." DeGress wrote this account on the night of July 22. "The enemy advanced in good order colors in the center," he recounted, "and when I saw the 2 companys on my right give way, I had the 2 left guns spiked and fell back about 10 yards to my right section with which I fired on them, until within 15 yards when I ordered the right gun spiked and waiting a short time I fired the left gun myself." When the fourth gun was spiked, DeGress and the few of his artillerists that had stayed with him made their escape.[49]

Stovall's Georgia men, under the command of Abda Johnson, were hard on the heels of Brown's successful attackers. While Manigault and Sharp had managed to drive four Federal regiments out of the trenches north of the Georgia Railroad, there was still significant work to be done south of the tracks. First Lieutenant William Smith, who was leading the Independent Rebels (B) of Gwinnett County that day, wrote that the Georgians marched at a right oblique "some distance against the enemy in the direction of the Georgia Railroad and a small section house." They "crossed the railroad and changed front(s)" so that they were perpendicular to the Georgia Railroad. Only companies A and B of the 42nd Georgia were north of the tracks, with

---

[47] Manigault, *A Carolinian Goes to War*, 227; Baxley, *Walk in the Light*, 115.

[48] Kimball, *History of Battery "A" First Illinois Light Artillery Volunteers*, 87.

[49] DeGress, report of Captain Francis Degress to Major T. Maurice, July 22, 1864.

the rest of the brigade lined up to the south of the tracks.[50]

With troop formations set and battle flags waving, the 1,200 men of Stovall's Brigade gave the Rebel yell and charged toward the Federal breastworks south of the Georgia Railroad. Waiting for them were three regiments of Federals under Colonel Samuel Mott of Ohio. The Midwesterners had thrown back an assault by Benton's Mississippi Brigade less than half an hour earlier. Now, to their right, the powder-stained Yankees could see Manigault's Rebels streaming over the breastworks north of the railroad. Even worse, some 500 of Sharp's Mississippians were now enfilading their lines with volleys from the railroad cut, "at a distance of not more than twenty-five paces." Mott's officers began ordering a right oblique to face this new danger, while, at the same time, a fresh brigade of Georgians was closing at the double-quick on their defensive line from the west. The veteran blue-clad soldiers knew when the gig was up, and they began to break and run for safety.[51]

In 1899, the editor and owner of the *Atlanta Constitution*, Evan Park Howell, wrote Colonel Lovick Thomas that as a twenty-five-year-old artillery captain posted between Hardee and Cheatham's Corps, "I was in position to see the charge of Stovall's brigade, under Colonel Johnson, and especially your regiment, the 42nd Georgia, which led the charge on the Decatur road and the Georgia Railroad that day." Howell had already lost half of his brave command that day dueling Federal artillery batteries across the battlefield. "I saw your brigade go into the fight across the valley where the Georgia Railroad runs along the Decatur road," he wrote, "and saw the line unwavering until it captured the battery." Howell, whose brother Albert was a captain in the 1st State Line, was referring to the two 10-pound Parrott guns in Company A of the 1st Illinois. These guns were posted behind the breastworks south of the railroad track and in the direct path of the 42nd Georgia. Howell added that in 1879 he had "rode over this battlefield" with General William Sherman, on the army commander's first post-war visit to Atlanta. "He told me personally," wrote Howell, "that he watched (from the Augustus Hurt House) the charge made by Stovall's brigade and remarked that there was not a more gallant charge made during the war than the one made on that occasion."[52]

---

[50] Goodson, *Letters and Diaries*, 77 (Rogers report).

[51] Report of Colonel Samuel R. Mott, *OR*, ser. 1, vol. 38, part 3, 217.

[52] Evan P. Howell to L. P. Thomas, July 19, 1899, vol. 4, pg. 530, Thomas collection, BRC.

Nineteen-year-old Henry Kitchens of Covington was a member of the Harper Guards (E) at the time of the charge. "I remember somewhere was a little brick building," he wrote Colonel Thomas in 1897. "We made kind of a halt and Col. Thomas ran in front of [the] regiments and said, 'come on, boys.'" Captain Joshua M. Mitchell of the Walton Tigers (H), in an unpublished letter found in the papers of Lovick Thomas, gave a detailed account of what happened next. "Our lines moved forward gallantly," he recalled, "but was halted for a few minutes in about twenty paces of the enemy, who was well entrenched. When we advanced again in a rush, the enemy fled." The former medical doctor claimed that he was one of the first of the Georgians to reach the top of the breastworks. "I could select my way better than the men, perhaps," Mitchell wrote, "and jumped upon the enemy's works several paces in advance of our line."[53]

Captain Mitchell's eye-witness account then continued. "I looked to the situation on each side the moment I mounted the enemy works. Part of my Regt. was to my left," he recalled, "and I think extended as far as the battery did. My right extended just to a deep hollow (that) the right of my regiment had to cross, then rise up a steep hill at the enemy's line." According to Mitchell, the Federal cannons were silent, "with no living men about either, unless wounded on the ground." This was due to Sharp's Mississippians, who had used the railroad cut to get behind the battery and fire a volley into the artillerymen. Fortunately for the men of the 42nd, they did not have to charge into Federal cannons loaded with double-canister. Since a single charge of canister weighed fourteen pounds and contained 240 projectiles, this was the difference between success and failure. Mitchell's company was down to nineteen men, but the thirty-eight-year-old captain formed them into a line "at right angles to the enemy, fired along the line, and drove the men as far as I could see across the woods."[54]

Resistance to the attacking Georgians proved greater the farther one got south of the Georgia Railroad. The 1st Georgia State Line had been to the immediate right of the 42nd Georgia. The two Parrott guns of Company A

---

[53] Henry Z. Kitchens to Colonel L. P. Thomas, March 19, 1897, Thomas papers, AHC; Captain Joshua Marion Mitchell to Colonel L. P. Thomas, from Greensboro, GA, January 18, 1897, vol. 4, pg. 530, Thomas collection, BRC. Mitchell regretted that many of the memories of the war had faded by 1897, "but the events of that day were so thrilling that many of them seem now right fresh in bloom."

[54] An experienced artillery crew could, under duress, fire four rounds of double-canister in a single minute (Mitchell to Colonel L. P. Thomas January 18, 1897, Thomas collection, BRC; Dodwell, "Bouton's Battery," 87).

of the 1st Illinois had been trained on their position, forcing the militiamen to lie prostrate for twenty minutes to avoid certain death. One of their first casualties was their commander, Colonel John M. Brown, brother of the Georgia governor. Seeing the success of the 42nd Georgia to their left, twenty-one-year-old Captain Albert Howell sprang to his feet and drew his sword. Running down the Confederate line, he roused the troops and took his place ahead of the color-bearer. The regiment followed their gallant young captain, with his full set of black whiskers, over the breastworks at the point of the bayonet. The 1st Georgia State Line took ninety-two casualties in their charge, including the loss of two of their color-bearers. Howell would earn a promotion to colonel for his bravery that summer afternoon, and, in honor of his men's furious bayonet charge, the new leader of the 1st Georgia State Line had the "screaming eagle" placed on their regimental banner.[55]

From the Federal perspective, the 57th Ohio posted closest to the railroad cut was the first to give way. Facing the 42nd Georgia to their front and hundreds of Mississippi troops on their flank, they "whirled back," and fled for the woodlands in the rear. Next to retreat by companies was the 55th Illinois. "Every intelligent soldier knew without orders," wrote a member of the 55th, "that liberty depended upon a rapid movement to the left rear." It was an uphill escape through a dense thicket as the Bluecoats hurriedly retreated half a mile eastward to the line of works that they had constructed only two days previous. There in the safety of their former trench line, the Midwesterners reformed by squads and were soon preparing a counterattack.[56]

Federal Counter-Assault

Around 4:30 P.M., there was a short pause as the Yankees fell back and the Confederates consolidated their gains. Rebel officers organized details to escort several hundred Union prisoners back to Atlanta. Captain Walker of the 10th South Carolina also used this opportunity to have the four captured Napoleon cannons and limbers on the Decatur Road pulled back to the Confederate lines. "The Battery our Regt. first captured was near a road running

[55] W. H. Bragg, *Joe Brown's Army*, 91. Transported to Milledgeville, Brown died on July 25. His funeral was held in the executive mansion, and he was buried in nearby Memory Hill Cemetery. Governor Joe Brown later erected a large monument over his brother's grave ("The Soldier Boys: Reunion of the Gallant Forty-Second Georgia," *Atlanta Constitution*, July 23, 1886, 5; "Thirty Years Ago, and Now," *Atlanta Constitution*, October 16, 1892, 21).

[56] Nourse, "Atlanta Campaign," 339.

through the works," wrote Walker on July 23, "so I had the guns brought round to our side, and they were subsequently carried into Atlanta. There was no break in the line of works near the DeGress Guns, and I could not do likewise with them." Fifty-five horses lay dead about the guns in Company A as the advancing Rebels had shot the horses to prevent the escape of the guns. The four Napoleons were eventually turned over to Douglas's Texas Battery with Brown's Division.[57]

South of the railroad, men of the 42nd Georgia were proud of their capture of the remaining two 10-pound Parrott guns belonging to Company A of the 1st Illinois Light Artillery. Captain William Smith remembered that Private Wesley Clower, the oldest man in Company B, was the first to reach the guns. Captain Mitchell remembered that the men turned one of the guns and "were directing on the enemy some distance away on the left." Adjutant Hugh Wylie "attempted to fire said piece with a match or cigar." Private John McWilliams of the Fowler Guards (D) remembered Wylie using an ember from a nearby campfire, but, in any case, the gun was spiked and remained mute. Artillery officer Evan Howell was "sent by the commanding officer with my horses to bring the guns out." Whatever pause there had been at 4:30 was short-lived, as Howell arrived to find Union artillery fire "so terrific that it was impossible to get the horses to the guns." Hoping that an opportunity would eventually present itself, Howell stayed close to the guns until nightfall.[58]

The Federal high command had watched the breaking of the Union line from the Augustus Hurt House atop Copenhill. They were 1,000 yards from the Troup Hurt House, separated by an impenetrable swampy valley. General Oliver Otis Howard rode over to Sherman's headquarters after the Confederates had broken the 2nd Division of the XV Corps. He offered to commit his troops to the battle, but Sherman thought better. "Let the Army of the Tennessee fight it out, this time," was Sherman's response. Soon a distraught and animated DeGress approached the commanding general. In his memoirs, Sherman wrote, "Poor Captain DeGress came to me in tears, lamenting the loss of his favorite guns." Howard watched as General John Schofield massed

---

[57] Captain Walker to his wife, Orie, July 23, 1864, in Baxley, *Walk in the Light*, 115; Walker, "1898 Report submitted by Colonel C. I. Walker," 9. According to Col. Walker, the four Napoleons taken on July 22 were recaptured by the Federals on the retreat from Tennessee, after the Battle of Nashville.

[58] Mitchell to L. P. Thomas, January 18, 1897, Thomas collection, BRC; Evan P. Howell to L. P. Thomas, July 19, 1899, Thomas collection, BRC.

his twenty pieces of artillery, "so located as to give a sweeping fire along the line of works at the interval held by the enemy." General Sherman took a personal interest in sighting each of the cannons. Everything near the Troup Hurt House soon became a target for canister balls and spherical case shot.[59]

Arthur Manigault marveled at the accuracy of the massed Federal artillery on Copenhill, recalling, "Their shells tore through the lines or exploded in the faces of the men with unerring regularity." Sherman had General Charles Woods draw up his division under the cover of the batteries. Despite being the "largest, tallest, stoutest officer on the ground," Woods was called "Susan" by Sherman and Howard, from their days as cadets at West Point. In columns of fours, Wood's Division, some 2,000 soldiers, set off to march around and through the swamp with the intent of regaining the occupied line of works. South of the railroad, near Leggett's Hill, it was the brilliantly energetic Major General John Logan who was organizing another group of Federal soldiers for the counterattack.

With the death of James McPherson three hours earlier, Logan, a former politician from southern Illinois, had been given command of the Army of the Tennessee. This change-in-command came as this proud Union army was in a fight for its very life. Mortimer Leggett said of Logan, "He seemed to have a power to awaken all the enthusiasm that was in the troops, to the extent that no other officer in the army seemed to possess." "Black Jack" Logan gathered August Mercy's Brigade, which numbered about 1,000 soldiers, and another 750 men from three regiments under Colonel James Martin and personally led them to the fight. They were joined by the routed troops of the 2nd and 4th divisions of the XV Corps, perhaps 1,800 infantry soldiers, who had regrouped after their initial collapse and retreat.[60]

Waiting for this hodgepodge of Federal units were roughly 2,500 Southern soldiers, who were still exuberant from their successful seizure of the Federal trench lines. Manigault, standing by the white framed widow Pope House, was pleased to see Colonel Bushrod Jones approach with 900 more officers and men. This Alabama brigade had been a quarter of a mile behind Abda Johnson's Georgia Brigade and was now ready to enter the fight. The

---

[59] The Federals referred to the Augustus Hurt House in their reports as the Howard House. A local distiller named "Colonel" Howard had occupied the vacant house as the Federals assumed control of the area, and they assumed he was the rightful owner of the house atop Copenhill (Howard, "The Battles about Atlanta," 395).

[60] Manigault, *A Carolinian Goes to War*, 228; Howard, "The Battles about Atlanta," 395; Ecelbarger, *The Day Dixie Died*, 175.

Charleston native described Jones as "one of the coolest fellows I ever saw, handling his command with great ease and judgement." With artillery shells bursting around them, Manigault, Sharp, and Jones quickly decided that Jones would deploy his troops north of the railroad tracks while Abda Johnson's Georgia Brigade would continue to hold the trenches south of the railroad. The Rebel officers could see that, only 800 yards to the north, Woods was massing his seven regiments for an attack. One immediate problem was that the Alabama brigade was facing to the east and Jones would have to realign his formations to the north, lining up near the captured DeGress Battery. "Not knowing where they were wanted," wrote Manigault, "their formation was unsuited for the occasion, and they were compelled to change fronts under a heavy fire."[61]

Private Kitchens of the Harper Guards (E) said that his company rolled bales of cotton onto the railroad tracks and "made breastworks of them. We fought from behind those bales of cotton so long as we held that battery." Well south of the Georgians there were another 700 Alabamians under Colonel John Coltart who, upon the Federal retreat, had entered the abandoned line closest to Bald Hill. With an unprotected right flank, Coltart's men were the first to be attacked by oncoming Federals. Despite killing Lucien Greathouse, the colonel of the 48th Illinois, the men from Alabama gave way, and they initiated the retreat back to the Atlanta fortifications.

Abda Johnson's Brigade, to Coltart's left, was soon pressed by three regiments under James Martin and by regrouped Federal regiments under Samuel Mott. Colonel Martin was bringing back the three regiments that had been called away to Bald Hill only hours before Cheatham's assault. The 55th Ohio, which had fled upon the Georgian's assault, had taken almost an hour to reorganize their lines in a ravine near their former trenches. Just as they were setting off to reenter the maelstrom, the familiar form of "Black Jack" Logan galloped up the Decatur Road toward them. Following behind Logan were Martin's three regiments. One member of the 55th Ohio remembered Logan "on a coal-black charger streaked with foam, hatless, his long black hair flying, his eyes flashing with wrath—a human hurricane on horseback."[62]

It did not take long for these spirited Midwesterners to push Abda Johnson's Georgians to the opposite side of the exposed trench line. Here the Rebels of the 40th, 41st, 42nd, 43rd, 52nd, and 1st State Line made a stand for

---

[61] Manigault, *A Carolinian Goes to War*, 228.

[62] Henry Z. Kitchens to Colonel L. P. Thomas, March 19, 1897, Thomas papers, AHC; Nourse, "Atlanta Campaign," 341.

thirty minutes. The remnants of the 47th Ohio, led by their four remaining officers, made the first charge on the trenches south of the railroad. Out front was Colonel Thomas Taylor, a twenty-seven-year-old lawyer from Georgetown, Ohio. The Unionists met a hot reception from the Georgia troops. "I advanced on the 'double quick' and got within a few feet of the works," Taylor wrote home a few days after the fight, "when such was the hail storm of fire and bullets which swept over us that both flag staffs were shot off, the regiment's standard was torn from the staff by the fragment of a shell, one color bearer killed, and a color corporal wounded, [and] others as a matter of course fell."[63]

According to Taylor, it took three charges before the Federals drove the Georgians "from their works and recapturing a section of artillery upon the left of the railway which the [Rebels] had turned upon us." Abda Johnson explained in his official report that his right flank was unsupported and subject to damaging enfilade fire. He finally ordered his brigade to retreat several hundred yards. "We shot the horses of the battery," wrote Henry Kitchens, "to keep them from using the guns on us as we retreated. Some of which were as fine as I ever saw." In the tree line, the Georgians were met by Gibson's reserve Louisiana brigade, and Colonel Johnson gave a brief thought to renewing the attack. Before anything could be decided, however, orders from General Clayton arrived, calling both brigades back to the Atlanta fortifications.[64]

North of the tracks, Arthur Manigault, Jacob Sharp, and Bushrod Jones were being attacked on two fronts. The Alabamians under Jones faced north to oppose the advancing troops of Woods's 1st Division. Manigault and Sharp were looking east as three regiments belonging to the "little Dutchman" August Mercy burst out of the woods 100 yards to the east. The 81st Ohio led the counterattack toward the Hurt House with the 66th Illinois on their right and the 12th Illinois on their left. The attacking column was 500 yards across. The 66th called themselves "the Western Sharpshooters," and they were armed with brass-framed, lever-action Henry rifles capable of shooting sixteen rounds before reloading. Mercy's Bluecoats pressed straight toward the four large guns of the DeGress Battery.[65]

At this critical junction, orders came down for Sharp and Manigault to

---

[63] Thomas Thomson Taylor to Orie Taylor, July 26, 1864, Taylor papers.

[64] Ibid.; Henry Z. Kitchens to Colonel L. P. Thomas, March 19, 1897, Thomas papers, AHC; Manigault, *A Carolinian Goes to War*, 293–94 (appendix report of the Battle of Atlanta).

[65] Ecelbarger, *The Day Dixie Died*, 183.

retreat, which they did in good order. Unfortunately, this sudden withdrawal exposed the right flank of Bushrod Jones's position. The surprised Alabamians retreated in disorder with many of their men being taken as prisoners. Mercy's men could claim the honor of recapturing the prized DeGress Battery as they arrived at the guns only minutes ahead of Wood's Federals. The exhausted Confederates remained near the front until just before dark when they withdrew to the inner works of Atlanta.[66]

This late afternoon see-saw affair at the Georgia Railroad was a bloody battle within the larger July 22 Battle of Atlanta. Captain Andrew West of the 41st Georgia described the carnage near the railroad cut when he wrote, "Muskets, canteens, knapsacks; in fact, everything that makes up the accoutrements of the soldier, was here, scattered in every direction." West remembered that "To the left of the Decatur road was the historic square Hurt house, which was filled with thousands of bullets and fragments of shell. At this point the dead lay thickest." To the young Confederate officer, "The wonder seemed not that there were so many dead, but that any lived." The Rebels could point to the brief capture of ten Federal cannons as the highpoint of the affair. The 2nd Division of the XV Corps lost close to 500 officers and men killed or wounded defending those guns. Another 500 had become prisoners, destined to spend the remainder of the war in Confederate prison camps. Battery A of the 1st Illinois Light Artillery had thirty-two men killed, wounded, or missing and fifty-five horses killed or captured. DeGress reported that his Battery H lost fourteen artillerymen and thirty-nine horses.[67]

The Confederate casualties were horrendous. Arthur Manigault entered the battle with 1,400 soldiers and lost 430 men including 35 of his officers. Brown's Division suffered about 1,400 casualties in all and Clayton's Division another 600. Abda Johnson and Bushrod Jones led the two brigades most heavily engaged in Clayton's command, losing 298 and 128 respectively. Each casualty had a story behind it. Private William Rogers of the 43rd Georgia wrote that his messmate Jesse Clement was wounded in the charge and lost a leg. Rogers lamented that he and Clement had "slept together from the time we left Cumming until now." The Milton Tigers (C) were stunned by the death of their captain, thirty-two-year-old Singleton Maxwell. Some

---

[66] Manigault, *A Carolinian Goes to War*, 293.

[67] West, "The Private Soldier," 1899 Speech at 42nd Georgia Reunion, Thomas collection, BRC; West, "Anniversary of the Battle of Atlanta: Thrilling Story of the Enormous Encounter," *Atlanta Georgian & News*, July 22, 1904; Ecelbarger, *The Day Dixie Died*, 190; Manigault, *A Carolinian Goes to War*, 292 (appendix report of the Battle of Atlanta).

Rebels remained bitter about the whole affair. Colonel Irvine Walker of Manigault's Brigade wrote in 1896, "We had built those breastworks, given them up to the enemy, retaken them at a very heavy sacrifice, and now we had to give them up again. The whole struggle of the afternoon, the lives lost, the suffering inflicted, had all been for nothing." He viewed it as a representation of what had befallen the Army of Tennessee at every turn, "marches, skirmishes, battles, all seemingly with no good result."[68]

John Bell Hood telegraphed Richmond at 10:30 P.M. with a brief report that conveyed an impression of victory. Despite inflicting 3,722 casualties on the Federal Army of the Tennessee, he had failed to crush Sherman's left wing. In fact, Hood had managed to wreck his Army of Tennessee in this battle, to the point that it would never again be able to launch tactical strikes on Sherman's forces with any hope of meaningful success. The fall of Atlanta now seemed inevitable. Albert Castell estimated total Confederate casualties in the largest battle of the Atlanta campaign at 5,500 soldiers. Gary Ecelbarger puts the figure between 5,700 and 6,300. Patrick Cleburne, the most tenacious of the Rebel commanders, lost 1,388 men or 40 percent of his division. The loss of 60 majors, colonels, and generals gives some indication of the havoc created among the Confederate chain of command. Many of their best officers were no longer in the field. "The fundamental reason for Hood's failure on July 22," Castell writes, "is that he tried to do too much with too little in too short a time."[69]

---

[68] Castel, *Decision in the West*, 412; recollection of Pvt. William H. Rogers in Goodson, *Letters and Diaries*, 76; Walker, "1898 Report submitted by Colonel C. I. Walker," 11; Captain J. P. C. Whitehead, the adjutant of Stovall's Brigade, was among the severely wounded on July 22, 1864. Hugh Wylie of the 42nd Georgia would take his position on July 24.

[69] Ecelbarger, *The Day Dixie Died*, 213–14; Castel, *Decision in the West*, 413.

Chapter 7

The Jonesborough Escape

"They got on their breastworks this morning and hollered to us to bring them some tobacco—They would give us a pound of coffee for a plug. We sent them an ounce of lead was our only reply."
—Captain Benjamin Weaver, describing the Siege of Atlanta

Atlanta Siege Lines

No one knew the ramifications of what had happened on July 22 better than Captain Lovick Thomas. In the darkness of that summer evening, he rode away from the Rebel lines and into the city of Atlanta. Living near the center of town was his father, Lovick Thomas Sr., his brother-in-law Joseph Porter, and his younger brother by fifteen years, Henry Thomas. According to an account later written by Henry (later known as "Heck"), Lovick pulled up on his horse, named Kennesaw, and told his relatives "to get what valuables we could haul, and to get down to Newton County at once—to save what we could, as Atlanta was going to fall." The teenage Henry had seen the ambulance carrying the body of General William Walker pass their house that afternoon and now his brother (nicknamed "Bud") was confirming their worst fears. "Atlanta was doomed" was the opinion of his respected older brother. The men asked Lovick for his account of the day's battle and were told that the officer had, "led the charge on the Yank's breastworks, but that he thought one man had beat him to the works a little."[1]

It took several days for the two armies to recover and reorganize their commands after the Battle of Atlanta. With McPherson's death, Sherman needed a new commander for his Army of the Tennessee. "Black Jack" Logan was the popular choice, but the Illinois politician was not a West Pointer and, in Sherman's eyes, lacked the allegiance ingrained from years in the old army. The promotion went to Oliver Otis Howard. Even more offended than Logan by the decision was General Joseph Hooker of the XX Corps, the senior

[1] "Letter from Heck Thomas of Oklahoma," Bruff's Column, *Atlanta Constitution*, August 24, 1910.

corps commander under Sherman and a former commander of the Army of the Potomac. Hooker resigned at the perceived insult, and Henry Slocum took his position.

On the Rebel side, the command structure was also reorganized. Arriving on July 25, thirty-one-year-old Stephen Dill Lee was brought from Mississippi and given command of Hood's old corps, with Benjamin Cheatham returning to his Tennessee division. The Georgia men knew the young Lee from his active role at the battles of Chickasaw Bayou and Baker's Creek. Barton's Brigade had been in the same division as General Lee's Alabama Brigade in the Vicksburg Siege. Stephen Lee's personal courage, therefore, was never in doubt, and he was well received by the troops of the 42nd Georgia. Since the Vicksburg exchange, General Lee had commanded cavalry and run military districts in Alabama and Mississippi. A nickname from the troops was soon pinned on Lee. While Hardee was "Old Reliable," Hood was "Old Peg Leg," and A. P. Stewart was "Old Straight," veterans now referred to Lee as "Old Temporary"—a reflection of how many different assignments he had been given since the war began.[2]

Sherman did not have the troops to surround and lay siege to Atlanta, and the failed assault on Kennesaw Mountain had taught him to respect the circle of strongly fortified lines protecting the city. A Federal victory would only come from cutting the last remaining railroad that supplied Hood's army with food and ammunition. This was the Macon and Western Railroad, running southwest to Macon by way of East Point. Six miles from Atlanta, East Point was the vital junction of the Macon & Western and the Atlanta and West Point Railroad. Sherman's next strategic objective would be the cutting of the railroad near East Point, thereby giving Hood no alternative but to abandon Atlanta.

On July 26, Sherman pulled the Army of the Tennessee out of the trenches east of Atlanta. Having thoroughly destroyed the Georgia Railroad to Augusta, these soldiers were now free to march completely around and behind the Union line north of Atlanta and begin advancing toward East Point from the western side of Atlanta. The next morning, July 27, battalions from the Army of the Tennessee began to extend the Federal line south of Proctor Creek, fortifying a ridge line that overlooked Atlanta. With no cavalry to screen his infantry, the new commander of the Army of the Tennessee was cautious as he moved his veteran troops into contested territory. Oliver

[2] Hattaway, *General Stephen D. Lee*, 127.

O. Howard had never forgotten May 2, 1863, when his XI Corps had been outflanked and smashed by Stonewall Jackson at Chancellorsville. Howard was determined to keep his right flank secure and prepared for a possible Rebel attack.

On the same morning of July 27, John Bell Hood ordered Stephen Lee to move two of his three divisions, those of Brown and Clayton, out of the works on the eastern side of Atlanta and march them to an area between Peach Tree Road and the Marietta wagon road. There they rested, with extra ammunition being distributed, so that each man would have sixty rounds in his cartridge box. After consulting with Hood, Lee began, in the late afternoon, to move his men west, beyond the existing left flank of the Confederate line. That evening Lee was to post his two divisions on Lick Skillet Road and block any further push south by Howard's advancing Yankees. The small community of Lick Skillet was near the Chattahoochee River and was connected to the southwest suburbs of Atlanta by a meandering crushed gravel road that roughly follows what is today I-20. Hood's plan was to have Lee defend the Lick Skillet Road and prevent Howard from advancing any further south in the direction of East Point. With Howard held in check, Hood would then have A. P. Stewart slip behind Lee on July 28, moving his corps to the west and then, on July 29, double-back into Howard's unprotected rear.[3]

John Bell Hood's battle plan never came to fruition as the young, inexperienced Stephen Lee decided to do more than just block Howard's advance. Upon arriving on the field on the morning of July 28, the newest general in the Army of Tennessee made a hasty and ill-conceived decision to attack Howard's right flank immediately. Lee had not reconnoitered the field but believed that Howard was still moving and was vulnerable. This assumption proved to be completely incorrect. Without informing Hood, Lee pushed his new command forward. For the next five hours, in a series of disjointed and mismanaged attacks, the young general sent Brown's four brigades and three of Clayton's four brigades across a shallow valley and up an open slope in the vain hope of dislodging Yankee infantry. The Bluecoats were fighting with spirit from behind newly built breastworks situated along a wooded ridgeline. A. P. Stewart, at the request of Lee, contributed two of his divisions to the fight.

---

[3] Hess, *The Battle of Ezra Church*, 39.

The principal Federal defenders were from Morgan Smith's 2nd Division of the XV Corps, the same men who had lost and then recaptured the Union line at the Railroad Cut on the afternoon of July 22. The Battle of Ezra Church, fought just north of what is now Westview Cemetery, was a complete disaster for the Confederates, producing some 3,000 casualties, about a quarter of the Southerners engaged. By contrast, Federal losses amounted to 632 men. These casualty figures belie the fact that it had not been an easy victory for the Federals as the Confederate charges had been made with vigor and an awe-inspiring recklessness.[4]

Fortunately for the men of Stovall's Georgia Brigade, they had not been involved in the slaughter at nearby Ezra Church. Clayton had been ordered to hold his brigade in reserve, tasked with defending the inner defenses west of Atlanta. Francis Nash of the Milton Tigers (C) wrote that they spent the day on the picket line in front of the Georgia militia. On the morning of the twenty-ninth, the 42nd Georgia marched through Atlanta and then southwest for three miles, on what is today Cascade Road, in the direction of Sandtown. Here they stopped and constructed breastworks on July 30. According to Arthur Manigault, who had lost another 150 men and twenty officers at Ezra Church, Clayton's Division was a mile from Lick Skillet Road, which was off to their right, and a half-mile from Sandtown Road to their left.[5]

The Rebel division was tightly concentrated on a high ridgeline overlooking a creek, with the Federals moving in on the opposite hill. There could be no retreat by the Southern defenders, if Atlanta, the supply base of the Confederacy, was to be held. Manigault's Brigade defended the important right flank and protected a redoubt of artillery. To his left was Stovall's Georgia Brigade, which Manigault simply commented, "a very mean one it was." Stovall's trenches stretched across a hill that is now part of present-day Greenwood Cemetery. Next along the line came Baker's Mississippians, the 36th Alabama from Holtzclaw's Brigade, and then Gibson's Louisiana Brigade.[6]

For the next four weeks, Clayton's Division would hold this trench line despite intense fighting in the no man's land between the opposing forces. Clayton estimated that, on "aggregate," he was losing almost 100 soldiers a

---

[4] Ibid., 160.

[5] Nash, *Civil War Diary of Francis H. Nash*, 32.

[6] Relics found by Bill Erquitt have confirmed these positions. Manigault's trench lines are still clearly visible today in the woods past the baseball field in Cascade Park (Manigault, *A Carolinian Goes to War*, 225).

day. "So severe was the skirmishing from 30th July to the 26th August," Clayton reported, "that it might be called a continuous battle." In the bigger picture, the 42nd Georgia was part of Hood's "railway defense line" that the Confederate commander kept extending westward and thereby managing to thoroughly frustrate every attempt by Sherman to reach East Point. In early August, a Federal push at Utoy Creek produced 1,000 Federal casualties but failed to dent Hood's defensive line. A heavy demonstration in front of Stovall's Brigade by members of Palmer's XIV Corps began General John Schofield's ill-fated assault.[7]

Adding to the stalemate was Sherman's decision to bombard Atlanta itself. The cannonade started on July 20 and did not end until August 25. The Union army had 254 field pieces, mostly 3-inch rifled guns, 10- and 20-pounder Parrott rifles, and 12-pounder smoothbore Napoleons. While the smaller weapons were only effective at pounding the Rebel fortifications, the 3-inch and Parrott rifled guns had the range to bombard the city itself. On August 10, two 30-pounder Parrotts were brought down by rail from Chattanooga to add to the destruction. Atlanta still had thousands of civilians who were forced to seek safety in underground bombproofs. According to historian Stephen Davis, at least 100,000 projectiles—about 500 tons of Northern iron—were fired into and around downtown Atlanta, damaging and destroying many buildings and killing some two-dozen people. Davis points out that this unprecedented destruction by Sherman's artillery was a violation of the rules of war while gaining no strategic benefit for the Union cause. Indeed, the front entrance to Atlanta remained locked and secured.[8]

The Siege of Atlanta was difficult on the infantry defending the Southern trenches. Arthur Manigault called the four weeks defending Atlanta "the most anxious, exhausting, and perilous of any period, during the whole course of my experience as a soldier." The defensive line was stout with deep trenches, head logs protected the men "whilst firing," and a double row of palisades covered the entire front. The Rebels found themselves trapped in their defenses challenged to even cook their meals or find fresh water for their canteens. Sleep was difficult to come by, as fighting on the picket line was incessant, and Federal artillery continually rained down on them.[9]

Confederate morale was clearly at one of its lowest ebbs since the beginning of the war. On August 4, 1864, Henry Clayton, a frustrated division

---

[7] Report of the Atlanta Campaign, in Clayton, military papers.

[8] S. Davis, *What the Yankees Did to Us*, 115, 168, 249.

[9] Manigault, *A Carolinian Goes to War*, 238.

commander, issued General Order #3 berating "the Pickets of this Division" for "retiring without a contest upon the first approach of the enemy, frequently without firing a gun, and still more often without making suitable resistance." Clayton urged his brigade commanders to hold the officers on picket duty strictly accountable. "The specious pretext that the *picket has been flanked*," the Alabama general wrote, "will no longer be received for the cowardly stampede of an entire picket line." Sounding much like Winston Churchill in 1940, Clayton expounded, "If we expect to succeed it must be understood that it can only be done by *fighting*. We must fight in the trenches, fight in the open field, and fight on the Picket line."[10]

Captain Benjamin Weaver was down to thirty men in his Company B when he wrote his wife that "We have to lie in the trenches day and night." The opposing trenches were some 600 yards apart, and sharpshooters forced the men to keep their heads down. Both sides dug an elaborate network of skirmish lines in front of their main positions. Lieutenant William Donaldson of the Gwinnett Independents (B) had his jaw shattered "all to pieces" and was shipped to a Macon hospital. On August 7, at least eight members of the 42nd were captured while on the picket line. In all, the regiment would have at least twenty-one of its soldiers fall into Union hands over the course of the month. They would spend the remainder of the war at Camp Chase prison in Ohio.[11]

Affairs were little better on the Union side of no man's land. Major John Widmer, the new commander of the 104th Illinois in the XIV Corps, wrote that his main line "was within point-blank range of the rebel skirmish line, which too, was hidden in a dense growth of pines." A sharpshooter had killed three of his men from this concealment. Widmer saw the sniping as pointless and sought a truce. At dawn on August 7, he stepped out of the Federal skir-

---

[10] Henry Clayton, report of the Atlanta Campaign, in military papers. Prime Minister Winston Churchill gave his "We shall never surrender" speech to the House of Commons on June 4, 1940, after the miraculous evacuation of Dunkirk.

[11] B. P. Weaver to home, August 18, 1864, Weaver letters. Charles Clayton, Charles Cox, Wilson Sorrells, Lewis Wood, William Barnett, David Faulkner, John Graham, and George Hays were all captured on August 7. Malon Doby and David Johnson were captured on the August 6 (L. Henderson, *Roster of the Confederate Soldiers of Georgia*, 4:563–608). Second Lt. Hill was captured on the picket line August 7 along with Sergeant Lewis Wood, William Barrett, George Camp, and Jacob Carter, all from Company I. On August 18–19, the Fulton Blues saw Clark Bryant, John Bryant, James Ramsey, Samuel Sewell, and Francis O. Smith all captured on the picket line.

mish pit and called out, "'Hello, Johnny!" In about two seconds, wrote Widmer, "a living fellow in gray clothes rose up in the other pit and replied, 'Hello, Yank.'" Major Widmer met a lieutenant from the 42nd Georgia about fifteen feet from the Rebel pits. He asked the Confederate officer to escort him to the captain in charge of the Rebel skirmish line. This turned out to be Albert Howell of the 1st Georgia State Line, "who seemed to be very much a gentleman." Widmer made his case for a truce along the two skirmish lines. "I told the captain," recalled the Illinois major, that "we were conducting a rather barbarous warfare on our part of the line; that we were now and then breaking an arm or a leg, or killing some poor fellow, but we were deciding nothing, only causing suffering." Howell agreed to a truce that would last until either side received orders to resume hostilities. A twenty-minute notice would be given, and "when firing began no attempt should be made to hit an opponent for the first two or three rounds; this to enable every soldier to get safely back to his pit."[12]

Minutes later, men in Blue and Gray were exchanging coffee, tobacco, cornbread, and hardtack. Widmer reported that a Georgia lieutenant called him aside to clarify the rules for treatment of Confederates who voluntarily came into Federal lines. "I thought it highly probable," Major Widmer responded, "that they would be sent North and set at liberty upon taking the oath of allegiance to the United States." His words affected the demoralized Rebels; two dozen "Johnnies," mostly from the 42nd Georgia, laid down their rifles and came across the lines in the coming nights. On August 13, Hood's chief of staff Francis Shoup wrote in his journal that Stovall's men had been trading tobacco for coffee the previous day and that several of Stovall's men had deserted. "Two of men ran away last night," wrote Captain Weaver. "I think they have gone home. T. M. Thomason is one of them." Some of the desertions were permanent, but others were temporary. Weaver reported, "A great many of our men have run away to see after their families but the most of them have returned."[13]

August 7, 1864, saw the wounding and capture of Captain Thomas Hill of the Fulton Blues (I). After the surrender of Vicksburg a year earlier, Hill had chosen to return to his native Bastrop County, Texas. He was soon the

---

[12] Strayer and Baumgartner, *Echoes of Battle*, 287–88.

[13] Ibid., 287; journal of Brig. Gen. Francis A. Shoup, July 25–September 7, 1864, *OR*, ser. 1, vol. 38, part 3, 691; B. P. Weaver to home, August 18, 1864, Weaver letters. Widmer proved to be incorrect, and those captured along the Atlanta siege lines in August were sent to Camp Chase, Ohio.

Inspector General of Confederate Cavalry in the Indian Territory. Ordered east of the Mississippi River, Hill returned to Georgia in spring 1864 as an honorary member of the fabled Terry's Texas Rangers. By summer 1864, Captain Hill was back in command of Company I. After his capture, the Texan would spend the rest of the war as a prisoner at Johnson's Island in Lake Erie.[14]

Spirits rebounded a bit when the men in the Southern trenches learned that Joseph Wheeler had taken almost all the Rebel cavalry on a raid into northern Georgia and Tennessee. Many predicted that Wheeler would cut Sherman's supply lines and force the Federal army to retreat all the way back to Chattanooga. Leadership in the Georgia Brigade improved with the return of Colonel Robert Henderson and General Marcellus Stovall, both of whom were back directing brigade operations by August 19. Their arrival corresponded with a Friday night lightning strike that disabled Sergeant George Knight of the Walton Tigers (H). Lieutenant James Gresham and J. R. H. Clegg were also shocked by the strike but carried on. After thirty men in Clayton's picket line deserted on August 13, General Hood clamped down on any communication between opposing soldiers. "I can see some of the Blue Devils from where I sit," Weaver wrote on August 18. "They got on their breastworks this morning and hollered to us to bring them some tobacco—They would give us a pound of coffee for a plug. We sent them an ounce of lead was our only reply."[15]

Furlough to Jonesborough

Acting Major Lovick Thomas was not present in the trenches during the Siege of Atlanta. On July 1, with Federal invaders nearing the Chattahoochee River above Atlanta, Jennie had left Lawrenceville with her two small children as well as enslaved house servants. Maude and Mary, bound for her grandfather's home four miles east of Jonesborough. The seventy-nine-year-old Samuel Lee and his third wife, Jane, operated a grist mill and lumber mill on Line Creek, close to where today the water exits the dam at Lake Spivey. His fine wooden house, in the Cotton Indian Creek District of Henry County, sat a

[14] "Death of Honored, Good, True Citizen Captain T. A. Hill," *Weimar (Texas) Mercury*, October 10, 1924, 1. Thomas Hill appears in a 1908 reunion photograph of the Terry's Texas Rangers taken in Houston (see photograph at Portal to Texas History, https://texashistory.unt.edu/ark:/67531/metapth995/).

[15] Sams, *Wayfarers in Walton*, 184; B. P. Weaver to home, August 18, 1864, Weaver letters.

half-mile north of the mill on the top of a ridge line overlooking the valley created by the large creek. By July 14, Lovick was aware that his family had "refugeed" from Lawrenceville and was now south of Atlanta. Though they didn't realize it at the time, Jennie would have been wiser to have stayed in Lawrenceville. Jonesborough was a railroad hamlet on the Macon & Western Railroad, and, in the coming months, a prize coveted by both armies. Lawrenceville, by contrast, was not favored by rail service and would, in that same time span, be a victim of only an occasional Federal raiding party.

Shortly after the Battle of Atlanta on July 22, Lovick found himself "so much broken down that, under direction of our surgeon, I hurried off for the hospital at Griffin." With a July 25 furlough in hand, Lovick probably had little intention of going as far as Griffin. When the train stopped in Jonesborough, the tired Confederate officer was "carried by friends to the old mill." According to Jennie's diary, Lovick would rest and regain his strength at the Lee home over the next four weeks. Colonel Thomas would, years later, describe this month with his wife and children in the most glowing terms. Since the "vandal hordes" had not reached Henry County, Lovick found the old Lee homestead "a veritable Egypt." His written descriptions were vivid: "The Negroes were sleek, fat and contented; the mules and horses were as sleek and fat as the darkies, and the cows had a satisfied look as they slowly wind their way up the long lane from the old mill. The bee-gums stood in long rows in the rear of the homestead, and the chickens and pigs were plentiful." Each morning, the famished officer devoured freshly prepared biscuits covered in butter and honey. Pleasant afternoons were spent at the mill teaching his children Lovick Pierce III and Martha, who were affectionately known as Bud and Mat, how to fish "for the little perch and horny-heads, so plentiful there."[16]

These pleasures were soon to vanish, another casualty of the advancing Federal juggernaut. On August 13, Sherman informed his top generals of an ambitious plan to bypass Hood's entrenched army. He called it a "grand movement by the right flank." Leaving one of his seven corps above the Chattahoochee to protect the W & A, the Federal commander planned to abandon his trench lines facing Atlanta and march some 60,000 men twenty miles to the south where they could find and destroy the Macon & Western Railroad. Plans changed the following day when Sherman learned that Hood had

---

[16] L. P. Thomas, "'Confed': The Story of the Capture and Escape of a Gallant Southern Soldier during the Fighting around Atlanta," Close Calls; A Thrilling Narrative of War Times, *Atlanta Journal*, December 15, 1900. The horny-head chub is a small fish often used for bait.

sent Joe Wheeler and almost all the Confederate cavalry, more than 5,000 troopers, north of Atlanta to destroy the W & A rail lines that supplied the Union army. Sherman saw this as an excellent opportunity to launch a cavalry raid of his own. On August 18, twenty-eight-year-old General Judson Kilpatrick set out with 4,700 cavalrymen, tasked with the duty of cutting the M & W railroad at the town of Jonesborough.[17]

Sherman's cavalry had failed the Union commander in the recent past, and it had become clear that the Yankee troopers were not up to the standards of Rebel horsemen. It all started on July 28 with the "Great Cavalry Raid" of generals George Stoneman and Edward McCook. In two columns, the 8,000 horsemen converged on Lovejoy Station, a village along the Macon and Western. According to Sherman's instructions, Stoneman and McCook should have been able to destroy the railroad track all the way south to Macon. There was even loose talk of rescuing Union officers imprisoned in Macon and a possibility of reaching the 20,000 Yankee soldiers held at the Andersonville Prison.

Stoneman, it turns out, was never interested in the carefully planned rendezvous with McCook at Lovejoy Station. The column under McCook reached the rendezvous point first with Joe Wheeler's cavalry in close pursuit. Soon surrounded and with Stoneman nowhere to be found, McCook's column fell apart in a desperate attempt to return to Federal lines. Stoneman's horsemen, after a half-hearted attack on the rail lines, rode straight for Macon by way of Monticello. Turned back at the Ocmulgee River by a motley force that included untested local militia and wounded soldiers taken from Macon hospitals, Stoneman turned back for Atlanta. His troopers were soon met by three brigades of Confederate cavalry under Alfred Iverson at the Battle of Sunshine Church. Forced to surrender some eighteen miles north of Macon, George Stoneman had ultimately been "responsible for the destruction of both his own and McCook's divisions, close to two-thirds of Sherman's available cavalry."[18]

By sending "Fighting Joe" Wheeler north on August 10 with 5,000 Rebel riders, Hood had just removed the rook that protected his queen. Sherman could now unleash upon middle Georgia the erratic and controversial Judson Kilpatrick with 4,700 troopers, the remainder of his cavalry, and two

---

[17] S. Davis, *What the Yankees Did to Us*, 195.

[18] Castel, *Decision in the West*, 440. Stoneman believed he would earn everlasting fame if he freed the Union officers held in Macon and then went on to liberate the notorious Andersonville Prison.

batteries of horse artillery. Sherman saw Kilpatrick, an 1861 West Pointer, as very much a "fool but a fighter." Kilpatrick's cavalrymen, who had no choice but to put their lives in the hands of this unstable man, saw in their general, as one officer wrote, "the most vain, conceited, egotistical, little popinjay I ever saw."[19]

With only 400 troopers, General Sullivan Ross's 3rd Texas Cavalry could do little more than harass Kilpatrick as the large Federal column moved southward. The Yankee raiders reached Jonesborough on the evening of August 19 with the ubiquitous Texans remaining in striking distance. Entering the railroad town at 5:00 P.M., the Federal troopers spent the next six hours tearing up the tracks of the M & W, burning warehouses and the train depot, stores, and cotton reserves, and, all the while, getting themselves ingloriously drunk. A heavy providential rain interceded that night, preventing the Union arsonists from heating and bending the rails over bonfires of piled-up ties. Kilpatrick knew by midnight that disparate detachments of Rebel cavalry would soon be converging on Jonesborough. He did not know that "Sull" Ross had also managed to secure Confederate infantry. Six miles to the south at Lovejoy Station, Patrick Reynold's Arkansas Brigade was disembarking from boxcars to protect the village on the M & A. At first light, Kilpatrick ordered his tired Federal troopers to head east, and upon reaching Lee's Mill, to turn south to capture Lovejoy Station.[20]

The Lee household had been informed on the evening of August 19 that Yankee cavalry was in control of Jonesborough, only four miles to their west. Without the benefit of reveille, an anxious Major Lovick Thomas was up at dawn. He strapped his old army blanket over his shoulder and was out the door before breakfast. Meeting a neighbor along the mill road, Lovick arranged for them to work together to watch for enemy movements. The men would return home for a quick breakfast and then meet at a nearby spring, "where we would secrete ourselves and watch Kilpatrick." As Lovick walked up the lane to the Lee house, a Confederate wagon train on the retreat from Jonesborough caught up with him. They were hauling army supplies and trunks filled with valuables belonging to the citizens of Jonesborough. Lovick told the officer of the train that they were in an exceedingly dangerous place and needed to move on at once. It was at this precise moment that five or six Federal horsemen, advance scouts for Kilpatrick, rode out in front of the wagon train and shouted for them to "Surrender, surrender, and hold up your

---

[19] D. Evans, *Sherman's Horseman*, 184.

[20] Castel, *Decision in the West,* 472.

arms."[21]

The sudden presence of Federal horsemen caught everyone on the road by surprise. There was no reasonable avenue of escape for the wagon train because tall trees and fences had them boxed in. The early morning light also revealed lines of Union horsemen approaching in seemingly every direction. Major Thomas found himself caught dead to rights in a situation that he had often sworn he would never allow. He was a captive, and, as a uniformed officer, he was quite the prize for the trespassers who were now pointing their seven-shooters in his direction.

As the Confederate prisoners proceeded under escort toward the Lee house, Lovick's mind raced as he considered the situation. His first thoughts were of his wife and little ones sleeping on the second floor of her parents' home. His next thoughts were of spending the rest of the war in a cold, dreary Yankee prison camp. He needed to find the opening that could reverse his fortune. Military surrender is a tricky thing at best, with its uncertainties, its helplessness, and the realization that you can no longer be of any help to the men you had pledged to support. But with Lovick Thomas, it went still deeper. He had been surrendered once before, a victim of General Pemberton's surrender of Vicksburg back in 1863. In that case, he held no vote and had no choice. This time, he was the author of his own surrender. This cut especially deep because, as an officer, he had often told his men that he would neither place himself in a situation that could lead to his capture, nor would he submit to a future which included time in a Yankee prison. "I had warned my men of the danger in lagging behind on the march," he wrote, "and the horrors of lying in gloomy Northern prisons." Yet here he was. He had slipped out of the Lee house to find the Yankees, and they had found him. He was the prisoner he had admonished his men never to become.[22]

Escape from Yankee Captors

Lovick Thomas described his brief experience as a Yankee prisoner in a well-received account that he penned for the *Atlanta Journal* in 1900. Readers had the opportunity to share his line of thinking in these tumultuous moments. It was his belief, Thomas noted, that his best and likely only chance of escape was to put his captors off guard by indicating full acceptance of his fate. The major surrendered as a gentleman and an officer and had apprised them of

[21] L. P. Thomas, Close Call narrative, *Atlanta Journal*, December 15, 1900.
[22] Ibid.

his name and rank and to what command he belonged. As the captured wagons pulled up at the Lee farm, a single Federal trooper was assigned to guard Thomas while the other half-dozen riders became consumed with smashing trunks, plundering wagons, and taking stock of the farm buildings.

Now with a lone sentry, Lovick believed he had a chance to cultivate a human connection from which an escape opportunity might arise. He confided to his guard that he was visiting his wife and children after a long separation and pled the courtesy of a final goodbye to "his wife and babies." The guard responded with a stern and resolute "No, sir!" That was the military response Lovick expected. "You cannot make me believe that you are so hard hearted, I will not believe it," exclaimed the major. "You are too generous for that. I would under the circumstances allow you at least to tell your wife and babies, good bye." Under this barrage of guilt, the guard seemed to soften and, while not giving permission, cast his eyes to the ground. Not hesitating, Lovick moved toward the gate that led to the back door of the house. The Confederate officer hoped he had read the guard's face correctly.[23]

Major Lovick Thomas won his high-stakes gamble and knocked on the door where his wife met him. He embraced Jennie, perhaps for the last time. Kissing her and holding her tightly, he pressed his treasured silver watch, a gift from his father, into her palm. She deftly secreted the timepiece in the billowing folds of her gown. It was a wise move; the Yankees rifled the house and carried off many items of value. In the midst of this confusion, one of the Lee family slaves walked into the yard, trying to make sense of the situation. Responding to the perceived threat, one of the Yankees ordered the slave to remove himself. The loyal servant refused and would not move an inch even though the Yankee soldier leveled his gun on him. At this point, Major Thomas intervened and ordered the young man to move away.

Having reassured Jennie that he would survive his awaiting ordeal, Lovick turned and passed through the big front gate and rejoined the other prisoners in the road. The Yankee horsemen remained busy, bursting open trunks and "scattering things in a lively way." With his guard "like a shadow at his back," Lovick moved about the area. "Major, ain't we in a hell of a fix?" asked the officer of the wagon train. At that, Lovick had an idea, and he asked the Rebel captain for the loan of one of his horses. Since the wagons were no longer his responsibility, the wagon master felt no need to deny the request, and he told Lovick to take his pick among the horses tied to a rail. Thomas

---

[23] Ibid.

unhitched a horse which, being fitted for wagon duty, lacked a saddle, bridle, and other riding accouterments.

To the amusement of the milling Yankees, the Confederate major began leading his new horse by the halter aimlessly down the line of wagons. As he walked past each wagon he called out in a loud voice, "Boys I need a bridle, who will loan me a bridle?" He would stick his head in one wagon and then the next, always with the same query: "Who will loan me a bridle?" Turning and making his way back to the house, he expanded on his needs: "Who's got an extra saddle? I'm going to need a saddle." By this point, the Yanks considered the behavior of the Rebel officer to be slightly unhinged. But as it soon turned out, the major's wanderings were not as pointless as they first seemed. Lovick Thomas had made a full circle and was now standing directly in front of the large gate which still stood slightly ajar. This was his moment to make good on his oft-repeated declaration that he would never wear the Union brand. In one quick motion, he turned the horse's great body, and, imposing the animal between him and his would-be captors, Thomas sprang on foot through the gate and in a few long strides disappeared behind the house.[24]

With the building offering further cover, Thomas sprinted through the field, jumping three low cross-fences, "which I think now, I never touched," before making it to the relative safety of a small copse of woods. Neither the guard nor the other horsemen had managed to get off a single shot at the fleeing Confederate. Hidden in a thicket of small trees that overlooked the house and the nearby road, Thomas watched the unfolding chaos that was erupting on the Lee farmstead. The infuriated and embarrassed Yankees fired their rifles as if facing a ghost army. So fast and perfect was his escape that the Federals believed the Rebel major was still on the property and proceeded to ransack the farmhouse while looking under it, around it, and in it. Pigs squealed, and terrified children added their voices to the shouted threats to burn all the farm buildings down. It was a heart-rending spectacle for the fugitive watching through the rail fence, especially knowing that two of these children were his. Soon the road toward Lee's Mill was filled with the blue-uniformed columns of Kilpatrick's main force.

Captain Thomas was hiding so close to the road that he could hear two troopers argue over a pocketknife. A section of artillery came rattling down the hill, and soon the men from Robert Minty's Michigan and Pennsylvania brigades were dismounting in the field around him to feed their horses and

[24] Evans, *Sherman's Horseman*, 434–36.

prepare breakfast. Kilpatrick and his officers were having their breakfast down the hill at Lee's Mill. The Federal officers promised Samuel Lee that they would not burn his mill, but they did empty his smokehouse and dispossess the old gentleman of his slaves and mules. Breakfast was soon interrupted by the sound of gunfire from the north. Sul Ross's hard-charging Texans had been pushing Kilpatrick's rear guard of Ohio horsemen since dawn when the Federal columns had left Jonesborough. Now, it seems, the two sides were going to square off in the field by the very thicket of small trees where Lovick had sought sanctuary.[25]

"I found an oak tree near at hand with grape vines and foliage for a hiding place," wrote Thomas, "and as nimbly as a squirrel I was soon up the tree and hid in the branches." He sat on a limb and hugged the trunk of the tree, as a Federal skirmish line formed along the fence line below him. The main force of Federals was further back near the mill. The Texans under the tall, twenty-four-year-old Sul Ross fought from the tree line to the north and began a lively skirmish. Ross ordered a light battery be unlimbered in the road, and soon the leaves and limbs around the Confederate fugitive were "disturbed considerably" by the discharge of Rebel cannon and the persistent whiz of errant minie balls. It would be several hours before Lovick saw an opportunity to improve his predicament.[26]

Thomas could not move until the main force of Yankees had crossed over the stream, but, once that had happened, he was ready to take his chances with the small number of horsemen still fighting along the fence line, even if his odds were not particularly good. Lovick slipped down the tree, keeping the trunk between himself and the enemy. "I placed my back against the tree and put on my shoes," he recalled in 1900, "and slipped off a few paces, where I had thrown my old army blanket on the leaves." No sooner had Lovick retrieved his bullet-torn blanket from the days at Resaca than the leaves at his feet began to dance and the "buzzing of bees" filled his ears. With Federal Springfield carbines firing in his direction, Thomas ran the gauntlet through the woods and made his "second dash for liberty."[27]

At full speed, Lovick Thomas headed for a company of cavalry that he was not entirely sure "were Confeds or Yanks." Fortunately for the breathless fugitive, the horsemen proved to be from the friendlier side of the Mason-

[25] Ibid., 435.

[26] L. P. Thomas, Close Call narrative, *Atlanta Journal*, December 15, 1900.

[27] Ibid.

Dixon. Exhausted but fighting mad, Thomas "ran right up among our cavalry" and asked for a gun and urged them "to move on the enemy at once." This unarmed Confederate officer, who had seemingly appeared from nowhere, was exhorting these tired Texans to advance on his tormentors and roll them up like full leaf Virginia cigars, capturing every Yankee this side of the stream in one earnest charge. Being a major in the infantry does not make one a major in the cavalry. "They looked at me a little doubtfully," recalled Thomas, and the Texans quickly sent the distraught officer, who was as mad as a stick-poked bear, to the rear to report to Colonel Ross. Behind the lines, Thomas found an old spent army horse that the Federals had left behind in a field. "Borrowing a plow line from a lone woman I found living in a small cabin by the road side," Thomas wrote, "I rigged up a bridle, threw my blanket over the horse, and joined the cavalry."[28]

The Battle of Lovejoy Station

It was now 11:00 A.M., but Lovick Thomas's eventful day was far from over. In the afternoon, he would be enveloped in one of the largest cavalry saber charges in the war. As the infantry major was finding a mount, Minty's Federal Brigade had reached Lovejoy Station from the northeast. Seeing few Rebels, the men from Michigan dismounted and approached the town on foot. They were not aware that waiting in the shelter of a deep railroad cut was Dan Reynold's Arkansas infantry Brigade. At fifty yards, the Razorbacks stood up and fired a volley into the surprised troopers. The Federals fell by the score, with the rest running back to their horses. Lovejoy Station was well defended, and Confederate infantry, formed in a line of battle, now began advancing from the railroad station toward the retreating Yankee horsemen. By noon, it was clear that the noose was tightening around Kilpatrick's sizeable raiding party.[29]

Rain began to come down in torrents. Believing in error that he faced thousands of Rebels in every direction, Kilpatrick regrouped his forces and made a retreat to the Federal lines by way of Decatur. On a high knoll (today, part of the Nash Farms Battlefield Park in Henry County), Kilpatrick consulted Colonel Robert Minty, a flamboyant former British cavalry officer. Any retreat back toward Atlanta would require going right through Sul Ross's indefatigable little band of Texans. Minty deployed his three regiments, the

---

[28] Ibid.

[29] Castel, *Decision in the West*, 472–73.

4th Michigan, 4th Regulars, and 7th Pennsylvania, in three lines, each man stirrup to stirrup in columns of four with sabers drawn. They would spearhead the charge with the rest of Kilpatrick's force following closely in a compact column.[30]

Sul Ross had his men dismounted and strung along a rail fence across an old field of corn stalks at the bottom of the knoll. A Federal bugler sounded "forward," and the mud-splattered lines of blue came thundering down the hill. The outnumbered Texans could not stop Minty's determined advance. Standing in the rain, Lovick Thomas was there to witness one of the largest cavalry charges of the war and the only successful mounted charge made in the whole Georgia campaign. Some of Ross's men were cut down or captured, and the flag of the Third Texas Cavalry lost to the Federals, but most of the Texans managed to escape and reform once the Yankees had passed. A lone Rebel howitzer in a stand of walnut trees fired canister into the charging horsemen. In minutes, Minty's troopers were behind the gun and mixed up in a wild melee among Ross's caissons, led horses, and ambulances. As Federals slashed at anything and everything with their sabers, the Rebels swung the howitzer around and fired a double load of canister into the backs of the troopers. "A withering blast ripped through the struggling mass," David Evans writes, "friend and foe alike." Amid this chaos was a distraught Lovick P. Thomas.[31]

As the newest recruit in Ross's cavalry, the unarmed Lovick was assigned to a group of men who were holding the reins of the horses for the dismounted troopers along the fence line. "With sabers drawn and horses at full tilt, they rushed upon us," Thomas remembered, "and over the line they came like a whirlwind, right into the horses held in the rear. Such a whacking and dashing among this little band of 'Confeds' you never saw." Thomas ran among the stampeding horses, "trying to catch one that I might save myself upon." He succeeded in catching the bridle of a small iron gray mule and joined the squad of former rein holders who were bounding for the rear. The demoralized Texans pressed their mounts without pause for ten miles. The small detachment crossed the swollen creek by Lee's Mill and continued north until they finally stopped and sheltered within a dark forest. The bone-tired men spent the night there without fire or rations. The next day they retraced their steps and stumbled into Jonesborough, where they found the rest of Ross's command. One trooper laid claim to the little mule that had

[30] Ibid., 473.

[31] Evans, *Sherman's Horseman,* 450.

saved Thomas on his third escape from Yankee captors during an eventful August 20, 1864.[32]

Captain Thomas's one-day flirtation in the service of the Confederate Cavalry had come to a merciful end. In a few more days, he would rejoin the 42nd Georgia in the Atlanta trench lines. Kilpatrick straggled back to the Federal lines on August 22, having lost 237 of his raiders. He reported to Sherman that he had destroyed enough track near Jonesborough to keep supplies from reaching Hood for ten days. But two days later, Sherman could hear whistles from the trains operating on the Macon & Western. Realizing the limitations of his depleted cavalry, the Union commander had no choice but to reconstitute his "grand movement" and use his infantry to permanently break the M & W Railroad.[33]

The Battle of Jonesborough

On the night of August 26, the XIV Corps pulled out of the breastworks in front of Clayton's Division. Francis Nash wrote in his journal that the 42nd were the pickets for the right wing and, at first light, ventured into the abandoned Federal trenches. Much to their pleasure, they found copious quantities of blankets, clothing, hardtack, and bacon. That day, Stovall's Brigade pushed out two miles to the front but did not find any Yankees. With Wheeler destroying the tracks of the Western & Atlantic north of Atlanta, the soldiers speculated that Sherman had been forced to withdraw to protect his line of communications. The citizens of Atlanta came out of their bombproofs, ecstatic that the shelling had stopped and their city had been saved from the "blue hordes."[34]

Sherman had, in fact, started his great "wheeling movement" and was swinging his army well to the south of Atlanta with the intent of cutting the Macon & Western Railroad. The XX Corps retreated to the bridges over the Chattahoochee and established a defensive position above Atlanta. The rest of the Federal army, with wagons holding a fifteen-day supply of food and carrying 100 rounds of ammunition per man, began marching toward Fairburn, which they reached on August 28. Sherman's army spent the next day destroying the tracks of the West Point Railroad. Hood was soon cognizant of Sherman's intentions and ordered Hardee's Corps to Rough & Ready, a

[32] L. P. Thomas, Close Call narrative, *Atlanta Journal*, December 15, 1900.

[33] S. Davis, *What the Yankees Did to Us*, 219.

[34] Nash, *Civil War Diary of Francis H. Nash*, 37; Castel, *Decision in the West*, 485.

village on the Macon railroad three miles southeast of East Point. Lee's Corps moved south to East Point. Private Nash wrote that, on the night of August 29, Stovall's Brigade relieved part of Cleburne's Division in the ditches near East Point.

At dawn on the thirtieth, the XV and XVI Corps, marching on parallel roads, set out from Fairburn in the direction of Jonesborough. Their progress was hindered by stiff resistance from Confederate cavalry under Armstrong and Ross. In the afternoon, the Federals crossed the Flint River and established a beachhead on the eastern side of the reddish-colored river less than a mile from Jonesborough. The capture of the little village would mean that the last railroad to Atlanta had been permanently cut. At a 9:00 P.M. meeting held in Atlanta, Hood ordered Hardee and Lee to march their corps through the night to Jonesborough and attack the Federals the following day. Hood considered it vital that Howard's men be driven back across the Flint River.

Hardee was closer to the new Federal position, and, with the army's provision and ordnance trains, reached Jonesborough in the morning. Lee's troops set out at midnight from East Point and, following rough roads for twelve miles, reached Jonesboro at 11:30 A.M. Henry Clayton called it "an exceedingly fatiguing march." By this point, Oliver Howard, the commander of the Army of the Tennessee, had brought all three of his corps across the Flint River. The Federals built breastworks as the day progressed. Ross's cavalrymen reported to Hardee that two more Federal corps were moving on Jonesborough from the north. Prospects for a successful Rebel attack diminished by the hour, but Hood's telegrams from Atlanta continued to pressure Hardee to launch an assault that afternoon.[35]

At 3:00 P.M., Confederate artillery opened on the Federal position, and ten minutes later skirmishers were sent forward. The Rebel front, some 20,000 men, extended for a mile and a half. Hardee's two divisions formed the left wing, with two of Stephen Lee's divisions, under Carter Stevenson and Patton Anderson, composing the right or north wing. Lee's third division under Henry Clayton was split up, with the Georgia brigades of Stovall and Cumming sent to form the second line behind Stevenson. The brigades of Manigault, Gibson, and Jones formed behind Anderson's Division. As the men fixed their bayonets, Arthur Manigault looked down his line of exhausted and apathetic soldiers and knew "there was not much fight in them." He wrote, "The troops were weary, dull, sluggish, and entirely without that

---

[35] Field report of General H. D. Clayton, September 16, 1864, *OR*, ser. 1, vol. 38, part 3, 821–22.

spirit which had hitherto characterized them on all such occasions as the present. They had long since lost confidence in their leaders and knew full well what was likely to be the consequence of an attack against superior numbers, protected by a complete set of field works." O. O. Howard had 12,000 Federals to oppose the Rebel charge and another 7,500 men waiting in reserve.[36]

The doomed assault was to be made in *en echelon,* from left to right, first by Hardee and then Lee. Albert Castel notes that Lee, "displaying the same aggressive spirit and talent for blundering that he revealed at Tupelo and Ezra Church," started his men forward as soon as Hardee's artillery went into action. The blue coated veterans under "Black Jack" Logan waited for the gray lines to draw close to their fortified position before dispensing rounds of punishing volleys. On the Confederate right, the first Confederate line got no closer than 200 yards before stopping and exchanging fire. The second gray line got within eighty yards before retreating. "The men did not behave as well as on any previous occasion," Manigault recalled, "and all efforts to carry them on, failed."[37]

Henry Clayton, who directed the second Rebel line, wrote in his official report that, "many of the troops in front, who had scarcely engaged the enemy, were coming back, and some of them were endeavoring to conceal themselves in the gullies of the old fields." Clayton had kind words for General Randall Lee Gibson, who seized "the colors of one of his [Louisiana] regiments and dashed to the front and up to the very works of the enemy."[38]

The 42nd Georgia was part of Carter Stevenson's second line charging at the center of the Federal line. Marcellus Stovall wrote that his Georgia Brigade formed behind the Alabama brigade of Edmund Pettus and to the left of Alfred Cummings's Georgia Brigade. Stovall reported that his command got "within forty paces of the enemy's works, when the fire became general on both sides." His men stood in the open field exchanging fire for thirty minutes. Hardee's Corps had attacked with little vigor on their left, and Stovall's troops experienced a galling cross fire from Yankees on their exposed flank. Stovall's only field officer (above the rank of major or colonel) on the field, Major Marcus Nall of the 41st Georgia, was killed "gallantly leading his regiment." To their right, brigade commander and West Point graduate, thirty-five-year-old General Alfred Cumming, was disabled by

[36] Manigault, *A Carolinian Goes to War*, 245.

[37] Castel, *Decision in the West*, 502; Manigault, *A Carolinian Goes to War*, 246.

[38] Field report of General H. D. Clayton, September 16, 1864, *OR*, ser. 1, vol. 38, part 3, 821–22.

wounds and taken from the field. Stovall wrote in his field report that a quarter of his brigade was killed or wounded that afternoon. The 42nd had several wounded, and five members of the Calhoun Guards (K) captured. Twenty-nine-year-old 2nd Lieutenant Benjamin Gholston of the Gwinnett Beauregards (A) was the only soldier killed from the 42nd Georgia. In all, Lee's Corps suffered 1,400 casualties in the ill-fated assault.[39]

The Surrender of Atlanta

In the late afternoon of August 31, Yankee raiders cut the telegraph line between Rough & Ready and Jonesborough, confirming that Sherman now had control of the last railroad line into Atlanta. A crestfallen Hood saw no option other than the immediate evacuation of Atlanta by his army. Stephen Lee's demoralized corps was ordered north from Jonesborough to cover the retreat. Setting out at 1:00 A.M., the sleep-deprived troops marched fourteen miles before camping. Lee's departure left Hardee's Corps alone to face most of Sherman's army. With the railroad in his hands, Sherman decided it was best to destroy Hardee's isolated corps before turning his large army on the Gate City. The second day of the Battle of Jonesborough was a reversal of the previous day, with XIV, IV, and XXIII corps all assailing Hardee's right flank. The Confederates, now fighting from a defensive position behind breastworks, somehow managed to hold Jonesborough and, in doing do, avoid annihilation. On the evening of September 1, Hardee pulled his battered corps out of harm's way and eight miles south to Lovejoy Station.

During that same evening of September 1, 1864, Stewart's Corps and the Georgia Militia under Gustavus Smith marched down Decatur Street on their way out of Atlanta. The downtown commissary warehouses were opened to all comers, and soldiers and civilians alike were soon helping themselves to bacon, coffee, whiskey, shoes, and clothing. The 42nd and the rest of Lee's Corps were six miles to the south, holding a defensive position on the McDonough Road. They were soon startled by the sound of terrific explosions coming from the railyard of Atlanta where five locomotives and eighty-one boxcars, many packed with gunpowder, shells, and cartridges, were set off by Hood's engineers. The light from the massive explosions reflected off the clouds and could be seen by Lovick Thomas and his regiment.

[39] Field report of General M. A. Stovall, September 16, 1864, *OR*, ser. 1, vol. 38, part 3, 825–26. William Cockrell, George Powell, William Scarborough, Asa Willis, and Silas Yarbrough were all captured at the Battle of Jonesborough on August 31, 1864.

Francis Nash reported that the 42nd set out for McDonough on the morning of September 2. They passed through the small village the following day en route to Lovejoy Station, where Hood hoped to concentrate his demoralized army.[40]

The first Yankees to enter Atlanta on September 2nd were members of the XX Corps coming south by way of Vinings. Mayor James Calhoun, with a delegation of leading citizens, rode out under a white flag from Marietta Street. In 1860, the mayor had been a delegate to the Constitutional Union Party Convention that had nominated John Bell of Tennessee for president. Now, four years later, he was surrendering his shelled and wrecked city. On a note that he handed to General William Ward, Mayor Calhoun had written, "The fortune of war has placed Atlanta in your hands. As mayor of the city I ask protection to non-combatants and private property."[41]

Sherman did not get confirmation of the fall of Atlanta until the morning of September 3. He had spent the afternoon and evening of the previous day continuing to make unsuccessful attacks on Hardee's tired men, who were now drawn up in a strong position along a low ridge one mile north of Lovejoy Station. A pleased William Sherman penned a dispatch to Halleck and the president in Washington: "Atlanta is ours, and fairly won." The following night under cover of darkness and heavy rain, the Federals pulled away from Lovejoy Station and headed twenty miles north to claim their prize. President Abraham Lincoln's re-election in November was now assured.[42]

Francis Nash wrote in his diary that the 42nd remained at Lovejoy Station, twenty-five miles south of Atlanta, until September 17, 1864. Clayton's Division was assigned to trenches closer to McDonough and held the right flank of the Confederate army while the main Rebel army rested and refitted at Lovejoy Station. Jennie, the wife of Lovick Thomas, wrote in her diary that, "I soon after went near McD [McDonough] to board as L. was there." Arthur Manigault noted that Sherman's army, down to 80,000 men, was, "as weary and exhausted as ourselves, required rest, re-equipping and re-organizing." Hood's Army of Tennessee was down to 30,000 men, but they had somehow survived the Atlanta campaign and were still a factor in the outcome of the war. Albert Castel notes that this Rebel army at Lovejoy Station still retained "a large core of veterans who, if they no longer have much hope of Confederate victory, nevertheless [were] prepared to fight out of pride, a

[40] Castel, *Decision in the West*, 522–23.

[41] Mayor Calhoun's son William was the capable captain of the Calhoun Guards (K).

[42] S. Davis, *What the Yankees Did to Us*, 288; Castel, *Decision in the West*, 528.

sense of duty, and honor, and sheer hatred of the Yankee invaders who are laying waste to their Southland."[43]

During this strange time, Sherman decided to make Atlanta a "pure Gibraltar" by ordering the expulsion of all civilians from the city, regardless of their sympathies for or against the Union. Between September 10 and 20, all civilians were compelled to register with Federal authorities. Those loyal to the North were allowed to take their possessions and board trains to Louisville. Another 705 adults, with 860 children and 86 servants, were given travel permits to pass out of Atlanta and into the Southern lines. The town of Rough & Ready, which is today referred to as Mountain View, was designated as the transfer point. The railroad hamlet had been named for President Zachary Taylor and was some fourteen miles by rail south of Atlanta. Under a flag of truce, Southern officers met the refugees at Rough & Ready and directed them south toward Macon. This arrangement between Sherman and Hood started on September 12 and lasted for ten days. It ultimately included a limited exchange of recently captured prisoners; 1,128 Federals were exchanged for 1,332 Confederates. James Russell Bowman of the Walton Blues (G), who had been captured near Jonesboro, was pleased to be part of this limited prisoner exchange.[44]

On the morning of September 17, the 42nd marched out of McDonough headed for Palmetto Station. It was a three-day march by way of Fayetteville. On the morning of the twentieth, Lovick Thomas had his regiment near Campbellton, where they constructed breastworks on the Campbellton & Petersburg Road. Hood was concentrating his army around Palmetto Station, some twenty-four miles southwest of Atlanta and directly on the West Point Railroad. On September 24, Charlie Rowland wrote Katie that the Georgia Brigade had melted away from 2,000 men to 500 over the course of the Atlanta Campaign of 1864. "Many are limping through the land without legs or arms and many more lie under the red hills between here and Dalton," he wrote. "It is a mysterious dispensation of Providence, but it is right as we shall see one day."[45]

President Jefferson Davis reached Hood and his Army of Tennessee on

---

[43] Nash, *Civil War Diary of Francis H. Nash*, 39; 1864 diary entries of Jennie P. Thomas in Huff, *Four Families*, 83; Manigault, *A Carolinian Goes to War*, 251; Castel, *Decision in the West*, 542.

[44] S. Davis, *What the Yankees Did to Us*, 299; Confederate service records of Russell Bowman, NARA, M226, roll 6.

[45] Rowland to home, September 24, 1864, in *Charles Alden Rowland*, 47.

September 25. The following morning, Lovick Thomas wrote a quick note to Jennie. Captain Benjamin Weaver was leaving for Gwinnett County "in a matter of minutes on business," and Lovick wanted her to know that he was safe. "We have a pleasant camp," he wrote her, with "plenty to eat, no fighting, [and] no enemy very near us." The expectation was that President Davis would review the troops that morning. "We are arranging to be under arms at 8 one half o'clock this morning and tis nearly that time now." The Army of Tennessee was drawn up behind their works for the grand review. Arthur Manigault thought President Davis looked "thin, care-worn, and angry. A scornful expression rested on it." The event made Hood's staff apprehensive as voices from different brigades called out for the Confederate leader to "Give us Johnston! Give us our old Commander." Manigault thought the whole affair was "cheerless and unsatisfactory" as President Davis "scarcely designed to lift his hat from his head as we saluted."[46]

[46] L. P. Thomas to home, September 26, 1864, in Huff, *Four Families*, 90; Manigault, *A Carolinian Goes to War*, 256.

Chapter 8

The Nashville Campaign

"We have been marching for days, so many that I cannot count them."
—Major Lovick Thomas to Jennie, October 29, 1864

Return to North Georgia

On September 29, 1864, with three days of rations in their haversacks, the infantry and artillery of the Army of Tennessee began crossing the Chattahoochee River using three pontoon bridges. Now a respectable 44,000 in number, the Southerners moved toward Marietta, intending to cut the Western & Atlantic in north Georgia and force Sherman to abandon his Atlanta prize and give chase. Despite the long odds against them, a month of rest had revived these veteran troops and, with morale improved, the Rebels were on the offensive and prepared to raise the price of what Sherman had declared "fairly won." On October 3, A. P. Stewart swung his corps right and captured Acworth and Big Shanty, thereby severing Sherman's supply line to Chattanooga. The Federal commander, intent on launching his "March to the Sea," was forced instead to respond to this threat by marching his army in the opposite direction into North Georgia. Leaving Slocum's Corps in Atlanta, Sherman directed his remaining 60,000 men to Marietta, where they could shadow Hood's Army of Tennessee.

For the men of the 42nd Georgia, heavy rains turned the roads to mud, slowing their progress. On October 3, Private Francis Nash wrote in his journal that the 42nd reached Lost Mountain after seventy-four miles of marching. They would rest there for the next three days. Charlie Rowland wrote to his wife, Katie, "I remember very well when we passed by here on the retreat in June, but had no idea we would see it again under such circumstances." From the heights of Lost Mountain, the men could see the smoke from the efforts of Stewart's men to burn the tracks near Big Shanty. Hoping to steadily pull Sherman away from Atlanta without provoking a general engagement, Hood gave orders to march west. For three days Clayton's Division marched through Dallas, Van Wert, and Cedartown, eventually reaching the mountain spa town of Cave Springs. They were now south of Rome and the Coosa

River, a few miles from the Alabama state line.[1]

Private Nash noted the arrival of General P. T. G. Beauregard to the Southern camp at Cave Springs. The former hero of Fort Sumter and First Manassas was now in charge of the new "Military Division of the West," and theoretically Hood's new superior. The strategy Hood laid out before the forty-six-year-old Louisianan was to move quickly into northern Georgia and destroy the Western & Atlantic from Dalton to Chattanooga. Hood would then fall back to Jacksonville, a small Alabama town near Gadsden. If Sherman came after him, Hood would fight him there. Beauregard was skeptical at first, but soon warmed to the complex and daring scheme. The Creole general did insist that Joseph Wheeler's cavalry be sent back to Georgia to watch and harass Sherman. In exchange, Nathan Bedford Forrest would bring his cavalry divisions down from west Tennessee to guard Hood's flanks. Orders were also given to the First Regiment of the Georgia State Line Troops to leave Stovall's Brigade and return to Georgia. Governor Joseph Brown needed his militia to defend their home state from Sherman's large army. The First Regiment departed having earned the respect of the Confederate regulars within Clayton's Division.[2]

On the morning of October 11, Stovall's Georgia Brigade marched into Alabama and camped near Cedar Bluff. "Gen. Lee ordered the smallest brigade in our Division to be detached," Rowland wrote, "and as ours was the smallest the agreeable duty fell on us." Now responsible for guarding the excess artillery and wagons, they drifted south to within five miles of Jacksonville, where they camped for several nights. "The ground is cold and wet and we are camped in the pine woods," complained Rowland, "the soil of which seems to have no bottom when it rains." Stovall had less than 900 men and was guarding about 300 wagons and 70 pieces of artillery. He worried about the Federal division under General John Corse posted in Rome, Georgia. Hood foresaw a battle with Sherman in Alabama and ordered supplies from Selma and other Confederate points to be sent in the direction of Jacksonville.[3]

Off to their east, Samuel French's Division from A. P. Stewart's Corps

---

[1] Nash, *Civil War Diary of Francis H. Nash*, 41; Rowland to home, October 4, 1864, in *Charles Alden Rowland*, 50.

[2] The State Line troops played a significant part in the Battle of Griswoldville, near Macon, on November 22, 1864. L. P. Thomas's half-brother, Pen Bedell, was killed in the Confederate charge that day as a member of the Georgia Militia.

[3] Rowland to home, October 14, 1864, in *Charles Alden Rowland*, 52.

fought a pitched battle for control of the railroad depot at Allatoona. It was an intense and savage fight at close quarters. Sherman managed to hold the Allatoona Pass, but the hot-tempered commander was in no position to defend a hundred miles of railroad "now that Hood, Forrest, Wheeler, and the whole bunch of devils, are turned loose without home or habitation." He renewed his plea with Grant and Lincoln to "make Georgia howl" by marching his large army southeast to Savannah. Sherman was convinced that destroying Georgia's ability to wage war was more beneficial for Northern war aims than chasing Hood's army across Alabama.

More disturbing still, from Sherman's perspective, was news that Hood's army had disappeared into the rugged mountainous forests around Rome. Sherman wrote his division commander in Rome, "I can not guess [Hood's] movements as I could those of Johnston, who was a sensible man, and only did sensible things." Hood was, in fact, using a gap west of Rome to reach the Chattooga River valley and sprint his army northward. He reached the Federal garrison at Resaca on October 12 and demanded its surrender. Much like Anthony McAuliffe at the Battle of the Bulge, Colonel Clark Weaver would have none of it. Hood decided to trudge on to Dalton, where he found a garrison commander more willing to surrender his fort and the 44th U.S. Colored Troops (USCT) who had been tasked with defending it. The 800 new prisoners were put to good purpose ripping up the tracks around Tunnel Hill. In all, twenty-five miles of tracks from Resaca to Tunnel Hill were destroyed.

With Sherman closing in, Hood marched southwest toward Alabama. His army passed the old Chickamauga battlefield and into the valley of the Chattooga River south of Lafayette. Morale in the Southern army had improved considerably, but not enough for his subordinate generals to endorse a general engagement with Sherman's much larger army. On October 16, John Bell Hood changed his strategic plans. Rather than fight Sherman in Alabama, the Army of Tennessee would march northward and cross into Tennessee. If they moved quickly, the Rebels might defeat General George H. Thomas's scattered forces and capture Nashville—and with it the vast Federal stores warehoused there. There was even loose talk of a resupplied Confederate army crossing the Ohio River and invading the North.[4]

---

[4] Groom, *Shrouds of Glory*, 73.

Northern Alabama

With the long Lookout Mountain Range standing off from their right shoulder, Hood's army pushed southwest down the valley toward Gadsden. Before any advance on Tennessee could begin, the Army of Tennessee would first have to cross over to the western side of this section of the imposing Appalachian mountain chain. Meanwhile, Stovall's men were marching east for twenty-one miles, with orders to help screen the movement of the Rebel army from Sherman's scouts. Back just inside of the Alabama-Georgia border, the 42nd spent October 20 in a line of battle just a half mile from Cave Springs. Resupplied at Gadsden, Hood's Rebel army began marching west on October 21. Being the farthest back in the line of march, Stovall's Brigade brought up the wagons and artillery under General Arnold Elzey and then covered the rear of the moving army. The Georgia Brigade marched forty-three miles in two days and were still miles short of Gadsden. Rowland penned Katy at 11:00 P.M. on the twenty-second that "Genl Hood has moved toward Guntersville (going I suppose into Tenn) leaving a few boats for us to cross upon." Worried that there were Federal cavalry in the area, the men were roused at midnight so that the rear guard could cross the Coosa River by daylight.[5]

Shortly before starting the night march, the soldiers were encouraged by the unexpected arrival of Captain Benjamin Weaver of the Independent Rebels (B). After recovering from illness, Weaver had left his Gwinnett home on October 8 and set out to find his regiment. It was not an easy task, and his journey made full use of all modes of nineteenth-century transportation, employing rail, steamboat, and foot to take him first on a long hike to Newnan and West Point, then a train to Montgomery, a riverboat (the *Toney*) to Selma, a train to Blue Mountain Station, a hike to Jacksonville and then on to Cave Springs. Only then did Weaver discover that the 42nd was now marching toward Gadsden, which meant a reversal of direction on his part. Weaver found his company "in fine health," though many of his men were now barefoot. On the night of October 23, Private Nash reported that the 42nd Georgia camped six miles west of Gadsden, close enough for the forty-year-old Alpharetta farmer to gain permission to fall out and spend one night with his uncle Frank Parks in Cherokee County, Alabama. It took the Georgians a day to cross the Coosa River and two and a half days to traverse Sand Mountain, a sandstone plateau at the southern tip of the Appalachians. Weaver called it, "a strange country, many places level and such rich land and

[5] Rowland to home, October 22, 1864, in *Charles Alden Rowland*, 56.

has large streams of water flowing high up on the mountain" The people, he thought, "are generally Tories and poor ignorant creatures." On the other side of the plateau they descended into the valley of the Tennessee River.[6]

What had once been fertile farmland on either side of the river was now, three years into the war, a wasteland of burnt out plantation houses and abandoned fields. The 42nd Georgia reached Somerville on October 26. The rear guard was now sixteen miles southeast of Decatur. The first Rebel plan had been to cross the wide Tennessee River at Decatur; however, Hood decided not to attack the sizable Federal garrison at Decatur but instead pushed west another fifty-five miles to Tuscumbia, Alabama. Here, past Muscle Shoals, the Army of Tennessee could cross the Tennessee River unimpeded by upriver Federal gunboats. Florence, Alabama, on the other side of the river, would be their new objective. Hood hoped that supplies could be brought to Florence on the railroad, once rebuilt, and that he would be in an easier position to rendezvous with Forrest's cavalry.

On October 28, an exhausted Major Lovick Thomas wrote Jennie from Courtland, Alabama, still twenty-eight miles short of Tuscumbia. They were following the Memphis & Charleston tracks to their new supply base. Ever since leaving their campsite at Cave Springs, the 42nd Georgia had been averaging some twenty-five miles a day. In all, the march from Palmetto to Tuscumbia covered approximately 400 miles through the rugged hill country. "The men are getting very tired, with sore feet, etc.," he wrote her. For days, meals had been eaten as they moved. "My horse is a little lame from the march," noted Lovick, "but hope he will be all right in a day or so. I could not well get along now without him." According to Rowland, "Many of our poor men are without shoes and have to walk over rocks and through the cold mud and water. I don't see how they stand it." The civilian supply officer reflected further on their long march across Alabama: "We have had all sorts of weather, sometimes a cold rain and then bright and pleasant. I never saw the forests look so beautiful, the leaves were colored from the darkest brown to the most beautiful crimson." He recalled that on one evening during the march, "some of the boys started a rabbit." Both the staffs of Stovall and Clayton were soon in full pursuit, and, to everyone's surprise, Rowland caught the hare. "He was fat and we had him for supper," reported Charlie Rowland. "The Genl [Stovall] got the lion's share but of course complained

[6] B. P. Weaver to home, November 6 and 13, 1864, Weaver letters.

that it was raw and not cooked enough, he eat it however with evident relish."[7]

As the Confederates approached the Tennessee River, General Hood ordered General Stephen D. Lee and his corps "to force a passage" near the vicinity of Florence, Alabama. On October 30, Lee implemented Hood's directive by sending Edward Johnson's Division several miles upstream to Muscle Shoals where they crossed the river using pontoon boats. Four miles to the east, Clayton, with Gibson's Louisiana Brigade in the lead, made a contested river crossing directly opposite Florence. The troops in small boats pushed off from shore alongside the burned-out railroad bridge. It became a spirited half-mile crossing to reach Florence on the opposite bank as Federal cavalry troopers peppered their approaching boats. Two Confederate artillery pieces responded with deafening covering fire. Stovall's men followed close behind the troops from the Pelican State and joined in pushing the Federals out of Florence, street by street. The following day a pontoon bridge was built across the Tennessee River so that the rest of the Army of Tennessee could cross. The people of Florence, long under Federal occupation, were surprised and relieved by the sudden arrival of the Confederate's western army. Ben Weaver witnessed "the ladies shouting and crying by turns." Being among the first to arrive north of the river, Lieutenant James Monroe Gresham, sergeants Jonathan Huff and George Knight, Corporal George Eason, and privates Beverly Blasingame and Asa Mitchell, all from Walton County, were sent scouting and scavenging. They managed to surprise a squad of enemy cavalry and put them to flight after a sharp engagement. The jubilant Walton men bore away with them a prize better than any Yankee captive: a side of bacon.[8]

William Sherman watched Hood's movements in Alabama impatiently from his headquarters in Kingston, Georgia. In mid-November, Grant finally wired him the long-anticipated permission to launch his "March to the Sea" toward Savannah. Major General George Thomas would be left in Nashville to deal with Hood and the Army of Tennessee. The "Rock of Chickamauga" would soon be reinforced by Federal troops coming from all directions. When and if they all arrived in time, Thomas would have a hodgepodge of twelve divisions of infantry and three divisions of cavalry, some 75,000 men. In a

---

[7] L. P. Thomas to Jennie, October 29, 1864 in Huff, *Four Families*, 91; Rowland to home, November 2, 1864, in *Charles Alden Rowland*, 60.

[8] Lee, "From Palmetto, GA, to Defeat at Nashville," 257; B. P. Weaver to home, November 13, 1864, Weaver letters; Sams, *Wayfarers in Walton*, 187–88.

furious ten days in November, Sherman sent the IV and XXIII corps to Tennessee, cut his communication with Chattanooga, and, with 62,000 veteran troops, marched south out of a very nearly destroyed Atlanta. The Army of Tennessee was no longer to be Sherman's problem.

Middle Tennessee

John Bell Hood's army did not set out from Florence for Tennessee until Sunday, November 20. It was a dangerous delay of twenty days that gave precious time for George Thomas to consolidate the Federal forces in Middle Tennessee. Hood, unfortunately for the Southern cause, had no choice but to wait for the railroad to be repaired so that badly needed ammunition, clothing, and provisions could be brought to his army. The Confederate officers in the Army of Tennessee took advantage of this delay to enjoy the hospitality provided by the ladies of Florence. "There has been a sort of ball in town every night since we have been here," wrote Rowland on the third, "and to night all hands have gone and I am staying in the Generals room in case any business occurs." Division ambulances were employed to carry the "gay parties to the dance." Benjamin Weaver wrote that the 42nd camped in town until November 10, when they moved two miles north of town and built breastworks. Waiting for the railroad to be repaired, the men were on half rations of cornbread and jerked beef, relieved only when a successful forage might produce a "goat or mutton occasionally." "We suffer cold here as we have no tents and but few blankets," Captain Weaver wrote on November 13. He complained about rheumatism in his back and hips, but complimented his servant Eli who is "of good service to me to make fires, wash, cook, etc."[9]

According to field reports submitted on November 6 and 7 for Hood's Army of Tennessee, there were 44,832 men present for the Tennessee campaign. Stephen D. Lee's Corps had 12,782 men, divided into three divisions under Henry Clayton, Edward "Alleghany" Johnson, and Carter Stevenson. Clayton's Division had three brigades: Georgians under Marcellus Stovall, Louisiana soldiers under Randall Gibson, and Alabama troops under James Holtzclaw. Stovall's five Georgia regiments numbered 871 men in all, with the 42nd, under Major William Hulsey, by far the largest regiment with 345 men present. Abda Johnson's 40th was the smallest regiment with only 76

[9] Rowland to home, November 3, 1864, in *Charles Alden Rowland*, 62; Weaver to home, November 13, 1864, Weaver letters.

men present. Even with these small numbers, many of the Georgia men no longer had rifles. Of the 345 men in the 42nd present at Florence, only 248 could respond with rifle in hand. Henry Clayton added to his report that, "Of those reported absent without leave in Stovall's brigade, 19 officers and 300 men are prisoners of war, and in Gibson's brigade 32 officers and 737 are prisoners of war."[10]

On November 17, the long-promised Nathan Bedford Forrest arrived with his vaunted troopers after successful raids at Johnsonville and throughout Middle Tennessee. Three days later, with bands playing, Stephen D. Lee's Corps led the advance out of Florence. Against long odds, the Tennessee winter campaign began on the morning of November 20, an operation that historian James Lee McDonough characterized as "the Confederacy's last great gamble." Charlie Rowland wrote of "a cold drizzling rain as gloomy and dismal a day as I ever saw. I hope it is not ominous of the result of our expedition." The 42nd crossed into the state of Tennessee on November 21 under a blinding snowstorm from the west. They only made eight miles that day and found, when finally making camp in the twilight, "our meal mixed up for baking bread froze before we put it in the oven." At night, Rowland reported that the soldiers "made huge fires, taking the whole trunk of a tree for a log."[11]

Thirty-two regiments in Hood's army hailed from Tennessee, compared to only fifteen from Georgia. The men from the "Volunteer State" had been absent from Tennessee for more than a year, and each man gave an enthusiastic shout upon stepping back on their home soil. Like most Southern soldiers, the Tennesseans had always been fighting as much or more for their home state than for their new country. Twenty-five miles to the northeast, in the town of Pulaski, 28,000 Federals were waiting under the command of Major General John Schofield. Hood, it seems, would be fighting his old friend and West Point classmate. The Rebel plan was to use rapid movement to surround the Yankees in Pulaski and catch the thirty-three-year-old Schofield somewhere south of Columbia and the Duck River crossing.

With Lee's Corps headed straight for Pulaski, Alexander Stewart marched to his right on the crushed-stone Lawrenceburg Pike, and Benjamin Cheatham marched his corps on the left of Lee, moving toward Waynesboro.

---

[10] Organization of Clayton's Division, Lee's Corps, November 7, 1864, *OR*, ser. 1, vol. 39, part 3, 897–98.

[11] Knight, *Hood's Tennessee Campaign*, 139; Rowland to home, November 19 and December 6, 1864, in *Charles Alden Rowland*, 66, 69.

The artillery and baggage wagons followed Stewart on the macadamized road. Brisk cavalry actions marked the Confederate advance as Forrest and "Red" Jackson screened the progress of the infantry. Had Schofield lingered near Pulaski, Hood's plan might well have worked, but the XXIII Corps made a fortuitous northward retreat to a defensive line on the southern outskirts of Columbia. Behind the little Tennessee town flowed the Duck River. On November 26, under heavy rain, the Federals completed a pontoon bridge over the river in preparation for another withdrawal toward Nashville. That evening lead elements of Lee's Corps deployed in front of the Federals defending Columbia.[12]

Charlie Rowland noted the difference between the first fifty miles of the march, which he called "a dreary wilderness of trees," and their arrival in the rolling valleys of Middle Tennessee. He noted that it was "the most beautiful country I have ever seen in the South, evidence of wealth and refinement are to be seen in almost every farm." Private Nash reported that the 42nd camped on the night of the twenty-seventh, "in line of battle in sight of the enemy's works before Columbia." By November 28, Cheatham had posted his divisions on Lee's right and Stewart had pulled up behind Cheatham. The besieged Schofield sent his wagon train north toward Franklin and withdrew his army to a strong position on the opposite side of the Duck River. With news that Forrest had managed to cross the Duck River in force to his east, Schofield began, on November 29, to march his army toward Franklin, by way of Spring Hill, a crossroads village twelve miles to the north. The division of Jacob Cox was ordered to hold Columbia until midnight and then bring up the Federal rear.[13]

A confident Hood was also moving at dawn on the twenty-ninth, crossing the Duck River three miles upriver at Davis Ford and heading for Spring Hill with both Cheatham's and Stewart's corps, as well as Allegheny Johnson's division from Lee's Corps. Stephen Lee would stay behind in Columbia with the divisions of Clayton and Stevenson, along with almost all the Rebel artillery. Lee spent the day bombarding the Federal rifle pits above and across the river and bringing out pontoon boats to hopefully deceive the Yankees

---

[12] Groom, *Shrouds of Glory*, 124. Winston Groom counted thirty-one regiments from Alabama, twenty-eight from Mississippi, fifteen from Georgia, thirteen from Arkansas, seven from Texas, seven from Louisiana and four from Missouri. Florida, five, South Carolina, four, and North Carolina, two, finished out the Army of Tennessee.

[13] Rowland to home, December 6, 1864, in *Charles Alden Rowland*, 69; Nash, *Civil War Diary of Francis H. Nash*, 47.

into believing that a river crossing was imminent.

At 3:00 P.M. on the twenty-ninth, generals Hood, Cheatham, and Cleburne pulled up on a low rise two miles from Spring Hill. They could see the advance division and wagons of Schofield's army on the Columbia-Franklin Pike. Here was the golden opportunity for the Confederates to block the pike and create panic in the Federal ranks. Lee had been successful in holding Schofield's attention during the morning, and now Schofield's army was strung out in a long line of retreat between Columbia and Spring Hill. What transpired in the next twelve hours has become one of the great "might-have-beens" of the Civil War. Winston Groom called it "one of the great mysteries of the war, ranking right up there with Lee's infamous 'lost orders' during the Antietam campaign." Miscommunication among Confederate generals, and maybe too much whiskey, resulted in the failure to seize the road in the failing light. Somehow, Schofield's entire Union army, roughly 20,000 soldiers, marched as quietly as possible past the glowing Confederate campfires that burned only a few hundred yards away. Hood's physical limitations had handicapped his ability to move about the battlefield on that pivotal evening and probably cost him his best opportunity for a major Southern victory.[14]

Battle of Franklin

Stephen Lee followed Cox's retreating Yankees with his two divisions, marching twelve miles in the darkness to reach Spring Hill at 9:00 A.M. They arrived to find a furious and tormented John Bell Hood, angry that Schofield had escaped his trap. Stewart's Corps led the chase that morning followed by Chatham and then Lee. Schofield's men were not out of the woods yet. The town of Franklin sits south of the modest but deep Harpeth River. The country bridge and railroad bridge were both gone, and the Federal pontoons that had served them on the Duck River were not present. Until the bridges were rebuilt, Schofield's men would have to hold the small town from what might become a determined Rebel attack.

In the afternoon, General Hood reached the summit of a small hill where he could see that the Federals had their backs against the Harpeth River. The young commander immediately decided to make a fight of it by charging his army across the open farm fields that led into the small town. The Yankees had only been there a few hours and were hastily constructing breastworks. There was still an opportunity, thought Hood, to make up for

[14] Groom, *Shrouds of Glory*, 137.

the debacle at Spring Hill. The tired Federals watched from behind their new trenches as 20,000 Confederates, 125 regiments in six divisions, aligned themselves across a three-mile battlefront on an open plain in front of Franklin. It was an Indian-summer afternoon with a gentle breeze to lift the regimental flags. Hood had only two six-gun artillery batteries to support the attack because Stephen Lee was still miles back with the rest of the guns.

A little before 4:00 P.M. on November 30, 1864, the Army of Tennessee launched the largest infantry charge of the war. Southern soldiers fell by the thousands as they attempted to cover a mile of open ground. In only a few places did they reach and carry the Federal works in terrible hand-to-hand fighting. The struggle went on for hours, with Alleghany Johnson's Division of Lee's Corps close enough to join the struggle at nightfall. In a rare night charge, three of his brigades managed to carry and hold the enemy works. General Arthur Manigault fell wounded at the head of the South Carolina regiments. "History may speak of Pickett's charge at Gettysburg or the charge of the Federal Irish Brigade at Marye's Hill," wrote Stephen D. Lee in 1876, "but I consider the charge of Johnston's Division at Franklin the most gallant feat of arms which I witnessed during the war." Alleghany Johnson was wounded in the battle, and his division lost 587 men.[15]

Francis Nash wrote in his diary that the fight at Franklin "continued nearly all night with great slaughter on both sides." Hood lost nearly a third of his army, with his number of Confederate dead being twice that of George Pickett's Division in the famed charge at Gettysburg. Casualties in Hood's best divisions approached 50 percent. The dead bodies of four Confederate generals, including the irreplaceable Patrick Cleburne, were laid on the veranda of the Carnton Plantation House. The officer corps of the Army of Tennessee was decimated, as the army lost twelve generals, thirty-four colonels, ten majors, and nine captains. In the aftermath of the charge, Confederate divisions had become the size of brigades, and brigades were reduced to the size of regiments.[16]

The two divisions under Stephen Lee, those of Clayton and Stevenson, arrived too late to be a part of the battle. "Our division escaped the awful fight at Franklin and thank God for it," Rowland wrote. Henry Clayton believed "night mercifully interposed to save us from the terrible scourge which our brave companions had suffered." Over the course of one evening, then,

---

[15] Lee, "From Palmetto, GA, to Defeat at Nashville," 258.

[16] Nash, *Civil War Diary of Francis H. Nash*, 48; Groom, *Shrouds of Glory*, 218. The other generals taken to Carnton were John Adams, Otho Strahl, and Hiram Granbury.

these two divisions suddenly became the only intact units left in Hood's army. Near midnight, Stovall's Brigade was sent forward across the dark battlefield to relieve Rebel troops lying prone in the forward positions just short of the Federal breastworks. In the night skirmishing that followed, Stovall reported the death of one officer and the wounding of a handful of soldiers. The officer was the much-admired Ben Weaver, captain of the Independent Rebels (B). His combat death, a mortal wound to the head, was deeply felt throughout the regiment.[17]

As dawn broke, the Rebels discovered that Schofield's Bluecoats had abandoned Franklin late in the night and were now retreating northward to Nashville. The Georgians were some of the first Confederates to enter the battle-scarred town and were assigned to guard stores and Federal prisoners. Around 1:00 P.M., Hood ordered Clayton and Stevenson to push ahead along the Columbia-Franklin Pike for three miles toward Nashville. On the evening of December 2, the 42nd stopped on Overton Hill within four miles of Nashville and two miles from the Federal lines. "We can see the spires of Nashville from our line and hear the trains coming and going," Rowland wrote.[18]

Battle of Nashville

The first Southern capital to fall to Yankee arms, Nashville had been in Federal hands since 1862 and, by December 1864, was "probably the most thoroughly and skillfully fortified city on the North American continent." Down to just 23,000 soldiers, Hood believed his "only remaining chance of success in the campaign at this juncture was to take position, entrench around Nashville and await Thomas's attack which, if handsomely repulsed, might afford us an opportunity to follow up our advantage on the spot and enter the city on the heels of the enemy." Having this military scheme come to fruition was

---

[17] Report of Maj. Gen. Henry D. Clayton of operations November 20–December 27, 1864, *OR*, ser. 1, vol. 45, part 1, 697; report of Brig. Gen. Marcellus Stovall of operations November 20–December 27, 1864, *OR*, ser. 1, vol. 45, part 1, 700–701. G. B. Thornton compiled a list of casualties in Clayton's Division, Army of Tennessee, in the Battle before Franklin, TN, November 30, 1864, that shows Captain B. P. Weaver, Company B in 42nd Georgia, mortal head wound; Private J. B. Westmoreland, Company B, 52nd Georgia, slight wound right leg; Private David A. Duck, Company G, 43rd Georgia, severe wound to left hand; Private J. Grace, Company G, 41st Georgia, severe head wound. Captain Weaver's grave is unknown, but he may be buried at the Federal Cemetery at Stones River, TN. There is a P. B. W., buried in Section L, grave number 448, from "the vicinity of Franklin."

[18] Nash, *Civil War Diary of Francis H. Nash*, 48; Rowland to home, December 5, 1864, in *Charles Alden Rowland*, 69.

a long shot at best. Colonel James Rusling, an inspector in Thomas's quartermaster department, was skeptical of Hood's prospects for success, writing, "the Rebs might as well butt their brains against the Rocky Mountains as attempt to take it [Nashville]." Charlie Rowland pondered, "What Gen. Hood intends doing, the Lord above us only knows." He wondered why the weary, poorly equipped Rebel army was not situated in "comfortable winter quarters."[19]

Hood deployed his army in a convex semicircle along a four-mile front atop the hills south of Nashville, with Stephen Lee's Corps centered on the Franklin Pike, Stewart to Lee's left and Cheatham to his right. The two divisions under Lee that had not been decimated at Franklin were given the important responsibility of holding the Confederate center. Stovall's Georgians were tasked with defending a strategic rise along the Franklin Pike. A mile north down this macadamized road stood Blockhouse Casino and Fort Negley, both heavily fortified anchors to the Federal defenses of Nashville. Hood established his headquarters at Traveler's Rest, the home of James Overton, a mile to the south of Stovall's position.

Both armies did their best to stay warm over the next two weeks in the face of bitterly cold winter weather, "diversified with rain, hail, sleet, and snow." Nashville had once been surrounded by beautiful woods and groves in every direction, but within days the soldiers' axes had brought down nearly every tree within five miles. On December 8 a severe ice storm paralyzed both armies for the next six days. "The breath would freeze upon my whiskers as soon as it left my mouth," Rowland wrote on the fourteenth. "We made huge fires but the wind would blow the smoke into our eyes, and our backs would freeze while we would burn in front." The Rebels suffered severely for lack of tents, proper shoes, blankets, and winter clothing. Despite their ill-clad appearance, 2nd Lieutenant William Smith of the Independent Rebels (B) remembered the boys of the 42nd sitting around bright campfires singing "Annie Laurie," "My Homespun Dress," and "Home Sweet Home." Major L. P. Thomas took an enthusiastic "part in the sweet vocal melodies that echoed and re-echoed throughout the valleys about Nashville." Henry Clayton called it a time of "almost incessant work upon lines of entrenchments, of which four were constructed by this division." There were fresh pork and meal for

---

[19] Horn, *Tennessee's War*, 321–23; Rowland to home, December 14, 1864, in *Charles Alden Rowland*, 72.

rations, and the men found frozen rabbits and partridges to add to the cooking pot.[20]

The slow, methodical Major General George Henry Thomas was ready to assault Hood's much smaller army on the morning of December 15. He assembled 55,000 soldiers for the attack and left the rest to defend the forts around Nashville. The tactical plan was to occupy Cheatham's Corps and the Confederate right with Steedman's Corps, challenge the center with Schofield and T. J. Wood's corps, and, in a great wheeling movement, crush the outnumbered Confederate left under Stewart with the newly arrived XVI Corps under A. J. Smith. Wilson's large cavalry corps of 12,000 men, armed with seven-shot Spencer repeating rifles, would finish off the Rebel army by sweeping around Hood's far left.

The Confederate commander had made things much worse for himself by sending Forrest and his cavalry away to Murfreesboro, hoping to capture the Federal garrison that was defending the town. A pea-soup fog covered the battlefield on the morning of December 15 and managed to mask much of the early movement of Federal forces. The attack on Cheatham's Corps by eight regiments of U.S. Colored Troops, one of the largest made by African-American soldiers in the war, failed to make any progress. This gave Hood time to shift troops to his left as Thomas began to turn the great wheel.

The vulnerable Confederate left flank had five little forts designed to prevent any flanking movements. The redoubts, numbered one to five, each held a battery of smooth-bore Napoleon guns as well as 100 or so infantrymen. The most distant Redoubt 5 was overwhelmed in a morning assault, but Redoubt 4 stubbornly held out until 3:00 P.M. By then, Hood had moved Manigault's and Deas's brigades from Lee's relatively inactive center to a stone wall by the Hillsboro Pike, located behind the redoubt. Attempting to cover the entire front that Lee's Corps had defended that morning, Stovall and Holtzclaw extended their lines in one rank to the left. Lee's reinforcements were barely in place behind the wall when they fell under a furious Federal cannonade, which was followed by the "irresistible blue flood" of thousands of Yankee infantry. The brigades of Manigault and Deas broke

---

[20] Horn, *Tennessee's War*, 323; Rowland to home, December 14, 1864, in *Charles Alden Rowland*, 72; William T. Smith, "Close Calls: The Second of the Journal's Thrilling War Stories," *Atlanta Journal*, c. 1900, in the Thomas scrapbook, Thomas papers, AHC; report of Maj. Gen. Henry D. Clayton of operations November 20–December 27, 1864, *OR*, ser. 1, vol. 45, part 1, 697.

from their cover and fled for safety.[21]

The Federal rout was on as Confederate defenders on either side gave way and the remaining redoubts were quickly seized. Wilson's cavalry began to swirl in the Confederate rear as bands of desperate Rebels attempted their escape from peril. General Hood and his Southern army were saved from complete disaster by the growing darkness and the careful retreat in good order of Lee's Corps. Clayton had seen the danger and pulled Gibson's Louisiana troops out of the trenches and thrown them "perpendicularly to check the advance of the enemy." This maneuver bought Lee and Cheatham just enough time to pull most of their men away from disaster. Thomas telegraphed an impatient Washington that he had routed Hood's army and pushed back the Confederate soldiers almost two miles from their original entrenchments. "I shall attack the enemy again tomorrow if he stands to fight," Thomas added. Hood, naturally, had every intention of continuing the fight.[22]

December 16, The Second Day

Winston Groom has described December 16, the second day of the Battle of Nashville, as "arguably the darkest day in the Southern Confederacy." The once proud Army of Tennessee collapsed in the late afternoon of December 16 and was virtually destroyed as the relentless Federal pursuit continued into the night. Interestingly, it was also a day when the 42nd Georgia Infantry Regiment and Stovall's Georgia Brigade did some of their finest fighting in the war, first defending Overton Hill and then in a desperate rear-guard action that effectively saved themselves and the surviving remnant of the Confederate army.[23]

On the night of the fifteenth, John Bell Hood had struggled to reform his scattered survivors into a capable defensive formation for the next day's battle. It was midnight before his three-mile front was firmly established and the troops could begin building breastworks.. Cheatham's Corps was transferred to the left wing with Stewart forming the new center. Bates's Division under Cheatham was given the responsibility for defending the poorly drawn

---

[21] Horn, *Tennessee's War*, 331.

[22] Report of Maj. Gen. Henry D. Clayton of operations November 20–December 27, 1864, *OR*, ser. 1, vol. 45, part 1, 698; Ambrose, "The Battle of Nashville: Fighting Gets Hotter," 31.

[23] Groom, *Shrouds of Glory*, 266.

military crest of Compton Hill, a point that would hold the left flank in place. As Hood's most intact and cohesive corps, Lee's troops would hold the right flank. The vitally important Franklin Pike ran through this section of the Confederate defenses. Any Federal breakthrough here would cut off the pike road, denying any possibility of retreat for the rest of the Southern army.

Carter Stevenson's Division formed the center with Alleghany Johnson on his left, with both divisions holding the west side of the pike road. Henry Clayton placed his men to the right of the pike on Overton Hill, referred to by locals as Peach Orchard Hill. Defending a hill that was an impressive 300 feet high, Clayton's small division would anchor the far right of the Rebel line. Holtzclaw's and Gibson's brigades faced forward (north), while the lines of Stovall's Brigade were refused southward for several hundred yards to protect the vulnerable Confederate right flank. Eldridge's Battalion of eight cannon, divided in three batteries, reinforced Clayton's position. Because of their late deployment, the men in the 42nd Georgia were not given an opportunity to sleep at any point that night as strong breastworks had to be built.[24]

George Thomas chose to make the first Federal assault the next morning on Clayton's Division defending Overton Hill. At best, thought Thomas, the charge would succeed and give him control of the Franklin Pike, and, at worst, the assault would draw Confederate reinforcements to Hood's right flank and thereby weaken the vulnerable Confederate left flank. The details of the morning attack were left to Thomas John Wood and his IV Corps. Brigadier General Wood understood that his men would be undertaking a difficult infantry charge. There were several hundred yards of open and ascending ploughed fields of mud and corn stubble. After slogging through this mess, the men would then have to cross a wooden fence and begin the steep climb up Overton Hill. Under concentrated fire, his men would then navigate the obstacles constructed by the Southern soldiers and finally fight their way into the enemy's breastworks. The sight of numerous Rebel flags flying from these breastworks indicated that the position was heavily defended. Wood, though, had seen his veterans scale Missionary Ridge a year earlier, and he believed that they would accomplish the impossible a second time.[25]

At about 9:00 A.M. massed Federal artillery, consisting of three 6-gun rifled batteries, opened fire on the men defending Overton Hill. The shelling continued for hours. Stephen Lee called it, "one of the most terrible cannonading of the war." Stanford's Mississippi Battery under Lieutenant James

[24] Ambrose, "The Battle of Nashville: Fighting Gets Hotter," 33.

[25] McDonough, *Nashville*, 224.

McCall, which was positioned between Stovall and Holtzclaw, was all but destroyed by relentless Yankee fire. Stovall, whose men suffered the worst punishment, called the shelling "as severe as any to which I have ever been exposed" and noted that a Federal battery "enfiladed my entire front." Holtzclaw considered it "the most furious I ever witnessed, while the range was so precise that scarce a shell failed to explode in the line." As J. P. Cannon, in Holtzclaw's 27th Alabama, wrote in his diary, "Shells and solid shot passed through the stone fence behind which we were sheltered, scattering stones in every direction and mangling men in a shocking manner."[26]

Leading the Federal assault would be the brigade of Philip Sidney Post. His troops were primarily Ohio regiments, with the 41st Ohio deployed as skirmishers. Behind Post would come the brigade of Abel Streight in three battle lines. Wood's formations would be attacking north to south directly at Holtzclaw's Alabama regiments and Gibson's Louisiana soldiers. Coming from the east and assaulting Stovall's Georgia regiments would be two full brigades from James Steedman's Division. Steedman's right flank, adjacent to Post's Brigade, would be the black brigade of Colonel Charles Thompson. The first battle line had the 100th U.S. Colored Troops Regiment on the right and the 12th USCT on the left. The 13th USCT made up the second line. Charles Grosvenor's Brigade constituted Steedman's left flank. They included the 18th USCT and several white regiments. The Federals had no choice but to form up in the open and begin their assault on a cold and rainy afternoon. The blue lines stepped off at about 3:00 P.M. Wood called it "a most splendid array in which the advance was made gave hopeful promise of success." Unfortunately for the Federals, almost from the beginning they were met "with a most terrific fire of grape and canister and musketry before which no troops could live."[27]

Skirmishers from Post's Brigade got to within 100 yards of the Rebel breastworks but were blocked by "a line of abatis of logs in which holes had been bored and long pins driven and sharpened." According to Captain J. T. Patton of the 93rd Ohio, "The logs were lashed together at the ends by grapevine witches and formed an unbroken barrier to our further advance." When the first line of blue infantry drew close to the abatis they were shot to pieces.

---

[26] Lee, "From Palmetto, GA, to Defeat at Nashville," 259; report of Brig. Gen. Marcellus Stovall of operations November 20–December 27, 1864, *OR*, ser. 1, vol. 45, part 1, 707; Horn, *Tennessee's War*, 338; report of Brig. Gen. James T. Holtzclaw of operations November 20–December 27, 1864, *OR*, ser. 1, vol. 45, part 1, 705; McDonough, *Nashville*, 222.

[27] McDonough, *Nashville*, 225; Horn, *Tennessee's War*, 339 (report of General Wood).

"Men could not live under such a fire," remembered Patton, and "the line began to waver, and then fall back." The second line was greeted with the same Confederate volleys, to be followed by Streight's three lines each meeting the same result. The aggressive Post was struck down from his horse by grape-shot and would later be awarded the Medal of Honor for his bravery in leading this unsuccessful assault upon Overton Hill.[28]

Henry Clayton thought that the attackers got no closer than thirty yards of his fortifications, but some Federals wrote in their field reports that a handful of their men had reached the Gray line only to be captured or killed. Clayton believed the Yankees, "suffered great slaughter," and he estimated their loss in killed and wounded at more than 1,500 soldiers. The Alabama lawyer and politician further noted in his battle report that five regimental colors had been "shot down in a few steps from the works." It was with great difficulty, a pleased Clayton reported, that his officers could contain "the enthusiasm" of their Southern troops and keep them from "going over the works in pursuit of the enemy." James Holtzclaw, holding Clayton's left, noted that a Negro brigade had gallantly charged up the hill, only to be killed by the hundreds. "Pressed on by their white brethren in the rear they continued to come up in masses to the abatis," Holtzclaw wrote, "only to die. I have seen most of the battle-fields of the West, but never saw dead men thicker than in front of my two right regiments."[29]

On Clayton's right, Steadman's Division assaulted Stovall's entrenched Georgia troops. The black soldiers charged with the most determination but also garnered the most casualties. The 13th USCT in Thompson's Brigade, in their first battle of the war, lost 40 percent of their regiment, some 220 officers and men including five color-bearers. "They put two or three lines of negro troops in front and one line of whites behind them," remembered Lieutenant William Smith of the Independent Rebels (B), "and with these several columns they assaulted us time and again, only to be driven back in utter confusion." A flag from the 18th USCT fell close enough to be brought into Stovall's Rebel lines. Future literary giant Ambrose Bierce, then a young Federal officer, witnessed the attack by black troops and wrote, "The front of the

---

[28] Patton, "Personal Recollections of Four Years in Dixie," 1:437.

[29] Report of Maj. Gen. Henry D. Clayton of operations November 20–December 27, 1864, *OR*, ser. 1, vol. 45, part 1, 698; Horn, *Tennessee's War*, 338; report of Brig. Gen. James T. Holtzclaw of operations November 20–December 27, 1864, *OR*, ser. 1, vol. 45, part 1, 704–707.

enemy's earthworks was protected by an intricate abatis of felled trees denuded of their foliage and twigs. Through this obstacle a cat would have made slow progress; its passage by troops under fire was hopeless from the first—even the inexperienced black chaps must have known that." Captain Henry Freeman of the 12th Colored saw his men stuck in the abatis "like flies in a spider web." The twenty-two-year-old Bierce watched as "their long lines swept into that fatal obstruction in perfect order and remained there as long as those of the white veterans on their right." Smith reflected, "The dark negroes, with their dark blue coats, lying dead in heaps on the beautiful snow made a strange contrast and a scene never to be forgotten."[30]

Though the Federal artillery had demolished his breastworks "in several places," Marcellus Stovall felt his Georgians had "stood true to their colors" and successfully repulsed the Federal advance. Wiley Baggett of the Calhoun Guards (K) wrote that it was in front of his company that "the enemy tore it [the breastworks] down with their artillery." He was among the dozen men from the 42nd Georgia who were wounded in the fight, and William Baker of the Fulton Blues (I) lay dead. Lieutenant Hugh Wylie, the popular longtime adjutant of the 42nd, was also killed while riding along the lines. Stovall had promoted the thirty-two-year-old Wylie to his staff as A.A.A.G. (appointed assistant adjutant general) on July 24, 1864. Rowland and Wylie were "bedfellows" on the Nashville campaign, and Rowland wrote home that the brave Wylie had been shot through the heart. "He dismounted from his horse but died almost immediately," reported Rowland. "I cut off some of his hair for his mother," Rowland went on, "and his body was buried in Columbia." General Stovall noted the loss of Wylie in his official report praising the staff officer as "conspicuous for gallantry as a soldier, efficiency as an officer, urbanity as a gentleman." For the men of the 42nd, this stand on Overton Hill was a most welcome victory as it had been months since they had been in a direct fight with Federal troops. Unfortunately, they were not able to savor their victory for long.[31]

---

[30] D. Smith, *In the Lion's Mouth*, 73; Freeman, "A Colored Brigade in the Campaign and Battle of Nashville," 2:418; Knight, *Hood's Tennessee Campaign*, 117; Smith, "Close Calls."

[31] Goodson, *Letters and Diaries*, 89 (account of Wiley Baggett); Rowland to home, December 22, 1864, in *Charles Alden Rowland*, 78. Lieutenant H. M. Wiley is buried in the Confederate section of Rose Hill Cemetery in Columbia, TN; report of Brig. Gen. Marcellus Stovall of operations November 20–December 27, 1864, *OR*, ser. 1, vol. 45, part 1, 701–702. Listed as wounded in the December 16 fight were Matthew Edgar (G), Jacob L. Warren (G), T. S. Edmonds (H), Thomas J. Holbrook (I), Nathan Lee (I), Thomas K. Watts (I), Wiley J.

Confederate Collapse

"After the repulse," wrote General Thomas Wood in his operations report, "our soldiers, white and colored, lay indiscriminately near the enemy's works at the outer edge of the abatis." While the career army officer acknowledged defeat at the hands of Clayton's soldiers, he took credit for drawing two brigades of Confederate reinforcements to Overton Hill and thereby weakening the left flank of Hood's overmatched army. At 4:00 P.M. the rumble of musketry and cannon signaled a major fight for control of Shy's Hill, two miles away on the Confederate left flank. Masses of Union infantry and dismounted cavalry simply overwhelmed Cheatham's thin line and the onrush of blue infantry continued down the other side of Shy's Hill. A chain-reaction of fear swept through the ranks of Stewart's Corps, which had been holding the Confederate center, and then down the line to the divisions of Johnson and Stevenson. "It was one scene of confusion and rout," remembered Sam Watkins. To make matters worse for the outnumbered Confederates, the best route of retreat for Cheatham and Stewart's men, the Granny White Pike, which was directly behind them, was now cut off by a brigade of Federal cavalry under General John Hammond. Only the Franklin Pike, which was directly behind Stephen Lee and the Rebel right flank, still lay open to the south.[32]

Henry Clayton and Marcellus Stovall were standing together on Overton Hill when an aide de camp from Stephen Lee rode up with an order to prepare a retreat. It was already too late for the Army of Tennessee as Clayton, who had an elevated view of much of the battlefield, could see "the troops on my left flying in disorder." Abandoning some guns, Clayton was able to get the Eufaula Light Artillery away. It was his belief that the entire Confederate army was now in complete undisciplined confusion, with the sole exception of Pettus's Brigade, his division, and the 39th Georgia, which had been sent to him and held in reserve. General Stovall was proud that "not a man of my brigade, so far as I could see, left the works without orders, and that I was able to march them out in good order and save the battery which I supported." It had become warm enough for the icy ground to turn to mud, and now, in a driving rainstorm, thousands of Confederates were attempting to

---

Baggett (K), James T. Calhoun (K), Edward T. Chupp (K), Alexander Ratterree (K), Moses G. Terry (K), and Green J. Watkins (K).

[32] Horn, *Tennessee's War*, 339 (report of General T. J. Woods); Watkins, *"Co. Aytch": A Side Show of the Big Show*, 224.

trudge through the grime on an uphill route over the hills that had formerly been behind them.[33]

Efforts by Hood and other officers to rally the fleeing men as they made for the Franklin Pike were futile. It was a rout, a mass of men running for their lives with Federal soldiers capturing any who hesitated. The collapse had been so sudden that half of the army's artillery—fifty-four guns—had been abandoned in the trenches. Stephen Lee managed to take a position on the Franklin Pike at the crest of the first major hill behind the former Rebel line. The thirty-one-year-old general implored the retreating men to rally around him and buy time for the army to escape and reform. Only a small handful accepted his invitation. Stovall was, in fact, losing men in his brigade at the very beginning of their desperate retreat. It was, "A race for half a mile for life and liberty," wrote J. P. Cannon, "through open fields to a range of hills." The 42nd Georgia lost at least twenty-six men in this desperate one-mile retreat. "I was wounded in the leg," wrote Wiley Baggett, "and had just gotten into the ambulance when the line gave way." The Atlanta native barely stayed ahead of his Federal pursuers.[34]

Most of the 42nd fell back to the right, but some, following Captain Joshua Mitchell of the Walton Tigers (H), attempted a shortcut over a hill to where the ammunition wagons were parked. As this group reached the brow of the elevation, they were confronted by a large group of colored Union troops and were immediately captured. Lieutenant James Monroe Gresham later claimed his captors were drunk and that one ordered him to remove the pair of Yankee pants that he was wearing. Given that the ground was frozen mud and the winter wind was constant, Gresham was reluctant to comply. "I

[33] Report of Brig. Gen. Marcellus Stovall of operations November 20–December 27, 1864, *OR*, ser. 1, vol. 45, part 1, 701.

[34] McDonough, *Nashville*, 260; report of Maj. Gen. Henry D. Clayton of operations November 20–December 27, 1864, *OR*, ser. 1, vol. 45, part 1, 699; Goodson, *Letters and Diaries*, 89 (account of Wiley Baggett). Listed as killed on the December 16 was Lt. Hugh M. Wylie (field staff), Cpl. William S. Baker (I), Pvt. Joseph Melton (I), Pvt. James David Garrison (K). Sgt. James Randolph St. John (F) was mortally wounded & died in Confederate hospital in Corinth, January 1, 1865. Pvt. William J. Turner (K) was mortally wounded and died in Federal Hospital, January 2, 1865. Listed as captured on December 16 were James Freeman (B), Eli J. Matthews (B), Robert A. Mills (B), Samuel Paden (B), Thomas Thomason (B), John L. Lawson (C), George W. McFarland (C), Elias M. Norton (C), William T. Roddy (C), John E. Fincher (C), James B. Austin (G), Eldridge Burton (G), James C. Davis (G), Calvin Patrick (G), Francis J. Thurmond (G), James Monroe Gresham (H), Jonathan W. Huff (H), Young A. Huston (H), John W. Johnson (H), James D. Malcom (H), Joshua M. Mitchell (H), Martin T. Sims (H), Joseph Blalock (I), William N. Sewell (I), and James F. Palmer (K).

began to fumble to unbutton," recalled the feisty lieutenant, "taking care not to raise my roundabout coat for fear they would discover my pistol in the belt underneath." Gresham was saved from a direct confrontation when a white Federal soldier escorting a squad of prisoners told him and his comrades to fall in line. Trampling over rows of fallen Federal dead, the Georgians were escorted to the rear.[35]

William Smith maintained that many of the men in Stovall's Brigade were captured before they ever got out of the trenches. "I had bought a ham of meat during the night before and laid it on a stump in rear of our line of battle, not having eaten anything for two days," he recounted. "As I started to run out with my regiment my eyes turned toward my ham of meat and what a blessed, loving, fond farewell look I gave that ham, while the Yankees shouted, 'Surrender, damn you!" Smith made a dash through a piece of wooded land and into a cornfield. With minie balls plowing up the snow around him, the young lieutenant began bounding "four or five corn rows at each jump." He eventually eluded the initial danger and found General Stephen Lee on the pike with twenty or more stands of colors and a few hundred men around him. "I reported to him with one man," recalled Smith, "I told him we were ready to die if necessary."[36]

Bitter Retreat

Having been awake for two straight days without sleep, the 42nd Georgia faced a Moscow-style Napoleonic retreat that was only just beginning. Henry Clayton stopped his harassed division after falling back a half-mile. Here the Eufaula Light Artillery under Lieutenant W. J. McKenzie unlimbered and along with the 39th Georgia of Cummings's Brigade deployed in support of the two guns. Darkness was beginning to set in, and the rain had tapered off. A silver winter moon was clearly visible. A young drummer boy beat the long roll in perfect time as the men formed. According to George Evans Brewer, a captain and Baptist minister in the 46th Alabama, there were sixteen stands of colors among the two or three hundred men drawn up to face the Yankee pursuers. Henry Clayton watched in admiration as the officers and men of the Eufaula Battery managed their guns with "coolness and deliberation."

---

[35] Sams, *Wayfarers in Walton*, 192–93. James M. Gresham, "Close Calls: Some Experiences at Nashville After Being Captured," *Atlanta Journal,* c. 1901, clipping in the Thomas scrapbook, AHC.

[36] Smith, "Close Calls."

Clayton would later claim in an unpublished account that he sent to General Hood, "with becoming modesty, that I was the last man to leave the Battle Field at Nashville."[37]

Clayton's small line, like that of Lee's first position, was soon flanked and in retreat. A half-mile farther back in the Brentwood Hills, a third line, this one made up from Stovall's reassembled brigade, was placed across the Franklin Pike. Upon seeing the Georgians, the pursuing Federals halted, took time to reform their lines, and pushed a skirmish line to their front. Clayton could not find Marcellus Stovall in this din of confusion, so he appointed Colonel Abda Johnson to take charge of the brigade. His instructions to Johnson were for the Georgians to contest the Yankee skirmishers, but to "break to the rear" when the main column of Yankees came into range.

Clayton rode up the Franklin Pike another half a mile, where he found Holtzclaw's Brigade already in a line across the road. "Soon," wrote Clayton, "Stovall's brigade, broken and flying, was by my order permitted to hop through the ranks of Holtzclaw's Brigade, which had meanwhile made preparations to receive the enemy." Disorganized groupings of Federals from T. J. Wood's IV Corps were leading the chase. They were as tired, intermingled, and confused as were the Southerners they were pursuing. Upon seeing Holtzclaw's men formed up, they again stopped to reform. Henry Clayton did not have the forces to stop the Federal pursuit, but he was effectively buying time for a Southern retreat.[38]

Four miles from the abandoned breastworks, in the village of Brentwood, Confederate officers unsuccessfully tried to reorganize their commands in the darkness. Sam Watkins recalled that it was "like trying to stop the current of Duck river with a fish net." For many of the confused Rebel survivors that milled about Brentwood that night, it seemed as if they were haunted by the specter of Missionary Ridge. Unable to reform by unit, orders were given to continue the retreat toward Franklin. Clayton's Division would continue to serve as the rear-guard of this disorganized mass of men.[39]

The division commander fell back slowly using troops from Holtzclaw's Alabama Brigade as the rear-guard of the rear-guard. Downpours of rain intensified, and the retreat was called a halt around 2:00 A.M. some seven miles

---

[37] Brewer, "Incidents of the Retreat from Nashville," 327; report of Maj. Gen. Henry D. Clayton of operations November 20–December 27, 1864, *OR*, ser. 1, vol. 45, part 1, 699; Henry Clayton, unpublished account of his military service, 22.

[38] Henry Clayton, unpublished account of his military service, 25.

[39] Watkins, *"Co. Aytch": A Side Show of the Big Show*, 224.

north of Franklin. Clayton's drowsy men collapsed in the rain puddles and sought rest. There would be no more fighting that night as the pursuit, led by Wood's Corps, had stopped and bivouacked about a mile from Brentwood.[40]

Union commander George Thomas had clearly won a resounding Federal victory in the two-day Battle of Nashville. His troops took 4,462 prisoners, including 287 Confederate officers. Division commander Allegheny Johnson was counted among those captured that day. When General Thomas rode across Overton Hill, he saw the long lines of dead black troops. Turning to his staff, he announced, "Gentlemen the question is settled; Negroes will fight." The Battle of Nashville finished the Southern Army of Tennessee. "Genl Hood will no doubt attribute the disaster at Nashville to the inconsistency of his troops and to some extent this is true," thought Rowland a few days later, "but he broke their spirit at Franklin where a fight was as unnecessary as it was insane at Nashville. If we had repulsed the enemy at Nashville we would have gained nothing."[41]

Hollow Tree Gap

The morning of December 17 broke dark and cloudy. The Federal commanders had their men up early as they saw a clear opportunity to destroy the rest of Hood's army that winter day. The Federal infantry was still several miles behind the Confederate rear guard, but Wilson's large and aggressive Federal cavalry was close enough to be nipping at the heels of the Rebels in a matter of hours. Clayton had the Confederate rear-guard up early as well and, in drenched and muddy uniforms, they resumed their retreat toward Franklin. There would be no rations to distribute. In two miles, they stopped at a small pass in a range of hills and formed another line of battle.

It was here at Hollow Tree Gap that the men under Stephen Lee would make the first of many stands and blocking actions against the Federal cavalry. Riding past them in the dawn light were Rebel cavalry going in the opposite direction to tangle with Wilson's troopers. According to George Brewer, the Alabama preacher, the Rebel horsemen, "twitted us about running away, stating that they were going back to show us how to whip Yankees,

---

[40] D. Smith, *In the Lion's Mouth*, 96.

[41] Ibid., 101; Morgan, "Reminiscences of Service with Colored Troops in the Army of the Cumberland," 5:114; Rowland to home, December 22, 1864, in *Charles Alden Rowland*, 77.

so we need not be afraid anymore." Lee posted Stovall's Brigade to the right of the road with Edmund Pettus and his Alabama Brigade on the left. Holtzclaw and Gibson drew up their brigades 600 yards behind them in support. A lost artillery battery from Cheatham's Corps, Bledsoe's Missouri Battery, deployed in support of Stovall's position.[42]

According to John Johnston of the 14th Tennessee Cavalry, a worn and dejected General Hood was also there sitting on his horse, "looking up the road as if expecting the appearance of the enemy in that direction." Historian Derek Smith writes that General Stovall was still separated from his men, and command of the brigade that morning had been given to Robert Henderson of the 42nd Georgia. If this was the case, then William Hulsey or L. P. Thomas directed the 42nd Georgia at Hollow Tree Gap. Stovall, in his operations report, makes no mention of not being at Hollow Tree Gap, but Clayton's field report has Henderson in command of the brigade later in the day. C. L. Bragg, the biographer of Marcellus Stovall, called the Georgian's actions during the Battle of Nashville and subsequent retreat "his finest hours as a Confederate general," but it may be that much of this credit belongs to Robert Henderson.[43]

At about 8:00 A.M. troopers from James Wilson's 7th Division gave chase to the Confederate horsemen they encountered in the dawn light. Though no cavalry was immediately visible, the Rebel infantry back at Hollow Tree Gap could hear the clatter of hundreds of galloping horses thundering toward them. In the vanguard was Joseph Knipe's 1st Brigade of Federal cavalry. Having passed over roads strewn with abandoned Rebel equipment and "every door yard filled with ill clad shivering prisoners," Captain Obediah Hayden of the 9th Indiana Cavalry believed his troopers would "walk over" what remained of Confederate resistance. With sabers swirling and carbines firing, the 19th Pennsylvania Cavalry and the 10th Indiana Cavalry charged after the retreating Rebel cavalry.

The problem for the waiting rear guard of Rebel infantry was how to distinguish friend from foe. At first Captain Brewer of the 46th Alabama saw that "it was our cavalry," the same ones "who was going to show us how to fight Yankees," who were barreling back toward them at full gallop. "Immediately upon their heels and even mixed with the rear were the pursuing cavalry of the enemy," remembered Brewer. Since most of the Rebel horsemen

---

[42] Brewer, "Incidents of the Retreat from Nashville," 327.

[43] D. Smith, *In the Lion's Mouth*, 116; Johnston, "Cavalry of Hood's Left at Nashville," 35; C. L. Bragg, *Distinction in Every Service*, 114.

were wearing captured blue overcoats, it was not clear to the foot soldiers which side each horseman had given his allegiance.[44]

As the Federal troopers passed into the Rebel lines, they were beaten back by musket fire and artillery blasts. Ordered to surrender, many of the Yankee troopers dismounted, but some soon thought better of it. They chose to remount and fall back through the gap at a gallop, despite the Rebel bullets whistling about them. When the smoke cleared, Stephen Lee's men saw a roadbed of dead and wounded horsemen "and several guidons scattered in the sullen woods and fields." Twenty-two troopers from these two cavalry regiments were killed or wounded, and another sixty-three were captured. Hayden's earlier confidence of easily bagging the retreating Confederate army was gone, as "Hollow Tree Gap undeceived us."[45]

As Marcellus Stovall reported, "My command captured between 75 and 100 prisoners, one stand of colors, with many horses, saddles, bridles, and such other things as constituted the outfit of a Federal cavalryman." It was not long, however, before more Federal cavalry arrived and began to lap around both of Henry Clayton's flanks. The 10th Indiana Cavalry charged from the east smashing through the 4th and 30th Louisiana regiments taking about seventy-five of the Louisiana troops prisoners. At 10:00 A.M. Stephen Lee ordered a vigorous retreat from the gap, down the Franklin Pike to the Harpeth River. Henry Clayton recalled, "as we emerged from the hills upon the plain, I think it no exaggeration to say there were no less than ten or fifteen columns of the enemy cavalry, at least four times our number."[46]

Upon reaching the Harpeth River, with Federal horsemen at their heels, Clayton posted Gibson's Louisiana Brigade, now down to 250 men, on a small rise near the abandoned Fort Granger, about 1,000 yards from the raging swollen river. They were joined on this elevated position by Bledsoe's Missouri Battery, and a contingent of Abraham Buford's cavalry. The thirty-two-year-old Gibson would have to hold his position until Stephen Lee could get the rest of his rear guard across the angry Harpeth River. William Smith

---

[44] Brewer, "Incidents of the Retreat from Nashville," 327.

[45] D. Smith, *In the Lion's Mouth*, 111; Hayden quote on the Civil War Trails marker at Hollow Tree Gap.

[46] Report of Brig. Gen. Marcellus Stovall of operations November 20–December 27, 1864, *OR*, ser. 1, vol. 45, part 1, 701. A historical marker for the December 17 Battle of Hollow Tree Gap was erected and dedicated by the Franklin Civil War Roundtable in December 2014. The interpretive sign includes a picture of Gen. Marcellus Stovall and is located at the southwest corner of Moore's Lane and Franklin Road (Henry Clayton, unpublished account of his military service, 26).

of the Independent Rebels (B), with two other officers, was instructed to take 150 men and hold the bridge "at all hazards." About seventy-five men from the 18th Alabama in Holtzclaw's Brigade appeared, having earlier been cut off from their command in the retreat from the gap. They were ordered to stand with and reinforce Gibson's men on the small rise.

Around noon, three Federal cavalry columns, more than 4,000 horsemen bore down on Gibson's overmatched outpost. It was one of the largest cavalry charges in American history and most of Stovall's men, who were now on the other side of the river, were helpless spectators as the Federal troopers were out of their musket range. Yale graduate Randall Gibson reported, "The enemy, five thousand strong, charged in three columns with squadrons covering the intervening ground and connecting them—one in front, one in rear upon the left flank, and one in rear upon the right flank." It was clear to all participants that the Confederate's posted above the river were in dire danger of being surrounded. Bledsoe's artillery fired ten rounds into the charging cavalry, but, according to Gibson, "it had no effect, except to increase the speed of his flanking columns." The general ordered Bledsoe's battery away and then had his brave men make a collective fighting sprint for the bridge into Franklin. The men from the Pelican State lost ten killed, twenty-five wounded, and five captured in their rush to safety. Some of Gibson's men were forced to swim across the swollen river to reach Franklin. Holtzclaw's detached group of Alabamians, still on the north bank of the Harpeth, helped to cover the retreat, but many were soon surrounded and forced to lay down their arms. Clayton laid the blame for the debacle on Rebel horsemen, "who, all day long, behaved in a most cowardly manner."[47]

William Smith had just deployed his men from the 42nd Georgia on the bridge into Franklin, but "before I could tell the men that we were expected to die there or hold the bridge, eight or ten thousand mounted infantry and cavalry came down at full speed along the pike and through the field." "The boys killed them by the score," Smith remembered, "many riderless horses came dashing across the bridge, but we held on like grim death." The longer Smith's contingent managed to hold the bridge, the longer Stephen Lee had to reorganize and push his rear guard out of Franklin and toward Spring Hill. Unfortunately for the Georgians defending the bridge, Federal

[47] Wade, "Historian recounts Hood's retreat on 149th anniversary"; report of Maj. Gen. Henry D. Clayton of operations November 20–December 27, 1864, *OR*, ser. 1, vol. 45, part 1, 699; Gibson, "Report of Brigadier-General R. L. Gibson of Operations in Vicinity of Nashville," 142.

cavalry under Richard Johnson had forded the Harpeth on the Hillsboro Pike to the west and were now charging into the streets of Franklin. "One of my men shouted to me, 'Great God, look in our rear,'" recalled Smith. The town of Franklin was suddenly full of Federal cavalry, and there were thousands more in front of Smith's band of defenders. As Lee's engineers worked frantically to destroy the bridge before it could fall into Federal hands, Smith shouted to his men "as best I could" that they should "save themselves if they could, or surrender if they desired to do so. I was going out or try."[48]

So began what Smith would declare in 1900 to be his "closest call" of the war. Running through gardens, over fences, and through houses, he worked his way through the war-ravaged town. "I ran on jerking off pickets," he wrote, "and dodging through enclosures with hundreds of Yankee bullets flying past my head." Smith crashed through a fence and into an alley. A horseman in a waterproof gum coat was immediately upon him. Was the rider Yank or Confed? "I was out of breath, tired until I could go no further," recounted Smith. The Rebel officer on the horse reached out his hand, but Smith did not have the energy to spring up on the mount behind the rider. As bullets began to pepper the pair, Smith "threw down all I had but my sword and pistol and company papers. Quick as lightning I was behind him on his horse." William Smith had been saved by one of Forrest's officers who had seen the Confederate fleeing through the town and determined that he would "save me or die in the attempt." Not everyone was as fortunate as Smith. Both Thomas Phillips (A) and Levi Victory (E) were both captured in the town of Franklin.[49]

By noon, Stephen Lee's rear guard had managed to both escape Franklin and destroy the railroad and footbridges across the Harpeth. In the coming days, supply trains for George Thomas's pursuing army would be held up waiting for the bridges to be repaired. In the early afternoon, the aggressive James Wilson reorganized his 10,000 cavalry troopers in the streets of Franklin and then sent them barreling down three separate roads hoping to envelop and cut-off Lee's rearguard before nightfall. "On either side of us great columns of cavalry were moving through the fields in parallel lines," wrote Daniel Comstock of the 9th Indiana Cavalry. "The entire cavalry corps was in sight." The winter afternoon would be filled with "a moving series of sharp but sporadic firefights" along the Columbia Pike as Wilson's horsemen continually harassed and fired upon what remained of Lee's retreating infantry

[48] Smith, "Close Calls."

[49] Ibid.

corps. Stephen Lee believed "a more persistent effort was never made to rout the rear guard of a retiring column."[50]

Three miles south of Franklin, where the pike road crossed Winstead Hill, General Lee had Carter Stevenson's Division replace Clayton's tired men as the rear guard. Stevenson's men deployed on the hill and beat back a cavalry charge but also watched as Stephen Lee was wounded in the foot by a bursting Federal shell. The proud South Carolinian would not relinquish his command of the corps to Carter Stevenson until darkness had fallen. Lee continued to give commands from an ambulance as the retreat continued. Clayton was marching slightly ahead of Stevenson: "I say nominally, for I was at no time out of sight of numbers of enemy cavalry and we were repeatedly charged upon one and the other side."[51]

Three miles further down the Columbia Pike, at about 4:00 P.M. in a cold drizzle and failing light, Stevenson was again forced to deploy his 700 soldiers in the open fields about a mile north of the West Harpeth River. James Chalmer's cavalry drew up to guard his flanks. As Wilson's columns bore down on the Confederates with Edward Hatch's Division on the left and John Hammond's brigade on the right, the Rebel horsemen immediately galloped off to safety, leaving, according to Stevenson, "my command to their fate." As many as 5,000 Yankee troopers were soon firing into Stevenson's left flank and rear. With his back to the West Harpeth River, Stevenson resorted to forming his men in a "hollow square" formation. This Waterloo-type configuration was rarely seen in the Civil War and only used when infantry found themselves surrounded by cavalry. "It was the most trying experience I had during the war," remembered George Brewer of the 46th Alabama, "the enemy were so numerous as to envelop us, and withal they were brave fighters." Staying in this configuration, Stevenson steadily pulled his depleted division back across the small river and toward Spring Hill.[52]

Upon learning that Stevenson was in serious trouble, Henry Clayton stopped his small division and deployed his men in a large beech grove along the pike. He placed "Stovall's brigade, Col. R. J. Henderson commanding on the right, Gibson in the center, and Holtzclaw's, Colonel Bushrod Jones commanding, on the left." As the sound of Stevenson's battle increased, Clayton ordered Holtzclaw's Brigade forward to assist Stevenson with instructions to

---

[50] Comstock, *Ninth Cavalry*, 38; D. Smith, *In the Lion's Mouth*, 119; Lee, "From Palmetto, GA, to Defeat at Nashville," 259.

[51] Henry Clayton, unpublished account of his military service, 28.

[52] Brewer, "Incidents of the Retreat from Nashville," 328.

keep their right "resting on the pike." Randall Gibson was told to send skirmishers forward, and then instructed to take command of the two deployed brigades, as Clayton and his staff rode forward to catch up with Bushrod Jones and his Alabama regiments. The arrival of Holtzclaw's reinforcements came none too soon for Carter Stevenson, as a large force of Federal cavalry was forming to charge his exposed left flank. Bushrod Jones deployed his men on the Confederate left and soon beat back the Federal advance.

After conferring with General Pettis about the disposition of his troops, Clayton rode back toward his other two brigades. Darkness had fallen, but Clayton and his staff were still able to recognize a sizable column of Yankee cavalry "moving obliquely and just entering the road a few paces in my front." These Federals had gone around the range of hills bordering the Columbia Pike, thereby putting themselves in position to hit Stevenson's beleaguered contingent from behind and force a final surrender. These troopers were not aware that there were more Confederates further up the Pike. They were, in fact, headed for a direct collision with the lines of Gibson and Stovall.[53]

Bucking their horses, Clayton's small entourage turned left and raced in front of Gibson's men barely avoiding being misidentified and shot down by their own troops. Upon learning of the Federal presence from Clayton, Gibson "promptly wheeled his brigade to the left and delivered a volley which shattered the enemy, killing many of them." Months of ceaseless drill had given the officers and men in this division the confidence to perform complicated maneuvers in the dark and under fire. "I changed front upon the left regiment," General Gibson recorded, "and ordered Colonel Henderson, temporarily in charge of Stovall's brigade, upon my right, to face by the rear rank and wheel to the right, so as to cover the road." The Georgians had swung themselves into line parallel to the pike, with their right on Gibson's refused line. The new front fired into two cavalry regiments under General Richard Johnson.[54]

The brigades of Gibson and Stovall then fell back behind a stone fence for protection. "This position was scarcely taken," reported Clayton, "when the enemy again began to move from the left upon the pike to our immediate front." The Rebels let the 10th Indiana Cavalry pass them and then sprung their ambush. According to Indiana trooper Daniel Comstock, a voice in the

---

[53] Report of Maj. Gen. Henry D. Clayton of operations November 20–December 27, 1864, *OR*, ser. 1, vol. 45, part 1, 700.

[54] Ibid.; report of Brig. Gen. R. L. Gibson of operations December 15–17, 1864, *OR*, ser. 1, vol. 45, part 1, 702–703; Salling, *Louisianans in the Western Confederacy*, 218.

darkness declared, "Halt! Who comes there?" Colonel George Washington Jackson answered, "Federal Cavalry." This response, according to Comstock, provoked a volley that lit up the night, "and down went man and horse." A second volley sent many of the stunned horses into a ditch and crippled many of the riders. Color-Sergeant Benjamin Franklin Ricks of the 9th Indiana was among the dead Hoosiers, and his colors were lost to the Rebels. "[T]he surprise, the darkness and disorder," Comstock wrote, "the impossibility of returning the fire, left but one thing to do." The surviving troopers spurred their horses and vanished into the night leaving saddles, arms, blankets, and other spoils of war to the defenders. The Georgians let out a yell and jumped the stone wall in a brief but spirited pursuit. The Hoosiers reformed in the forest, with Comstock making a summation of his experience: "The chagrin of failure, the separation from comrades who might be dead or wounded, and in the hands of the enemy, added tenfold to the horror of the night, itself wet, cold, comfortless." There would be no more fighting on December 17. Describing Stephen Lee's rear-guard on that long winter day in 1864, historian Stanley Horn had nothing but praise "for a prodigious performance under the greatest difficulties."[55]

Napoleonic Retreat

Francis Nash recorded in his journal that, on Sunday, December 18, the 42nd Georgia marched south and camped in the old elevated Yankee breastworks north of the swollen Duck River that looked down on the town of Columbia. "On the morning of the 18th," Henry Clayton reported, "the army began to assume shape, and a feeling of hopeful security was felt." The Federal IV Corps had been held up at the Harpeth River for eighteen hours and did not cross into Franklin until the morning of the eighteenth. They camped three miles south of Spring Hill that night. The Federal cavalry was played out after the fighting on December 17, and needed rations, forage, ammunition, and rest. As such, they used December 18 to refit and reorganize.[56]

Private Frank Nash wrote that, on the nineteenth, his regiment marched through Columbia and out the Pulaski Road for another three miles. That same day Cheatham's Corps assumed rear guard responsibilities at Columbia,

---

[55] Comstock, *Ninth Cavalry*, 39; Brewer, "Incidents of the Retreat from Nashville," 328; Horn, *Tennessee's War*, 159.

[56] Henry Clayton, unpublished account of his military service, 30.

and, late that evening, General Nathan Bedford Forrest arrived after a two-day forced march from Murfreesboro. Hood briefly thought of making a stand behind the Duck River, but his shattered army was in no condition for battle. Instead, the remnants of the Army of Tennessee would have to make a nightmarish eighty-five-mile winter retreat to the Tennessee River.[57]

The irascible Forrest met Hood in Columbia and was assigned command of the rear guard. Employing his cavalry and a special 2,000-man infantry contingent under General Edward Walthall, Forrest managed to keep Wilson's resupplied cavalry at bay and behind the retreating army. Lee's Corps, now under Carter Stevenson, led the retreat from Columbia southward. The 42nd camped two miles short of Pulaski on December 20 and three miles past the town the following night. They were now in the "barrens" of Tennessee. It was so cold that the march had to be stopped every hour to light a fire and warm their frozen bodies. On the night of December 22, twelve miles south of Pulaski, C. A. Rowland wrote a long and remarkable letter home. "I hardly know where to begin to tell you of our misfortune," he began. "Everything has gone wrong," he wrote, "and we present today the most melancholy spectacle you can imagine in regard to an army. The road covered with ice, snow and mud is filled with poor sick and wounded soldiers struggling along toward Alabama, not knowing where they are going or where they will find a resting place for their weary feet." Stovall's Brigade had left Florence and entered Tennessee with nearly 700 men, but now only 290 remained. Rowland believed that Hood's "insatiable thirst for ambition" had destroyed the Army of Tennessee. "Our retreat resembles and is only surpassed by that of the French from Moscow," he went on. "My heart sickens at the sight, the poor fellows came to our fire along the roadside last night, with snow frozen to their naked feet, sick and without food."[58]

The army camped on Christmas Day in Bainbridge, a few miles above Florence, Alabama, on the Tennessee River. Forrest's ambushes at Anthony Hill and Sugar Creek would send Wilson's cavalry back in disorder and buy just enough time for an eighty-boat pontoon bridge to be assembled across the wide Tennessee River. The 42nd crossed on December 27 with the last of the Confederates crossing the bridge by daybreak on the twenty-eighth. Henry Clayton characterized the 100-mile retreat from Nashville as "a most

---

[57] Nash, *Civil War Diary of Francis H. Nash*, 51.

[58] Ibid.; Rowland to home, December 22, 1864, in *Charles Alden Rowland*, 76–77.

painful march, characterized by more suffering than it had ever been my misfortune to witness."[59]

John Bell Hood had entered Tennessee with a bit more than 35,000 effective soldiers and artillerymen and suffered some 17,000 casualties in the thirty-four-day campaign. The Army of Tennessee was reduced to about 18,000 men. The survivors marched to Tupelo, Mississippi, where a winter camp was established on January 7. The controversial Hood asked to resign as Commander of the Army of Tennessee on the thirteenth and was replaced a few days later by General Richard Taylor. A system of furloughs was announced on January 15, which resulted in one in fifteen soldiers being allowed to go home, most never to return. Gibson's Louisiana Brigade, or what was left of them, was ordered to the defenses of Mobile on January 18. There was little food for the men in Tupelo, and their clothes were in tatters. January proved to be unusually cold and brutal. Many of the Southern soldiers deserted after crossing the Tennessee River and others received ad hoc furloughs. Captain Lovick Thomas was one such soldier. Jennie Thomas records in her diary that her husband reached their Gwinnett home on January 5. "He had lost his horse and blankets," she wrote. There would be little time to rest for the few still under the flag of the 42nd Georgia who made it to Tupelo. They had orders to move east to Augusta, Georgia.[60]

Johnson's Island Prisoner of War Camp

A Northern prisoner of war camp awaited the considerable number of Southerners captured at the Battle of Nashville and in the subsequent retreat. Brought into the Tennessee capital after dark on the sixteenth, the officers and common soldiers were separated into two groups. The enlisted ranks were bound for Camp Chase in Columbus, Ohio, while officers were sent to Johnson's Island Prison on Lake Erie. By February 1865, Camp Chase held 9,400 Confederates, most of whom had arrived starving and in poor health. They were issued starvation rations and one blanket before being assigned to wooden huts with no heat. Two hundred sixty Confederates are buried in the Camp Chase Cemetery, including eleven from the 42nd Georgia. More than half of those who perished at Camp Chase during the war died in 1865, over

---

[59] Report of Maj. Gen. Henry D. Clayton of operations November 20–December 27, 1864, *OR*, ser. 1, vol. 45, part 1, 699.

[60] Knight, *Hood's Tennessee Campaign*, 140; D. Smith, *In the Lion's Mouth*, 261; Jennie P. Thomas, diary, March 1865, in Huff, *Four Families*, 91.

the last four months of the conflict. In all, sixty-eight members of the 42nd Georgia spent time at Camp Chase.[61]

Captain Joshua Mitchell and Lieutenant James Gresham, both members of the Walton Tigers, were headed for Johnson's Island three miles from Sandusky, Ohio. Though the prison had a lower death rate than Camp Chase, it was swept by the winter wind coming across the great lake. Gresham kept a handful of notes from his experience as a prisoner. He began on December 17, as he and Mitchell, along with 500 captured Rebel officers, were placed in freight cars bound for Cincinnati. They had spent the previous night in the Nashville penitentiary. Huddled in the dark box cars, the prisoners were occasionally given a slice of bread and a bit of meat, but they remained extremely hungry. In Louisville, the captured men were marched through town under heavy guard. Southern ladies in upstairs windows waved handkerchiefs at the passing prisoners. The hungry men crossed the Ohio River in railcars that were carried across the water by flatboats. "En route I stole a Yankee guard's rations," he scrolled, "or rather his meat—no knife to divide with Captain Mitchell. We would bite by turn."[62]

The collection of captured Southern officers crossed the Buckeye State by rail and were finally hustled out of the freight cars at Sandusky. The dirty and famished men walked, slipped, and often fell as they traversed three miles of frozen lake to reach the island and the prison camp. At the gate, they were stripped and searched. Their money and valuables were seized—in exchange for a receipt. Old prisoners yelled "fresh fish" as they eyed the new arrivals. Gresham and Mitchell were some of the first to enter the prison yard and the first to reach the waiting bread wagon. Gresham asked for bread to feed six and returned to Mitchell with six loaves. The two men from Walton County sat on the frozen ground and, in a matter of minutes, ate three loaves each despite having nothing to drink. Gresham wrote that he was still hungry, but not nearly as hungry as the several hundred Rebel officers who came in the gate behind him and found the bread wagon empty.

Fortunately for the two Georgians, they were soon recognized by Major Henry Dickerson McDaniel, also of Walton County. As an officer in the 11th

---

[61] The following soldiers are buried at the Confederate Cemetery at Camp Chase, Ohio: William M. Carr (A), Charles Green (B), James A. Miller (B), James M. Rutledge (D), L. Victory (E), Leonard Bass (G), Elisha Lake (G), Sgt. J. W. Huff (H), J. L. Johnson (H), W. H. Sorrells (H), William F. Barrett (I), Noah B. Smith (I); Ingmire and Ericson, *Confederate P.O.W's Buried in Northern Cemeteries*, 58–93.

[62] Gresham, "Prison Life at Johnson's Island, Ohio," 428.

Georgia, the twenty-nine-year-old McDaniel had been wounded and captured in the retreat from Gettysburg. Before the war, he had graduated first in his class at Mercer Law School and established a practice in Monroe in 1857. Four years later he was selected by the voters of Walton County to be the youngest delegate to the Georgia Secession Convention in Milledgeville. Major McDaniel led the new prisoners to his ten-by-fifteen-foot room in Block Four and provided them with a warm meal. Gresham remembered it as their "first square meal" in six days. Many of McDaniel's seven roommates were lawyers. Gresham found the conversation in their mess to be lively and divorced from the reality of their hardship. He was also pleased to see Colonel Charles D. Phillips of the 52nd Georgia, who had been wounded in four places and left for dead at the Battle of Baker's Creek. Also a prisoner at Johnson's Island was Thomas Hill, who had been captain of the Fulton Blues (I) and captured during the Atlanta siege in August 1864.[63]

Johnson's Island had thirteen two-story barracks within its fourteen-foot stockade walls. Lieutenant Gresham wrote that in the morning they were assigned quarters. At first, they slept on the cold floor with one thin blanket under them and one around them. When other prisoners were exchanged, the pair confiscated the extra blankets and "slept warm afterwards." The prisoners were permitted to go to the lake at 9:00 A.M. and 3:00 P.M. to cut the ice for water. Rations were spartan, and there was no tobacco to smoke. "I had never loved to cook till then," wrote Gresham, "boiled our mite of beef in quart oyster can, wooden top with nail to lift—never permitted eye of grease to swim on top, burnt bread crust in stove to make black coffee in oyster can." They were not above catching and eating an unlucky rat or cat. Cooper Nisbet wrote that their rations of firewood "were not sufficient in quantity to keep us warm for more than a few hours a day." It was a bitter winter, with "the Thermometer standing below zero most of the time." Fiddling and dancing helped to keep the prisoners warm. Gresham sized up the German guards as cruel, but they were eventually sent to the front and replaced by crippled veterans who were kinder to the 3,000 prisoners at Johnson's Island. Letters from home came by way of Richmond through Old

---

[63] Nisbet, *Four Years on the Firing Line*, 328–30. The eight roommates were Major Henry D. McDaniel, Colonel William Dixon Mitchell of the 29th Georgia from Thomasville, Colonel Nathan Louis Hutchins of Lawrenceville, Colonel Cooper Nisbet of the 66th Georgia, Captain Dennis Sanders of the 3rd Georgia, Robert H. Couper from St. Simon's Island, Lieutenant Alvin Freeman from Newnan, Captain "Hamp" Wilkins, Captain Columbus Heard of the 3rd Georgia, and Captain Cuyler King of the Georgia 1st Battalion.

Point Comfort under a flag of truce.[64]

On January 8, 1865, Henry McDaniel, a future governor of Georgia, wrote his fiancée Hester Felker. "James Gresham and Josh Mitchell are here," he wrote, "both well and likely to make themselves comfortable. From them I have learned many items of news from Walton but not all the particulars I wish to know." In February, Ira McDaniel received a letter from Lieutenant Robert Couper of Ware County, a recently released prisoner from Johnson's Island. "Your son was in good health when I left," Couper wrote. He instructed Ira where to send a box of supplies that would reach his son in the Ohio prisoner of war camp. "What he wants is something substantial," wrote Couper, "flour, corn meal, peas, dried fruit, molasses, bacon, sausages, a little red pepper, or anything of the kind that may occur to you."[65]

There was always a bit of tension in the camp. Several prisoners accepted the offer of full rations and more privileges if they signed an Oath of Allegiance. These men were separated from the other prisoners. Northern newspapers were circulated about camp and the POW's read the editor's speculation that all commissioned officers would be hung after the final surrender. Upon Lincoln's death in April 1865, Gresham wrote that the "Yankees came in armed to quell riot." It was not necessary, "for prisoners preferred him to Andy Johnson." Colonel Charles W. Hill of the 128th Ohio apologized to his prisoners for the over-reaction of his guards. After Lee's surrender at Appomattox, prisoners were finally allowed to receive packages from home. Captain Mitchell received a box with tobacco, flour, and two sides of meat. Gresham got a package of #16 Virginia Twists. "Such smoking and chewing," he reminisced.[66]

Junior officers were released after taking the Oath of Allegiance on an alphabetical basis. On June 10, Henry McDaniel wrote Hester, "James Gresham will be released in a few days, also your old friend Dr. Mitchell. He will be apt to visit you, as you are one of his favorites." He went on, "Receive him kindly for my sake. The War has improved him in many respects. Barring a little oddity, I think him a very worthy man. He has done much good here in a professional way, working for others." James Gresham left the island prison on June 16, 1865. He waited "day & night" for Joshua Mitchell in Sandusky, as the captain still had ten dollars in his pocket. The two Georgians

[64] Ibid., 336.

[65] Sams, *With Unabated Trust*, 214–15.

[66] Sams, *With Unabated Trust*, 221; Gresham, "Prison Life at Johnson's Island, Ohio," 429.

went through Louisville, and Gresham received ten dollars from a Kentucky aunt. In Nashville he sold his #16 tobacco for a pair of shoes and a linen duster. It was a difficult hike from Cartersville to the Chattahoochee River, and Captain Mitchell struggled to keep up. James Gresham called on "my best girl at the first opportunity." She had refused to marry him until the war was over, but now, three years later, it "didn't take long to arrange matters." James Gresham and Mattie Atkinson were married in Morgan County on November 14, 1865.[67]

[67] Ibid., 221; Gresham, "Prison Life at Johnson's Island, Ohio," 429.

PHOTOS

The 42nd Georgia drew their companies from the counties of Milton, Gwinnett, Walton, Newton, DeKalb, and Fulton. *Source: Libary of Congress*

A hand-tinted ambrotype of Captain Lovick Thomas of the Gwinnett Beauregards (A). He was probably in training at Camp McDonald when this photograph was taken. *Source: L. P. Thomas collection, Bell Research Center, Cumming, Georgia*

The 42nd Georgia stayed two days at Camp Van Dorn outside of Knoxville. They managed to avoid an epidemic of disease that swept through the 40th and 52nd Georgia. This historical marker was erected in 2013 through the efforts of Gary Goodson and the General Barton Stovall Association.

The 42nd Georgia served in Smith's Army of East Tennessee in the Cumberland Gap and Kentucky campaigns of 1862. *Source: University of Kentucky Special Collections Research Center*

Union General George Morgan outmaneuvered Confederate General Carter Stevenson for control of Cumberland Gap. *Source: University of Kentucky Special Collections Research Center*

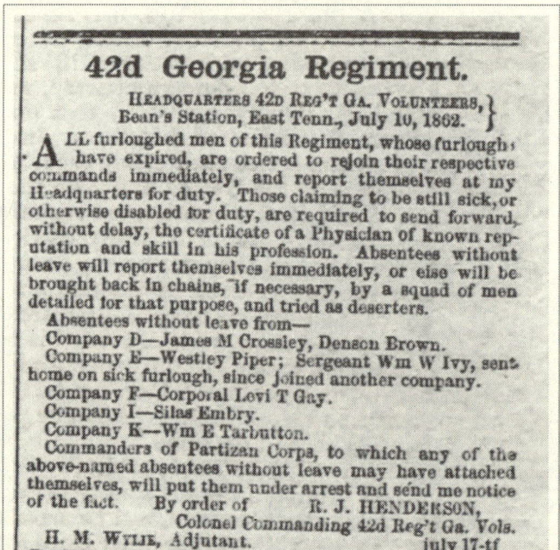

42d Georgia Regiment.

HEADQUARTERS 42D REG'T GA. VOLUNTEERS,
Bean's Station, East Tenn., July 10, 1862.

ALL furloughed men of this Regiment, whose furloughs have expired, are ordered to rejoin their respective commands immediately, and report themselves at my Headquarters for duty. Those claiming to be still sick, or otherwise disabled for duty, are required to send forward, without delay, the certificate of a Physician of known reputation and skill in his profession. Absentees without leave will report themselves immediately, or else will be brought back in chains, if necessary, by a squad of men detailed for that purpose, and tried as deserters.

Absentees without leave from—

Company D—James M Crossley, Denson Brown.

Company E—Westley Piper; Sergeant Wm W Ivy, sent home on sick furlough, since joined another company.

Company F—Corporal Levi T Gay.

Company I—Silas Embry.

Company K—Wm E Tarbutton.

Commanders of Partizan Corps, to which any of the above-named absentees without leave may have attached themselves, will put them under arrest and send me notice of the fact. By order of R. J. HENDERSON,
Colonel Commanding 42d Reg't Ga. Vols.

H. M. WYLIE, Adjutant. july 17-tf

Illness quickly depleted the ranks of the Georgia regiments sent to East Tennessee. This appeal appeared in each edition of Atlanta's *Southern Confederacy* in August 1862. *Source: Georgia Historical Newspapers, Digital Library of Georgia*

Noah Brogdon was the 2nd Lieutenant of the Gwinnett Beauregards (A). As disease swept through the 42nd Georgia, Brogdon died of mumps in Cumberland Gap. His wife and brother retrieved his body and brought it home by wagon.

Lieutenant William Akridge of the Walton Tigers (H) resigned in poor health in November 1862. The 42nd averaged a death a day from disease in 1862.

L. P. Thomas, Sr. was the "Fighting Quartermaster" of the 35th Georgia. *Source: Frederick Ware Huff of Kennesaw, Georgia*

This unpublished ambrotype of a teenage Heck Thomas in his homemade Confederate uniform was found in the Thomas papers at the Bell Research Center. *Source: Bell Research Center, Cumming, Georgia*

Dr. Ira Lawson Gunter served in the 42nd Georgia before being transferred to the Confederate Medical Corps where he served as an Assistant Surgeon in Atlanta area hospitals. *Source: Dr. Jim Latimer of Jonesboro, Georgia*

A view of George Morgan's 7th Division occupying Cumberland Gap. Four days later General Morgan would abandon this position and flee across Eastern Kentucky to the Ohio River. *Source: Library of Congress*

Sergeant John Douhit of the 52nd Georgia left a detailed account of how the men in Barton's Georgia Brigade managed during the Kentucky Heartland Campaign in the fall of 1862.

This beautifully carved wooden canteen depicts the Cotton Mill Prison in Madison, Georgia. In 1862, Captain Lowndes Calhoun's detachment guarded more than 800 Federal prisoners at the mill. On the front of the canteen is the name of Lieutenant Ira Ford of the 18th Wisconsin. On the back of the canteen are the initials of 41-year-old James Green Berry Jones of the Fowler Guards (D), a member of Captain Calhoun's detachment of Madison prison guards. *Source: Bill Erquitt of Atlanta, Georgia*

Statue of General Stephen Dill Lee at Vicksburg National Military Park. *Source: Library of Congress*

Corporal David Burson McCord and his older brother Jesse David Burson, both soldiers in the Walton Blues (G), were killed at the Battle of Chickasaw Bayou, December 29, 1862.

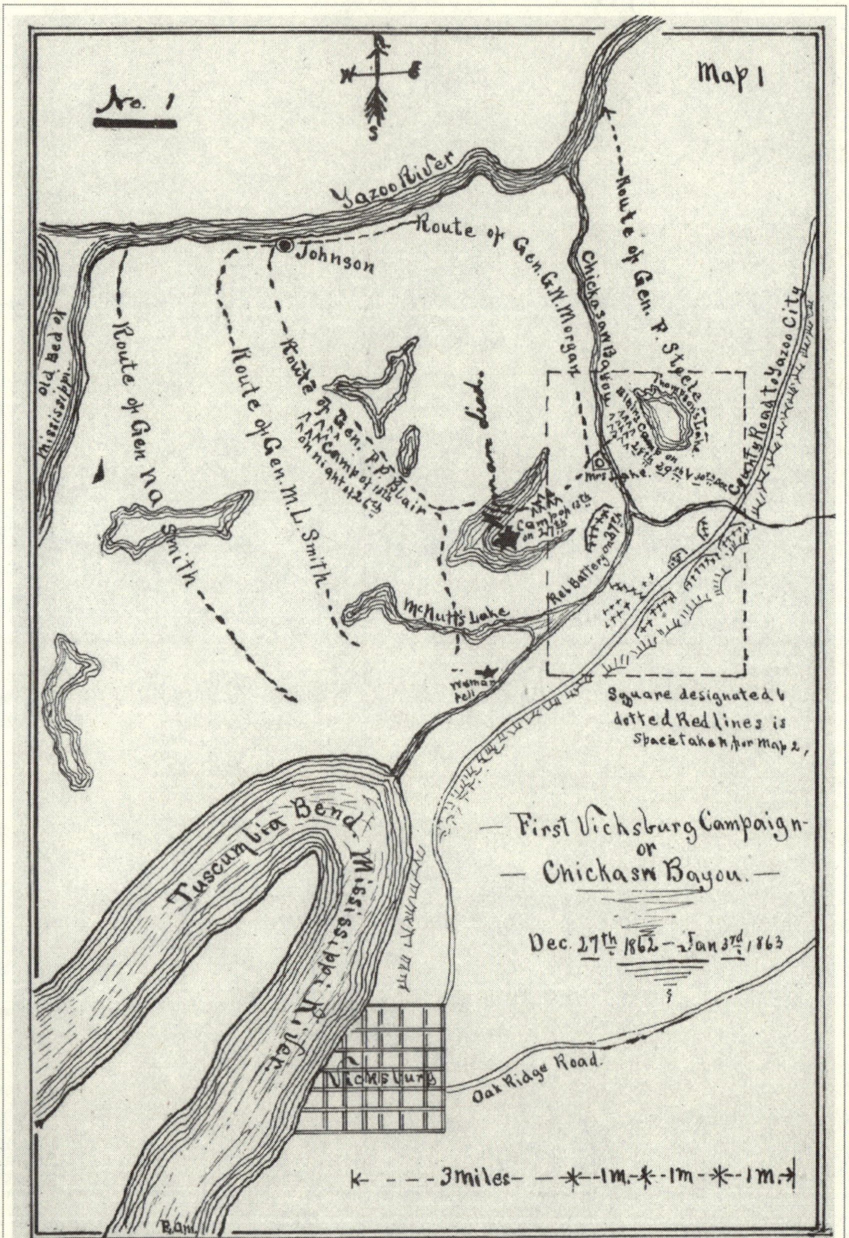

Map of the Battleground of Chickasaw Bayou, December 29, 1862. Drawn by Edward A. Munn in 1892 for the veterans of the 13th Illinois Infantry. *Source: Library of Congress*

The 42nd served in the division of Carter Stevenson from Cumberland Gap through the Vicksburg campaign. General Stevenson was from Virginia, a West Point graduate, and had served with distinction in the Seminole Wars and the Mexican War. *Source: Alabama Department of Archives and History*

Sergeant Jonas Cain of the Milton Tigers (C) was one of dozens of soldiers to receive medical discharges while serving in Mississippi.

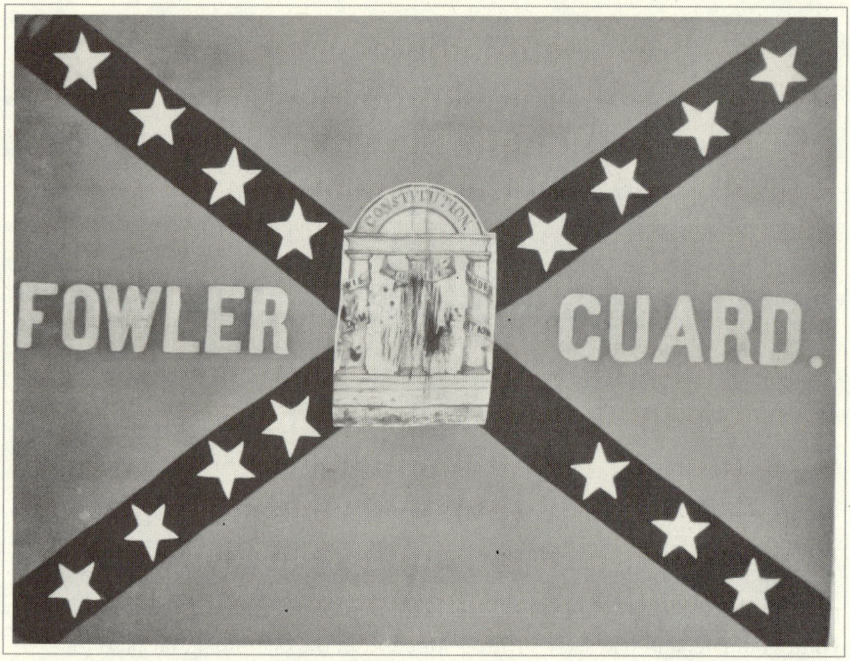

One of two flags from the 42nd Georgia that survive today. The Fowler Guard flag was captured in hand-to-hand fighting at the Battle of Champion's Hill on May 16, 1863. At the center the flag is a hand painted silk "tombstone" panel in the form of a Georgia State Seal. The company designation "FOWLER GUARD" was painted in white paint after the flag had been completed. *Source: Collection of Joseph Cashia, Franklin, Tennessee*

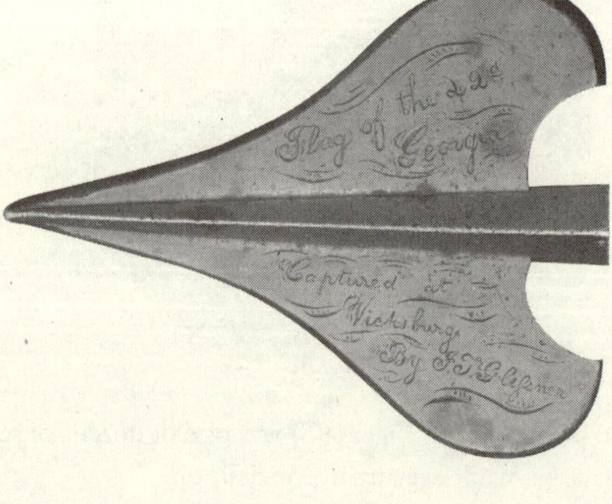

This spear finial belonged to the flag of the 42nd Georgia and was captured at the Vicksburg surrender on July 4, 1863. *Source: Collection of Kerry Elliott of Atlanta*

The 38 men from the 42nd Georgia captured at the Battle of Baker's Creek were exchanged at Fort Delaware on July 4, 1863. *Source:* Harpers Weekly, *June 27, 1863*

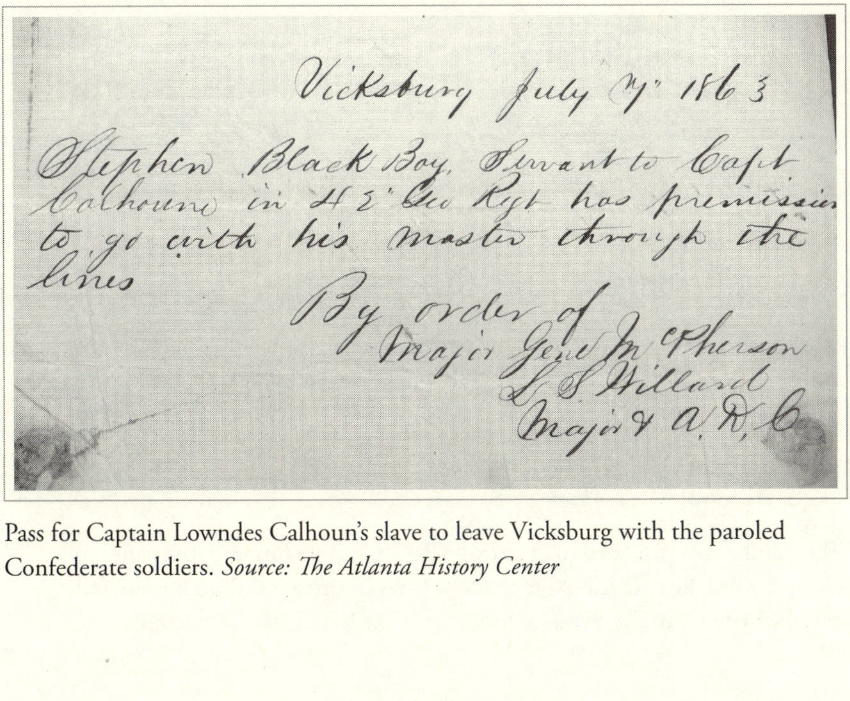

Vicksburg July 7th 1863

Stephen Black Boy, Servant to Capt Calhoun in 42d Geo Regt has premission to go with his master through the lines

By order of
Major Genl McPherson
L. S. Willard
Major & A.D.C.

Pass for Captain Lowndes Calhoun's slave to leave Vicksburg with the paroled Confederate soldiers. *Source: The Atlanta History Center*

Efforts were made in the fall of 1863 to promote Colonel Robert Henderson of the 42nd Georgia to Brigadier General. *Source:* History of the 42nd Regiment, Georgia Volunteers, *published by Judge Lowndes Calhoun in 1900.*

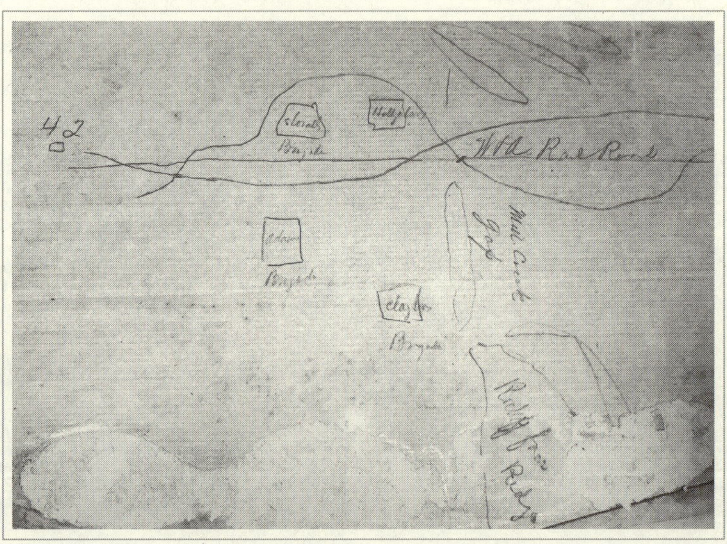

This hand-drawn undated map of Mill Creek Gap was found in the papers of General Marcellus Stovall. Note that the 42nd Georgia is well to the left and detached from Stovall's Brigade. *Source: Augusta State University Archives*

Colonel Christopher John Dickerson (1828-1872) of the 10th Michigan was the highest-ranking Federal officer directly captured by the 42nd Georgia.
Source: The History of Hillsdale County, Michigan, *Everets & Abbott, 1879*

A view of the Battle of Mill Creek Gap from the Union lines as depicted in this Alfred Waud drawing from *Century Magazine*, 1887. The 42nd Georgia defended the gap from May 5 until May 12, 1864.

The second day of the Battle of Resaca was the bloodiest day of the war for the 42nd Georgia. 117 men from the regiment fell in an ill-fated charge against the New York brigade of Adolphus Williams. *Source: Library of Congress.*

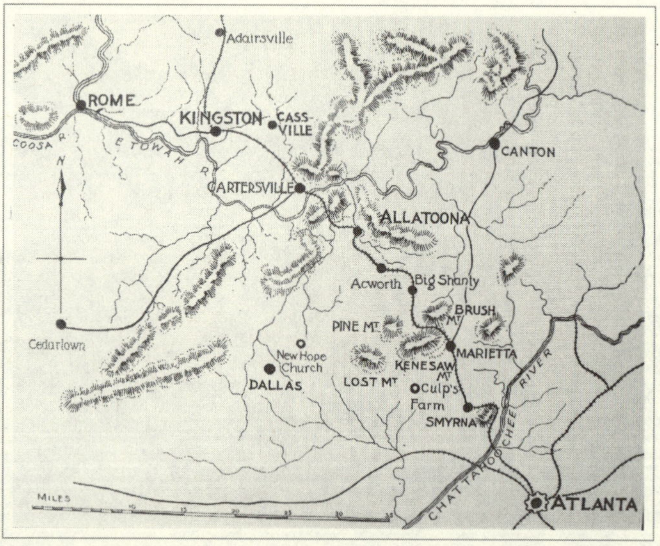

This excellent map first appeared in Union Major William Ludlow's 1891 account of the Battle of Allatoona. *Source: William Ludlow,* The Battle of Allatoona, 1864

Thomas Jefferson Mercer led the Harper Guards (E) of Newton County. His company was the center company in a line of battle and carried the regimental flags. *Source: Private collection of Robert Elliott of Grayson, Georgia.*

Jennie Peeples Thomas was the resourceful wife of Captain L. P. Thomas. *Source: Bell Research Center, Cumming, Georgia*

Private Asbury Fletcher Smith of the Harper Guards (E), with his wife Jane. A native of Talbot County, Asbury was killed in the skirmishing of July 20, 1864 at age 39.

This spirited but inaccurate 1890 engraving by Alfred Waud shows the capture of the DeGress Battery. *Source: This first appeared in Joseph M. Brown's* Mountain Campaigns of Georgia

Captain Cornelius Irving Walker of the 10th South Carolina used sharpshooters firing down from the second-floor porch of the Widow Pope House to break the resistance of Federal soldiers holding the trench line near the Troup Hurt House. *Source: Confederate Museum of Charleston, South Carolina*

Dr. Vick Thomas VI holding the sword that his ancestor and namesake carried at the Battle of Atlanta. The weapon is believed to have been manufactured by the College Hill Arsenal of Nashville, Tennessee, from the same mold as those made at the Nashville Plow Works. *Source: W. Clifford Roberts*

Captain Singleton Maxwell of the Milton Tigers (C) was killed in the Battle of Atlanta, July 22, 1864. *Source: Private collection of Fritz Neuschel of Wilmette, Illinois*

Private James Willis, Jr. of the Fulton Blues (I) found the Federal Army between his home and the trenches held by the 42nd Georgia, a distance of less than a mile. *Source: Private collection of Bill Erquitt of Atlanta, Georgia*

Charles Rowland served as the civilian quartermaster of Stovall's Brigade in the Atlanta and Tennessee campaigns.

Captain Benjamin Weaver was the popular leader of Gwinnett County's Independent Rebels (B). He received a mortal head wound late in the Battle of Franklin, Tennessee.

Lieutenant James Monroe Gresham of the Walton Tigers (H) was captured at the Battle of Nashville on December 16, 1864. He was sent to Johnson Island Prison Camp on Lake Erie, Ohio. *Source: Private collection of Annette Wade of Monroe, Georgia*

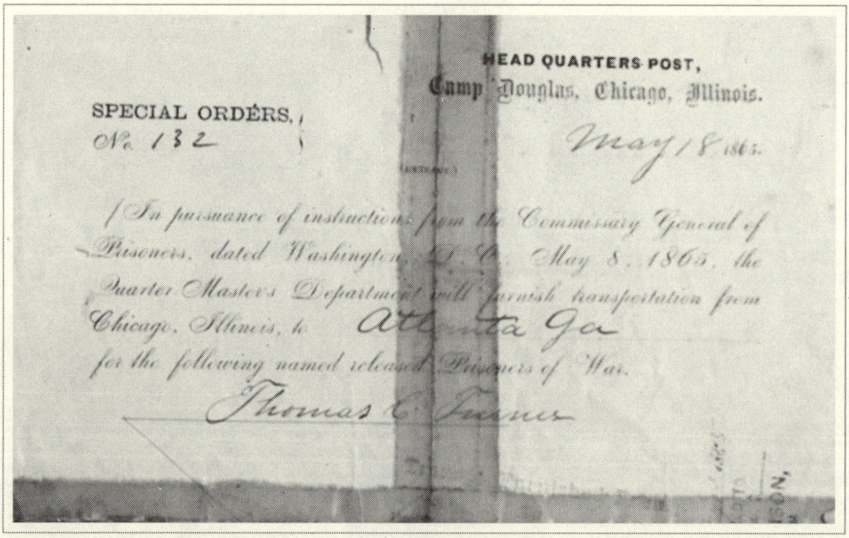

HEAD QUARTERS POST,
Camp Douglas, Chicago, Illinois.

SPECIAL ORDERS,
No. 132

May 18 1865.

In pursuance of instructions from the Commissary General of Prisoners, dated Washington, D. C., May 8, 1865, the Quarter Masters Department will furnish transportation from Chicago, Illinois, to Atlanta Ga for the following named released Prisoners of War.

Thomas C. Turner

A railroad pass to Atlanta given to Private Thomas Turner of the Fowler Guards (D) upon his release from Camp Douglas in Chicago. *Source: Private collection of Lisa Yarn of Athens, Georgia*

GEN. R. J. HENDERSON — GEN. R. F. HOKE — LIEUT. COL. L. P. THOMAS

BATTLE OF BENTONVILLE.

BY L. P. THOMAS. — Continued from Page 3.

xteen, fell mortally wounded when gal-
ntly charging in the front rank.
Finding during the night that Schofield
d reached Goldsboro and that Sherman
as moving towards Cox's brigade and
at all our wounded who could bear
ansportation had been removed, we
oved to the neighborhood of Smithfield
ation. General Johnston says further in
[illegible] "We took about 15,000 men

which I did. He and his staff were gath-
ered around a small fire partaking of
their scant supper, of which I was invited
to partake. Whether I refused from the
apparent scarcity of their rations, or
over-awed by being in the presence of an
officer so superior in rank, I do not now
remember, but I entered into conversation
with the general and his staff officers,
and was soon informed that an order was

three; fire! he would pay the penalty. A
second effort had been made to induce
General Johnston for pardon; but he stood
firm for discipline and the execution must
proceed. Just as the young soldier was
being escorted to the ground, one of the
general's aid was seen rushing with
lightning speed towards them—the execu-
tion was stayed; the commanding gener-
al had just received confirmation of the
reports that General Lee had surrendered,
and the poor fellow was saved. Our
army encamped around and about Greens-
boro—our brigade at High Point, N. C.,
where we stacked our arms for the last
time. Sherman and Johnstin agreed to a
truce on April 18, 1865, and all was over,
our pay rolls were furnished us, and our
army paid from the specie saved. It was
run out from Richmond under guard, and
was through the quartermaster of our
[illegible] each receiving

L. P. Thomas' 1900 article on the Battle of Bentonville kicked off a series of successful articles in the *Atlanta Journal* known as Close Calls, each written by a Confederate veteran. *Source: Thomas scrapbook at the Atlanta History Center*

Private William Mayo of the Walton Tigers (H) was one of 237 soldiers from the original 42nd Georgia present at the final surrender in Greensboro, North Carolina, April 26, 1865.

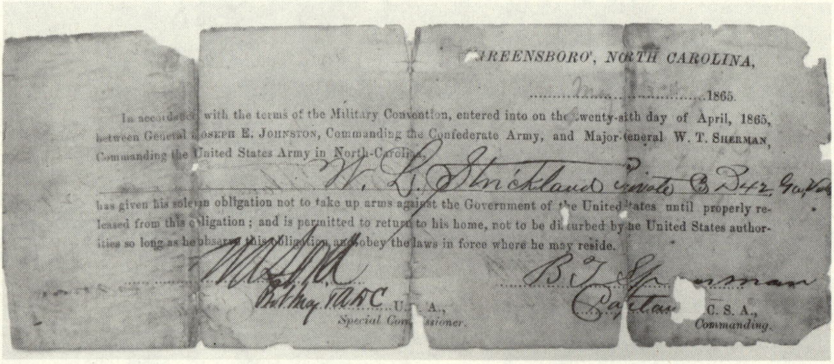

Greensboro', North Carolina,

..........1865.

In accordance with the terms of the Military Convention, entered into on the twenty-sixth day of April, 1865, between General Joseph E. Johnston, Commanding the Confederate Army, and Major-General W. T. Sherman, Commanding the United States Army in North-Carolina,

W. L. Strickland

has given his solemn obligation not to take up arms against the Government of the United States until properly released from this obligation; and is permitted to return to his home, not to be disturbed by the United States authorities so long as he observes this obligation and obey the laws in force where he may reside.

.... U. S. A., Special Commissioner.

.... C. S. A., Commanding.

Few of the original parole documents from the Greensboro Surrender have survived. W. L. Strickland from Heard County had been a member of the 56th Georgia before he became a soldier in the Consolidated 42nd Georgia Regiment only weeks before the final surrender. *Source: Private collection of Robert Elliott of Grayson County*

Heck Thomas, the much younger brother of Lovick Thomas, became a famous U.S. Marshall after he killed Western outlaw Bill Doolin in 1896.

Wiley Floyd Dennard of the Newton Rifles (F) was only 17 years old at the Greensboro Surrender. Like many veterans, he moved to Atlanta after the war. Wiley served in the Atlanta police force under L. P. Thomas.

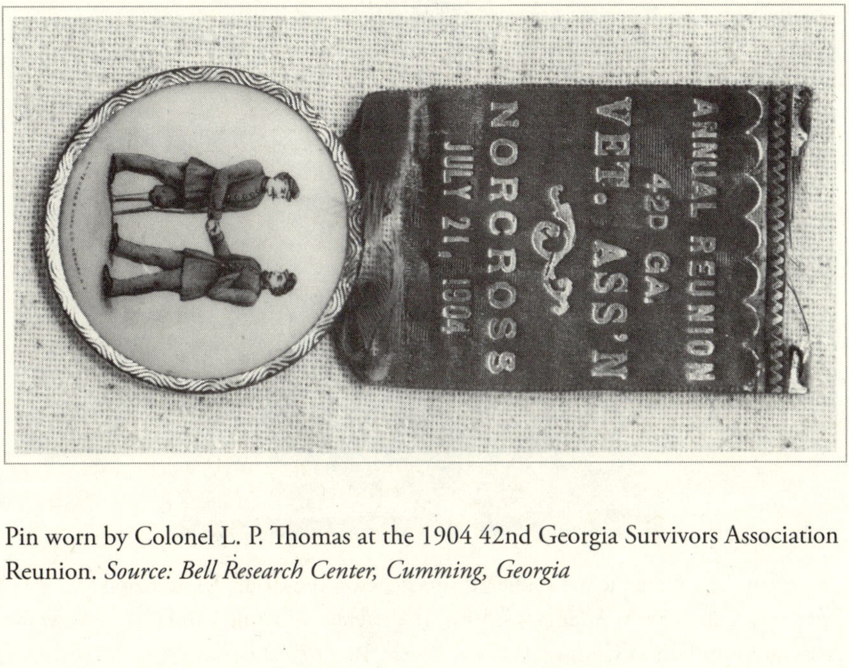

Pin worn by Colonel L. P. Thomas at the 1904 42nd Georgia Survivors Association Reunion. *Source: Bell Research Center, Cumming, Georgia*

Cabinet Card of Colonel L.P. Thomas in 1888 when he was Sheriff of Fulton County. *Source: Private collection*

This statue of a Confederate soldier at rest sits atop the Confederate Knoll at Westview Cemetery in Atlanta, Georgia. It was erected by the Confederate Veterans Association of Fulton County in 1889. *Source: W. Clifford Roberts*

Sketch of Judge Lowndes Calhoun when he was the Commander of Atlanta's United Confederate Veteran's Camp #159.

Leonidas Felix Livingston, the former Quartermaster of the 42nd Georgia, was elected to ten terms in the U.S. House of Representatives from the 5th Congressional District. "Uncle Lon" was a populist politician from the red hills of Newton County. *Source: Oct 19, 1900 edition of the* Atlanta Constitution

Many of the Reunions of the 42nd Georgia were held in the basement of the Fulton County Courthouse. Several veterans of the 42nd Georgia spent their post-war careers working at the Courthouse.

This is part of the "Saucer Drawing" made by Theodore Davis in the 1880s to assist the artists in their painting of the Cyclorama in Atlanta, Georgia. Davis interviewed local soldiers as part of this project which may explains why "Howell's Wyard Guns," the "1st Georgia State Capt A. Howell," and the "42nd GA Capt L. P. Thomas" are specifically noted. *Source: The Atlanta History Center*

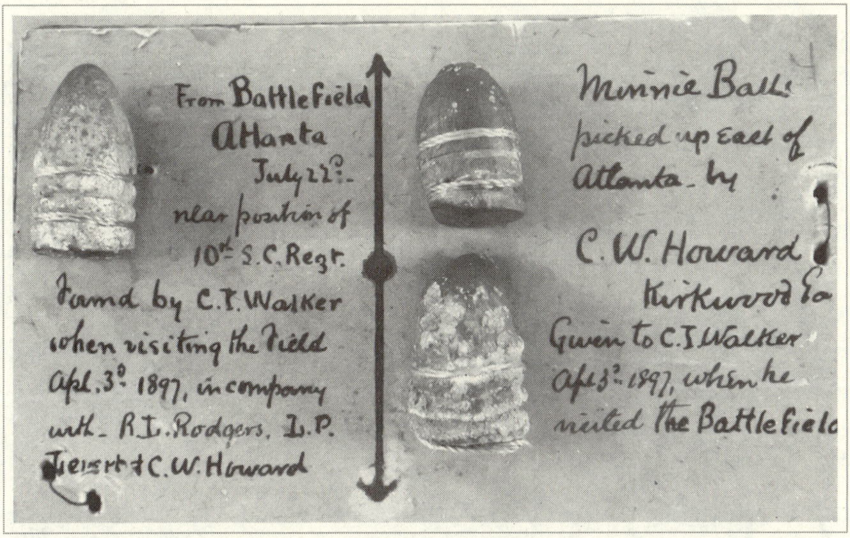

Collection of minié balls given to Colonel Irving Walker of South Carolina on his tour of the Atlanta Battlefield on April 3, 1897. Walker and L. P. Thomas were unable to resolve their dispute over which regiment had captured the DeGress Battery on July 22, 1864. *Source: The Confederate Museum of Charleston*

Sergeant William Edwards of the Newton Rifles (F) holding the tattered remnants of the battleflag of the 42nd Georgia. Edwards enjoyed marching with the flag at Confederate Reunions. *Source: Mike Couch of Cleveland, Georgia*

United Confederate Veteran ribbons worn by Colonel Lovick Thomas at various National Reunions. *Source: Bell Research Center, Cumming, Georgia*

James Madison Brown of the Walton Blues (G) wearing a Southern Cross of Honor on one lapel and, on the other, his United Confederate Veteran ribbon for R. E. Lee Camp 1055 in Monroe, Georgia.

Andrew Jackson West employed Lovick Thomas in the wholesale grocery business after the war. Colonel West, formerly a member of the 41st Georgia, became wealthy investing in the Atlanta real estate market. He was also a major leader in the United Confederate Veterans in the state of Georgia. *Source:* Atlanta Constitution, *Oct 15, 1899*

Lieutenant James Monroe Gresham became a successful farmer and a representative from Walton County in the General Assembly. Like many former officers of the 42nd Georgia, he requested that only Confederate veterans serve as pallbearers at his funeral. *Source: Private collection of Annette Wade of Monroe, Georgia*

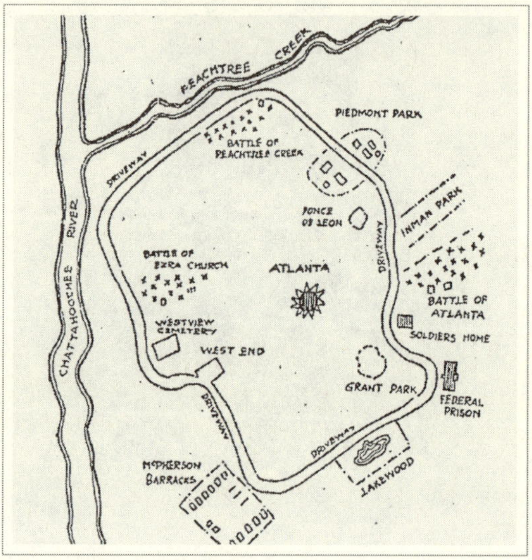

A 1905 map in the *Atlanta Constitution* touting a parkway through the proposed Atlanta Military Park. Atlanta would have become a very different city if this project had been financed by the U.S. Government.

New battle flags were issued to regiments in the Army of Tennessee in the spring of 1863. The rectangular design was favored by General Joseph Johnston. Color Bearer Sergeant William Edwards smuggled the regimental flag away with him at the Greensboro surrender. He kept the flag at his Covington home until his death in 1926. *Source: Georgia Capitol Museum*

Chapter 9

Carolinas Campaign

"You will never be ashamed I am sure, to claim once to have belonged to the 42nd Geo. Regt."
—Colonel L. P. Thomas, addressing his officers at the final surrender at High Point, North Carolina

South Carolina

General William Sherman's March to the Sea ended with the capture of Savannah shortly before Christmas 1864. He then spent the month of January refitting and preparing his army for an invasion of South Carolina, the first state to secede from the Union. The people of the United States, he wrote supreme commander in chief General Ulysses S. Grant, will "rejoice to have this army turned loose on South Carolina, to devastate that State in the manner we have done in Georgia." It was truly an audacious campaign to push an army of 60,000 men in the dead of winter across the many swamps and rivers of South Carolina. His army was supplied in part by units of "bummers" who, like hordes of locust, fanned out ahead of the advancing columns of troops. Destruction of civilian homes, farms, and villages was rampant. The state capital of Columbia was destroyed in a colossal fire. Beleaguered Rebel cavalry provided what little resistance there was to the advancing Yankees.[1]

The remnants of the Army of Tennessee were ordered east in a desperate attempt to stop Sherman's advance into the Carolinas. On January 23, 1865, Private Francis Nash of the Milton Tigers (C) reported that the 42nd Georgia, after a two-week rest, left their winter camp at Tupelo, Mississippi, by rail car for Meridian. By January 28, they had reached Selma, Alabama, where they boarded a steamboat for Montgomery. By January 30, they were in Columbus, Georgia. Desertion plagued the Army of Tennessee in its retreat from Nashville, and many in Stovall's Georgia Brigade left their regiments upon their return to their home state. The diehards of the Stovall Brigade,

---

[1] Sherman, *Memoirs*, 2:224.

those who were still willing to fight, traveled by rail to Milledgeville, by way of Macon. Disembarking two miles from the state capitol, they marched through the deserted streets on February 1, 1865.

It was a twenty-four-mile march from Milledgeville to the old plantation town of Sparta and another twenty-eight miles moving northeast to reach the village of Camak. The Georgia Railroad, which connected Atlanta and Augusta, had a station in Camak. Boarding a train, Private Nash's regiment reached Augusta on the evening of February 4. They marched through the city in the growing darkness and boarded another train at the South Carolina Depot. A desperately sick Marcellus Stovall, deemed medically unfit to command, parted ways with his troops upon reaching his hometown of Augusta. Command of the brigade subsequently fell to Colonel Abda Johnson. On January 29, Major L. P. Thomas put his uniform back on and ended his leave of absence and rejoined his regiment in Milledgeville. By February 6, General Sherman had alerted General Oliver O. Howard, one of his wing commanders, that Stovall's Brigade, some 400 strong, was defending Binnicker's Bridge over the South Edisto River. The newly arrived Georgians were now part of the Confederate Edisto Line some sixty-two miles east of Augusta.[2]

Sharp skirmishing commenced on the evening of February 9 for control of the bridge. Overmatched by the oncoming XVII Corps of Sherman's right wing, Stovall's men fell back toward Orangeburg Court House. Three miles south of the town they stopped and constructed breastworks. The Georgians camped on this North Edisto Line until February 11, when they were driven across the river and into Orangeburg. "Beyond the bridge was a battery in position," Sherman recalled, "covered by a cotton and earth parapet, with wings as far as could be seen." The Rebel infantry in these Orangeburg fortifications were elements of General Stephen Lee's Corps and the Alabama brigade of Edmund Pettus. While the Federal division of Giles Smith menaced the Confederates from across the partially burnt bridge, Ohio and Illinois men from Colonel Wells Jones's Brigade went two miles north of the town. By "wading water three to five feet deep for more than a mile," the Midwesterners were able to cross the river swamp and begin an attack on the Rebel defenders from the flank.[3]

The fight was a rather one-sided affair. The outnumbered Confederates made a hasty retreat, and Yankees entering Orangeburg soon managed to

[2] C. L. Bragg, *Distinction in Every Service*, 165; Nash, *Civil War Diary of Francis H. Nash*, 57–58.

[3] Sherman's report, Campaign of the Carolinas, *OR*, ser. 1, vol. 47, part 1, 20.

burn half of the town to the ground. After a night march of twelve miles up the Columbia Road, the tired men of the 42nd boarded railroad cars for Kingsville, South Carolina. Complicating the affair at Binnicker's Bridge had been the conduct of Colonel Abda Johnson of the 40th Georgia, who, as senior officer, had entered the fight commanding the five regiments of Stovall's Brigade. By the evening of February 9, it appeared that Johnson had become utterly intoxicated, "being in stupid slumber over a fire." The officers of the brigade had no choice but to appeal to Colonel James Gordon, commander of Jackson's Brigade, to assume command of both his and Stovall's Brigade.[4]

General P. G. T. Beauregard elected not to defend Columbia with the small number of troops available to him. Stovall's Brigade was instead sent north into the Piedmont region of South Carolina. By February 17, the day the state capital fell to the Federal "sons of Vulcan," the Georgia men were camping outside of Camden. Private Nash recorded, "We marched 20 miles and camped at Fairfield Courthouse on February 18, 1865; and on the morning of February 19, 1865, we marched 16 miles farther and we were that night in the Fairfield District of South Carolina." By the twenty-first they were wading across the Catawba River and marching into North Carolina. Two miles from Charlotte, what was left of Stovall's Brigade was reunited with the remnants of Stephen Lee's Corps of the Confederate Army of Tennessee. Encouragement came with the news that Robert E. Lee, now commander in chief of all Confederate armies, had brought General Joseph E. Johnston back from oblivion. "Old Joe" was to head all operations in the Carolinas and drive back Sherman's advancing army. Johnston received raucous cheers from his loyal veterans as the general reviewed his troops in Charlotte on February 27, 1865.[5]

Three days earlier, on February 24, acting major Lovick P. Thomas of the 42nd Georgia pressed written charges against Colonel Abda Johnson. There had been a history leading up to these allegations. Johnson had been accused of exhibiting poor leadership at both the Battle of Resaca and the fighting retreat on November 17 near Nashville, and his drunkenness at Binnicker's Bridge had been the final straw. Major Thomas charged Johnson with intoxication and "conduct unbecoming of a commanding officer in time of danger." Specifications for each charge were listed, coming from seven of-

[4] Confederate service records of L. P. Thomas, NARA, M226, roll 60.

[5] Nash, *Civil War Diary of Francis H. Nash*, 61–62.

ficers and the sergeant major of the 41st Georgia. On March 16, 1865, General Daniel Harvey Hill, who had taken command of Stephen Lee's Corps, issued Special Orders No. 21. Abda Johnson's case would be heard by a Board of Examiners for Stovall's Brigade made up of five officers. Because it was so late in the war, the formal court-martial for Abda Johnson never took place, but the Cartersville lawyer was immediately removed from command of Stovall's Brigade upon the filing of charges by Lovick Thomas. In the subsequent battles, Colonel Henry Kellogg of the 43rd Georgia, a native of Forsyth County, would lead Stovall's Brigade into action.[6]

Battle of Wise's Forks

Ever since his army of 60,000 men had left Savannah at the beginning of February, Sherman had been able to thoroughly confuse the Confederate high command about his intentions. Was "Old Billy" moving on Augusta, Charleston, or Columbia? The dispersed Rebel defenders were, thus, only able to provide a token defense at any given point as Sherman's veteran army smashed through the Palmetto State. Upon entering North Carolina, though, Sherman now "knew that my special antagonist, General Jos. Johnston, was back with part of his old army; that he would not be misled by feints and false reports, and would somehow compel me to exercise more caution than I had hitherto done." Johnston was indeed skilled at discerning Sherman's intentions.[7]

Johnston correctly surmised that Sherman intended to move next on either Raleigh, the capital of the Old North State, or Goldsboro, which was further to the east and the inland junction of two coastal railroads, the Atlantic & North Carolina and the Wilmington & Weldon. The Rebel leader ordered his dispersed Confederate units to assemble in the town of Smithfield, halfway between Goldsboro and Raleigh. What Johnston had at his disposal,

[6] The original Special Orders #21 is in the Thomas collection, BRC. The five officers named to the Board of Examiners by General Hill were Maj. L. P. Thomas (42nd Georgia), Maj. William C. Lester (43rd Georgia), Cap. J. F. Groover (40th Georgia), Capt. Thomas Leigh (41st Georgia), and Lt. John J. Stephenson (42nd Georgia). The officers prepared to testify against Abda Johnson were Lt. Col. Charles Gordon, 66th Georgia, commanding Jackson's Brigade; Capt. James Summers, Company F (42nd Georgia); Capt. Elijah Clarke, Company D (42nd Georgia); Capt. J. F. Groover, Company E (40th Georgia); Capt. Stephen Clements, Company F (41st Georgia); Lt. John Stephenson, Company F (42nd Georgia); Lt. Henry Snow, Company K (42nd Georgia); and Serg. Major Robert Douglass (41st Georgia).

[7] Bradley, *This Astounding Close*, 13.

at this late point in the war, was a polyglot of forces. The Army of Tennessee numbered no more than 5,000 troops, among whom, Johnston estimated, were 1,300 soldiers without arms. Everything—artillery, supply wagons, daily rations—was in short supply. Retreating from Wilmington after the fall of Fort Fisher was Robert Hoke's veteran division of 5,000 men from the Army of Northern Virginia and a contingent of teenage soldiers, the North Carolina Junior Reserve, both under the overall direction of General Braxton Bragg. Coming north from Charleston was General William Hardee with the garrison troops of Savannah and Charleston. Three thousand of these men were "Red Infantry," former heavy artillerists now marching on foot. Hardee's best division was composed of infantry veterans under General Lafayette McLaws. Hardee estimated that a quarter of his force deserted before reaching Cheraw, South Carolina. Johnston's most reliable and mobile force would be his cavalry under Wade Hampton and Joseph Wheeler. The new Confederate commander called these four separate field forces, in all less than 25,000 men, his "Army of the South."

Not only must General Johnston deal with Sherman's advancing army, but there was also a significant threat moving in his direction from the Carolina coast. General John Schofield had landed thousands of fresh troops at both New Bern and Wilmington and was progressing inland. The most direct challenge was Schofield's Union Provisional Corps under General Jacob Cox, which was advancing on Goldsboro from the port of New Bern. Cox's taskforce of 13,000 soldiers was under orders to open the Atlantic & North Carolina Railroad from New Bern to Goldsboro, where Sherman hoped to resupply his advancing army. The Federals did not think much of the flat Carolina coastal region, mostly covered in longleaf pine trees. Captain Dwight Fraser of the 128th Indiana wrote home, "I cannot say much concerning the country, only that it is good for darkies, cotton, and turpentine."[8]

Upon the request of General Bragg, Johnston dispatched most of Stephen Lee's Corps from the Army of Tennessee, 2,218 men in all, including the 42nd Georgia, in the direction of Camden, North Carolina. This relief contingent was under General Daniel Harvey Hill and made up of Hill's own division under Colonel J. G. Coltart, Henry Clayton's Division, and from Carter Stevenson's Division, the brigade of General Edmund W. Pettus. Private Nash reported that the men of the 42nd reached Kinston, North Carolina, by rail on March 8, two hours before daylight. Henry Clayton's Division

[8] Sokolosky and Smith, *"To Prepare for Sherman's Coming,"* 28.

was down to 416 men in two brigades, Stovall's Georgians under Colonel Henry Kellogg of the 43rd Georgia, and Jackson's Brigade, four depleted Georgia regiments under Colonel James C. Gordon. The Battle of Wise's Forks, the second largest Civil War battle in North Carolina, began that morning four miles east of Kinston. With Hill behind trenches holding the attention of the advancing Union Provisional Army, Robert Hoke's North Carolina division, some 4,500 infantry, "burst through like a torrent" on the left flank of General Samuel Carter's Division and managed to capture an entire regiment of New Englanders. Hill joined the advance, but the North Carolina Junior Reserve panicked and refused to move forward. The veterans of the Army of Tennessee left the young boys behind amid a bit of jeering and then charged into Colonel Charles Upham's beleaguered Federal brigade from the right flank. The Confederates took 890 prisoners, most from the 15th Connecticut and the 27th Massachusetts. It was the last mass capture of Yankee officers and men in the Civil War. This Rebel success lost momentum in the afternoon as Hoke's men became bogged down in the difficult swampy terrain and the Federals had time to reform and solidify their lines.[9]

Until this point, Clayton's two Georgia brigades had been held in reserve near Jackson's Mill on the Confederate right flank. At 2:00 P.M., Bragg ordered Clayton forward to Wise's Forks. Moving down the Dover Road, Henry Clayton found the Federal main line. The Alabama major general prudently pulled up and was content to engage in a lively skirmish. Hill's entire line was soon ordered back in a heavy rain mixed with sleet. They spent the night building breastworks along the British Road.[10]

Cox's badly whipped Federals retreated to Wise's Forks and erected earthworks as well as bringing up and placing twenty-one artillery pieces among their infantry lines. A Confederate attack on March 9 did not materialize. Heavy skirmishing consumed the day, and three men in the 42nd, including Lieutenant John O. Medlock the commander of Company A, fell wounded. That night, Confederate reinforcements arrived by rail for Hill's Army of Tennessee contingent. Bragg was still convinced that a Southern victory was possible and ordered a frontal attack about midday on March 10. Hoke's Division with about 4,400 men struck the left side of the Union line. The Tar Heels fought their way into the Federal works only to be beaten back

---

[9] Broadwater, *Battle of Despair*, 26; Hughes, *Bentonville*, 27; Battle of Wyse Fork Historical Marker, erected 1991 by C. S. S. Ram Neuse Camp No. 1427, Sons of Confederate Veterans, Kinston, North Carolina.

[10] Sokolosky and Smith, *"To Prepare for Sherman's Coming,"* 123–24.

by Union counter attacks.[11]

Harvey Hill's contingent from the Army of Tennessee, eight brigades totaling just 1,800 men, formed ranks around noon into two battle lines. The left flank was held by the Alabama brigade of Edmund Pettus, and to their right marched the Georgia brigades of Jackson and Stovall. Unit cohesion quickly dissolved as the Rebel infantry negotiated low-lying swamps, fallen timbers, and thick underbrush. The Georgia men managed to get well in front of the Alabama men to their left and Mississippi troops to their right. Stovall's soldiers confronted Colonel John Orr's Hoosier Brigade fighting from behind earth-and-log parapets. In a sharp exchange of fire with the Federal infantry, Kellogg's men were at a significant disadvantage as they were standing in the open. Stovall's Brigade took the largest loss—more than forty casualties—of any brigade in the Army of Tennessee. The 42nd reported twenty-five casualties at what they termed the Battle of Camden. First Lieutenant Tyre Adolphus Starr, of the Newton Rifles (F), was seriously wounded. The twenty-four-year-old Starr had been an 1862 graduate of Emory College in Oxford, Georgia.

John Orr ordered a Federal counterattack, and his men from Indiana took several prisoners before being driven back by the late arriving Alabamians under Pettus. Colonel Orr reported taking 103 casualties that afternoon. The Confederate attack did not have much of a prospect of success and ended in less than an hour. George Knight of the Walton Tigers (H) was captured, ending his streak of good fortune, which included surviving a lightning strike only eight months earlier. Clayton blamed his defeat on "the want of proper management in the officers who happened to be in command." After the battle, he relieved Colonel Gordon and put Jackson's Brigade under the command of Colonel Osceola Kyle. On the night of March 10, the men of the 42nd Georgia slipped away from their breastworks and retreated through the town of Kinston. The abandonment of this town meant that Confederate sailors had no choice but to destroy the ironclad *CSS Neuse* and join the retreat. Private Nash reported that the 42nd passed through Goldsboro and reached Johnston's rendezvous point of Smithfield, North Carolina, on the morning of March 14, 1865.[12]

---

[11] Ibid., 155.

[12] Ibid., 180–83; Nash, *Civil War Diary of Francis H. Nash*, 65.

Battle of Bentonville

Joe Johnston's only hope of Southern success this late in the war was to somehow hit and cripple one of Sherman's two wings before the other wing could reach the battlefield. Unfortunately, the ill-equipped Rebels were facing a veteran foe, one of the most experienced armies in American history. With high morale and strong leadership from senior officers down to company captains, this "western" army had captured Atlanta and swept through Georgia and South Carolina. Now deep into North Carolina, Sherman's men were a gruff-looking bunch with long beards, blackened faces from tending campfires, shredded uniforms, and worn shoes. They were also fighters and welcomed any opportunity to tangle with "Johnny Reb."

Sherman believed from faulty intelligence reports that Johnston was setting up defensive lines to his northwest to defend Raleigh, the state capital. This meant that the roads from Fayetteville northeast to Goldsboro, which had been Sherman's campaign objective since leaving Savannah, would be unprotected. With the weather improving, it was time to spread out his federal columns and make a two-day sprint for Goldsboro—where communications and a supply route could be established between a coastal seaport and the North.

By the evening of March 17, 1865, Confederate General Joe Johnston saw an opportunity to hit the left wing of the Federal army because the "Mingo Swamps" separated the two wings of Sherman's large army. Two miles south of the village of Bentonville, cavalry commander Wade Hampton chose the battlefield. Pine forests on the edge of large, flat, and empty tobacco fields would offer concealment and the opportunity to ambush an unsuspecting Federal army. Major L. P. Thomas recalled that on the morning of Sunday, March 19, his men were marching along the main road from Smithfield Station to Bentonville. "Firing could be heard in the distance," Thomas remembered later, "and the movements of couriers and aides rushing here and there indicated a battle on hand." Wade Hampton's dismounted horsemen were stubbornly giving up ground as the Yankees advanced toward the concealed Rebel infantry.[13]

The Left Wing, which Sherman called his Army of Georgia, was com-

---

[13] Thomas, "The Fight at Bentonville," 216. L. P. Thomas published this article in both the *Atlanta Journal* in July 1901 and in the *Southern Historical Society Papers* of 1901. Citations are to the *Southern Historical Society Papers.*

manded by a thirty-eight-year-old New Yorker named Henry Warner Slocum, a West Point graduate and a veteran of every eastern theater battle from Bull Run to Gettysburg. Under his command were almost 30,000 men in the XX and XIV corps. The XX, with their "bloody five-pointed star," was an oddity in Sherman's Army in that they were Easterners. The XIV Corps, on the other hand, had a reputation as ill-disciplined fighters, especially now that they were under the command of volcanic General Jefferson C. Davis. Under their acorn badge, they had stood heroically in 1863 with George Thomas on Snodgrass Hill at the Battle of Chickamauga. General Sherman took the firing in his front that morning to be nothing more than cavalry skirmishers and urged Slocum to "push them out of the way." With those words, he rode off with his staff to find the right wing. As Nathaniel Hughes would write in *Bentonville*, "It was an overconfident, impatient Union army—from the commanding general himself to the crack-of-dawn forager that stumbled into Johnston's trap."[14]

The XIV's advance division on the morning of March 17 was led by thirty-five-year-old William Passmore Carlin, a West Point graduate who was known as a steady combat leader. Meeting stiff resistance on this fine spring day, the Yankees deployed into lines of battle and pushed ahead. Coming upon an open field to their left, with a country road going in a north-easterly direction, Slocum ordered Carlin to send a brigade under George Pearson Buell, cousin of Don Carlos Buell, the Federal leader at Perryville in 1862, to move up this country trail and probe for the Confederate right flank. Meanwhile, Jeff Davis ordered Carlin to bring forward his other two brigades and advance head-on into whatever resistance lay in front of them.

The position of the awaiting 10,000 Confederates has been described as a sickle. The Rebel left under Braxton Bragg straddled the Goldsboro Road and represented the handle. This was Hoke's Division and the North Carolina Junior Reserves. At the top of the sickle was Confederate artillery, and then the cutting blade would be the remnants of the Army of Tennessee under the overall command of A. P. Stewart. The veterans of the Army of Tennessee were still divided into three, now tiny corps. At the road was Stewart's Corps, now under William Loring, with only 900 men present. In the woods to the west of the road was Stephen Lee's Corps of 2,700 men. They had been under the command of General Harvey Hill since they arrived in North Carolina. Further to the west, holding the right flank, were the 900 men of

[14] Hughes, *Bentonville*, 94.

Cheatham's Corps being led by General William Bate. Behind the sickle, the other half of the Confederate army, William Hardee's two divisions, those of William Taliaferro and Lafayette McLaws, constituted the reserve.[15]

There were two divisions in Harvey Hill's Corps, one under Carter Stevenson and the other under Henry Clayton. There were only 867 effectives (about the size of an early war regiment) divided among three brigades in Clayton's small division. Two of the brigades, Stovall's and Jackson's, were deployed on the front of the wood line, with the brigade of Alpheus Baker behind them in the forest. Henry Kellogg of Forsyth County, the colonel of the 43rd Georgia, was in command of Stovall's Brigade that morning. More than half of his brigade's 376 troops, though, were in one regiment—the 42nd Georgia under the direction of Major Lovick Thomas. "Having been informed that the line would probably be changed," wrote Clayton in his official report, "I delayed constructing breastworks." As the skirmishing in his front grew heavy, the call was given to dig-in as best they could. The Georgians lacked entrenchment tools but for the next hour, "they worked like beavers, and the scene was the busiest upon which I ever looked. It was a veritable human bee hive." Using sharpened sticks, they dug a shallow trench, and the soldiers used their bayonets to cut limbs of pine saplings which they placed in front of the trench. Hugging the ground in their concealed position behind blackjack oaks and pine trees, the Georgians watched as the Yankee battle line began pushing across the open flat field toward them. The Confederate skirmishers from the 39th Alabama gradually fell back into the woods and took their places in the Rebel line of battle.[16]

Six Federal regiments, all uncertain as to what exactly lay in front of them, hopped a rail fence and began their reconnaissance in force. Thomas recalled, "It was a grand sight to see them moving on us, 'Old Glory' floating in the breeze so proudly." The veterans of Sherman's March to the Sea, some 1,300 men in all, had 600 yards of Willis Cole's open farm field to cross. On the Union left was Buell's three regiments, the 21st Michigan and the 69th

---

[15] Ibid., 55–56.

[16] Clayton in his battle report said that he began the battle with 323 in Stovall's Brigade, 109 in Jackson's Brigade, which was under the command of Colonel Osceola Kyle, and 435 in Baker's Brigade. In October 1887, Wade Hampton published an article on the Battle of Bentonville in the *Century Magazine*. According to Col. Thomas and Gen. Henderson, there were passages that they disagreed with, and a reporter from the *Atlanta Journal* was summoned to take down their thoughts ("The Last Charge of the War, Made by Genl R. J. Henderson," *Atlanta Journal*, 1889, clipping, vol. 1, pg. 27, Thomas Collection, BRC); Bradley, "Two Confederate Views of Bentonville," 94.

Ohio out front, with the 13th Michigan in close support. They were moving at an oblique angle searching for the Confederate right flank. The further Buell advanced into the heart of the Confederate position, however, the more of his left flank was exposed to the men of Cheatham's Corps, who lay hidden in the woods. To the right of Buell's three regiments were the 94th Ohio, 88th Indiana, and the 33rd Ohio under wing commander Colonel Cyrus Briant. These Bluecoats were far less exposed, working their way through a dense forest. The 94th Ohio was headed straight toward Stovall's Georgia Brigade. Major Thomas wrote, "I never was more particular and careful in giving officers and men orders to hold their fire. My orders had gone up and down my line repeatedly, instructing the men and officers to keep down, and await a sign, or orders; even threatening those who should first disobey." About 1:00 P.M. Briant's men came to "within 3 rods" of the Rebel line.[17]

Major Thomas remembered "the sharp click-clack of muskets being cocked" along his trench line and "the excited breathing of the man next to you." He wrote, "When not over forty or fifty paces from us, the order so anxiously awaited was given, and a sheet of fire blazed out from the hidden battle line of the Forty-second Georgia." A Rebel yell immediately followed the volley. General Buell reported that the "Rebs opened fire from all sides & gave us an awful volley." The first line of Midwesterners "halted, reeled, and staggered," recalled Thomas, "while we poured volley after volley into them, and great gaps were made in their line." The two Ohio regiments soon fell off to either side, leaving the 88th Indiana standing alone. The Hoosiers took frightful casualties in a manner of minutes. Lieutenant Allan Dougall, who won the Medal of Honor that day, reported that every officer on the right wing of his regiment fell in front of Stovall's Brigade.[18]

To the left of the 88th Indiana, the 13th Michigan advanced through the decimated blue line of the two regiments in front of them and charged into a hail of bullets. Major Willard Eaton, who led the 13th Michigan, was killed in the assault. Buell wrote, "We went for them with a yell & got within 5 paces of their works." "The brave Federals fell everywhere—," Thomas recalled, "their colors would rise and fall just a few feet from us." With Major Eaton falling dead in the assault, Buell's broken regiments retreated across the field and took cover in a wooded ravine. Henry Clayton counted twenty-

---

[17] Thomas, "The Fight at Bentonville," 217; Bradley, *Last Stand in the Carolinas*, 172.

[18] "The Last Charge of the War, Made by Genl R. J. Henderson," *Atlanta Journal*, 1889; Hughes, *Bentonville*, 73; Thomas, "The Fight at Bentonville," 217; Bradley, *Last Stand in the Carolinas*, 186.

five Federals "killed dead," and scores more wounded or captured. "My own loss did not exceed ten killed and wounded," he reported.[19]

It was time for the Federals to fall back and, in a role reversal, go on the defensive. The XIV Corps other division, three brigades under Bostonian James Dada Morgan, moved to Carlin's right and began to fortify their position. Slocum's other half of his wing, the XX Corps, began to rush reinforcements forward as well as set up a defensive line of artillery positions one mile in the rear at the Reddick Morris farm. At 2:30 P.M., General Slocum sent a dispatch rider to Sherman. He noted that strong forces were in his front, adding, "It is reported by prisoners that Johnston and Hardee are here. I think a portion of the Right Wing should be brought forward at once." The Confederates, at this time, were clearing the field of the wounded in their front and "our boys were picking up guns thrown down by the enemy." Enough, claimed Thomas, to equip the entire North Carolina Reserve with modern rifles. There were scores of wounded among the 42nd, none more serious than Moses Martin. Thomas noted that Sergeant Martin, who had enlisted with him at the beginning of the war in Lawrenceville, was carried off and later had to have his leg amputated above the knee. General D. H. Hill and his staff rode down the Rebel line and complimented the men on their valor. It was now time for a Gray counter-assault.[20]

At about 3:00 P.M., the Army of Tennessee under the direction of "Old Reliable" William Hardee stepped out from the woods onto Willis Cole's tobacco field. They were lean, bearded men in gray and brown garments of every shade and cut. Rebel yells echoed across the battlefield as the men dressed their lines "conforming to the right." Every felt hat came off as the men caught sight of General Joseph Johnston and cheered him loudly. Colonel Charles Broadfoot of the 1st North Carolina Junior Reserves had a wonderful view of the ordered lines, "but it was a painful sight to see how close their battle flags were together, regiments being scarcely larger than companies and a division not much larger than a regiment should be." For the first 400 yards, Hill's Corps marched at common time behind mounted officers, in two lines some 300 yards apart. When they were 200 yards from the Union fieldworks, Harvey Hill ordered a charge. The first volley from Carlin's Federals staggered the Confederates, but there was not a second volley as the Yankees turned and ran for safety. "Onward we moved for perhaps half a mile or so," wrote Lovick Thomas, "carrying everything before us." The Army of

[19] Hughes, *Bentonville*, 73; Bradley, "Two Confederate Views of Bentonville," 94.

[20] Hughes, *Bentonville*, 80; Thomas, "The Fight at Bentonville," 218.

Tennessee captured three artillery guns and several hundred prisoners. As the Confederates chased the Federals back to the Goldsboro Road, things became confused as units mixed and men began to stop to plunder Union supply wagons.[21]

The Army of Tennessee finally reorganized itself along the Goldsboro Road. Thomas recalled that while the 42nd rested on the pine-covered ground as it neared dusk, "the sunbeams were slanting through the pines and gilding them with mellow light." It was then that Thomas "saw the grandest sight I ever witnessed on the battlefield." About 4:30 Major Thomas had a front row seat for the sweeping advance of Hoke's "gallant and dashing" division against the Federal division of James Morgan. "Hoke's division was composed of fresh men, freshly uniformed, and a grand sight it was as they made the charge. Sweeping grandly through the open pine woods, they bore down upon the enemy like a hurricane, and as the lines met, it was like the meeting of clouds in a summer thunderstorm." It was a fierce fight that continued into darkness. Morgan's men, with an impenetrable swamp behind them, mostly held their lines.[22]

At this late point of the battle, Confederate units moved in different directions. The left of Hardee's line assaulted and almost overran Morgan's left flank while Hardee's right flank continued to move down the Goldsboro Road. Morgan's third Federal brigade under Benjamin Dana Fearing made a determined stand to hold the road. Fearing's men were forced to retreat only after Henry Clayton's two Georgia brigades—Jackson's and Stovall's—hammered them back. The last of the Confederate troops, the divisions of Taliaferro and McLaws, were brought up to make a final assault. By now, however, Henry Slocum had positioned his entire XX Corps on defensible high ground with artillery support. With the sun setting, the Rebel reserves pushed forward but were repulsed at every point. These attacks constituted the Confederate high-water mark at the Battle of Bentonville. About midnight, General Johnston ordered a retreat to the positions his weary men had held earlier in the morning.

That night, most elements of Otis Howard's Federal right wing began to arrive on the battlefield. Sherman expected Joseph Johnston to retreat across the Mill Creek Bridge, his only route of escape back to Smithfield. Surprisingly, Johnston dug in and held his position for two days. Hopelessly

---

[21] Thomas, "The Fight at Bentonville," 218.

[22] "The Last Charge of the War, Made by Genl R. J. Henderson," *Atlanta Journal*, 1889; Thomas, "The Fight at Bentonville," 218.

outnumbered, the Confederate commander wanted time to get his wounded off the field. He may also have wished for another Kennesaw Mountain, with Sherman launching a bloody frontal assault on his defensive positions. Historian Mark Bradley speculates that Johnston might also have believed that there was a moral victory to be gained for his beleaguered troops, as they stood for two days toe-to-toe with Sherman's much larger force. General Sherman, though, declined this opportunity to destroy the much smaller Rebel army. Instead, he focused on removing his wounded to Goldsboro and establishing supply lines all the way to New Bern on the coast. Thomas recalled that "for the next two days the two armies stood facing each other like game cocks, neither caring to make another advance."[23]

It rained for much of March 21, the second day. Sherman's Corps commanders tightened the noose around the Confederate positions but were instructed not to bring on a general engagement. A. P. Stewart's Army of Tennessee held the center of the line facing Jeff Davis and his XIV Corps. While no frontal assaults materialized, skirmishing occurred up and down the line. According to D. H. Hill's report, Major Thomas was "the corps officer of the day," and personally led the skirmish line on a sortie to burn the Willis Cole house, thus stopping Federal sharpshooters from using it. Everything changed at 4:00 P.M., when a Union division in the XVII Corps under General Joseph Mower, a bold leader and Sherman favorite, aggressively probed the far left of the Confederate line. With their rifles held above their heads, Mower's two brigades waded across a wide swamp and stormed the hill beyond. His skirmishers trekked on at a steady trot and managed to capture the log cabin headquarters of Joseph Johnston. Mower had pushed away the cavalry screen, trying to defend the Confederate army's left flank, and now threatened to cut off any Rebel retreat by seizing the bridge over Mill Creek.

Disaster beckoned for Joseph Johnston's army if a fierce counterattack could not be launched at once. The only infantry present was Cumming's Brigade, just crossing the Mill Creek Bridge into Bentonville. The Georgia Brigade "was a shadow of itself," with just 200 soldiers present. Alfred Cumming had been severely wounded at Jonesboro, and half the brigade, made up of the 34th, 36th, 39th, and 56th Georgia, had been captured at the Battle of Nashville. The newly arrived brigade now also included men from the 52nd Georgia, and a few soldiers from the 42nd Georgia, troops who had

[23] Bradley, *This Astounding Close*, 23; "The Last Charge of the War, Made by Genl R. J. Henderson," *Atlanta Journal*, 1889.

been detached in the Nashville retreat for guard duty near Columbus, Mississippi. Leading the tiny brigade into the fray was Colonel Robert Henderson, who had been the widely respected leader of the 42nd Georgia from Camp McDonald to the Battle of Nashville.[24]

It was an audacious Rebel counterattack that came in the nick of time. Henderson's men charged forward at a double-quick in two lines. Henderson would later say, "to me belongs the mournful honor of having led the last charge made in behalf of the Confederacy." On their left, coming at a dead run, was Wade Hampton with 300 dismounted Georgia troopers. To Henderson's right, 500 cavalrymen under Joseph Wheeler, some mounted and others dismounted, pounced on Moyer's left flank. The furious Rebel attackers seemed to hit the Federal line at the same moment. Henderson's men carried two Federal lines and pushed the Bluecoats off the hill and back to the swamp. An appreciative Joseph Johnston would later promote Henderson to brigadier-general for his gallantry on the field that afternoon.[25]

In the mix of this battle were the last eighty members of Terry's Texas Rangers. They stampeded into the Federal lines, emptying their carbines and then their pistols at point-blank range. Forty-nine-year-old William Hardee even participated in the charge. Finding Wade Hampton after the fight, the lieutenant general exclaimed, "That was Nip and Tuck, and for a time I thought Tuck had it." Unfortunately, minutes later glory would become a tragedy. General Hardee spotted the newest recruit of the Texas Rangers, his sixteen-year-old son Willie Hardee, slumped in his saddle, mortally wounded in the chest. Thus, the Battle of Bentonville ended. Johnston reported his casualties to Lee as 224 killed, 1,470 wounded, and hundreds missing. Though Johnston believed "The enemy's losses are far greater than ours," they were not. The Federals lost 194 killed, 1,112 wounded and 221 missing.[26]

Surrender at Greensboro

On the night of March 21, General Hill had Major Lovick Thomas report to his headquarters. "He and his staff were gathered around a small fire partaking of their scant supper, of which I was invited to partake," recalled Thomas. The leader of the 42nd declined the invitation. "Whether I refused from the

---

[24] Hughes, *Bentonville*, 200.

[25] "The Last Charge of the War, Made by Genl R. J. Henderson," *Atlanta Journal*, 1889.

[26] Hughes, *Bentonville*, 207.

apparent scarcity of their rations, or overawed by being in the presence of an officer so superior in rank," he wrote, "I do not now remember." Hill told Thomas to expect a general retreat that evening. The 42nd had held the bridge at Baker's Creek in 1863, and they would do the same at Bentonville. Major Thomas was ordered to provide the last skirmish line for Hill's departing corps, which began to pull out of the trenches at 2:00 A.M. It was a difficult retreat made on muddy roads in complete darkness. On the morning of March 24, the 42nd Georgia marched through the town of Smithfield. They established camp three miles down the Raleigh Road and stayed there until April 10.[27]

On March 31, Joseph Johnston's army, which he now called the Army of Tennessee, numbered 25,011 present for duty. So that it might function better as a cohesive fighting force, the senior general reorganized his entire army into two corps in early April. Divisions were dissolved, and shrunken brigades and regiments were consolidated. The intent was for each surviving infantry regiment to have ten companies, with roughly forty men in each company. For some of the Confederate officers present, this meant that they would no longer have any troops to command. Most of these leaders chose to return home, as, for them, their War of Southern Independence was effectively over. In a letter written to Jennie from Hillsboro on April 13, Thomas listed many of these now displaced officers whom he had fought and served with—General Henry Clayton, Colonel William Hulsey, Adjunct James Henry Griffin, Colonel Henry Kellogg, and Captain William Lester. In Stovall's Georgia Brigade, the 52nd Georgia Regiment was down to less than forty men, and was, therefore, assigned to the 39th Consolidated Georgia Infantry as a "detachment." The 40th, 41st, and 43rd Georgia regiments were consolidated into a single regiment, the 40th Georgia Battalion, under the command of Colonel Stephen D. Clements. A native of Meriwether County, Clements had started the war as a 1st lieutenant in Company F of the 41st Georgia.[28]

Lovick Thomas was designated to lead the consolidated 42nd Georgia Regiment, as he had been promoted on April 9, 1865, to lieutenant colonel. "In this reorganization," recalled Thomas, "I was promoted to lieutenant

[27] Thomas, "The Fight at Bentonville," 220; Nash, *Civil War Diary of Francis H. Nash*, 65.

[28] L. P. Thomas to Jennie P. Thomas, April 13, 1865 in Huff, *Four Families*, 92; "May 1, 1865 final surrender documents of the 42nd Georgia," *Muster Rolls and Lists of Confederate Troops Paroled in North Carolina*, NARA, M1781, roll 3.

colonel and J. J. McClendon was made the major." The forty-year-old Jonathan Jackson McClendon was a Georgia politician and had, until this promotion, been serving as the adjutant of the 34th Georgia. According to Lovick, Jackson "represents his county [Coweta] in convention, whenever he chooses to run." There were enough veterans of the old 42nd Georgia to make up six of the ten companies in the newly consolidated regiment. Only the Harper Guards (E) and the Newton Rangers (F), both of Newton County, had enough men to remain, after the reorganization, as single companies in their own right. Captain T. J. Mercer's Harper's Guards became the new Company A, and the Newton Rangers became Company B. For the rest of the Georgia men, combinations had to be made. Companies G and K combined to become Company C under Captain William Calhoun. Companies B and D combined to become the new Company E under Captain Elijah Clarke, and I and H merged to be the new Company F under Captain John Stubbs. Finally, Captain James Moore would lead Company H, which was a combination of the old A and C companies. Benjamin Toombs Spearman of Heard County had held his Company K of the 56th Georgia together enough for them to become the new Company D of the consolidated 42nd Georgia. Remnants of the 36th Georgia became companies G and I, and, finally, Company K, under Captain Daniel Brewster, would be an awkward combination of men from 34th and 56th Georgia. Perhaps most impressively, the Consolidated 42nd Georgia would still feature one of the best musical bands in the Army of Tennessee, with eleven musicians under the capable leadership of Captain Otto Spahn. Thomas and his consolidated 42nd Regiment would be in Robert Henderson's brigade in Carter Stevenson's division, in Stephen D. Lee's corps of 6,191 men.[29]

One of the surprising departures among the Confederate officers was Henry Clayton of Alabama. Among the letters published in the 2015 book *The Lost Papers of Confederate General John Bell Hood* is an 1879 letter from Clayton to Hood. Knowing that General Hood was soon to publish his war memoir *Advance and Retreat*, the former division commander asked that Hood, in his new book, correct a slight made by Stephen Dill Lee against

[29] Thomas, "Reminiscences of Forty-Second Georgia," 14; L. P. Thomas to Jennie P. Thomas, April 13, 1865, in Huff, *Four Families*, 92; Confederate Service records of Colonel L. P. Thomas, NARA, M226, roll 60; "May 1, 1865 final surrender documents of the 42nd Georgia," *Muster Rolls and Lists of Confederate Troops Paroled in North Carolina*, NARA, M1781, roll 3.

Stovall's Georgia Brigade. "Genl. Lee did me and my command great injustice in regard to the Battle of Nashville," wrote an upset Clayton. "Before leaving the army he published a general order complimentary to my command, except one Brigade. I felt this such an undeserved reflection on that Brigade that I refused to publish it." Clayton gathered statements about the meritorious conduct of the Georgians from generals Gibson and Holtzclaw, and along with his own statement, submitted them to Lee. "Still he refused to make the correction," said Clayton. When Lee returned to the army in North Carolina, "I still being in command of Stovall's Brigade, promptly asked to be transferred, stating as a reason that I could not in my judgment again command Stovall's Brigade under him." Stephen Lee, it appears, had never quite forgiven the Georgians for their contribution to the disaster on Champion Hill in May 1863. "Now, General, can't you do me justice?" asked Clayton of Hood. On page 304 of John Bell Hood's 1880 book, Henry Clayton and his division would be acknowledged and complimented for their performance as the army's rear guard during the initial stages of the retreat from Nashville.[30]

General Joseph Johnston ordered a review of the newly reorganized Army of the South. Lovick Thomas recalled his impression of Old Joe on his horse that day, "He seemed from that piercing look to give each soldier as he passed a most scrutinizing look." The three-week halt at Smithfield allowed the reorganized units to drill and maneuver, but there was also leisure time to rest, enjoy card games, horse racing, and the resumption of revival meetings.[31]

The Confederate retreat began on a rainy April 10. Private Francis Nash wrote in his diary that the Consolidated 42nd marched through Raleigh on April 11. Women lined Fayetteville Street that day, handing out meat, bread, and tobacco to the ill-clad soldiers. As regimental bands played "Dixie," the butternut columns passed by the statehouse and bronze statue of George Washington. It was on this second day of the march that the 42nd encountered a soldier from General Lee's Army of Northern Virginia. Thomas recalled that this Confederate "had in some way gotten away in advance of his comrades and no doubt was moving as rapidly as he could toward his loved ones and home, disheartened, sore-footed, and hungry." His report that Lee had surrendered to Grant on April 7 did not seem plausible. "This seemed so

---

[30] S. M. Hood, *The Lost Papers of Confederate General John Bell Hood*, 121, 238; J. B. Hood, *Advance & Retreat*, 304.

[31] Thomas, "The Fight at Bentonville," 221; Bradley, *This Astounding Close*, 66.

absurd," wrote Thomas, "that we could not believe it, and the poor fellow was put under arrest and held for a day." As "news of the surrender came thick and fast," the Southern prisoner was released by Carter Stevenson with apologies.[32]

Stephen Lee's Corps camped at Hillsborough on the evening of April 13. Lovick wrote his wife that evening explaining that he had "given Col. Kellogg who lives near Cumming, Forsyth Co., two hundred dollars to send you, which he promises to do as soon as he gets home." The next day they marched eighteen miles through the rolling countryside before reaching the railroad bridge over the flooded Haw River. On the afternoon of Easter Sunday, April 16, the 42nd was ordered to halt and camp two miles from Greensboro. The soldiers took this order to mean that their surrender was forthcoming. That evening, President Jefferson Davis slipped out of Greensboro with a cavalry escort, moving south toward Charlotte. General Johnston was left to begin negotiations with General Sherman for final surrender terms. Complicating these military negotiations was the fatal wounding of President Abraham Lincoln on the evening of April 15. Thomas wrote, "Among officers and men there was an alarming demoralization in all of the commands."[33]

Wholesale desertions now plagued the Confederate army, particularly among Carolina men who were only a few days' walk from their homes. Discipline within the ranks collapsed as speculation abounded as to what lay next for everyone involved. On April 27, 1865, the men of the 42nd moved on to High Point, a short distance from Greensboro. On that same day, General Joe Johnston issued General Orders No. 18 announcing the surrender of all Confederates in North Carolina, South Carolina, Georgia, and Florida. Colonel Lovick Thomas called the men together and read the order to his troops. Johnston wrote, "Events in Virginia, which broke every hope of success by war, impose on its general the duty of sparing the blood of this gallant army and saving our country from further devastation and our people from ruin." After reading the order, Thomas folded the order and slipped it into his pocket. The following day long lines of troops stacked their arms, and officers completed their muster rolls in duplicate. Each man received one Mexican silver dollar as a final payment from the Confederate Treasury.[34]

---

[32] Bradley, *This Astounding Close*, 95; Thomas, "Reminiscences of Forty-Second Georgia," 2; Nash, *Civil War Diary of Francis H. Nash*, 69.

[33] L. P. Thomas to Jennie P. Thomas, April 13, 1865 in Huff, *Four Families*, 92; Bradley, *This Astounding Close*, 146; Thomas, "The Fight at Bentonville," 221.

[34] In 1901, L. P. Thomas said that he still owned his Mexican silver dollar "and prized

It was a very different surrender than what had happened at Appomattox Courthouse a few weeks earlier. Lee's soldiers had marched into the Virginia village and formally surrendered their implements of war in front of a line of saluting Union troops. Johnston's army, by contrast, was spread out over a wide area from Salisbury to Greensboro. Their surrender would not be a formal ceremony in front of Sherman's victorious Union army. Instead, each Confederate regiment stacked their weapons in camp, received their written pardon, and then began the slow journey home on foot. "There in an old field," recalled Colonel Thomas, "I had our regiment to stack arms, about four hundred in number, and when that was done, amid a silence that could almost be felt, many a tear was shed by brave officers and brave men while standing over those guns." Thomas was then presented a "beautiful address," signed by every officer and noncommissioned officer in the regiment. They tendered their "heart-felt thanks for that kindness and courtesy, which has always characterized your conduct." They noted that his deportment on the battlefield was "gallantly unsurpassed and heroism never faltering." They wrote that "We have always approached you satisfied that our grievances would be repaired and our wrongs addressed."[35]

Thomas also secured a copy of Special Field Orders No. 65 from General Sherman. "Foraging will forthwith cease," and disbursing officers would pay cash or provide vouchers for purchases. Sherman offered a quarter-million ration to the former Confederates and placed General Schofield in charge of parole arrangements. Johnston's quartermaster issued the men cloth,

---

it as one of the most valued relics of the war" (Thomas, "The Fight at Bentonville," 222; "General Orders 18," General J. E. Johnston, April 27, 1865, *OR*, ser. 1, vol. 47, part 3, 843–44).

[35] Dunkerly, *The Confederate Surrender at Greensboro*, 3; Thomas, "Reminiscences of Forty-Second Georgia," 2; "42nd Georgia Officer's Appreciation Letter," High Point, NC, April 28, 1865, notebook 1, 35, Thomas collection, BRC. Officers signing the letter of appreciation were Jonathan J. McLendon, Major, 42nd Georgia; Joseph Heard, 1st Lt., Co. E; J. S. Stubbs, Captain, Co. F (formerly in 32nd GA); William L. Wilson, 1st Lt., Co. E; Leonidas F. Stephenson, 2nd Lt., Co. F; Benjamin Toombs Spearman, Capt., Co. D (formerly of 56th GA); J. J. Stephens, 1st Lt., Co. D; J. B. Smith, 1st Lt., Co. I (formerly in 56th GA); John B. Steward, 2nd Lt., Co. I (formerly in 36th GA); Henry Britton Mikel, 1st Lt., Co. G; James Moore, Cap., Co. H; John O. Medlock, 1st Lt., Co. H; Henry Harris Burgess, 2nd Lt., Co. D; Elijah H. Clarke, Captain, Co. E; Richard M. Morris, 1st Lt., Co. E; William T. Smith, 2nd Lt., Co. E; James Pendleton Brewster, Captain, Co. K (formerly in 56th GA); E. L. Thomas, 1st Lt., Co. K; James T. Lovelass (?), 1st Lt., Co. J; William A. Deweese, 2nd Lt., Co. G (formerly in 56th GA); Henry Y. Snow, 2nd Lt.; William M. Durham, adjutant, 42nd Georgia.

thread, and cotton yarn, which they could barter for food on their journey home. On May 1, 1865, Confederate brigade commanders presented themselves at the Britton House Hotel in Greensboro. Schofield's inspector general, General William Hartsuff, reviewed their muster rolls and counted out the number of parole forms needed for each respective brigade. Each parole form required the former Confederate soldier to give "his solemn obligation not to take up arms against the Government of the United States," and he would in return "be permitted to return to his home, not to be disturbed by the United States authorities so long as he observes this obligation."[36]

On the morning of May 3, the men broke camp and headed for home, going south on the Salisbury Road. Company supply wagons hauled Federal rations. Most commands carried out their battle flags in violation of the convention. Color bearer Sergeant William Edwards took the flag of the 42nd home with him to Covington, where he kept it until his death in 1926. "With one wagon and the old regimental ambulance, we moved out," remembered Lovick. Upon reaching Salisbury, the Georgians headed south on the Charlotte Road, while men from Texas, Tennessee, and Arkansas headed west on the Morganton Road. Upon reaching Augusta, Georgia, the men of the 42nd began to scatter in different directions. Colonel Thomas, riding his splendid bay horse Kennesaw, headed for Greensboro, where several of his men lay in an army hospital. There he found "Mose" Martin of Company A, who had lost a leg at Bentonville. As the men shook hands, the sergeant said, "Colonel, if I ever get home, and should have a 'baby boy,' I will name him after you." Lovick Richmond Martin was born in Gwinnett County in 1866.[37]

Assessment

In his 2008 book *General Lee's Army*, historian Joseph Glatthaar wrote that "no words could prepare the men of 1862 for the disasters they would endure." Unlike the "Boys of '61," who had months to meld into military service, the volunteers of 1862 were thrown into the war with little preparation. Glatthaar notes, in his extensive study of soldiers who served in the Army of Northern Virginia, that among the 1862 recruits, "Three of every four soldiers were killed, wounded, taken prisoner, or died of disease." If the casualty figures of the 42nd Georgia are any indication, the numbers for the

[36] Bradley, *This Astounding Close*, appendix E, 27, 226.

[37] Georgia Capital Museum, *Hallowed Banners*, 42; Bradley, *This Astounding Close*, 242; Thomas, "The Fight at Bentonville," 222.

Army of Tennessee are virtually identical. Of the 1,355 men who served in the 42nd Georgia, 514 (38 percent) were dead by 1867. It is impossible to know how many men in the regiment were wounded, but the records of division surgeon Gustavus Thornton (see appendix) show that it had to be in the hundreds. Finally, 182 men (13 percent of the regiment) would serve time in Federal prisoner of war camps.

That Lovick Thomas survived the war with only the one slight wound at Kolb Farm is remarkable. Fifty-nine of the 153 Gwinnett County men (39 percent) who served in his Company A lost their lives during the conflict although only twelve were combat deaths. Eleven men were too sick to return to the company after the Vicksburg exchange. Twelve men deserted in the trying campaigns of 1864, all taking the oath of allegiance and agreeing to stay north of the Ohio River until the war had ended. Fifteen men were captured and taken to Northern POW camps. Two did not return. William Carr died of smallpox at Camp Chase, Ohio, and John Singleton died of inflammation of the lungs at Camp Douglas, near Chicago. Only thirty men (20 percent) were present at the final surrender at Greensboro on April 26, 1865. Five other men were serving in other units by 1865, and they were formally surrendered at Kingston, Georgia, on May 12, 1865. Poorly supplied in difficult conditions in the windswept mountains around Cumberland Gap and then in the malaria-infested Mississippi Delta around Vicksburg, the 42nd Georgia took tremendous losses from disease in their first eighteen months of service. A soldier perished nearly every day from disease until the regiment reformed in Decatur, Georgia in October 1863.[38]

The soldiers of the 42nd Georgia Volunteer Infantry fought in the War Between the States for exactly three years and were involved in twenty-two battles. As participants in the western theaters of war, they marched in the heat of summer and cold of winter an estimated 2,538 miles across seven states and traveled another 1,998 miles on top of windswept boxcars pulled by rickety trains. Historian Larry Daniel believed "the western army's bonding and survival mentality were molded in part during these lengthy and arduous trips." They certainly did not have a leader of the caliber of Robert E. Lee to rally around, nor the confidence that comes from winning large-scale battles. Surprisingly, the 42nd Georgia was not present for three of the largest

[38] Glatthaar, *General Lee's Army*, 206. Federal records show that members of the 42nd Georgia were sent to the following prisons: Camp Chase, sixty-eight men; Camp Morton, forty-four men; Camp Douglas, forty men; Rock Island, ten men; Point Lookout, ten men; Alton, six men; Johnson's Island, four men.

battles of the West: Shiloh, Stones River, and Chickamauga. They were not atop Missionary Ridge when it was scaled by Federal infantry but were, instead, in a periphery fight that day at Rossville Gap. The soldiers of the 42nd were also fortunate that they arrived on the field at Franklin too late to participate in Hood's doomed charge. Their best fighting came in fierce but lesser known affairs, with place names such as Chickasaw Bayou, Buzzard's Roost, New Hope Church, and Hollow Tree Gap.[39]

The worst performance for the 42nd Georgia, and certainly for Barton's Brigade, was on May 16, 1863, at the Battle of Champion Hill. It was their first major battle, and the brigade was swept back in confusion off the hill, taking significant casualties and allowing their artillery support to be captured. It would take more than a year for Southern generals to regain confidence in the fighting abilities of the Georgians in Barton's Brigade. The bloodiest day for the 42nd Georgia was the second day at the Battle of Resaca when they charged into overwhelming Federal numbers and took more than 100 casualties. The order to charge was an afterthought to Joseph Johnston, but the poorly conceived assault undid the good work that the officers had undertaken in Dalton all winter and spring to rebuild the 42nd Georgia in both numbers and morale. The regiment had been a wreck after the forty-seven-day siege and surrender at Vicksburg in July 1863. Yet, it was a confident, robust regiment that emerged from the tree line and charged across the railroad tracks of the Western & Atlantic above the village of Resaca.

Much credit must go to Lovick Thomas, who would have to rebuild the 42nd a second time and hold it together during the ninety-day Atlanta campaign. The culmination of his efforts came in the most famous charge in the regiment's history at the Battle of Atlanta. The moment is captured to this day in the Cyclorama painting at the Atlanta History Center. Finally, the men of the 42nd Georgia did some of their best fighting near the end of the war. They beat back a large Federal assault on Overton Hill in the Battle of Nashville and, the following day, became the fighting rear-guard that saved Hood's retreating and demoralized army. The remnants of the 42nd were also in the middle of the last major fight for the Army of Tennessee at the Battle of Bentonville in North Carolina. The cumulative record of the 42nd Georgia Volunteer Infantry is a good one and worthy of the respect and admiration its survivors would receive in the Confederate reunions that followed the War Between the States.

---

[39] Daniel, *Soldiering in the Army of Tennessee*, 28.

Chapter 10

Beginning Life Again

"The country is blessed in having as its foremost citizens the remnant of the lives spared from the Forty-Second Georgia regiment."
—Colonel Andrew Jackson West, July 22, 1899
Speech to the 42nd Georgia Survivors' Association

"Wherever men have been needed, in this grand work, in the farm or the shop, in the office or mart of trade, in legislative halls or in the courts, in city, town, or country, veterans of the 42nd Georgia have been conspicuous."
—Georgia Governor Henry McDaniel, 1888

Surviving Reconstruction

Writing in 1910, Robert E. Lee biographer Douglas Southall Freeman pointed out that "Most of the men who came back from Appomattox and other fields of sorrow were young men." So, it was then that the generation chosen by fate to determine by force of arms those issues that the preceding generation had failed to settle by legislative action, was now responsible for the recovery and restoration of a "New South." Freeman explained, "All depended on the boys. How they did it, how they surmounted obstacles and repaired their broken fortunes, how they struggled to make life worth living and home worth having; how they rose and toiled and finally succeeded is no less heroic than the tale of their military prowess."[1]

These veterans of war returned to the farms they had not seen since spring 1862. Pleased to be reunited with their families, they confronted destitution at every turn. Sergeant Leonidas Stephenson was one of three brothers from the Newton Rifles (F) to return to their Georgia county from the Greensboro surrender. Because he had an injured foot, Leonidas was given an army mule named "Peter" to ride on his journey home. Returning to his wife Jeannette and baby son, the twenty-five-year-old L. F. Stephenson found

[1] Johnson, *Douglas Southall*, 90.

that he had become the owner of the only mule in the Rocky Plains District of Newton County. Peter, the lone mule, became the foundation for rebuilding Stephenson's financial wherewithal. Peter's eventual death at the age of thirty-six would warrant an article in the local *Georgia Enterprise*.[2]

Colonel Lovick Pierce Thomas was thirty years old when he returned to Lawrenceville, Georgia, in May 1865. Though he had "staked my all and lost," he was ready to begin anew. Indeed, fifty years of a fruitful life lay in front of him. His wife, Jennie, age twenty-nine, shared his desire to put the war behind them. She wrote in her diary, "We are subjugated, just as I had long believed would be the case, and I think Lovick will not willingly engage in another war soon." They now had three children: daughter Mattie, who was seven; son Lovick, who was five; and daughter Fannie, who was two.[3]

Decisions would need to be made quickly. While many of the men in the 42nd Georgia were returning to their neglected farms, Thomas was a merchant without a store, stock, or capital. Rebuilding their lives in Lawrenceville seemed out of the question. The large textile mill near Lovick's old general store had burned on May 30, 1864. This meant that most of the 250 machinists, operators, repairers, and overseers had left the Gwinnett county seat for other opportunities in other mill towns.[4]

Lovick and Jennie, like many other Georgians, saw opportunity in Atlanta, which was thirty-five miles southwest of Lawrenceville. The "Gate City" was already home to many veterans of the 42nd Georgia, and hundreds more would move there in the decades to come. Though much of this important city lay in ruin in 1865, there was an optimism that, like the mythical phoenix, Atlanta would rise again from its own ashes. Once rebuilt, four railroads would serve Atlanta: the Western & Atlantic to Chattanooga, the Atlanta & West Point Rail Road to LaGrange, the Georgia Rail Road to Augusta, and the Macon & Western to Macon. Whatever potential there was for renewed prosperity in the South, it would almost certainly run on steel rails through Atlanta. A correspondent for a Northern newspaper described 1866 Atlanta as "a never-ending throng of pushing and crowding and scrambling and eager and excited enterprising men, all bent on building and trading and swift fortune making."[5]

---

[2] "Donald Grier Stephenson," in Newton County Historical Society, *History of Newton County*, 925.

[3] Jennie P. Thomas, diary, November 4, 1865, in Huff, *Four Families*, 93.

[4] Holman, "Factory Town," 63–64.

[5] In 1873, the Atlanta and Charlotte Air Line Railroad added the fifth railroad into

"Just after the war ended," wrote Lovick, "I commenced, as we all did, to look about me and see what I had in the way of worldly goods to commence life over again." He was not completely destitute, as he came home with a horse, two mules, an ambulance wagon, and a Mexican silver piece. There was also the vacant city lot in Atlanta, on Crap's Alley, that he had bought, sight unseen, in December 1862. Jennie contributed one twenty-dollar gold piece that she had faithfully held on to throughout the war. Lovick began his new civilian life with a financial transaction. He recounted, "I traded a note on a prominent gentleman still living in the grand old county of Gwinnett—a note I had carried through the war—for a house in Lawrenceville, once used as a store house for the Lawrenceville Factory."[6]

The $150 note from William Simmons was traded to the manufacturer Enoch Steadman for a small one-story storehouse in Lawrenceville. With the help of a carpenter, Lovick dismantled the building, "with a view of hauling it to Atlanta." Transporting the lumber some thirty-five miles over wagon roads and then use the boards to build a new house on his town lot seemed like an impossible undertaking. "But members of my grand old regiment came to my rescue," he noted, as "they came in from all parts of the county with their horses, mules and wagons, and by their timely assistance the house was hauled to Atlanta."[7]

Fifteen veterans, "any one of whom would have died for their colonel," provided the workforce to load the wood from the old storehouse onto wagons and begin the journey to Atlanta. The well-worn wagons seemed to break down with each hill they climbed, and the caravan left a trail of shingles from Lawrenceville to Atlanta. It would be a two-day journey to reach their destination. Lovick rode ahead wearing his old military blue jacket, gray pants, military cap, and a red sash about his waist. Against the backdrop of a setting sun, he stopped on the hill near Oakland Cemetery, not far from the war-scarred battlefield of 1864, and waited for his ramshackle caravan to catch up to him.[8]

Speaking at a regimental reunion in Social Circle in 1899, Lovick, who was an inveterate story-teller, recounted the rich visual scene, "I sat on my old army horse, with the same soiled uniform on that I had worn during the

---

Atlanta (Shavin and Galphin, *Atlanta*, 82).

[6] "Col. Lovick P. Thomas," in *Memoirs of Georgia*, 1:943.

[7] Huff, *Four Families*, 26.

[8] Winn, "A Sketch of the Life and War Record of Colonel Lovick Pierce Thomas II," 60.

war, near the Kimball House corner of Decatur Street and watched that long wagon train as it came slowly toward me up the street." There were exquisite details to recount: "My old weather-beaten friends, with poor old horses and mules, hitched to dilapidated wagons, over-loaded with lumber belonging to that old house; wagons some of them that had stood neglected of no use all during the war; here they came, winding and moving along as well as they could hardly walk for the dirt, brick, and debris from the walls that had fallen during the siege of Atlanta." The gold coin could only pay some of the expenses, and his army comrades would have to wait until Lovick could earn the money to pay the balance. It was an important new beginning, as housing was scarce and expensive in the resurgent city of Atlanta.[9]

With the reconstructed house on Crap's Alley, the Thomas family had a significant head start by having a modest home to call their own. It took Lovick and one worker forty days to sort out the wood, secure a pile of old bricks for a foundation, and build a house on the half-acre town lot. Much later, at the age of sixty-five, Lovick would fondly recall his first Atlanta home: "Comrades; that house stands in the same place today, and although I have built quite a number of others, I love it best of all. It stands as a monument to the love and devotion, of members of my old regiment, and there it shall stand without change of alteration until time shall carry it off."[10]

Jennie Thomas was not nearly as pleased as Lovick about their new Atlanta home, believing the small house on Crap's Alley was isolated and in a poor part of Atlanta. "I felt quite low spirited when I landed at my new home and took a good cry when I went to bed," wrote Jennie in her diary. Thankfully, in 1866, the street name was changed to Trebursey Street. Jennie had a new baby, Sarah Clara Thomas, born in September 1865. Particularly important to the family, in their early days in Atlanta, was the presence of former slaves Jiles and his wife, Mandy. Jiles used the Confederate ambulance to bring in much-needed money as a dray driver operating from the nearby train station. Mandy did the cooking for the family, and her cornbread was the

---

[9] Thomas, "1899 Reunion Address at Social Circle," vol. 1, pg. 14, Thomas collection, BRC. Part of the 1865 surrender was a requirement that all buttons and insignias be removed from uniforms that would be worn in public.

[10] Introductory remarks of L. P. Thomas, 42nd Georgia Reunion, Social Circle, July 22, 1899, vol. 1, pg. 14, Thomas collection, BRC; Winn, "A Sketch of the Life and War Record of Colonel Lovick Pierce Thomas II," 60; Joseph Porter's sale to L. P. Thomas of a one-half acre lot, with 100 feet facing Craps Alley, took place on December 22, 1862, and was recorded in the Fulton County Deed Book on May 24, 1875. The house would eventually be given an address of 42 Tatnall Street.

centerpiece of each meal. A year into their new life in Atlanta, L. P. managed to purchase a barrel of flour. The family story is that Jennie asked Lovick where they should put the barrel. "Put it!" exclaimed the colonel, "where should we put it, but in the parlor." So, there it sat in the empty parlor, more prized than a plantation settee until finally exhausted.[11]

The Bush Arbor

Lovick's father, "the Fighting Quartermaster," also called Atlanta home in the years after the war. Lovick Thomas Sr. was named 1st lieutenant of Atlanta's new police force after the December 1865 local election. Thomas Sr. hired thirty-two officers, all, with hardly an exception, former Confederate soldiers. This small command represented what law and order there was in this crazy railroad terminus. After the December 1866 election, Thomas Sr. was made chief marshal of the police department. The elder Thomas was paid $1,500 to the mayor's annual income of $2,000. The youngest patrolman on the force was L. P. Sr.'s seventeen-year-old son, Heck Thomas.[12]

Valentine Barnwell's City Directory of Atlanta for 1867 announced a population of 20,228, with 9,288 black residents. The city limits were defined as everything within a mile and a half radius from the centrally located passenger depot. The directory lists Lovick Thomas Jr. as a grocery merchant living on Trebursey Street, and his father, Lovick Thomas Sr., as the chief marshal of Atlanta. Both men and their families were members of Trinity Methodist Church, which stood on Mitchell Street, across from the city hall. The 257-member congregation had just replaced their minister, Atticus Haygood, with Reverend William Crumley.[13]

In spring 1868, the Georgia state capitol was moved from Milledgeville to Atlanta. This move by Federal authorities reflected Atlanta's increasing importance to the state of Georgia. With a dearth of public buildings, the General Assembly took over the half-finished Atlanta Opera House owned by Hannibal Kimball. Post-war politics was a free-for-all in Atlanta and across the state. On July 23, 1868, a Democratic rally on a vacant plot of land by

---

[11] Jennie P. Thomas, diary, November 4, 1865, in Huff, *Four Families*, 94; Garrett, *Atlanta and Environs*, 1:705; Winn, "A Sketch of the Life and War Record of Colonel Lovick Pierce Thomas II," 60.

[12] Garrett, *Atlanta and Environs*, 1:703, 733.

[13] Ibid., 1:703; Meeks and Speer, *Heck Thomas*, 8; Barnwell, *Atlanta City Directory of 1867*, 261. Unfortunately, there were no street numbers in Atlanta until the 1870s.

the passenger depot drew 20,000 people to hear five hours of fiery anti-Reconstruction speeches, including those of Robert Toombs, Howell Cobb, and Benjamin Hill. Northern radicals and Southern carpet-baggers were denounced for plundering the public treasury. An immense bush arbor podium and hard plank seats for thousands had been constructed for the event. The day's affair closed with a spectacular torchlight parade led by marching bands. The "Bush Arbor" would remain and consequently be used by both factions as a staging area for political demonstrations.[14]

By 1868, the newly enfranchised black population faced intimidation from the Ku Klux Klan to abstain from voting, and violent deeds from the Union League militias if they did not vote for the Republican ticket. It was in this atmosphere of nightly political demonstrations that police officers Heck Thomas and Jack Smith were patrolling on Decatur Street near the Bush Arbor. Cursing and antagonizing a torch-lit parade of 200 black Republicans was Andrew Whitaker, a black man well-known for his Democratic stance. Sensing the inherent danger in this combustible situation, the two officers arrested Whitaker for his own safety and began a journey to the station house. Unfortunately, the small party soon ran into another torchlight procession of radical Republicans. Shouts and demands were immediately given to the two white policemen to release their prisoner. Heck refused and, according to his account, "the battle was on right then and there."[15]

Both the prisoner, a black Democrat, and Smith were knocked to the ground in the ensuing melee. The angry Republican mob beat them both with clubs and sticks. Heck remained on his feet and blew furiously on his police whistle, summoning help from the nearby station. Officers Ed Murphy, John Johnson, and Sid Holland ran to join the fight, which was taking place under a gas light post on Decatur Street. "Sid got mixed up with a big negro for possession of an axe," recalled Heck in 1910. In one account of the riot, Heck used his nightstick to take out the troublemaker with a heavy blow to the head, and, in another account, with a blow to the temple from the butt of his Colt revolver. Heck remembered that "a little negro—he wasn't much bigger than my fist—ran up on my right side and shot me square through the right thigh. But I never fell and stayed through the fight." Once the crowd

---

[14] Garrett, *Atlanta and Environs*, 1:783, 787. The outdoor venue has also been referred to as the "Brush Arbor."

[15] Meeks and Speer, *Heck Thomas*, 3; "Bruff's Column," *Atlanta Constitution*, August 24, 1910, 8.

had chosen gunplay and elevated the stakes, Heck fired into the surging attackers with his Colt, eventually disbursing the Republicans in every direction.

Thus, began the noted and sometimes violent law enforcement career of Heck Thomas, one in which he would be wounded eight times and take the lives of eleven criminals. Whether wearing the badge of an Atlanta police officer, a Texas railroad detective, or a Federal marshal in Oklahoma Indian Territory, the younger brother of L. P. Thomas would always call for his suspect to surrender, and if they did not comply, their death, was by their own volition.[16]

Wholesale Groceries

Upon his arrival in Atlanta, Lovick Thomas found it difficult to find meaningful employment in a city filled with unemployed Confederates. He began by clerking at several different wholesale grocers for nominal pay. This large-scale distribution of food and grocery supplies was a new but growing business in the South. With the plantation system giving way to sharecropping and tenant farming, country stores were opening at every crossroads. Atlanta merchants were sending jobbers and salesmen across the region to canvass these stores for business. The *Hawkinsville Georgia News* reported in 1882 that "Everybody has seen the Atlanta drummer. He skips with his little sample bag all over the face of the earth from Richmond to New Orleans."[17]

Salvation for L. P. Thomas came in the form of a fellow Confederate officer named Andrew Jackson West. The owner of West, Edwards & Company, A. J. West, who was eleven years younger than L. P. Thomas, ran one of the more substantial wholesale grocery concerns in Atlanta. Lovick quickly became a trusted salesman and traveling agent for Mr. West. The men had several things in common, starting with an upbringing in a farming community in Georgia. A. J. West was from La Grange, the county seat of Troup County on the Alabama line. The two friends were proud of their Revolutionary War ancestors and had enlisted in Confederate Service confident they were defending similar principles of liberty. Both were recognized for their bravery under fire and had received battlefield promotions. Finally, both former Confederate officers viewed Atlanta as their best opportunity to build

[16] Ibid.

[17] Russell, "Atlanta, Gate City of the South, 1847 to 1885," 174. By 1871, there were fifty-five wholesale dry goods and thirty-three wholesale produce firms in Atlanta.

their postwar careers.

When the Civil War started, Andrew West, only seventeen, had enlisted with the Troup Light Guards, later to become Company E of the 41st Georgia. At the Battle of Perryville, on October 8, 1862, Sergeant West had been the first to scale a hill and reach the Yankee battery in a dramatic charge. His captain, John Curtright, was cut down at Perryville, and West was severely wounded himself. The young Georgian was "left on the field as a dying soldier and attributes his recovery to the kindness of a Kentucky girl, Miss Lizzie Everhart." Upon regaining his health, West returned to Bragg's army and was promoted by special distinction to captain. He went on to fight from Vicksburg to the final surrender at Greensboro, North Carolina. Ragged and shoeless, Captain West may have accompanied Colonel Thomas on the slow walk back to Georgia. Lovick said of A. J. West, "It was from this aid of his that I first began life again, and soon began to see the dark clouds disappear, and bright tints and golden linings where once all was dark and gloomy."[18]

These arduous early postwar years in Atlanta also saw an emergence of ambitious political "movers and shakers" that read like a roster of Confederate officers from the Army of Tennessee. On December 2, 1868, the "boy colonel" William Hulsey of the 42nd Georgia was elected mayor of Atlanta. Only twenty-nine, the former attorney had returned from the war and gone into partnership with his former preceptor, William Ezzard. Hulsey then served eighteen months as the solicitor-general of the Coweta Circuit, before being removed by Republican Governor Rufus Bullock for partisan reasons. Mayor Hulsey would, during his term in office, be responsible for the creation of Oglethorpe Park on the northwest side of town. Colonel Hulsey would go on to be elected to the Georgia House of Representatives for Fulton County in 1878 and 1882. The *Atlanta Constitution* called the lawyer "a deep thinker, a careful student and fluent talker."[19]

Hulsey's fellow regimental staff officer, Robert Maddox also entered politics after the war. In 1866, he was elected to the State House, then sent by Governor Jenkins to the northwest to buy $200,000 worth of grain for the starving citizens of Georgia. Maddox was back on his feet quicker than most, thanks to the success of his tobacco and liquor store on the first floor

---

[18] "Col. A. J. West," in *Memoirs of Georgia*, 1:950; introductory remarks of L. P. Thomas, 42nd Georgia Reunion, Social Circle, July 22, 1899, vol. 1, pg. 14, Thomas collection, BRC.

[19] "Col. W. H. Hulsey," in *Memoirs of Georgia*, 1:828–29; "Thirty Years Ago, and Now," *Atlanta Constitution*, October 16, 1892, 21.

of the National Hotel, which was next to the Kimball House and a short walk from the Union Depot. In the census of 1870, he had real estate and personal property valued at $16,000.[20]

Chief of Police

The brokered national election of 1876 marked Redemption for the South and the withdrawal of Federal military forces. Former Confederates were now welcomed to participate in all levels of government. Georgia had been well ahead of the curve, having been ruled by the Democratic Party since 1872. By the time Georgia delegates wrote a new state constitution in 1877, Confederate general Alfred Colquitt was serving as governor, and the state was represented in the United States Senate by Redemption statesman Benjamin Harvey Hill and Confederate general John B. Gordon.

On April 1, 1877, Atlanta's board of police commissioners selected Lovick Thomas as the city's new chief of police. At a yearly salary of $1,500, the forty-two-year-old Thomas "would be held responsible for the good order of the city, and general good conduct of the officers and men of the force." Lovick's force of forty-three men operated from a general headquarters at 47 South Pryor Street, about six blocks south of the Western & Atlantic Train Depot. "Lovick has been traveling most of the time," Jennie Thomas wrote in her diary on April 9, "but has made so little and was so tired of it that he accepted the office of Chief of Police. I do not like it, chiefly because it looks like a coming down to accept such an office, and then it takes him out of nights and some of Sundays." Her assessment of the situation turned out to be wrong. What success Lovick had enjoyed in the wholesale grocery business had been crushed by the Panic of 1873. Law enforcement, on the other hand, turned out to be a true calling for Lovick Thomas. He proved to be an inspiring leader to his patrolmen, used his military background with the 42nd Georgia to organize every aspect of his department's duties and responsibilities, and proved to be a natural at communicating with the board of police commissioners, the newspapers, and the public.[21]

Only four days into office, Lovick Thomas was profiled by a "Georgia Soldier" in a column for the *Atlanta Constitution.* The unnamed writer was

---

[20] "Colonel Robert F. Maddox," in Pioneer Citizen's Society of Atlanta, *Pioneer Citizen's History of Atlanta*, 249–53; Reagan, *H. I. Kimball*, 125.

[21] "The Baton Brigade: Election of the New Police Force Yesterday," *Atlanta Constitution*, April 1, 1877, 3; Jennie P. Thomas, diary, April 9, 1877, in Huff, *Four Families*, 99.

most certainly the editor of the newspaper, Evan Howell, a former officer in the Georgia militia and a good friend of Colonel Thomas. The "Georgia Soldier" quoted from General William Sherman's new memoir about the loss of the DeGress Battery on July 22, 1864. He then wrote, "Doubtless, few of our readers know, except those with the army at the time, that one of the Confederate officers most conspicuous in this battle was Colonel L. P. Thomas, Jr. who commanded the 42d Ga. Regiment, and was on the extreme left of the attacking line on that day." This passage was the first of what would be numerous mentions in the *Atlanta Constitution* of the 42nd Georgia capturing the DeGress Battery at the Battle of Atlanta. Equally conspicuous in L. P.'s department was police captain "Tige" Anderson. A former brigadier-general in Longstreet's Corps, the fifty-three-year-old Anderson was a contemporary of Lovick's uncle, General Edward Thomas, and a native of Covington.[22]

The ambitious L. P. Thomas only served one term as police chief. In the election of December 4, 1878, Thomas was a successful candidate for deputy sheriff of Fulton County on a ticket headed by Judge William Asbury Wilson. The county sheriff and his handful of deputies were often well compensated because they were paid a percentage of the fees and proceeds that they collected. Thomas would earn more in this new position, with responsibilities that included providing security for judges, juries, and witnesses during trial sessions of the Circuit Court, serving all legal papers such as summonses, warrants, and writs, maintaining the county jail and workhouse, transportation and custody of prisoners, and, by court order, the seizing of personal property and real estate and selling it at auction. L. P. would serve as a deputy sheriff for the next six years, serving under Wilson and Ang Perkerson.[23]

The new mayor of Atlanta, William Lowndes Calhoun, was also elected on December 4, 1878. He won in a tight race against Captain James English. The son of former war-time mayor James Calhoun, Captain Calhoun returned from the surrender of the 42nd Georgia at High Point to practice law with his father. From 1873 to 1876, he served two terms representing Fulton County in the state legislature. Mayor Calhoun was successful in refinancing

[22] "Col. L. P. Thomas, Jr.; Our New Police Chief—A Scrap of History," *Atlanta Constitution*, April 4, 1877, 4; Warner, *Generals in Gray*, 6. General Anderson had been severely wounded at the Battle of Gettysburg.

[23] Keeping a percentage of the fees collected was common practice in Gilded Age America.

the city debt at a 6 percent rate of interest and hosting General William Sherman on his inspection trip to Atlanta in January 1879. After his term as mayor, W. L. Calhoun was elected judge of the Court of the Ordinary (Probate) for Fulton County. He served sixteen years in this position, often running without opposition in the bi-annual elections.[24]

International Cotton Exposition

On October 5, 1881, the grand opening of the International Cotton States Exposition took place at Oglethorpe Park on the northwest outskirts of Atlanta. With bands playing, President Chester Arthur and Georgia governor Alfred Colquitt led the dignitaries under a fifty-foot high entrance into the exposition featuring 1,113 exhibits powered by three large steam engines. It was the first World's Fair ever held in the South and an impressive achievement for a city of 42,000 people. The emphasis was on improved methods in the cultivation of cotton, all symbolized by the unveiling of James Henry Moser's mural, *The New South Welcoming the Nations of the Earth*.[25]

The three-month fair drew large crowds and excellent reviews from national newspapers, thereby helping to reshape a positive image of the South. Lowndes Calhoun was the mayor of Atlanta as the planning and preparations were undertaken, while James English, his former opponent, served as the mayor during the three months the fair was in operation. Virtually every important businessman and lawyer in the city gave their time and financial support toward the endeavor. The names include Henry Grady, Samuel Inman, Evan Howell, Joseph E. Brown, Robert F. Maddox, and Sidney Root. Colonel Maddox served on both the audit and construction committees.

Lovick Thomas was made chief of the Department of Protection, with the responsibility of supervising the Exposition Guard, the fire department, as well as the "general protection for person and property." Some 250,000 visitors would explore the exposition without incident. Chief Thomas, in his final report, noted that there were two attempts to break into the exposition at night. On both occasions, the marauders were chased away by the night watch under Captain Richard Clark and the force's trusted bloodhound, "Old Lee."[26]

---

[24] Massey, "William Lowndes Calhoun," 165.

[25] King, "International Cotton Exposition," 186–88.

[26] The International Cotton Exposition of Products, Machinery, and Manufactures; Prospectus, 1881, pg. 5; King, "International Cotton Exposition," 188.

Lovick was also pleased to see his old Georgia Brigade commander, General Marcellus Stovall of Augusta, Georgia. As secretary-treasurer of the Georgia Chemical Works, Stovall had a building constructed at the exposition that showcased his company's fertilizer products and included a model of the Augusta works built along the Savannah River. In the 1870s, Stovall had been involved in the give and take popular among former Confederate generals as to where the blame lay for poor tactical decisions made during the war. He was a sympathetic supporter of his former superiors Daniel Harvey Hill and Joseph E. Johnston. In an 1874 letter to General Johnston, Stovall wrote that his Georgia men had "perfect confidence" that, if allowed, Johnston would have prevailed over Sherman's Federals. "The morale of my Brigade," he wrote Johnston, "was never better than when it crossed the Chattahoochee." As for Hood, he recounted the day when President Davis visited the Army of Tennessee at Palmetto Station. Stovall wrote, "He was greeted along the line with a universal demand for 'old Joe,' and although his was preceded by a staff officer of the commanding general with the request that each demonstration be suppressed—the troops had caught the enthusiasm which that name never failed to inspire."[27]

The 42nd Georgia Veterans Survivor's Association

What had begun in 1884 as a one-time regimental reunion became a yearly gathering that ran until 1914. Organized by Colonel L. P. Thomas, the annual meeting was held every July 22, the anniversary of the Battle of Atlanta. The location of the gathering would change each year between Atlanta, Buford, Covington, Decatur, Monroe, and other hometowns of the veterans. Through the years these reunions also became community events. The reporter at the 1888 reunion in Monroe optimistically estimated the crowd to be around 6,000 people.[28]

The second reunion was held in Decatur in 1885, and it would establish a pattern for future gatherings. A special train from Atlanta was reserved to bring the veterans, with their family and friends, to and from the reunion. Baskets of food were loaded onto the train for the big picnic dinner to come. Upon the arrival of the train at the Decatur platform, the 200 or so veterans

---

[27] C. L. Bragg, *Distinction in Every Service*, 127; Marcellus Stovall, "June 28, 1874, letter to Joseph E. Johnston," Stovall papers, ASU.

[28] "A Grand Reunion: Monroe the Scene of a Notable Gathering," *Atlanta Constitution*, July 21, 1888, 5.

were formed by companies. Led by the Buford Brass Band, the column marched up to the Courthouse Square and down to the arbor. They followed color-bearer W. F. Edwards, who still carried the bullet-riddled regimental flag that had been smuggled home from the Greensboro surrender. Upon gaining their seats, they were welcomed by George Hillyer, mayor of Atlanta.

So, a custom was thus established: A short parade to a shaded arbor, where the master of ceremonies introduced the different speakers and guests. Among other considerations, men were recognized for traveling great distances to be present. In 1886, for example, Captain Thomas Hill of the Fulton Blues (I) traveled some 900 miles from his home in Weimer, Texas. Each reunion had a guest speaker who would make a speech, followed by a response made by a member of the 42nd. A poem was invariably read, and representatives would report on the status of the survivors in their respective companies. A capacious afternoon dinner would be followed by "social intercourse and hand shaking, music and rejoicing." There were often foot races and a rifle shooting contest in the afternoon. A group photograph was sometimes taken. The Atlanta newspapers usually had reporters present to garner colorful quotes.[29]

For most of the association's history, L. P. Thomas served as commander, W. L. Calhoun as the historian, W. M. Durham as secretary, and G. B. Strickler as chaplain. A notable list of dignitaries addressed the veterans over the years. Governor Henry McDaniel spoke in 1888, General Clement Evans in 1890 and 1893, General Stephen D. Lee in 1898, and Congressman Lon Livingston in 1900. There were many years where joint reunions were held with another regiment. They included the 7th, 16th, 35th, and 43rd Georgia.[30]

Sheriff of Fulton County

In the race for sheriff of Fulton County in fall 1884, Lovick Thomas ran unopposed in both the Democratic primary and the general election. Lowndes Calhoun also enjoyed a "clean walk over" on his way to becoming the judge of the ordinary court. Both veteran officers of the 42nd Georgia

[29] "The Soldier Boys: Reunion of the Gallant Forty-Second Georgia," *Atlanta Constitution*, July 23, 1886, 5; "Today's Reunions: The Veterans of the First & 42nd," *Atlanta Constitution*, July 22, 1885, 5.

[30] The executive committee of the association included W. H. Hulsey, W. L. Calhoun, R. F. Walker, T. M. Armstead, J. M. Wilson, H. Y. Snow, R. F. Maddox, J. R. Simmons, J. D. Cameron, and F. D. Stier.

proved to be popular with Atlanta voters for years to come. Calhoun would win eight two-year terms, and the charismatic Thomas won three two-year terms. Thomas was never opposed in his runs for sheriff, and Calhoun rarely had opposition. Lovick worked each day from a high, standing desk. The *Atlanta Constitution* noted that "The colonel likes his work and sticks to it well, but he will drop his pen at any second to swap a war story." Thomas worked with three experienced deputy sheriffs during his terms in office.[31]

In 1877, Jennie Thomas confided in her diary that Lovick "has been a most excellent husband, better than I have deserved." They were an affectionate couple, as she confessed, "I find I am most as foolish about him as when we were first married, still require a deal of petting; and fortunately, he is as foolish about me as I am about him." Jennie was deeply religious and "made it the habit of my life to ask God's direction in everything." Lovick seemed to have backed away from his Methodist faith after the war, but, according to Jennie's diary, returned to the fold by 1880. Lucy Hunt, the orphaned daughter of Lovick's sister Mattie Hunt, lived in the Thomas household in the 1870s. Jennie found her to be a difficult teenager and a challenge to raise.[32]

On December 31, 1885, L. P. and Jennie's second daughter, Fannie, was wed at the Trinity Methodist Church in Atlanta to Courtland Simmons Winn of Lawrenceville. The *Atlanta Constitution* called the afternoon ceremony "one of the most beautiful and impressive ever seen in Atlanta." The auburn-haired Courtland, then twenty-three, was a graduate of Emory College, had already passed the Georgia bar, and was serving the first of two terms as mayor of Lawrenceville. His grandfather, James P. Simmons, had owned the Gwinnett Manufacturing Company, where L. P. had run the general store before the war. Courtland's father, Samuel Winn, was a contemporary and friend of Lovick's from their days in antebellum Lawrenceville. The two Confederate veterans, in short, could not be more pleased with the union.[33]

Sadly, Lovick's world was thrown asunder only six months later by Jennie's sudden death. On July 6, 1885, the fifty-year-old Jennie collapsed after

---

[31] "Who Were Elected: A Big Vote at the Democratic Primary Yesterday," *Atlanta Constitution*, October 30, 1884, 4; "Those Who are in Office," *Atlanta Constitution*, March 29, 1886, 7; "Not Many Changes," *Atlanta Constitution*, January 6, 1889, 15.

[32] Jennie P. Thomas, diary, April 9, 1877, in Huff, *Four Families*, 99.

[33] "The Social World: Events of the Week on the Social Side of Life," *Atlanta Constitution*, January 3, 1886, 11; Huff, *Four Families*, 33.

breakfast. According to her daughter Fannie, "her entire right side was paralyzed," and within hours she was dead. "Papa is heartbroken," wrote Fannie two days later, "and so stunned by the shock of dear Mama's death we are all unable to comfort him in the least." The *Atlanta Constitution* noted that Lovick and Jennie had been devoted to each other since they were school children. Her letters to him in the California goldfields had cheered him in his search for "Eldorado." As a married couple, they had known, "nothing but the sunshine of devotion."[34]

On September 7, 1886, L. P. Thomas, at age fifty-one, a widower with five children, married for a second time. His new wife was a thirty-five-year-old widow named Jeannette R. Payne. Jennie brought to this second marriage a nine-year-old daughter, Jesse Belle Payne, and a demanding temperament. Jennie and Lovick would have three children of their own. The Thomas family moved to 367 Whitehall Street in 1890 and would live there until 1907 when they moved to a large house on North Avenue.

Fulton County Confederate Veterans Association

Until 1886, Federal military rule in the South had made the formal organizing of Confederate veterans legally impossible. Expressions of sympathy and commemoration of the Southern veteran had, to this point, been confined to the activities of the Ladies' Memorial Association. The first tentative steps in Georgia had taken place in 1884 with a handful of regimental reunions.

This changed with an appeal in the *Atlanta Constitution* on April 15, 1886, for former Confederates to meet in the basement of the Fulton County Courthouse on the evening of April 20 for the purpose of organizing "a Camp of Confederate Veterans." Thirteen prominent men were behind the announcement, including William A. Hemphill and Evan P. Howell, proprietors of the *Atlanta Constitution.* Representing the 42nd Georgia were Lovick Thomas and Lowndes Calhoun. As sheriff of Fulton County, Thomas had been able to secure the use of the courthouse for the inaugural meeting.[35]

The impetus behind the establishment of the camp was the expected arrival in Atlanta in two weeks of the former president of the Confederacy,

---

[34] Fannie Thomas to May Kirby Thomas Huff, July 8, 1885, in Huff, *Four Families,* 102; "A SUDDEN DEATH.: Mrs. L. P. Thomas Dies Suddenly at Her Home of Paralysis—A Sad Event," *Atlanta Constitution,* July 7, 1885, 1.

[35] Rodgers, *History of the Confederate Veterans Association,* 6.

Jefferson Davis. One hundred eighty-two veterans came to the inaugural meeting. They had mostly served in Georgia units, but nine Southern states were represented. Present from the 42nd Georgia was L. P. Thomas, W. M. Durham, T. M. Butt, and W. L. Calhoun. Colonel William Ambrose Wright was elected president of the association and W. L. Calhoun, and L. P. Thomas were elected to the five-person executive committee.[36]

Things moved quickly after the first meeting. Roughly 200 men from the new association marched for the first time in the Memorial Day parade on April 26. The highlight of 1886 would be a week later with the unveiling of the statue to the recently deceased Georgia Senator Ben Hill in downtown Atlanta. On May 1, several thousand Confederate veterans from across Georgia marched from the Court House to the starting point on Marietta Street. The Fulton County Confederate Veterans' Association (FCCVA), now 300 strong, led the procession. The veterans were placed in position with open ranks so that Jefferson Davis, riding in a white carriage, could once again review his "Boys in Gray." Atlanta schoolchildren strew flowers in front of the old war heroes as they marched down Pryor and Peachtree streets. Newspaper editor Henry Grady formerly introduced Jefferson Davis to the thousands present, and Georgia's two most famous generals, James Longstreet and John B. Gordon, appeared on the podium as the veterans voiced their approval with spontaneous Rebel yells.[37]

At the second meeting of the FCCVA, held on May 17, 1886, the members approved a twenty-three-point constitution. The purpose of the new association was "the conservation of Confederate memories, the promotion of fellowship, and the cultivation of friendship between surviving officers and soldiers." Other important objectives included "the exhibition of loyal respect to the recollections and impulses of a Confederate past," as well as "the extension of reasonable aid and sympathy to fellow-members in seasons of sickness and distress, and, in case of death, the rendition of suitable funeral honors."[38]

Calls for financial help from distressed veterans initially overwhelmed

[36] William Ambrose Wright (1844–1929) served on his father's staff in the famous Wright's Brigade. The young officer lost a leg at the Battle of Second Manassas and was later captured in the march to Gettysburg. In 1886, he was serving as the comptroller general of Georgia under Governor Alfred Colquitt. Colonel Wright is buried in the generals' section of Oakland Cemetery.

[37] Rodgers, *History of the Confederate Veterans Association*, 11, 106–107.

[38] Ibid., 108–12.

the young organization. Money for the relief fund was raised each fall with a weeklong Veterans Fair supervised by L. P. Thomas. It was estimated in 1886 that more than 3,000 former Confederates lived in Fulton County alone. In a time when women could not vote and the rights of freemen to the ballot were being increasingly restricted, the Southern veterans, like their counterparts in the North, came to realize that they were a potent political force capable of determining the outcome of state and local elections.

In 1888, Lowndes Calhoun, the former Captain of the Calhoun Guards (K), became president of the FCCVA. Establishing himself as a competent and popular leader, he went on to be reelected in 1890. One of the first tasks facing Calhoun was securing a permanent meeting place. For twenty-five dollars a month, the association rented the second floor of a large building at 26 Broad Street, one block north of the current Five Points MARTA station. Not surprisingly, the space was renamed Confederate Hall and served as headquarters for the veterans for years to come. The auditorium was packed for the dedication on September 17, 1888. After an orchestra had warmed up the crowd, President Calhoun introduced three speakers: Governor John B. Gordon, "honorary" member Henry Grady, and newspaper publisher Evan P. Howell. Introduced as the captain of the Howell Artillery Battery, Howell rallied the new association to, "Keep up the lick we've got now and I'll stand by you. Your purpose is grand, for it goes further than social reunions, and it is a friend to help Confederate soldiers that have had no friend since the war." There was an element of fairness at play. "The United States (I do not say this complainingly) has paid more to her veterans since the war," noted Howell, "than the Confederacy paid her soldiers during the war in Confederate money, and the old Confederate veteran has helped pay for it."[39]

Under Calhoun's inspired leadership the association soon secured a fifty-foot square plot on an elevated piece of land in the new Westview Cemetery. In exchange for the property to be used for the burial of its members, the association agreed to erect a monument in the center of the plot. Dr. Amos Fox, treasurer of the association, spearheaded the selection, design, and funding of the monument. The marble shaft arrived in New York harbor from Italy in March 1889. The twenty-foot monument is capped by an infantry soldier carved from Carrera marble. A handsome work of art, it remains today, "Erected by the Fulton County Veterans Association in Memory of their Dead Comrades." Another priority for the men was to assist

[39] Rodgers, *History of the Confederate Veterans Association*, 137; "The Men in Gray: Dedication of the Confederate Hall Last Night," *Atlanta Constitution*, September 18, 1888, 5.

the Atlanta Ladies Memorial Association (ALMA) in raising money to replace the rotting wooden markers above the 7,000 Confederate graves at the Oakland Cemetery with permanent marble markers. L. P. Thomas chaired the headstone committee.[40]

An important new tradition was also established on October 21, 1888, as the first memorial service was held at the Central Presbyterian Church. Honored in the service would be the Confederate veterans who had died in Fulton County the previous year. In attendance were hundreds of veterans and friends, members of the O. M. Mitchell post, Grand Army of the Republic, and ministers from throughout the city. President Lowndes Calhoun was particularly eloquent that night: "Remembering my return to this city after the war—the home of my boyhood and of my love—wounded in body, broken in health, with spirits depressed but not conquered, property destroyed and friends gone, and that I met here some of the men whose names will be called tonight, it is natural that I should be filled with solemn memories." This yearly tradition of holding a memorial service in October would continue for another twenty-five years.[41]

United Confederate Veterans

Founded in New Orleans in 1889, the United Confederate Veterans (UCV) managed in less than four years to unify the various veteran groups scattered across the former Confederate states. In 1890, the Fulton County Veterans Association became UCV Camp #159.

Georgia General John Brown Gordon was elected the first commander-in-chief of the national UCV and served in that capacity until his death in 1904. General Clement Evans, a Methodist minister and historian, headed the Georgia division for twelve years. Andrew West, the close business friend of L. P. Thomas, served as the adjutant general and chief of staff of the Georgia division. By 1896, there were 850 UCV Camps in the United States with fifty-five camps in Georgia. Former members of the 42nd Georgia were principal members of Camp 159 (Atlanta), Camp 305 (Covington), Camp 440 (Jefferson), Camp 665 (Decatur), Camp 736 (Cumming), Camp 973 (Gainesville), Camp 982 (Lawrenceville), Camp 1092 (Dawsonville), Camp 1115 (Newton), Camp 1300 (Buford), and Camp 1684 (Rockdale

[40] Rodgers, *History of the Confederate Veterans Association*, 17, 141.

[41] "Memorial Service: Of the Fulton County Confederate Veterans' Association," *Atlanta Constitution*, October 22, 1888, 4.

County).[42]

One of the highlights for members of this new veteran organization were annual reunions held at the state (division) level in the spring and the five-day national convention held each summer. The power and breadth of these gatherings were majestic in their stirring presentation. Host cities of the national reunion went to elaborate lengths to entertain and accommodate the thousands of former Confederates and their guests. Colonel L. P. Thomas's scrapbook, now on permanent display at the Bell Research Center in Cumming, Georgia, holds various medals and ribbons from Confederate reunions, including delegate badges to the 1897 national reunion in Nashville, Tennessee, and the 1899 reunion in Charleston, South Carolina.

Death of Colonel Robert Henderson

General Robert Johnson Henderson, the first commander of the 42nd Georgia, was also the first of the regiment's principal staff officers to expire in the postwar years. The sixty-nine-year-old Georgian died in Atlanta on February 3, 1891, after a protracted illness. Lovick Thomas and Lowndes Calhoun made the arrangements for the funeral. The day before the general's death, L. P. had received a telegram from William Smith of Buford expressing his concern over Henderson's deteriorating condition. "A brave, noble spirit is passing away," lamented Smith. The former captain of the Independent Rebels (B) went on to describe a private conversation that he and surgeon John S. Wilson had had with Henderson two years earlier on Alabama Street. "He said he did not think he would live long," Smith said, "and expressed a wish that when the final dissolution should come that the survivors of the old Forty-second should bury him." So it was that the pallbearers of General Robert Henderson would be five colonels, three captains, and one adjutant, all former officers of the 42nd Georgia.[43]

One of the challenges for the funeral party was getting Henderson's body from Atlanta to Covington and then on to City Cemetery. Lowndes Calhoun, as president and commander of UCV Camp 159 in Atlanta, sent out a directive asking seventeen veterans to escort the remains to the Union

---

[42] Luana Henderson, *Inventory of United Confederate Veterans Records*, 23–54.

[43] "A Soldier's Death," *Atlanta Constitution*, 1891. According to the *Atlanta Constitution*, the pallbearers were Colonel W. L. Calhoun, Colonel L. P. Thomas, Colonel R. F. Maddox, Colonel W. H. Hulsey, Dr. W. M. Durham, Captain H. Y. Snow, Captain B. F. Walker, and Colonel W. T. Smith.

Passenger Depot. A special car took the funeral contingent to Covington, where the train was met by a sizable crowd of citizens, veterans of the 42nd Georgia, and "a large number of the old plantation family servants." Henderson was laid to rest in a simple ceremony led by Presbyterian minister Dr. G. B. Strickler.[44]

The 1892 reunion of the 42nd Georgia was held in Covington in honor of Colonel Henderson. After appropriate remarks, the battle flag of the 42nd Georgia was laid across his marble tombstone. With heads uncovered, the veterans of the 42nd Georgia, in a mark of respect and reverence, marched around his cemetery plot. The stirring tribute moved Lowndes Calhoun to write, "Grand old regiment! Patriotic, true, steady, brave—it did its duty in defense of the South."[45]

Confederate Soldiers Home

In 1887, the Fulton County Veterans Association established a three-person subcommittee to explore the feasibility of building a home for indigent Confederate veterans. Samuel Inman, Lovick Thomas, and Lowndes Calhoun made up the committee, and it was Inman's idea to lobby the state assembly to sell the old capitol and use the proceeds to purchase a suitable facility for the care of disabled and needy Confederate veterans. The newspaper editor Henry Grady was an enthusiastic supporter through his editorial "Come Home, Major Joe Stewart" in the *Atlanta Constitution.* Grady was elected president of the enterprise, but his sudden and unexpected death meant that Lowndes Calhoun and Amos Fox would serve as co-presidents of the Board of Trust. With $35,000 raised through subscriptions, the committee purchased the Shultz estate near Grant Park in 1889.[46]

On Confederate Memorial Day 1890, the Grand Lodge of Masons of Georgia laid the cornerstone of the new home in an elaborate ceremony. After placing a collection of Confederate money and other southern mementos in the cavity of the stone, Grand Master John Davidson poured corn, wine, and

---

[44] "General Henderson Buried," *Atlanta Constitution*, February 6, 1891, 3; The funeral escort was made up of General Phil Cook, James A. Anderson, John A. Berry, K. C. Divine, W. D. Ellis, John A. Fitten, George Hillyer, Henry Jackson, T. B. Neal, W. A. Russell, F. T. Stier, Howard Van Epps, A. J. West, John H. Matthews, J. H. Porter, Thomas M. Armistead, and Green B. Adair.

[45] Calhoun, *History of the 42d Regiment*, 44.

[46] Rodgers, *History of the Confederate Veterans Association*, 124. Samuel Inman was reputedly one of the wealthiest men in Georgia.

oil on the stone and then struck it three times with his mallet. A grand parade of Confederate veterans from every county in Georgia followed the ceremony. The men from the FCCVA, with their camp badges and high white hats, led the procession. Thousands of citizens came out to see generals Kirby Smith and Joseph Johnston march from Confederate headquarters on Broad Street to a program at the sold-out DeGive's Opera House, and then on to the state capitol and then forward to Oakland Cemetery. The procession stretched from the capitol to the cemetery as thousands cheered their advance along Hunter Street. Hundreds of women in black greeted the marchers at Oakland Cemetery and proceeded to lay garlands and fresh flowers on each Confederate grave.[47]

Unfortunately, all did not go well from there. While construction of the Victorian-style home was completed in 1891, the Georgia Legislature failed to appropriate funds for staffing and maintenance of the facility for another ten years. The first Confederate Veteran was not admitted until June 3, 1901. Tragedy followed a few months later when fire destroyed the retirement home. The state legislature quickly appropriated funds to build a second Veterans Home at a new location at the south edge of the Ormewood Park Neighborhood. This retirement home for former Confederate soldiers on Confederate Avenue operated from 1902 to 1941. According to the official registers of the institution, thirteen members of the 42nd Georgia lived their final days in the Atlanta home.[48]

Congressman Leonidas Felix Livingston

In summer 1894, sixty-two-year-old Leonidas Livingston was running for his third term to the United States Congress from the 5th Congressional District of Georgia. The chairman of his re-nomination campaign was Colonel Robert F. Maddox of Atlanta. A lifetime farmer and resident of Newton County,

[47] Ibid., 185–90.

[48] Brooke, *Into the Gray Vale*, vii–viii. The Bell Research Center in Cumming holds the original set of *Official Records of the Rebellion* from the library of the Atlanta Confederate Home. Many of the old soldiers wrote comments in the pages. The following soldiers from the 42nd Georgia lived their last years in the Georgia Confederate Home on Confederate Avenue in Atlanta: R. P. Ferguson (A), J. M. Flarity (B), Shelton Prator (B), James W. Jackson (D), Thomas N. Tanner (D), William A. Persall (F), Reuben A. Aaron (F), George M. Herndon (H), William F. Smith (H), B. G. Neal (I), Frank T. Stier (K), J. H. Hilton (K), J. E. Robinson (K).

Livingston rose to political prominence as president of the Georgia Agricultural Society and the president of the Georgia State Farmer's Alliance. His primary opponent was Colonel Milton Anthony Candler, a Decatur lawyer who had been a former U.S. Congressman from the Fifth District. At a political rally in August 1894, Atlanta lawyer John Clay Smith claimed that Livingston, as quartermaster of the 42nd Georgia, had speculated in the goods of the commissary and had been thrown out of the regiment by his comrades.[49]

Newspaper writers immediately sought out L. P. Thomas, a former commander of the regiment, for comment. "It is true," said Thomas, "that Mr. Livingston was a commissary officer in the 42nd Regiment and served in that capacity until the Kentucky campaign when he was transferred to another regiment." Thomas thought Livingston had performed his important duties "faithfully" and dismissed the charges made by Smith. "Mr. Livingston," remembered Colonel Thomas, "was always a good soldier and bore an excellent reputation for personal bravery." Thomas M. Armistead, also a veteran of the 42nd, recalled a time when Livingston left his post behind the lines and, with a borrowed gun, joined a line of sharpshooters to beat back a Federal advance.[50]

Known as "Uncle Lon," Livingston personified the political influence enjoyed by many veterans of the 42nd Georgia. Beginning in 1892, the 6'2" blue-eyed Lon won ten consecutive terms to the U.S. Congress from his district that represented the counties of Campbell, Clayton, DeKalb, Douglas, Fulton, Newton, Rockdale, and Walton counties. Described by the *Atlanta Constitution* as "fat, plump, and hearty," the former farmer was a bigger-than-life populist who disliked banks, tariffs, and monopolies. Fighting against deflation on a national scale, Livingston advocated from the stump for the free coinage of silver at a ratio of 16 to 1.[51]

"Colonel" Livingston had the close support of Clark Howell and his

---

[49] "Northen vs. Livingston: The Governor Writes a Letter Asking for Information," *Atlanta Constitution*, March 26, 1891, 4; "Loyal to Leonidas: Rousing Meeting of his Supporters in the Fifth Ward," *Atlanta Constitution*, July 15, 1894, 19.

[50] "From His Comrades," *Atlanta Constitution*, July 16, 1894, 5; L. F. Livingston spent three months in Cobb's Legion before resigning to join the 42nd Georgia. On February 1, 1864, at the age of thirty-two, Livingston joined the Atlanta Arsenal Battalion and served in that 2nd Battalion state militia unit (Confederate Service Records of L. F. Livingston, NARA, M266, roll 37).

[51] "16 to 1: That is the Declaration of the Fifth Congressional District," *Atlanta Constitution*, August 2, 1894, 5.

Atlanta Constitution but was strongly opposed by Hoke Smith and his *Atlanta Journal.* In 1895, American and British diplomats jostled over trade with South America, with Congressman Livingston taking a leading role in asserting the rights of Venezuela. So impressed was the small nation with the congressman from the 5th District of Georgia that they sent a warship to New York City to personally bring Uncle Lon back to their Latin nation. "They gave me a baseball game—simply because it was American," he told reporters. "The government band was present and played 'Dixie' between the innings, and the crowd kept up a constant shouting of 'Hurrah for Cleveland,' and 'Hurray for the Monroe Doctrine.'"[52]

By 1908, Uncle Lon was the ranking Democrat on the Appropriations Committee, the senior member of the Georgia delegation, and was being profiled by the *Saturday Evening Post.* The national newspaper marveled at his ability to bring patronage back to the Empire state. "Lon is the greatest getter the Democrats have," they wrote. "He has more people in government jobs than any four other Democrats." Livingston's crowning achievement was bringing the new Federal prison for the South to Atlanta. The congressman always considered himself a man of the "plain people" and welcomed every opportunity to return to his wife and children on their ancestral farm in the red hills of Newton County. Livingston's long tenure ended in 1910 when he was defeated in the Democratic primary by an eloquent and fiery young lawyer named William Schley Howard.[53]

Deputy Sheriff Lovick Thomas

After three terms as sheriff of Fulton County, L. P. Thomas stepped back to pursue the purchase and management of real estate in the growing city of Atlanta. The change of vocation did not entirely suit the former Confederate, and he soon reentered Atlanta politics. On December 4, 1895, the sixty-year-old Thomas was elected from the First Ward to the Atlanta City Council by the slim margin of eighty-six votes. There were seven wards in Atlanta, with two slates competing for control of city government. City voters cast their ballots for all seven ward seats and two alderman positions, thus making every

[52] "Loyal to Leonidas: Rousing Meeting of his Supporters in the Fifth Ward," *Atlanta Constitution*, July 15, 1894, 19; "Colonel Livingston Talks," *Atlanta Constitution*, June 30, 1895, 15; "Colonel Livingston Talks: He Is Interviewed concerning the Situation in Venezuela," *Atlanta Constitution*, June 30, 1895, 15.

[53] "'Plain Uncle Lon' as Seen by the Saturday Evening Post," *Atlanta Constitution*, June 11, 1910, 2.

race a citywide contest. This was the first municipal election without a preceding primary, and, as a result, the "colored vote" mattered in this tight election. Of the 6,398 male registered voters, 971 were counted as "colored." Lovick did not show up for the Colored Voter Meeting at the county courthouse two nights before the election. He did send a letter to the meeting in response to a series of civil rights questions addressed to the candidates. The former Confederate colonel stated that he was willing to let the black voters judge him from his past record. Since Thomas had not directly addressed any of the group's questions, activist Henry Lincoln Johnson moved that his "letter be laid on the table." Cries from the crowd came back to "Put it under the table."[54]

Thomas only completed one year of his two-year term on the city council. In fall 1896, he chose to join the collective ticket of Dr. John Nelms, a former Confederate captain who had ridden under John Hunt Morgan. Nelms became the new Fulton County sheriff, with L. P. Thomas serving as his senior deputy. On January 1, 1897, Dr. Nelms and his new deputy sheriffs, having won the Fulton County election, were sworn in for the upcoming two-year term by the new judge of the ordinary court, Colonel William Hulsey. Captain Lowndes Calhoun had finally retired from this position to practice probate law, and he was pleased to be replaced by another officer of the 42nd Georgia.[55]

John Willis Nelms would be the capable sheriff of Fulton County for the next twelve years, and deputy sheriff L. P. Thomas would be by his side for the duration. Lovick Thomas was always there to add spirit and color to the daily duties of the Sheriff's office. L. P. established an impressive collection of walking sticks in his office, each carved from wood salvaged from a different Civil War battlefield. In 1907, his younger brother Heck Thomas, then the sheriff of Lawton, Oklahoma, sent him a strange looking cactus stick to add to the collection.[56]

Union veterans visiting Atlanta often sought out Lovick Thomas to exchange war stories, and any battlefield relics found around Atlanta were brought by the courthouse for the deputy sheriff's inspection and comment. In 1895, David Davis, the fifty-one-year-old postmaster of Conesville, Ohio, waited for Lovick Thomas at a Sunday service at Trinity Methodist Church.

---

[54] "Here Are the Men," *Atlanta Constitution*, December 5, 1895, 7; "Negroes and the Candidates," *Atlanta Constitution*, December 3, 1896, 4.

[55] "Dr. Nelms and his Deputies," *Atlanta Constitution*, April 29, 1896, 8.

[56] "People and Things," *Atlanta Georgian & News*, October 19, 1907.

As the congregation departed the church, Davis found Colonel Thomas at the door. "I haven't seen him since the day I met him at the old brick house, near Atlanta, on July 22, 1864," recalled the former member of Company D, of the 76th Ohio. He went on, "He captured me that day and I shall never forget him or the occasion."[57]

The L. P. Thomas Chapter of the United Daughters of the Confederacy was organized in Norcross on August 9, 1899. The group held monthly meetings and appropriate exercises on Lee's birthday, Confederate Memorial Day, and the birthday of Jefferson Davis. Every Memorial Day, Confederate graves in the Norcross Cemetery were wreathed in flowers by members of the chapter. "In naming the Chapter for Col. L. P. Thomas," the ladies wrote in their charter, they "desired not only to compliment a brave officer but also to pay a well-deserved tribute to the gallant Forty-Second Regiment, which numbered among its members so many of Gwinnett's bravest soldiers."[58]

---

[57] "Talk of the Town," *Atlanta Constitution*, May 1, 1895, 4.

[58] McNab, "L. P. Thomas Chapter, U.D.C., Norcross, Ga.," 15.

Chapter 11

A Setting Sun

"Do you feel much older now, old friend? The battle fields over which we fought look lonely; small trees that grew on them then, have changed and grown into giant oaks, and the whispering breezes only tell of our efforts to save our beloved Southland."
—Colonel L. P. Thomas to Captain Joshua Mitchell, January 28, 1896[1]

"It is with much pleasure that, as a native Georgian, born in Atlanta, as Mayor of this city, and as a son of the regiment, I cordially greet each and every one of the surviving members of the Forty-second Georgia upon this reunion, after forty-seven years have passed since your organization at Camp McDonald."
—Mayor Robert F. Maddox, Jr. addressing the 1909 Survivors of the 42nd Georgia Reunion[2]

Who Captured the DeGress Battery?

Since its inception in 1884, the 42nd Georgia Veterans Survivor Association held their annual reunion every July 22, on the anniversary of the Battle of Atlanta. The reunions would continue until at least 1912 when the ranks had fallen to a handful of graying warriors. Contemporary newspaper accounts of the gatherings invariably mentioned that by capturing the DeGress Battery, the 42nd Georgia had provided one of the few high points in the Battle of Atlanta. The veterans even erected a wooden sign near the railroad cut to signal their achievement.

Another commonly voiced sentiment at the reunions was an acknowledgment of Lovick Thomas's bravery on the afternoon of July 22, 1864. By putting himself well in front of the battle lines of the 42nd Georgia, Thomas

---

[1] L. P. Thomas, "Reminiscent," (letter to Joshua Mitchell, January 28, 1896), Joshua M. Mitchell Collection, AHC.

[2] "Reunion Is Held by Old Veterans: Survivors of the 42nd Georgia Regiment," *Atlanta Constitution*, July 23, 1909, 7.

compelled his men to follow him to victory or death. Former Federal officers, while visiting Atlanta in later years, would seek out the Fulton sheriff to tell him they had seen him that day well ahead of his men and that they admired his valor. One 1895 visitor, probably Colonel Thomas Thomson Taylor of 47th Ohio, told Courtland Winn that Thomas was "on foot with his hat in his left hand and his sword in his right-hand waving defiantly at the Federal Lines." The former Federal officer recalled that he had knocked the gun up and away from a Yankee private who had trained his weapon on Thomas as the Confederate captain scaled the breastwork. To have shot such a brave officer, thought Taylor, would have been "wanton murder."[3]

In 1895, the United Confederate Veterans of Atlanta published a detailed fifty-page pamphlet, complete with a fold-out map, titled *Battles of Atlanta.* Proceeds from the sale of the twenty-five-cent booklet went to their local UCV Camp 159. The five men on the committee who wrote and produced the work were all distinguished Atlanta citizens and had been full participants in the Battle of Atlanta. L. P. Thomas and W. L. Calhoun were former officers of the 42nd Georgia while brothers Albert Howell and Evan P. Howell had been young officers in the Georgia State Militia. By 1895, they owned and operated the *Atlanta Constitution* and were noted civic leaders in their "Phoenix City." Robert M. Clayton was the city engineer of Atlanta and a veteran of the 16th North Carolina. His regiment had been one of the few Tar Heel regiments in the Georgia campaign. As a skilled railroad engineer and city planner, he drew up the fold-out map that outlined infantry battle lines and carefully numbered the seventy-eight points of interest described in the camp brochure. Not surprisingly, full credit for the capture of the DeGress Battery was given in the booklet to the brave men of the 42nd Georgia Infantry. This was an unshakable belief that the authors would carry to their graves.[4]

In 1896, after reading accounts of the battle in Atlanta newspapers, General Cornelius Irvine Walker, leader of the South Carolina UCV, wrote the *Atlanta Constitution* editorial page to insist that it was the men of Manigault's Brigade, not the men of the 42nd Georgia, who had heroically captured the four 20-pound Parrott cannons of the DeGress Battery. With the

---

[3] Winn, "A Sketch in the Life and War Record of Colonel Lovick Pierce Thomas II," 53–54.

[4] Committee of the Atlanta Camp, United Confederate Veterans, *Battles of Atlanta* (Atlanta, Bergstrom Printing Company, 1895). In 1888, Atlanta adopted as its official symbol a phoenix rising from the ashes.

death of Arthur Manigault in 1886, it would be up to Walker to lead the fight to defend and restore the honors won that day by the South Carolina veterans.

L. P. Thomas was indignant that such a claim would be made some thirty-two years after the battle. The National Reunion of the Confederate Veterans was to be held in Atlanta in 1898, and it was important that this new controversy be put to rest before the delegates from across the South arrived. Colonel Robert Rodgers, the Atlanta camp historian, was given the task by General Clement Evans of determining precisely what happened near the Georgia Railroad on the afternoon of July 22, 1864. Rogers's definitive 1898 report was a front-page story in all the Atlanta newspapers. "In 1896 a controversy arose," so began his report, "in the matter of the capture of a Battery of General Sherman's army, known as 'Captain De Gress's Battery.'"[5]

Historian Rodgers read the letters submitted by the participants (including one by L. P. Thomas), walked the battlefield, and interviewed those he could find. The Georgia veterans, led by Thomas, based their claim primarily on William Tecumseh Sherman's 1875 memoirs. The Union commander had named the battery captured near the Troup Hurt House as the DeGress Battery. When the Georgians began to hold annual reunions in the 1880s, there were few published accounts to consult. What was indisputable was the fact that the 42nd Georgia had, with courage and grit, captured two Federal cannons and Sherman, the commander of Federal forces on the field, had identified the guns at that battle as being those under Francis DeGress. Colonel Thomas, speaking for the 42nd Georgia, considered the case closed.

Walker and Thomas also began a correspondence during this period. On August 12, 1896, after looking at Clayton's fold-out map in the new *Battles of Atlanta* pamphlet, Walker wrote Thomas, "On the map which my friend, Gen'l C. A. Evans, sent me, De Gress' Battery is placed to the south of the Railroad. If that battery referred to above, north of the brick house, was not De Gress' whose was it?" Nine months later, on April 2, 1897, General Walker stopped in Atlanta, having traveled by train from Nashville en route to his home in Charleston, South Carolina. Staying at the Kimball House Hotel, the commander of the South Carolina division of the United Confederate Veterans was visited by a delegation of Atlanta veterans in an extraordinary effort to resolve the question of who captured the DeGress Battery. A long conference followed in the parlor of Atlanta's finest hotel. Present

[5] Rodgers, *Who Captured the DeGress Battery?*, 3.

was Colonel Thomas of the 42nd Georgia; Colonel Robert Rodgers, the historian responsible for the final report; General Clement Evans, the commander of the Georgia division of the UCV; and Colonel Albert Howell, who had led the 1st Georgia State Line in the Confederate charge on July 22, 1864. According to the newspaper reporter also present, Thomas asserted that DeGress's artillery, "was scattered and extended on the south side of the Georgia railroad." General Walker, on the other hand, maintained that the four guns were, "stationed in regular battery order on the north side about one-eighth of a mile from the railroad." Walker also pointed out that there were two batteries deployed that day—Illinois batteries A and H—and that H belonged to Francis DeGress. It was a cordial meeting, but the issue would remain unresolved. The former Confederate officers couldn't agree on conclusions that day in the parlor of the Kimball House.[6]

The following day General Walker, accompanied by Le Vert and Howard, former members of Manigault's Brigade, walked the battlefield with Thomas and Rogers of Atlanta. According to Walker, they found the site where the four guns of the DeGress Battery were placed on the property of former artillery captain Max Corput.[7]

Twelve days later, General Walker wrote Thomas, again insisting that "DeGress was not south of the Georgia Railroad, but about 500 yards to the north of it." According to Walker, it was Colonel Thomas's responsibility to admit his mistake in writing so that the South Carolinian could publish it, as "this will settle the unfortunate dispute and save any newspaper warring, or, what is more important, unpleasantness between comrades in a cause in which we struggled side by side." Lovick Thomas responded a week later, writing, "I see that it would be entirely useless and out of place to controvert your position and I believe you are perfectly sincere in your statements and conclusions and hope you will give me equally the same credit for my position." The Georgian noted that, "Captain DeGress was the ranking officer that day and was in command of both batteries, battery A and battery H, and that some of the guns were south of the Ga. R.R."[8]

Lovick Thomas and the members of the 42nd Georgia had first made

---

[6] C. Irvine Walker, August 12, 1896 letter to L. P. Thomas, Thomas collection, BRC; "General C. I. Walker is Here: Believes that Manigault's Brigade did it," *Atlanta Constitution*, April 3, 1897, 2.

[7] "Taking of DeGress Battery," *Charleston Weekly News & Courier*, April 14, 1897, 7.

[8] C. I. Walker to L. P. Thomas, April 14, 1897; L. P. Thomas to C. I. Walker, April 22, 1897, Thomas collection, BRC.

their claims of capturing the DeGress Battery in the 1880s, well before the first volumes of the *Official Record of the Rebellion* were published. Historian Rodgers, using volume 38, was able to consult the detailed battle reports from the officers involved in the Battle of Atlanta, as well as two powerful and detailed narratives, those of Arthur Manigault and C. I. Walker. Rodgers ultimately and correctly concluded that it was the South Carolina troops under Arthur Manigault who had first captured the four rifled Parrott guns belonging to Francis DeGress. Rodgers's account notwithstanding, Colonel Lovick Thomas was never convinced that his men had not, in their finest hour, seized the DeGress Battery so identified by General Sherman. At one subsequent reunion, he remarked, "Perhaps it would be as much credit to us if that [the capture of a section of Battery A] was true, as one Yankee Battery at that time was as objectionable to us as any other, but in this our friend General C. I. Walker of South Carolina is Mistaken."[9]

Atlanta historian and artist Wilbur Kurtz would reexamine the entire DeGress controversy in a series of newspaper articles published in the early 1930s. Kurtz ultimately found himself in full agreement with the conclusions reached by Robert Rodgers in 1898. Kurtz was helped by eighty-seven-year-old John William McWilliams, who walked the battlefield with the historian in October 1930. McWilliams was a twenty-one-year-old private in the Fowler Guards (D) on July 22, 1864. The retired storekeeper explained to Kurtz in 1930 that the 42nd Georgia had started the hot summer day in reserve near the Atlanta Medical College. The soldiers were called into formation and marched southwest into the raging battle. They crossed the tracks of the Georgia Railroad near the present-day intersection of Waverly Way and DeKalb Avenue.[10]

After crossing the railroad tracks, the brigade continued marching southwest until they finally stopped, turned west, dressed their ranks, and studied the Federal lines, which was in a haze of smoke in front of them. As the regiment anchoring the far left of the formation, the men of the 42nd were on either side of the actual railroad tracks. According to McWilliams, the regiment followed the tracks toward the Federals and soon came to several large piles of bridge timbers lying on the north side of the track. This was near the present intersection of Elizabeth Street and DeKalb Avenue. The

---

[9] Volume 38 of the *OR* was published in 1891; Lovick Thomas, "Introducing R. L. Rodgers at Social Circle, Georgia, July 22, 1899," vol. 4, pg. 594, Thomas collection, BRC.

[10] Wilbur G. Kurtz, "The Broken Line & the DeGress Battery," *Atlanta Constitution*, February 8, 1931, SM4. DeKalb Avenue was called the Decatur Road in 1864.

lumber was cut and marked and ready for use in replacing or repairing railroad bridges. McWilliams remembered that companies A and B went to the left of the bridge timbers while the rest of the regiment drifted to the right.[11]

The private recalled that, among his company, he was a close second to Orderly Sergeant Bob Hollingsworth in scaling the Federal breastworks. McWilliams did not see the capture of any artillery by the 42nd Georgia, but neither could he see the railroad cut from his position further. McWilliams remembers being severely bruised by a minie ball during the Federal counterattack and then almost killed by an exploding Federal shell during the withdrawal. He credits the heavy blanket roll strapped across his shoulder with saving his life on that occasion.[12]

Confederate Parades

In a time before radio and television, nothing could surpass a city parade for excitement, pomp, and pageantry. The annual Confederate Memorial Day Parade held in late April was always one of Atlanta's biggest events. At 1:00 P.M. stores and businesses would close for the day. According to the *Atlanta Constitution*, every citizen "the high, the low, the rich and the poor, vied with each other in showing their loyalty and devotion to the day." There was great excitement even before the 3:00 parade had stepped off from the customs house. A reporter in 1888 noted "the soldiers, bedecked in their flaming uniforms, members of various secret orders, attired in their uniforms, the young cadets, in their modest suits of gray." Leading the procession in 1888 was Judge Lowndes Calhoun and Colonel L. P. Thomas, both mounted and wearing sashes. Behind them marched 100 or so members of the Fulton County Confederate Veterans Association. They were followed by two tribes of the fraternal order of Red Man, brass bands, the Gate City Guard, and other marching units. The parade moved down Marietta Street to Broad Street, on to Whitehall, and on to the cemetery by way of Hunter Street.[13]

By 1895, the scale of the parade had dramatically increased. That year, L. P. Thomas served as marshal of the day. Marchers were so numerous that they were divided into five divisions, with every military and civic group in

[11] Ibid.

[12] Ibid.

[13] "The Gallant Dead: Beneath the Grassy Hillocks in Oakland Cemetery," *Atlanta Constitution*, April 27, 1888, 1. The Improved Order of Red Men was a fraternal order established in 1834.

the city participating. Leading the parade on horse were L. P. Thomas, W. L. Calhoun, and W. H. Hulsey, all former officers of the 42nd Georgia. Prominent men in the parade were Governor William Atkinson and Confederate General James Longstreet. Lowndes Calhoun introduced the orator of the day, the Georgia historian Lucian Lamar Knight.[14]

Special occasions also warranted massive parades down the streets of Atlanta. Some of the biggest were the unveiling of the Henry Grady Monument (1891), the 400th anniversary Christopher Columbus parade (1892), the national United Confederate Veterans Reunion (1898), and the unveiling of the General W. H. T. Walker Memorial (1902). Jefferson Davis would ultimately warrant three parades: his 1888 visit to Atlanta, his 1889 parade in sympathy to his New Orleans funeral procession, and a giant 1893 parade when his funeral train stopped in Atlanta on the way to his final resting place in Richmond, Virginia. Colonel Robert Maddox was one of the five men constituting the Georgia delegation at Davis's final funeral at Hollywood Cemetery.[15]

The 1898 UCV Atlanta Reunion

In 1896, Lovick Thomas was elected commander of the Atlanta UCV Camp #159. He led a delegation to the seventh annual national reunion held in Nashville, Tennessee, in June of the following year. Thomas was chairman of the credentials committee for the convention that accredited 2,061 voting delegates from 1,031 camps. One of the Georgian's primary goals was securing the national reunion for Atlanta the following year. Colonel Andrew West made the principal speech to the convention, inviting them "to a city, at whose very threshold Hood fought, and Walker fell." Now a city of 120,000 with "thirteen great railroads, furnishing sharp competition from every direction," West was confident that the "Gate City" could accommodate some 15,000 veterans and 50,000 guests.[16]

Upon returning to Atlanta in a special train dedicated to Confederate veterans, Thomas announced that Atlanta had defeated Louisville for the right to hold the next reunion. "When the vote was counted, and the result

---

[14] "Flowers & Tears; Memorial Day Will Be Beautifully Observed Today in Atlanta," *Atlanta Constitution*, April 26, 1895, 9.

[15] "Finally Laid to Rest," *Raleigh News and Observer*, June 1, 1893.

[16] Minutes of the Seventh Annual Meeting and Reunion of the United Confederate Veterans (1898), 61, 86.

announced," explained Thomas to an *Atlanta Constitution* reporter, "the wildest scene followed. 'Dixie' was played, the drum corps rattled, and the thousands shouted themselves hoarse."[17]

The leaders of the Georgia Division and especially the host camps, Atlanta camps 159 and 925, now faced the daunting task of planning the logistics of holding the largest UCV national convention produced to date. They would work closely with George Moorman, the adjutant general and chief of staff, headquartered in New Orleans. General Clement Evans of Georgia would serve as president of the newly formed Confederate Reunion Association. Committees were quickly organized for finance, entertainment, quarters, badges, music, decorations, and receptions. The officer staff of the 42nd Georgia—L. P. Thomas, W. L. Calhoun, R. F. Maddox, W. M. Durham, W. L. Hulsey, and H. W. Snow—were all assigned to multiple committees. In April 1898, three months before the reunion, Camp 159 held their annual election of officers at the armory of the Gate City Guard. Based on the bylaws of the camp, L. P. Thomas was no longer eligible to serve as commander, and he stepped aside in favor of his good friend Colonel A. J. West. Thomas was soon elected brigadier general to command the North Georgia Brigade of the UCV.[18]

Atlanta successfully hosted the eighth annual national reunion in July 1898. It was the highwater mark for the UCV with 1555 camps sending delegations. Dozens of newly formed Sons of Confederate Veterans camps also sent nonvoting delegations. In a packed auditorium in Piedmont Park, the delegates assembled under Confederate and American flags and the suspended life-size portrait of Jefferson Davis. Loud applause greeted each Southern general as he took his seat on the stage. The "building fairly shook" when General John Brown Gordon, with his battle-scarred face, entered the hall to the tune of "Dixie." The Rebel yells and cries of "Gordon! Gordon!" drowned out the best effort of the band. The 25,000 veterans present passed a resolution in favor of the War with Spain and endorsed the building of a Confederate Battle Abbey in Richmond, Virginia. John B. Gordon was reelected as commander, with General Stephen D. Lee heading the Army of the Department of Tennessee and General Wade Hampton in charge of the

[17] "Atlanta Veterans Back from Fair" *Atlanta Constitution*, June 26, 1897, 6.

[18] "Four Weeks More Till the Reunion," *Atlanta Constitution*, June 23, 1898, 4.

Army of Northern Virginia. Charleston, "the Old City by the Sea," was selected for the next convention.[19]

With every hotel in Atlanta booked to capacity, auxiliary buildings around Piedmont Park fed and housed thousands of the old veterans. The 42nd Georgia held their annual reunion in the Costa Rica Building on the exposition grounds. Members present proudly received new regimental badges. The large parade down Peachtree Street on July 23 commenced under torrents of rain. A bare-headed General John B. Gordon, water streaming down his face, sat on his magnificent black horse by the reviewing stand as the thousands of determined veterans marched past.

Southern Cross of Honor

Conceived by the United Daughters of the Confederacy chapter in Athens, Georgia, in 1896, the ladies proposed to award every living Confederate soldier who had been honorably discharged from Southern service with a handsome bronze medal to be worn on the soldier's chest. Four years later, 9,500 medals had been manufactured, with the Atlanta chapter receiving 300 medals for distribution on Confederate Memorial Day, April 26, 1900. Veterans had to complete a certificate of eligibility before the medal would be conferred upon them in an elaborate ceremony.[20]

The first Atlanta ceremony took place in the hall of the House of Representatives in the state capitol. Mrs. Helen Plane, president of the Daughters of the Confederacy, awarded the first medal to Samuel Gibbs French of Mississippi, the only Confederate major-general still alive in 1900. Georgia Governor Allen Candler, the last Confederate veteran to serve as the state's chief executive, received the second cross of honor. The name of each of the 300 veterans was called out, and each received their medal from Mrs. Plane. The band struck up "Dixie" after the last name was called. Six members of the 42nd Georgia were present in the capitol that day. Similar ceremonies took place across the state. Though the majority of medals were given out in 1900, smaller ceremonies would continue on each Memorial Day until 1917.[21]

---

[19] Minutes of the Eighth Annual Meeting & Reunion of the United Confederate Veterans (1899), 11, 59.

[20] "Southern Cross of Honor Ready for the Veterans," *Atlanta Constitution*, April 21, 1900, 11.

[21] Ibid. Members of the 42nd Georgia to receive a Cross of Honor at the Atlanta ceremony were W. L. Calhoun (K), W. M. Durham (K), W. L. Grubbs (I), W. H. Hulsey (Staff), L. P. Thomas (A), and W. H. William (B).

Blue & Gray Reunions

The first significant gathering in Georgia of GAR and UCV members was at the Blue & Gray Reunion in July 1894, an event to commemorate the thirtieth anniversary of the Battle of Atlanta. The event in Grant Park drew 2,000 people and was the inspiration of Dr. Charles D'Alvigny, Captain George B. Forbes, and Colonel L. P. Thomas. Members of the 7th, 42nd, and 43rd Georgia made up most of the former Confederates present while the Federals were well represented by members of the Atlanta-based O. M. Mitchel Post of the Grand Army of the Republic. The various speakers on the platform that afternoon emphasized the valor of both armies and the spirit of brotherhood that now allowed them to "clasp hands under the banner of a reunited country."[22]

Spurred by calls for reconciliation by President William McKinley, the city fathers of Evansville, Indiana, visited Atlanta in August 1899 to promote a major reunion of North and South in their city that fall. A delegation of former Confederates from Atlanta made the trip to the four-day event. Two of the most prominent places on the program were given to Colonel William Hemphill and Colonel Andrew West, both from Atlanta. Assigned to elegant rooms in the St. George Hotel in Evansville, they had an opportunity to meet with President McKinley and his cabinet, but not before each had addressed the crowd of 50,000 in the fairgrounds' amphitheater. "No two men on any single occasion," according to columnist Wallace Putnam Reed, "made a better impression upon the nation at large as the representatives of Georgia and Atlanta."[23]

Atlanta returned the favor the following year, hosting one of the most significant national Blue and Gray Reunions ever held in July 1900. Again, the leadership of local UCV was called upon to coordinate the event. Colonel William Hemphill served as chairman, with Colonel Robert Maddox in the important role of head of invitations. As usual, the generals from both armies would be the major celebrities of the event. Present were generals John C. Breckinridge, Oliver O. Howard, Alexander P. Stewart, Willard Warner, Albert Shaw, Thomas J. Wood, John B. Gordon, John Coburn, and Stephen

[22] "The Blue & Gray: They Celebrated the Battle of Atlanta Yesterday," *Atlanta Constitution*, July 23, 1894, 5.

[23] "Blue and Gray Committee Gets Gov. Chandler as Reunion Guest," *Atlanta Constitution*, July 23, 1894, 7; Wallace Putnam Reed, "Two Atlanta Orators Win National Fame," *Atlanta Constitution*, October 15, 1899, 17.

French. Governor James Mount of Indiana agreed to make the keynote speech. The special guests were divided into groups and escorted to the Atlanta battlefields by local dignitaries. Lovick Thomas and Lowndes Calhoun were selected to host General A. P. Stewart, their former commander at the battles of Resaca and New Hope Church.[24]

On July 19, 1900, some 5,000 members of the Blue and Gray gathered at the Aragon Hotel on Peachtree Street to register for the three-day Atlanta reunion. The event began with an evening campfire that drew thousands to hear speeches by Gordon, Lee, Howard, and other distinguished officers. The following day was devoted to visiting the battlefields and sharing recollections. Guides and carriages were at each of Atlanta's three battlefields so that the veterans could retrace and mark their steps. The special guests of the reunion gathered on the first morning at the McPherson Monument for impromptu speeches from John C. Breckinridge and Albert Shaw, the commander in chief of the Grand Army of the Republic. From the commanding knoll where the monument rises, the veterans had a comprehensive view of the Battle of Atlanta.[25]

Congressman Leonidas Livingston, who had started the war as a quartermaster in the 42nd Georgia, used his cane to draw the lines of battle in the sand. Livingston had been there that day, July 22, 1864, and had helped place the mortally wounded General W. H. T. Walker in the medical litter. At noon, the dignitaries strolled down the hill arm in arm to a shaded glade where once soldiers defended dirt trenches. Debris of battle still worked its way to the surface after a heavy rain, and bullet holes still scarred the trees. On this day, the staff of the Aragon Hotel had set up tables "covered with snowy napery and bright silver." With Colonel William Hemphill serving as master of ceremonies, various generals stood to give complimentary speeches praising the bravery and good intentions of their former enemies.[26]

On the afternoon of July 21, a mammoth barbeque was held at the vast exposition building in Piedmont Park. Three long tables ran parallel down the hall and were covered in American and Confederate flags. Much to the delight of the old veterans present, a short but spirited controversy arose in

---

[24] "Vets of the Blue & Gray Meet Here Wednesday," *Atlanta Constitution*, July 17, 1900, 1.

[25] C. R. S. Horton, "Reunion of the Blue & Gray," July 1900, 27–28, magazine clipping, vol. 1, pg. 69, Thomas collection, BRC.

[26] Horton, "Reunion of the Blue & Gray," 28; "Blue and Gray, Arm in Arm, Fight the Battles of Atlanta All Over Again," *Atlanta Constitution*, July 20, 1900, 5.

the speeches between General Shaw of the GAR and General Gordon of the UCV. The Georgian objected to Shaw's assertion that it was time for "patriotic teaching for all and by all," where "the keeping alive of sectional teachings as the justice and rights of the cause of the South in the hearts of the children is all out of order, unwise, and unjust." General Gordon would have none of it, declaring, "If we cannot teach them that their fathers were right, then it follows that these Southern children must be taught that they were wrong. Are we ready for that? For one I am not!" With tact, General Shaw, the first Federal soldier to cross the Chattahoochee River in 1864, conceded the point and harmony was restored.[27]

On July 22, 1900, the anniversary of the Battle of Atlanta, reunions were held by the 10th, 30th, and 42nd Georgia regiments. It was on this momentous occasion that Judge Lowndes Calhoun presented his long-awaited history of the 42nd Georgia Regiment. The forty-five-page book opens and closes with Smith Clayton's poem "Faded Jacket of Gray." The book includes a short overview of the history of the 42nd, a roster broken down by company (often with year of death listed), Calhoun's account of the Siege of Vicksburg, various war dispatches, L. P. Thomas's account of the Battle of Bentonville, and a summary of previous regimental reunions.[28]

A National Park for Atlanta

On December 5, 1899, U.S. Congressman Leonidas Livingston of Georgia introduced House Bill H.R. 946 calling for the establishment of the Atlanta Military Park. The legislation would set aside some 1,275 acres for battlefield preservation at a cost to the Federal Treasury of $300,000. Georgia Senator Alexander Clay made a similar proposal in the Senate. A military park in Atlanta seemed like a logical next step after the creation of similar national military parks at Shiloh, Chickamauga, Vicksburg, and Gettysburg. Livingston's initiative would be the crowning achievement for the veterans of the 42nd Georgia, who considered July 22 to be a sacred day.[29]

The proposal had originated from Colonel William Hemphill, who visited the Gettysburg Battlefield in 1898, some thirty-five years after he had been wounded there in battle. The Georgia veteran presented his ideas at a

---

[27] Horton, "Reunion of the Blue & Gray," 27–28.

[28] Calhoun, *History of the 42d Regiment*, 3–42. Some 250 veterans of the 42nd Georgia were still alive in 1900.

[29] Atlanta Business Men's League, The Proposed Atlanta National Military Park, 27, 37.

meeting of the Atlanta Business Men's League. As founder and publisher of the *Atlanta Constitution*, Hemphill was extremely influential in all aspects of Atlanta's economy and politics. His proposal was enthusiastically received, and representatives of the league quickly and quietly secured options on 1,200 acres, at $100 an acre, of what had been the battlefield around Peachtree Creek. It was decided that the lands around the Battle of Atlanta and Ezra Church were, by this time, too developed and therefore too expensive to purchase.[30]

On January 31, 1900, two delegations from Atlanta, one from the Chamber of Commerce and one from the Atlanta Business Men's League, addressed the House and Senate committees on military affairs. The Atlanta men were armed with the Business Men's League's fifty-five-page illustrated brochure, which touted the proposal, the importance of the Civil War battles, and the virtues of Atlanta as a rising business city. Written endorsements from former Federal officers, including former President Benjamin Harrison, as well as support from UCV Camp 159, were included. Speaking for the legislation was Hoke Smith, generals John B. Gordon, A. P. Stewart, and O. O. Howard, Captain Evan P. Howell, and Mr. Forrest Adair. The delegations were well received, and General Joseph Hawley of Connecticut, who headed the Senate committee, enthusiastically endorsed the proposal.[31]

Private meetings with members of Congress and war department officials followed. Colonels W. A. Hemphill and A. J. West and generals O. O. Howard and A. P. Stewart met with President McKinley at the White House. "He received us with a great deal of friendliness," reported Hemphill, "and told us in as many words that he favored the project to establish a military park near our city." Secretary of War Elihu Root, on the other hand, was noncommittal about the proposed legislation, and David Henderson of Iowa, speaker of the house, warned that "he wanted to keep the appropriations for this congress down to the lowest possible limit."[32]

---

[30] "Captain Howell's Part in National Military Park," *Atlanta Constitution*, September 28, 1902, 18.

[31] Atlanta Business Men's League, *The Proposed Atlanta National Military Park*, 8–57; Joseph Ohl, "Atlanta Military Park Has No Chance This Year," *Atlanta Constitution*, February 21, 1901, 6; Supporting the proposal was President Benjamin Harrison, General J. J. Finley, General Smith Atkins, General A. P. Stewart, General William Bate, General Stephen D. Lee, General Oland Smith, General Charles Grosvenor, General Thomas Wood, General John Palmer, General John Parkhurst, General George Harrison, General John Coburn, and General Daniel Butterfield.

[32] "Talk with Col. Hemphill," *Atlanta Constitution*, February 2, 1900, 5.

In March 1900, Atlanta's Confederate veterans hosted a large reception for General Albert Shaw, commander in chief of the Grand Army of the Republic, at the Kimball House Hotel. Sitting between W. A. Hemphill and A. J. West, Shaw rose and endorsed a national military park for Atlanta. Two days later the House Committee on Military Affairs reported favorably on the bill. The ebullience and momentum, however, had peaked. Months later, Speaker Henderson would kill the appropriation in a tight budget year. He did keep the candle lit by promising the Georgians five times the amount originally requested at some future date. Talk soon floated about establishing three battlefield parks in Atlanta, each connected to one another by a touring boulevard.

The legislation was reintroduced in 1902, but, by then, some sixteen battlefields were under consideration as national parks. Fearing that allowing any one of these appropriations to pass would open a floodgate, Henderson "put his foot on the whole lot, and the Atlanta bill goes down with the others." Secretary of War Root urged Congress to create one National Park Commission as opposed to a separate commission for each park. It was 1904 before Atlanta's proposal was again before Congress. The general belief was that money would be appropriated during that session for establishing parks at Appomattox and Atlanta. This time, however, Root effectively ended Atlanta's chances by coming out against the creation of a military park at Appomattox.[33]

In 1905, L. P. Thomas, president of the 42nd Georgia Confederate Survivors Association, would take up the fight in the Atlanta newspapers. The revised legislation called for the appropriation of $250,000 for 1,202 acres in the three Atlanta battlefields, all to be connected by a thirty-mile "great driveway to be known as 'The Boulevard.'" Colonel Thomas, who had been successful in Atlanta real estate throughout his life, envisioned "a splendid succession of summer or winter homes" fronting the boulevard, making it "the most beautiful adornment possessed by any city in this country." The city council, chamber of commerce, veterans' groups, the governor, and the Georgia congressional delegation gave their endorsement with the hope that President Theodore Roosevelt, with Georgia roots on his maternal side, would become an advocate for an Atlanta military park. Later that year, the presi-

[33] Joseph Ohl, "Atlanta Military Park Has No Chance This Year," *Atlanta Constitution*, February 21, 1901, 6.

dent would pass through Atlanta, and banker Robert F. Maddox, former lieutenant colonel of the 42nd Georgia, hosted a dinner in his honor.[34]

Ultimately, this legislation, like all previous efforts, failed to pass through Congress in 1906. There were no natural advocates in a Republican-controlled Congress for a bill sending funds to Democratic Georgia. The steady growth around Atlanta only made the cost of a national military park more difficult and expensive with each passing year. Writing for the Civil War Trust in 2014, author Ian Michael Rogers would call the events at the turn of the century "a missed opportunity."[35]

Passing of the Old Veterans

The veterans of the 42nd Georgia passed away in due course, with Lovick Thomas being one of the last to answer the summons. The *Atlanta Georgian & News* noted at his death in 1910, "the gaps in the ranks are growing wider and wider. With each dip of the sun behind the western horizon there follows an ominous silence when the familiar name of some veteran is called, and in the awful stillness we shudder to think how fast the muster rolls of the dead are lengthening."[36]

On June 6, 1899, Colonel Robert Flourney Maddox passed away from a stroke at the age of seventy. He was buried at Oakland Cemetery in the handsome Maddox family mausoleum. L. P. Thomas, Lowndes Calhoun, A. J. West, William Hulsey, General John B. Gordon, and the Howell brothers were among seventy city dignitaries who escorted his casket to the cemetery. Maddox had returned to Atlanta after the war penniless but, with an enterprising spirit, spent the rest of his life working to build a personal fortune, as well as help his city, state, and Methodist church. He was a prosperous tobacco and liquor dealer and then, in 1879, formed a bank and cotton broker business with another war veteran, Captain Jett W. Rucker. The Maddox-Rucker Bank, at the corner of Broad and Alabama streets, would continue to prosper after the colonel's death. By 1916, it was known as the American National Bank and was the largest bank in the city with $15 million in deposits. Robert F. Maddox Jr. succeeded his father as the president of the bank

34 "Plan for National Park Has Gained New Headway," *Atlanta Constitution*, October 15, 1905, A7; Edwards Gallery, *Dinner for President Theodore Roosevelt*, silver gelatin print.

35 Rogers, "A Missed Opportunity: Preservation at Peachtree Creek."

36 "Colonel Lovick P. Thomas," *Atlanta Georgian & News*, May 4, 1910, 6.

and would be elected mayor of Atlanta in 1908.[37]

On January 9, 1904, Major General John Brown Gordon, the most beloved of all Georgia Confederates, passed away at the age of seventy-two. Gordon had a splendid war record with the Army of Northern Virginia, ending his service as one of Lee's three corps commander at the age of thirty-three. He became a formidable Georgia politician after the war; he served as governor and was elected twice to the United States Senate. In 1890, he became the first commander in chief of the United Confederate Veterans, a position he would hold until his death. The *Atlanta Constitution* eulogized him "as being, beyond a doubt, the greatest and most conspicuous southerner."[38]

General Gordon's death symbolized the end of a Rebel era in Georgia, and the old veterans of the 42nd Georgia fully participated in the efforts to preserve his memory. For two days, his coffin, covered in American and Confederate flags, lay in state at the state capitol. Services were held in the hall of the House of Representatives and then across the street at the Central Presbyterian Church. The Confederate veterans were given priority seating in the church. The procession to Oakland Cemetery was under the command of General Clement Evans and divided into six large divisions. With heads uncovered, the dense throng of well-wishers stood in complete silence as the immense black hearse, pulled by four jet-black horses, passed to the cadence of muffled drums. One thousand Confederate veterans, each carrying a flower, brought up the rear of the procession. Judge Lowndes Calhoun served as chief of staff for the Georgia Division of the UCV. Upwards of 75,000 people viewed and took part in the memorial ceremonies on January 14, 1904.[39]

One of the Confederate marchers that day was Captain William Smith of Buford, Georgia. Smith had been sick for weeks before the funeral but could not be dissuaded from attending. Fatigued by the experience, the sixty-one-year-old veteran of the Independent Rebels (B) caught the first train home after the parade. Smith was carried to his home from the Buford train depot and died five days later, on January 19, 1904. William T. Smith had

---

[37] Jett W. Rucker (1823–1900) was a native of Wilkes County and a "49er" during the California Gold Rush. He was a tobacco merchant (Garrett, *Atlanta and Environs,* 2:686).

[38] "The Death of Gen. John B. Gordon," *Atlanta Constitution*, January 11, 1904, 6.

[39] "Great Funeral Procession Arranged in Great Detail," *Atlanta Constitution*, January 13, 1904, A1; "To Sweet and Solemn Strains of Music Gordon's Corse was Borne to Oakland," *Atlanta Constitution*, January 15, 1904, 4.

married shortly after the 1865 surrender, and with his new bride set out for the Lone Star State, where he had won two terms in the Texas legislature. Returning to his native state in 1881, Smith set up a successful mercantile business in the growing railroad town of Buford. In 1888, he won the first of two terms in the state legislature. In 1902, he became a state senator representing Gwinnett County. Ten days after his unexpected death, a letter appeared in the *Atlanta Constitution* from Hiram Parks Bell, the former colonel of the 43rd Georgia, who was the state senator from Forsyth County. Bell had worked with Smith in the state Senate, where they were both served on the Confederate Pension Committee, a topic they both cared about deeply. Bell considered Captain Smith to have been a born orator, "with a fine, commanding presence, a good voice and an extensive vocabulary of well-chosen Anglo-Saxon words, which he used with the ardor of conviction and the intensity of enthusiasm, he rarely failed to convince his auditors."[40]

Within a week of General Gordon's death in 1904, each of the Confederate groups in Atlanta assembled in the hall of the House of Representatives at the state capitol. Urged forward by General Clement Evans, the group invited 100 prominent Georgians to organize the Gordon Monument Association. The 42nd Georgia would be represented by W. L. Calhoun, L. P. Thomas, and R. F. Maddox. Five current and former Georgia governors were among the new members of the association. Judge Lowndes Calhoun was made the president of the group, explaining,

> Twenty-three centuries ago ancient Athens placed a funeral monument over the graves of her warriors who for her glory died. This custom has prevailed to the present time and will continue down the ages forever. My countrymen and countrywomen, is it not time that we should more fully realize the duty which we owe to the past, the present, and the future, in preserving and perpetuating this sacred custom of the ages?[41]

The association unanimously decided that General Gordon would be remembered with an equestrian statue on the grounds of the state capitol. Subscriptions totaling $10,000 poured in from citizens across the South. The committee on design selected the sculptor Solon Borgium of Connecticut to

---

[40] Hiram P. Bell, "Letters from the People: The Late Hon. W. T. Smith," *Atlanta Constitution*, January 28, 1904, 4.

[41] "Shaft to Gordon Will Be Erected: Meeting Will Be Held this Afternoon," *Atlanta Constitution*, January 20, 1904, 7.

design and build the giant bronze statue. A large granite stone from Stone Mountain was secured for the base. In 1906, the state legislature appropriated the final $15,000 for the statue and established a commission with Governor Joseph Terrell as chairman. On May 25, 1907, a parade down Peachtree Street was followed by a noon ceremony on the capitol grounds in front of thousands of well-wishers. The Gordon statue, standing twenty-five feet from the Georgia clay to the top of his hat, was unveiled by his two daughters to the sounds of "Dixie" and Rebel yells.[42]

On November 17, 1908, L. P. Thomas served as a pallbearer for his dear friend and comrade Lowndes Calhoun. The judge was buried at Oakland Cemetery surrounded by many of his Confederate friends. Described as "broad-minded, genial, and always agreeable," Lowndes Calhoun had emerged as an important leader among Confederate veterans in Atlanta. He was a founding member of the Fulton County Veterans Association, became a commander of UCV Camp 159, and was a trustee of the Confederate Home in Atlanta. As historian of the 42nd Georgia Veterans Association, Judge Calhoun published a fifty-six-page *History of the 42nd Georgia* in 1900.[43]

The next to cross the river was seventy-two-year-old Colonel William Henry Hulsey. His death was reported on the front page of the *Atlanta Constitution* on May 18, 1909. "Judge Hulsey's life was spent in the service to his state and native city, Atlanta," reported the paper of record. Hulsey's home was situated "about the very point where Colonel Thomas with the Forty-second and Albert Howell with his command on his right swept over the federal works like a whirlwind." His nine pallbearers, including L. P. Thomas, A. J. West, George Hillyer, and W. M. Durham, were all Confederate veterans. The uniform jacket of the "boy major" was sold by the James D. Julia Auction House in March 2016 for $40,000.[44]

There were many highlights and a few low points in the last years of L. P. Thomas's life. He was financially secure and able to enjoy his passion for

---

[42] Colonel W. Lowndes Calhoun, "History of Movement of Gordon Monument," *Atlanta Constitution*, May 26, 1907, E6. The Gordon statue was cleaned and restored by the Sons of Confederate Veterans in 2008 ("As in Battle Gordon Sits on His Charger," *Atlanta Constitution*, May 26, 1907, E4).

[43] "Calhoun Funeral Notice," *Atlanta Constitution*, November 17, 1908, 14. Among the present were Dr. W. M. Durham, Col. A. J. West, Col. W. H. Hulsey, Dr. Charles D'Alvigny, Gen. Clement Evans, Dr. Amos Fox, Captain R. M. Clayton, Donald Bain, and Clark Howell.

[44] "Hulsey Answers Call of Death," *Atlanta Constitution*, May 18, 1909, 1; "The Old Veterans," *Atlanta Constitution*, July 27, 1886.

poetry and to pursue his family genealogy. His scrapbooks in the Bell Center Collection are filled with clippings of favorite poems, including those of his first cousin Sidney Lanier. But in June 1907, the angel of death visited sixteen-year-old Edward Lloyd Thomas, the youngest child of Lovick and his second wife, Jeannette Payne Thomas, and he was buried in the Thomas plot in Oakland Cemetery.[45]

On January 24, 1910, seventy-four-year-old Lovick Thomas suffered a stroke that paralyzed half of his body. A second stroke on the evening of May 2 ended his storied life. L. P.'s death was reported on the front page of the *Atlanta Constitution.* Its editorial page poetically pointed out, "in that most withering battle around Atlanta, July 22, 1864, when gaping lines were thinned by grape and cannon ball, and when the red soil was drenched a deeper hue by the life-blood of contending Americans, it was Lovick P. Thomas whose deeds of almost super human valor added fresh laurels to the brow of the Confederacy."[46]

Lovick Thomas's funeral was held at St. Mark's Methodist Church. Colonel Thomas had chosen his pallbearers before his death, and they were all members of the 42nd Georgia. They were followed by an honorary escort composed of members of Atlanta Camp 159, members of his old regiment, official members of St. Mark's Church, and prominent men in the Atlanta community, including eight judges, Sheriff John W. Nelms, and many former deputy sheriffs.[47]

L. P. Thomas did not live long enough to see his son-in-law Courtland Winn become mayor of Atlanta in January 1911. Winn succeeded Robert F. Maddox Jr. to the important post. Considered a progressive, Winn raised teacher salaries and oversaw the building of a city garbage incinerator, a project initially conceived of by L. P. Thomas in 1896. On February 12, 1912, Mayor Winn joined friends and Confederate veterans at the Terminal Station to meet the "Birmingham Special," which held the body of Congressman Leonidas Livingston. "Old Lon" was returning from Washington to be buried in his King community in Newton County. Courtland Winn directed that the national flag over city hall be lowered to half-mast until after his

---

[45] Thomas proudly displayed several pages of their shared genealogy in his scrapbook, vol. 3, pg. 326–329, Thomas collection, BRC.

[46] "Lovick P. Thomas," *Atlanta Constitution,* May 5, 1910, 4.

[47] "Funeral Services of Col. L. P. Thomas," *Atlanta Constitution,* May 5, 1910, 9. The pallbearers for the funeral of L. P. Thomas were O. F. Owen, J. L. Bell, T. M. Armistead, Col. A. J. West, Dr. W.M. Durham, and Captain W. H. Harrison.

internment.[48]

On July 22, 1910, Heck Thomas wrote a lengthy letter to his niece Fannie Winn, the oldest living daughter of Lovick Thomas. Heck told some old stories from the past and then noted that, for decades, he had sat down every July 22 and written a long letter to his older brother. Heck viewed July 22, the anniversary of the Battle of Atlanta, as the most important day in Lovick's life. The old lawman, who would be dead himself in two more years, asked if he could continue the tradition with a yearly letter to Fannie Thomas Winn.[49]

---

[48] "'Uncle Lon' Comes Home: Body Lies in State Today," *Atlanta Constitution*, February 13, 1912, 7; "Courtland S. Winn: The Man; A Typical Citizen of Atlanta," *Atlanta Georgian and News*, September 17, 1910, 3.

[49] Edward C. Bruffey, "Bruff's Column," *Atlanta Constitution*, August 24, 1910, 8.

Epilogue

Commemorations such as the recent sesquicentennial of the War Between the States (2011–2015) have moved with little notice from the role of the unwelcomed visitor to the finality of the banished. Where once there were grand parades and venerated old soldiers, now there is only silence. The heroes of that conflict have, for the most part, known the silence of the grave for 100 of those 150 years. The thin gray line has broken the bonds of the earthly plane. Time is a foe that no army can defeat. Death is a promise that will always be kept.

Whether these men of the 42nd Georgia died in the desperate charges at Resaca, Atlanta, or Bentonville; of disease at Bean Station or Vicksburg; from a sniper's bullet along the Atlanta siege lines; or from malnutrition at Camp Morton and Camp Chase, they fulfilled a destiny that was thrust upon them. These volunteers died in service to their young nation, and they lived on in the memories of those they left behind. Those who survived the war and stubbornly rebuilt their lives in the fractious years that followed were afforded the luxury of being escorted from life in a bed surrounded by a generation of loved ones born since the close of the war.

Mortality is inescapable, but it is not the final arbiter of a person's, or, in this case, a soldier's life. For as long as he is remembered, he has attained a certain degree of immortality. Only when his name is no longer spoken and his deeds are no longer recounted, does he truly slip into the shadows of the dead.

The men of the 42nd are now remembered by those of us who never knew them, spoke to them, broke bread with them, or shared a church pew with them. We know them as names on the headstones found at community or battlefield cemeteries such as Cedar Hill, Oakland, or Westview. Their markers can also be found in the numerous church cemeteries where they worshiped in life. Many were brought home to the family plot to be buried in the same soil that they had worked as vibrant young men. Saddest of all are those who lie buried, unmarked, in forgotten corners of battlefields, unknown to all but God and sharing their graves with some of the same men they knew in life and fought with in their struggle for independence.

Those who survived carried a two-fold responsibility to their communities and to their fallen comrades. Their first debt was addressed by Colonel

L. P. Thomas at the Greensboro surrender when he complimented the behavior of his reconstituted regiment and gave them the following guidance: "May your positions as Citizens in Society be as honorable as the responsible position you have occupied in a gallant but over powered army."[1] In this charge to his officers, Colonel Thomas directed them to become leaders in the community as they had been leaders of men in battle. The transformation of the City of Atlanta from a burned-out hull of brick storefronts to a dominant force in the New South was testament to how they fulfilled that charge.

The second duty, a debt owed to the fallen, was also honored by the men of the 42nd Georgia and by their Survivor's Association. In the second meeting of the FCCVA, the members voted to replace the deteriorating wooden crosses at Oakland cemetery with stone monuments. What they accomplished in that action was to assure that the names of the Confederate dead, fading in the space of one generation, would survive so long as stone would endure.

As the 42nd was charged, so are we all charged to keep alive these men through the tangible head stones and monuments but also through books, family accounts, and the proud tracing of our family genealogies. So long as we remember, they will continue to live in the pages of historical accounts. So long as we stop to listen, they will still speak to us through their diaries and letters.

It is with humbled reverence that we add this account of the men of the 42nd Georgia to the written record that their voices not be stilled and that we might continue the process of putting flesh back on their bones and humanity back to their story.

[1] Loose letter to his fellow officers, April 29, 1865, High Point, NC, Thomas collection, BRC.

Appendix 1

Detached 1862 Madison, Georgia Prison Guards
October 31, 1862 Muster roll of the W.L. Calhoun Detachment of prison guards from the 42nd Georgia stationed in Madison, Georgia. Their discipline, clothing and instruction were rated "good," but their arms and accoutrements were rated "poor." The names are listed with rank and company.

NAME	RANK	COMPANY
William L. Calhoun	CPT	K
Franklin W. Kendrick	CPL	D
John C. Crow	CPL	B
Daniel J. Barnett	PVT	A
Clark H. Bryant	PVT	I
William L. Bates	PVT	I
William H. Cole	PVT	A
John F. Chaffin	PVT	C
Wilson C. Carter	PVT	I
George W. Camp	PVT	I
Oliver P. Campbell	PVT	D
Sanford M. Davis	PVT	A
James W. Darby	PVT	D
George W. Elliott	PVT	I
Ezekiel Farris	PVT	D
David T. Garner	PVT	A
Union F. Gholston	PVT	A
John B. Gordon	PVT	I
Wilson L. Grubbs	PVT	I
John T. Hanson	PVT	A
Andrew J. Holbrook	PVT	I
Thomas J. Holbrook	PVT	I
John W. Jones	PVT	C
James Johns	PVT	I
James G.B. Jones	PVT	D
Amos N. Kelly	PVT	A
Franklin H. Kent	PVT	C
Columbus J. Kent	PVT	C
Nathan Lee	PVT	I
Moses Martin	PVT	A
John V. Mewborne	PVT	A
James Melton	PVT	I
Joseph Melton	PVT	I
Sylvester L. Nunnally	PVT	A
Calvin Owen	PVT	I
Thomas J. Peacock	PVT	I

John T. Pittard	PVT	A
Sanford Roberts	PVT	A
William T. Roddy	PVT	C
Alexander Roddy	PVT	C
James V. Roddy	PVT	C
Ira Sanford Rainey	PVT	C
John Smith	PVT	I
Henry L. Sims	PVT	I
John A.J. Spear	PVT	I
William W. Tarrent	PVT	I
Unnamed – Died in service		
-Possibly Robert Vaughan		

Appendix 2

Vicksburg Parole Records – 42nd Georgia Infantry
Taken from the 2015 transcription of Confederate Parole Records made by the Vicksburg National Military Park. Spelling of names has been corrected by the authors.
July 4-8, 1863

Regimental Staff – Colonel Robert Henderson – 641 Soldiers

Benson W. Adams C	PT-AQM	Field	
J.H. Griffin	SGTM		Field
Robert J. Henderson	COL		Field
William H. Hulsey	MAJ		Field
James M. Myers	CHAP		Field
J.R. Simmons	CPT-ACS		Field
P.H. Stallings	QM SGT		Field
W.B. Tindall	CSGT		Field
H.M. Wylie	ADJ		Field

Company A – Gwinnett Beauregards – Captain L.P. Thomas – 68 soldiers
1 Captain, 3 Lieutenants, 5 Sergeants, 4 corporals, 55 privates

J.Q. Adams	PVT	Field
Wiley F. Armstrong	PVT	Field
James A. Arnold	PVT	Field
John F. Baker	PVT	Field
Joseph T. Baxter	SGT	Field
Findall Bell	PVT	Field
Marshall Bolton	PVT	Field
A.M. Bramblett	SGT	Field
John W. Brantley	PVT	Hosp
Daniel G. Brogdon	PVT	Field

S.J. Brogdon	PVT	Field
Samuel J. Burrell	PVT	Field
James H. Calver	PVT	Field
William M. Carr	PVT	Field
William H. Cole	PVT	Field
C.C. Colwell	PVT	Field
William A. Craig	PVT	Hosp
Basil W. Davis	PVT	Field
W. Everett	PVT	Hosp
James M. Flowers	CPL	Field
R.P. Furgerson	PVT	Field
A.J. Furgerson	PVT	Field
David T. Garner	PVT	Field
Benjamin Gholston	2LT	Field
D.T. Gomer	PVT	Field
W. Googan	PVT	Field
H.M. Gunter	PVT	Field
John T. Hanson	PVT	Field
David M. Hogan	PVT	Field
William J. Ingraham	PVT	Hosp
William F. Johnson	PVT	Field
Robert D. Jones	PVT	Field
Amos N. Kelly	CPL	Field
Andrew J. King	PVT	Field
John C. King	PVT	Field
Daniel Kehelay	PVT	Field
S.R. Marks	PVT	Field
Abraham F. Martin	PVT	Field
Elisha Martin	PVT	Field
Darling P. McDaniel	1LT	Field
Wash L. McDaniel	SGT	Field
John O. Medlock	2LT	Field
L.D. Meddock	PVT	Field
Joshua W. Morgan	PVT	Field
Robert H. Munday	PVT	Field
J.V. Mewborn	PVT	Field
George W. Noel	PVT	Field
Thomas W. Phillips	PVT	Field
John T. Pitard	PVT	Field
S.C. Pittman	PVT	Field
John M. Pittman	PVT	Field
William J. Pruett	SGT	Field
James W. Roberts	PVT	Field
Cynes T. Shelnutt	PVT	Field
Erwin Smith	CPL	Field
John I. Spears	PVT	Field

N.R. Stevenson	PVT	Field
Lovick P. Thomas	CPT	Field
Wiley S. Townley	PVT	Field
Marion J. Townley	PVT	Hosp
Russell Whaley	CPL	Field
James B. Whitworth	PVT	Field
Hugh F.M. Wilson	1SGT	Field
J.B. Withworth	PVT	Field
William B. Worsham	PVT	Field

Company B – Independent Rebels (Gwinnett County) – 73 soldiers

2 Lieutenants, 4 Sergeants, 2 Corporals, 65 Privates

Asa D. Aderhold	PVT	Field
Ellsberry B. Andrews	PVT	Field
Andrew P. Beaver	SGT	Field
J.L. Bennett	PVT	Field
Daniel J. Brand	PVT	Field
Ransom Brand	PVT	Field
John C. Brownlee	PVT	Field
Warren Brownlee	PVT	Field
J.G. Bryan	1SGT	Field
Francis M. Buchanan	PVT	Wash Hosp
J.S. Burnett	PVT	Hosp
John W. Clower	PVT	Field
J.C. Davis	PVT	Field
W.P. Donaldson	1LT	Hosp
James Elleson	PVT	Field
James M. Flarity	PVT	Field
George W. Freeman	PVT	Field
James P. Freeman	CPL	Field
James Freeman	PVT	Field
Joseph F. Garner	PVT	Field
Joseph Garner	PVT	Field
Joseph Garnes	PVT	Field
Charles Green	PVT	Field
William D. Gresham	PVT	Field
Mathew M. Hale	PVT	Field
Milton N. Harris	PVT	Field
J.M. Henry	PVT	Field
Alexander A. Hewett	PVT	Field
Barry Hollingsworth	PVT	Field
William P. Hutchins	PVT	Field
George Jones	PVT	Field
John E. Kenerley	PVT	Field
Jessie Lancaster	PVT	Field
Francis M. Lanier	PVT	Field

F.M. Lauier	PVT	Field
Daniel J. Liddell	PVT	Field
Jesse L. Long	PVT	Field
D.W. Martin	SGT	Hosp
J.F. Martin	PVT	Field
Samuel C. Martin	PVT	Field
Eli J. Matthews	PVT	Hosp
James A. Miller	PVT	Field
Francis M. Mills	PVT	Field
Joseph B. Mills	PVT	Field
R.A. Mills	PVT	Field
William B. Mills	PVT	Field
T.W. Mitchell	PVT	Field
David P. Moore	PVT	Field
William C. Moore	PVT	Field
James B. Nash	PVT	Field
John T. Nash	PVT	Field
Wilson D. Nichols	PVT	Field
W.M. Oliver	PVT	Field
Samuel W. Paden	PVT	Field
Daniel J. Peters	PVT	Field
William P. Phillipps	PVT	Hosp
W.R. Roach	PVT	Field
Oliver H.P. Rollins	PVT	Hosp
John L. Stapp	PVT	Field
George C. Smith	PVT	Hosp
George I. Smith	PVT	Hosp
Robert P. Smith	PVT	Field
Silas A. Starr	CPL	Field
William S. Starr	SGT	Field
Dickerson M. Steele	PVT	Wash Hosp
G.M. Thomason	PVT	Hosp
Benjamin C. Thomason	PVT	Field
B.C. Thornton	PVT	Hosp #2
Andrew J. Williams	PVT	Field
James T. Williams	PVT	Field
Newton T. Williams	PVT	Field
William H. Williams	1LT	Field
Thomas F. Worthey	PVT	Field

Company C – Milton Tigers – Captain Henry Parris – 44 Soldiers

1 Captain, 2 Lieutenants, 2 Sergeants, 2 Corporals, 37 Privates

Elisha Thomas Allen	PVT	Field
Matthew L. Andrews	PVT	Field
James M. Barron	PVT	Field
John T. Barron	PVT	Field

Daniel W. Bennett	PVT	Wash Hosp
David D. Brewer	PVT	Field
Daniel Bruce	PVT	Wash Hosp
Giles M. Burnett	PVT	Field
Calvin Burton	PVT	Field
Lindsay Conner	PVT	Field
Augustus C. Dickert	PVT	Field
Francis M. Dunbar	LT	Field
James M. Eason	PVT	Field
William Fitch	CPL	Field
John M. Fowler	PVT	Field
James M. Green	PVT	Field
Lewis M. Harmon	PVT	Field
David F. Harris	PVT	Field
George W. Jameson	PVT	Wash Hosp
J.G. Jones	PVT	Field
William K. Jordan	PVT	Field
J.G. Kemp	PVT	Field
Francis H. Kent	PVT	Field
Isaac Langley	PVT	Field
Isaac Newton Lanier	PVT	Field
James Lee	PVT	Field
Charles D. Mann	PVT	Field
Singleton Maxwell	LT	Field
William T. Mills	CPL	Field
William R. Morris	PVT	Field
Elias M. Norton	PVT	Field
Francis H. Nash	PVT	Field
Thomas A.B. Panell	SGT	Field
Henry M. Paris	CPT	Field
Harvey A. Rainey	PVT	Field
Sanford Rainy	PVT	Field
William T. Roddy	PVT	Field
Simeon Scoles	PVT	Field
Miles Waddy Spence	PVT	Field
Allen Stanfeel	PVT	Field
Gideon B. Thomason	PVT	Field
John Thomas	PVT	Field
Milton Trammell	PVT	Field
Henry W. Vaughn	SGT	Field

Company D – Fowler Guards (DeKalb County) – Captain Elijah Clark – 71 Soldiers

1 Captain, 1 Lieutenant, 3 Sergeants, 3 Corporals, 63 Privates

James A. Adcock	PVT	Hosp
J.J. Armstrong	PVT	Field
S.M. Armstrong	PVT	Hosp

W.P. Atkins	PVT	Field
Benton B. Ayers	PVT	Hosp
J.D. Blackstock	PVT	Hosp
Terry T. Blackstock	PVT	Hosp
William A. Brown	PVT	Field
Albert C. Burgess	PVT	Field
Henry Harris Burgess	2LT	Field
James K.P. Cagle	PVT	Field
C.P. Campbell	PVT	Hosp
S. Campbell	PVT	Field
Benjamin F. Cardin	PVT	Field
Elijah Henry Clark	CPT	Field
Cleveland A. Clay	PVT	Field
James W. Clay	PVT	Field
Joseph Clay	PVT	Field
David F. Clotfelter	PVT	Field
J.W. Cobb	PVT	Field
Joel J. Crockett	SGT	Field
James W. Darby	PVT	Field
Samuel T. Eads	PVT	Hosp
William Ellison	PVT	Field
W. Elson	PVT	Field
Ezekiel Farris	PVT	Field
William C. Fargeson	PVT	Field
James T.W. Green	PVT	Field
Robert Halingsmith	SGT	Field
Elbert F. Harris	PVT	Field
Cyrus Henry	PVT	City Hosp
Henry C. Holcomb	PVT	Field
Daniel E. Jackson	PVT	Field
John A. Jackson	PVT	Field
James W. Jackson	PVT	Field
T.J. Jackson	PVT	Field
James Green Jones	PVT	Field
Robert J. Kelley	PVT	Hosp
Franklin Kendricks	CPL	Field
Adiel S. Lacy	PVT	Field
A.J. Lee	PVT	Field
Benjamin McCanless	CPL	Field
Alexander McGoha	PVT	Field
Angus McLeod	PVT	Field
J.G. McWilliams	PVT	Field
James L. McWilliams	PVT	Field
John W. McWilliams	PVT	Field
Benjamin G. Mitchell	PVT	Field
Daniel Moore	PVT	Field

James Moore	1SGT	Field
John Moore	PVT	Field
Alexander L. Morgan	CPL	Field
Richard M. Morris	PVT	Field
George S. Ozmer	PVT	Field
James M. Rutledge	PVT	Field
David Sheppard	PVT	Field
John T. Sheppard	PVT	Field
N.F. Smith	PVT	Hosp
William S. Spain	PVT	Field
S.T. Tanner	PVT	Hosp
A.L. Thomas	PVT	Field
B.T. Thomas	PVT	Field
George M. Thomas	PVT	Field
John H. Thomas	PVT	Field
Elijah F. Turner	PVT	Field
Thomas C. Turner	PVT	Field
Lewis H. Tweedell	PVT	Field
James W. Twilley	PVT	Field
John H. Warren	PVT	Field
Thomas C. Webb	PVT	Field
Nick White	PVT	Hosp

Company E – Harper Guards (Newton County) – 51 Soldiers

2 Lieutenants, 5 Sergeants, 2 Corporals, 1 Musician, 41 Privates

William Allen	PVT	Field
Isham M. Aycock	PVT	Field
Rufus K. Bennett	PVT	Field
Drewry J. Brewer	SGT	Hosp
C.W. Bishop	PVT	Field
Henry J. Boggus	CPL	Field
Robert W. Boggus	PVT	Field
George S. Cook	PVT	Field
James F. Dick	SGT	Field
Thomas L. Edwards	PVT	Field
George R. Elliott	PVT	Field
James J. Elliott	CPL	Field
William B. Ellis	PVT	Field
John E. Fincher	PVT	Field
J.M. Fincher	PVT	Field
William T. Gilpin	SGT	Field
William F. Green	PVT	Field
J.W. Hays	PVT	Field
John L. Hinton	PVT	Field
Seth K. Huff	PVT	Field

John W. Ingram	PVT	Hosp
William G. Johns	PVT	Hosp#2
John Asberry King	PVT	Field
John B. King	PVT	Field
Henry Z. Kitchens	PVT	Field
J.W. Leach	PVT	Field
Augustus J. Lee	2LT	Field
Joseph E. Lowry	MUS	Field
William G. Lowry	PVT	Field
Perry O. McCullough	PVT	Field
James M. Mercer	2LT	Field
William Morgan	PVT	Field
Isaac Newton Nolen	PVT	Field
William P. Noland	PVT	Field
John C. Parker	PVT	Field
James M. Parker	PVT	Field
Charles E. Polk	PVT	Hosp
Thomas W. Polk	PVT	Field
Henry B. Pope	PVT	Field
John Clark Pope	PVT	Field
W.B. Pope	PVT	Field
William Pope	PVT	Field
Steven D. Reagan	PVT	Field
David L. Rodgers	PVT	Field
John R. Sams	1SGT	Field
A. Tire Smith	PVT	Field
John O. Smith	PVT	Field
C.L. Stone	PVT	Field
Richard Taylor	PVT	Field
William E. Tucker	SGT	Field
Thom M. Wilkerson	PVT	Field

Company F – Newton Rifles – Captain J.M. Summers – 65 Soldiers

1 Captain, 2 Lieutenants, 3 Sergeants, 59 Privates

Francis M. Argo	PVT	Field
Bertram J. Bellah	PVT	Field
William A. Berny	PVT	Field
E. Vance Bradshaw	PVT	Field
William Bradshaw	PVT	Field
Elisha N. Brown	PVT	Field
John M. Cason	PVT	Field
Reuben M. Christian	PVT	Field
William B. Christian	SGT	Field
James D. Clealand	PVT	Field
William Crompten	PVT	Field
James R. Daniel	PVT	Field

David R. Dial	PVT	Field
Arthur Dobbs	PVT	Hosp
E.W.R. Dobbs	PVT	Wash Hosp
J.M. Dry	PVT	Field
William F. Edwards	PVT	Field
John D. Ellis	PVT	Field
William W. Freeman	PVT	Field
John B. Graham	PVT	Field
Lindsey G. Granger	PVT	Hosp
William J. Green	PVT	Field
George L. Hamilton	PVT	Field
Samuel H. Hardin	PVT	Field
T.H. Hardin	PVT	Field
William M. Harper	PVT	Field
George F. Hays	PVT	Field
Joseph W. Hays	PVT	Field
George M. King	PVT	Field
Alfred B. Lane	PVT	Hosp
Richard M. Lunsford	PVT	Field
William J. Lunsford	PVT	Field
Joel Marks	PVT	Field
James B. McCord	PVT	Field
Ferdinand McCullough	PVT	Field
Marcus A. McDonald	PVT	Field
Oscar V. McDonald	PVT	Field
Ephraim A. Melton	PVT	Field
Henry P. Mitchell	PVT	Field
John P. Murdock	PVT	City Hosp
H.M. Neal	PVT	Field
Green K. Odom	PVT	Field
John W. Odom	PVT	Field
John T. Ogletree	PVT	Field
James A. Parker	PVT	Wash Hosp
William A. Pergall	PVT	Field
Richard F. Phillips	PVT	Field
Minor H. Plunkett	PVT	Field
P.L. Plunkett	PVT	Field
Wiley P. Plunkett	PVT	Field
Isham B. Pritchett	PVT	Field
James Joseph Reagan	PVT	Field
Samuel R. Rodgers	PVT	Field
A.M. Rowden	PVT	Field
Sidney Snow	PVT	Field
Tyre A. Starr	2LT	Field
John J. Stephenson	2LT	Field
L.F. Stephenson	SGT	Field

John T. Stewart	SGT	Field
James M. Summers	CPT	Field
Marion G. Townsend	PVT	Field
William J. Wallace	PVT	Field
James Watson	PVT	Field
John Watson	PVT	Hosp
William Watson	PVT	Field

Company G – Walton Blues – Captain H. Wickland – 69 Soldiers

1 Captain, 3 Lieutenants, 3 Sergeants, 4 Corporals, 58 Privates

James G. Adams	PVT	Hosp
T.J. Bains	PVT	Field
Cicero M. Baxter	PVT	Field
J.J. Baxter	PVT	Field
William P. Bone	PVT	Field
Leonard Boss	PVT	Field
Charles M. Brantley	PVT	Field
James M. Brown	PVT	Field
Reuben T. Bulloch	PVT	Wash Hosp
Eldridge M. Burton	PVT	Hosp#2
J.P. Conner	PVT	Hosp
Andrew J. Cooper	PVT	Field
Ark Cooper	PVT	Field
William A. Craft	PVT	Field
Isaac Dildy	PVT	Field
Elisha Edgar	PVT	Field
John Edgar	PVT	Field
Samuel A. Edmonds	CPL	Field
Wiley W. Edwards	PVT	Field
Henry H. Evans	PVT	Field
James C. Forrester	PVT	Field
Samuel W. Forrester	PVT	Field
John G. Fuller	PVT	Field
Henry H. Green	PVT	Field
James J. Gunter	PVT	Field
William H.H. Gunter	PVT	Field
James J. Haney	PVT	Wash Hosp
John D. Harrison	PVT	Field
James H. Harrison	PVT	Field
Gideon Hendrick	PVT	Field
Nathan S. Hendrick	PVT	Field
Rufus P. Hughes	PVT	Field
W.J. Hughes	SGT	Field
John Jackson	PVT	City Hosp
R.J. John	1LT	Field

William L. Johnson	PVT	Field
Robert A. Landress	PVT	Field
Eli L. Mayfield	1SGT	Field
Robert Mayfield	CPL	Field
G.J. McDaniel	PVT	Field
George W. McDaniel	CPL	Field
H.B. Michael	1LT	Field
J.W. Michael	PVT	Field
W.D. Michael	PVT	Field
Austin Middleton	PVT	Field
Robert W. Miller	PVT	Hosp
R.F. Moore	PVT	Field
Addison W. Norwell	PVT	Field
J.C. Odum	PVT	Field
J.W. Odum	PVT	Field
Robert R. Patrick	PVT	Hosp
James B. Peeler	PVT	Field
William C. Peeler	PVT	Field
William Ridgeway	PVT	Field
Jefferson Sheppard	PVT	Field
Jesse H. Smith	CPL	Field
William J. Smith	2LT	Field
William H. Terrell	PVT	Field
Alfred G. Thompson	PVT	Field
Hilliard Thompson	PVT	Hosp
Thomas Thompson	PVT	Field
James W. Tuck	PVT	Field
Silas W. Vickers	PVT	Field
Jacob L. Warren	PVT	Field
John C. Weatherford	SGT	Field
Francis Weatherford	PVT	Field
John W. Weatherford	PVT	Hosp
H. Wickland	CPT	Field
Billington A. Wilson	PVT	Field

Company H – Walton Tigers – Captain J.M. Mitchell – 74 Soldiers

1 Captain, 3 Lieutenants, 4 Sergeants, 4 Corporals, 62 Privates

Seaborn Adcock	PVT	Field
Joseph E. Andrews	PVT	Field
Willis M. Argo	CPL	Field
James H. Armistead	SGT	Field
James A. Blackenship	PVT	Hosp
Beverly F. Blasingame	PVT	Field
J.T. Blasingame	PVT	Field
W.H. Boomer	PVT	Hosp

S.J. Brogdon	PVT	Field
George W. Cawley	PVT	Field
J.R.H. Clegg	PVT	Field
Joel B. Clegg	PVT	Hosp
Thomas M. Clegg	PVT	Hosp
Alexander M. Cox	PVT	Field
Charles A. Cox	PVT	Hosp#2
Charles Cox	PVT	Hosp
John J. Cox	1LT	Field
Wilson D. Durden	PVT	Field
George W. Eason	CPL	Field
William L. Eason	PVT	Field
Ed Edmonds	PVT	Field
Andrew J. Fowler	PVT	Field
James J. Fowler	PVT	Field
Elisha Garrett	PVT	Field
Amos H. Greene	PVT	Field
John H. Greene	PVT	Field
Edmond B. Gresham	PVT	Field
James M. Gresham	2LT	Field
John H. Grier	PVT	Field
Charles H. Griffith	PVT	Field
John D. Griffith	PVT	Field
F.M. Hanston	PVT	Hosp
James W. Harris	PVT	Field
John D. Herring	PVT	Field
Elisha M. Herndon	PVT	Field
George M. Herndon	PVT	Field
David M. Hogan	PVT	Wash Hosp
George M. Herndon	PVT	Field
James M. Huff	PVT	Field
Jonathan W. Huff	1SGT	Field
Young A. Huston	PVT	Field
Brazil G.H. Ivy	PVT	Field
Edmond M. Jackson	PVT	Field
John T. Jackson	PVT	Field
S.C. Jackson	PVT	Field
William M. Jackson	PVT	Field
J.R. Klegg	PVT	Field
G.A. Knight	PVT	Wash Hosp
John W. Lemmons	PVT	Field
George W. Lyon	PVT	Field
James D. Malcom	PVT	Field
William T. Mays	PVT	Field
Henry J. McAdams	PVT	Hosp
James M. McAfee	SGT	Field

F. Jerome McDowell	PVT	Hosp
J.D. McDowell	PVT	Field
A.M. McGaughey	PVT	Hosp
Asa C. Mitchell	PVT	Field
J.M. Mitchell	CPT	Field
Obediah F. Owens	PVT	Field
D.U. Phillips	PVT	Hosp
Charley C. Queen	PVT	Field
E.D. Queen	PVT	Hosp
William H. Serrels	PVT	Field
John N. Simmons	PVT	Field
William F. Smith	CPL	Wash Hosp
John W. Stark	PVT	Field
John A. Stephens	PVT	Wash Hosp
Frederick T. Stier	CPL	Field
David Williams	PVT	Field
E.W.H. Williams	PVT	Hosp
James F. Williams	2LT	Field
William J. Williams	SGT	Field
James H. Winter	PVT	Field

Company I – Fulton Blues – Captain T.A. Hill – 66 Soldiers

1 Captain, 1 Lieutenant, 5 Sergeants, 4 Corporals, 55 Privates

Wiley F. Baker	CPL	Hosp
W.F. Barrett	PVT	Field
W.F. Berry	PVT	Field
Joseph T. Blalock	PVT	Field
Clark H. Bryant	PVT	Field
William G. Bryant	PVT	Field
S.A. Caldwell	PVT	Field
George W. Camp	PVT	Field
Jacob Carter	PVT	Field
Wilson C. Carter	PVT	Field
J. Debenport	PVT	Field
G.S. Downes	PVT	Field
William Downs	PVT	Field
George W. Drake	PVT	Field
George W. Elliott	PVT	Field
David V. Faulkner	PVT	Field
Tarrent C. Gilstrap	PVT	Field
James G. Glynch	PVT	Hosp
John B. Gordon	PVT	Field
B.T. Grubbs	SGT	Field
Wilson L. Grubbs	PVT	Field
George M. Hancock	PVT	Field
Augustus L. Harwell	PVT	Field

Benjamin J. Herndon	PVT	Field
T.A. Hill	CPT	Field
F.J. Holbrook	PVT	Field
W.S. Holbrook	PVT	Field
F.C. House	1LT	Field
H.H. House	CPL	Field
Henry C. House	PVT	Field
Henry J. Lee	PVT	Field
J.H. Lee	PVT	Field
Nathan Lee	PVT	Field
Thomas M. Lee	PVT	Hosp
William J. Lee	PVT	Field
W.M. Manlain	PVT	Field
W.A. Mauldin	PVT	Field
W.C.J. McDonald	PVT	Field
James Melton	PVT	Field
Joseph Melton	PVT	Field
Calvin Owen	PVT	Field
William L. Payne	PVT	Field
Joel Queen	PVT	Field
James E. Ramsey	PVT	Field
M.M. Ray	PVT	Field
J.C. Sewell	PVT	Field
William N. Sewell	CPL	Field
James T. Shireling	PVT	Field
Henry L. Simms	PVT	Field
J.F. Smith	PVT	Field
J. Smith	PVT	Field
John A.J. Spear	PVT	Field
John S. Stubbs	1SGT	Field
Washington W. Tarrent	SGT	Field
I.T. Tarrent	SGT	Field
Louis Traber	PVT	Field
Benjamin F. Tuck	PVT	Field
Daniel D.M. Waits	CPL	Field
Thomas K. Waits	PVT	Field
John G. Wallace	PVT	Field
Joseph D. Willis	PVT	Field
William L. Wilson	SGT	Field
William A. Winters	PVT	Field
Lewis J. Wood	SGT	Field
Michael Woolf	PVT	Field

Company K – Calhoun Guards (Fulton County) – Captain W.L. Calhoun -51 soldiers
1 Captain, 2 Lieutenants, 4 Sergeants, 1 Corporal, 43 Privates

John Bailey	PVT	Field
Andrew J. Bass	PVT	Field
James H. Brown	2LT	Field
Jasper N. Bruce	PVT	Hosp
James T. Calhoun	PVT	Field
William L. Calhoun	CPT	Field
Jiles R. Chapman	SGT	Field
Edward T. Chubb	PVT	Field
George W. Clark	PVT	Wash Hosp
W.C. Cockerell	PVT	Field
James A. Collier	PVT	Field
A.J. Dickerson	PVT	Field
William M. Durham	PVT	Field
William J. Frederick	PVT	Field
Jessie J. Frederick	PVT	Field
James D. Harrison	PVT	Field
William J. Harrison	PVT	Wash Hosp
John Hall	PVT	Hosp
Barry Hardy	PVT	Field
Ben Harris	PVT	Hosp
Thomas E. Henderson	PVT	Field
John H. Hilton	PVT	Field
J.F. Hollingsworth	PVT	Field
J.N. Hornsby	PVT	Field
John W. Hornsby	PVT	Field
Jefferson J. Howell	PVT	Field
Richard Humphrey	PVT	Field
J.B. Jackson	PVT	Hosp
J.R. Jett	PVT	Field
James Johns	PVT	Field
David Johnson	PVT	Field
Henry Lawrence	SGT	Field
Joseph W. Maddox	PVT	Field
James F. Palmer	PVT	Field
Alexander J. Ratterree	PVT	Wash Hosp
J. Rattesee	PVT	Field
John Rattesee	PVT	Field
Cornelius L. Roberts	PVT	Field
William J. Scarborough	PVT	Field
James P. Shaw	PVT	Field
Kelly Sims	PVT	Field
John H. Smith	PVT	Hosp
Pinckney Smith	PVT	Hosp
Henry Y. Snow	1SGT	Field

Joshua Spillman	PVT	Field
J.E. Sullivan	PVT	Field
Moses G. Terry	PVT	Field
Benjamin F. Walker	2LT	Field
Green J. Watkins	CPL	Field
Joseph A. Williams	PVT	Field
Asa M. Willis	SGT	Field

Appendix 3

Muster Roll of Company H, Walton Tigers, October 21, 1863, in Eastern Tennessee, before being recalled to Missionary Ridge: 56 Soldiers

J.M. Mitchell private papers, Microfilm, Georgia Archives

John M. Mitchell	CPT
John J. Cox	1LT
James M. Gresham	2LT
James F. Williams	3LT
Jonathan W. Huff	1SGT
William J. Williams	2SGT
Thomas M. Armistead	3SGT
James M. McAfee	4SGT
George W. Eason	3CPL
William M. Argo	4CPL
Joseph E. Andrews	PVT
Seaborn Adcock	PVT
Zachariah J. Armistead	PVT
Beverly F. Blasingame	PVT
James P. Blasingame	PVT
Harrison H. Broach	PVT
George W. Cawley	PVT
J.R.H. Clegg	PVT
Joel B. Clegg	PVT
Alexander M. Cox	PVT
Charles A. Cox	PVT
John J. Dawes	PVT
William Durden	PVT
T.S. Edmonds	PVT
Edmond Edmondson	PVT
W.L. Eason	PVT
J.S. Eason	PVT
John H. Grier	PVT
Elisha Garrett	PVT
Edmond B. Gresham	PVT
James M. Huff	PVT

Egbert M. Herndon	PVT
Elisha M. Herndon	PVT
James W. Harris	PVT
Allen Ivey	PVT
William M. Jackson	PVT
Edmond M. Jackson	PVT
Scott A. Johnson	PVT
Lewis T. Knight	PVT
George W. Lyon	PVT
W.M. Lyon	PVT
John W. Lemmons	PVT
Asa C. Mitchell	PVT
David Malcom	PVT
U.J. Owen	PVT
Obediah F. Owen	PVT
John N. Simmons	PVT
W.H. Simmons	PVT
Francis M. Smith	PVT
John W. Stark	PVT
Nathaniel R. Thompson	PVT
Andrew J. Towler	PVT
David M. Towler	PVT
James J. Towler	PVT
William W. Turnbull	PVT
David Williams	PVT

Appendix 4

Soldiers Present at the Final Surrender, Greensboro, NC, May 1, 1865

The 42nd Georgia was formally surrendered by Colonel L.P. Thomas at High Point, NC. The names below were taken from the Federal rolls now on reel 3 of Microfilm M1781 at the National Archives in Washington, D.C. The original spelling of the last names that appeared on the rolls has been kept. The "Original Command" notation has been added by Cliff Roberts using Lillian Henderson's *Roster of the Confederate soldiers of Georgia, 1861-1865* as the reference.

Consolidated 42nd Georgia Staff Officers (5 Men)

Name	Rank	Commission Date	Original Command
L.P. Thomas	Lt. Colonel	9 April 1865	42nd GA, Company A
J.J. McClendon	Major	9 April 1865	34th GA, Adjutant
W.M. Durham	Adjutant	6 December 1864	42nd GA, Company K
H.S. Cox	Surgeon	11 September 1863	42nd GA, Regimental Staff
L.H. Tignes	Asst. Surgeon	13 June 1864	

Non-Commissioned Staff & Band (15 Men)
Name Rank Enlisted Where Original Command
H.B. Lane SGT Major 12 May 1862 Franklin 56th GA, Company K
E.V. Branham QM SGT 4 May 1862 Covington 42nd GA, Company F
J.M. Wilson ORD SGT 3 July 1861 Atlanta 42nd GA, Company H
B.Y. Neal Stewart 11 Apr 1862 Atlanta
Otto Spahn Musician CPT 10 Jan 1864 Macon 42nd GA
J.F. Brown Musician 4 Mar 1862 Decatur 42nd GA, Company D
L.A. Caldwell Musician 4 Mar 1862 Atlanta 42nd GA, Company I
D.F. Clotfelter Musician 4 Mar 1862 Decatur 42nd GA, Company D
W.G. Lowry Musician 4 Mar 1862 Covington 42nd GA, Company E
J.E. Lowry Musician 4 Mar 1862 Covington 42nd GA, Company E
G.W. Lyon Musician 15 May 1862 Atlanta 42nd GA, Company H
W.M. Lyon Musician 4 Mar 1862 Decatur 42nd GA, Company H
B.F. Moore Musician 4 Mar 1862 Monroe 42nd GA, Company G
E.E. Townsend Musician 24 Mar 1864 Henry Born in 1848
W.H. Williams Musician 12 May 1862 Monroe 42nd GA, Company H

Company A - Captain Thomas Jefferson Mercer (28 Men)
Name Rank Enlisted Where Original Command
Jos Heard 1LT 4 Mar 1862 Covington 42nd GA, Company E
H.J. Boggus 1SGT 4 Mar 1862 Covington 42nd GA, Company E
J.F. Dick 3SGT 4 Mar 1862 Covington 42nd GA, Company E
W.E. Tucker 4SGT 4 Mar 1862 Covington 42nd GA, Company E
J.T. Boyd 5SGT 4 Mar 1862 Covington 42nd GA, Company E
T.A. Boyd 1CPL 4 Mar 1862 Covington 42nd GA, Company E
J.B. King 2CPL 4 Mar 1862 Covington 42nd GA, Company E
J.W. Smith 3CPL 19 Aug 1862 Calhoun 42nd GA, Company E
R.K. Bennett PVT 4 Mar 1862 Covington 42nd GA, Company E
J.P. Biggers PVT 19 Aug 1862 Calhoun 42nd GA, Company E
R.W. Boggus PVT 12 May 1862 Covington 42nd GA, Company E
W.N. Brewer PVT 12 May 1862 Covington 42nd GA, Company E
G.J. Cook PVT 12 May 1862 Covington 42nd GA, Company E
C. Davis PVT 10 Aug 1864 Huntsville
J.L. Green PVT 25 Apr 1864 Covington
M. Jones PVT 11 Feb 1865 Covington
W.C. Jones PVT 20 Mar 1864 Covington
Charles Kitchens PVT 1 Feb 1865 Covington 42nd GA, Company E
Henry Kitchens PVT 25 Feb 1863 Covington 42nd GA, Company E
Moses Kitchens PVT 12 May 1862 Covington 42nd GA, Company E
J.W. Leech PVT 4 Mar 1862 Covington 42nd GA, Company E
F.M. Morris PVT 1864 Covington 42nd GA, Company E
I.N. Nolen PVT 12 May 1862 Covington 42nd GA, Company E
J.M. Nolen PVT 15 Apr 1864 Covington 42nd GA, Company E
Wm. H. Nolen PVT 4 Mar 1862 Covington 42nd GA, Company E
Burrell Pope PVT 16 May 1864 Covington 42nd GA, Company E

J.O. Smith PVT 1 Aug 1862 Calhoun 42nd GA, Company E
Richard Taylor PVT 4 May 1862 Covington 42nd GA, Company E
Company B – No officer listed (34 Men)
Name Rank Enlisted Where Original Command
J.T. Stewart 1SGT 4 May 1862 Covington 42nd GA, Company F
J.T. Ogletree 2SGT 4 May 1862 Covington 42nd GA, Company F
W.F. Edwards 3SGT 4 May 1862 Covington 42nd GA, Company F
W.H. Rowden 4SGT 15 May 1862 Covington 42nd GA, Company F
T.M. Hicks 5SGT 15 May 1862 Covington 42nd GA, Company F
J.M. Cason 3CPL 4 May 1862 Covington 42nd GA, Company F
W.F. Wilson 4CPL 11 Mar 1862 Decatur 42nd GA, Company F
Thomas Aithey PVT 4 Mar 1862 Covington 42nd GA, Company F
E.W. Aithey PVT 23 Jan 1864 Dalton 42nd GA, Company F
E.N. Brown PVT 12 May 1862 Covington 42nd GA, Company F
W.A. Brown PVT 19 Apr 1864 Covington 42nd GA, Company F
J.B. Beard PVT 9 Feb 1864 Dalton 42nd GA, Company F
F.M. Chandler PVT 4 Mar 1862 Covington 42nd GA, Company F
G.W. Chandler PVT 15 Oct 1864 Dalton 42nd GA, Company F
E.W.R. Dobbs PVT 4 Mar 1862 Covington 42nd GA, Company F
W.F. Denard PVT 15 July 1864 Covington 42nd GA, Company F
J.D. Ellis PVT 8 Aug 1862 Jones Sta 42nd GA, Company F
W.W. Freeman PVT 12 May 1862 Covington 42nd GA, Company F
J.N. Jones PVT 29 Jan 1864 Dalton 42nd GA, Company F
J.F. Knight PVT 20 Aug 1862 42nd GA, Company F
W.F. Knight PVT 25 Jan 1864 Dalton 42nd GA, Company F
W.J. Lunsford PVT 4 Mar 1862 Covington 42nd GA, Company F
Silas Morgan PVT 23 Apr 1864 Covington 42nd GA, Company F
J.P. Murdock PVT 12 May 1862 Covington 42nd GA, Company F
O.V. McDonald PVT 12 May 1862 Covington 42nd GA, Company F
W.P. Plunkett PVT 1 Mar 1864 Covington 42nd GA, Company F
W.S. Puckett PVT 4 Mar 1862 Covington 42nd GA, Company F
P.L. Puckett PVT 4 Mar 1862 Covington 42nd GA, Company F
Sidney Shaw PVT 4 Mar 1862 Covington 42nd GA, Company F
M.D. Smith PVT 4 Mar 1862 Covington 42nd GA, Company F
J.W. Stephenson PVT 20 Mar 1864 Covington 42nd GA, Company F
P.H. Stallings PVT 4 Mar 1862 Covington 42nd GA, Company F
M.G. Townsend PVT 4 Mar 1862 Covington 42nd GA, Company F
I.V. Wells PVT 20 Aug 1862 Woodsons 42nd GA, Company F

Company C – Captain Lowndes Calhoun (42 men)
Name Rank Enlisted Where Original Command
H.B. Mikel 1LT 4 Mar 1862 Monroe 42nd GA, Company G
H.Y. Snow 2LT 4 Mar 1862 Atlanta 42nd GA, Company K
W.T. Farrar 1SGT 4 Mar 1862 Atlanta 42nd GA, Company K
W.L. Johnston 2SGT 23 Aug 1862 Calhoun 42nd GA, Company G
J.T. Calhoun 3SGT 20 Aug 1862 Atlanta 42nd GA, Company K

J.C. Weatherford 4SGT 4 Mar 1862 Monroe 42nd GA, Company G
William Cooper 5SGT 4 Mar 1862 Monroe 42nd GA, Company G
J.J. Gunter 1CPL 4 Mar 1862 Monroe 42nd GA, Company G
Elisha Edgar 2CPL 4 Mar 1862 Monroe 42nd GA, Company G
Vinson Davis 3CPL 11 Mar 1862 Atlanta 42nd GA, Company K
C.H. Smith 4CPL 13 Sep 1862 Woodsons 42nd GA, Company G
J.E. Austin PVT 12 May 1862 Monroe 42nd GA, Company G
C.N. Baxter PVT 4 Mar 1862 Monroe 42nd GA, Company G
W.P. Bone PVT 4 Mar 1862 Monroe 42nd GA, Company G
J.M. Brown PVT 12 Mar 1862 Monroe 42nd GA, Company G
J.A. Collier PVT 11 Aug 1863 Calhoun 42nd GA, Company K
E.T. Chupp PVT 28 July 1863 Calhoun 42nd GA, Company K
J.W. Davis PVT 4 Mar 1862 Monroe 42nd GA, Company G
J.S. Davis PVT 25 Jan 1864 Dalton 42nd GA, Company G
J.F. Dennard PVT 10 Feb 1864 Dalton 42nd GA, Company K
J.W. Edwards PVT 4 Mar 1862 Monroe 42nd GA, Company G
J.W. Forrester PVT 4 Mar 1862 Monroe 42nd GA, Company G
N.P. Harrison PVT 26 April 1864 Atlanta 42nd GA, Company G
E.R. Howell PVT 5 April 1864 Dalton 42nd GA, Company K
E.B. Johnson PVT 31 Dec 1863 Dalton 42nd GA, Company K
Matthew Landoefs PVT 4 Mar 1862 Monroe 42nd GA, Company G
H.J. McAdams PVT 4 Mar 1862 Monroe 42nd GA, Company G
J.W. Nowell PVT 12 Mar 1862 Monroe 42nd GA, Company G
James B. Peeler PVT 4 Mar 1862 Monroe 42nd GA, Company G
W.C. Peeler PVT 4 Mar 1862 Monroe 42nd GA, Company G
A.J. Ratteree PVT 8 Jan 1863 Vicksburg 42nd GA, Company K
C.L. Roberts PVT 18 Sep 1862 Madison 42nd GA, Company K
W.J. Smith PVT 18 Feb 1864 Atlanta
F.T. Stier PVT 4 Mar 1862 Atlanta 42nd GA, Company K
T.C. Thompson PVT 4 Mar 1862 Monroe 42nd GA, Company G
A.G. Thompson PVT 12 Mar 1862 Monroe 42nd GA, Company G
D.H.C. Thompson PVT 18 Feb 1864 Atlanta 42nd GA, Company G
E.L. Turner PVT 28 July 1862 Calhoun 42nd GA, Company K
J.T. Wiggins PVT 4 Mar 1862 Monroe 42nd GA, Company G
F.M. Weatherford PVT 12 May 1862 Monroe 42nd GA, Company G
J.L. Warren PVT 4 Mar 1862 Monroe 42nd GA, Company G

Company D – Captain Benjamin Toombs Spearman (29 Men)
Name Rank Enlisted Where Original Command
B.T. Spearman CPT 30 Apr 1862 Atlanta 56th GA, Company K
I.J. Stephens 1LT 10 May 1862 Franklin 56th GA, Company K
W.W. Strickland 1SGT 10 May 1862 Franklin 56th GA, Company K
J.B. Strong 3CPL 10 May 1862 Franklin 56th GA, Company K
W.T. Albright PVT 10 May 1862 Franklin 56th GA, Company K
W.R. Cook PVT 10 May 1862 Franklin 56th GA, Company K
Dempsey Cook PVT 10 May 1862 Franklin 56th GA, Company K

W.W. Cobb PVT 10 May 1862 Franklin 56th GA, Company K
Joseph Cash PVT 1 Apr 1865
W.A. Dunson PVT 10 May 1862 Franklin 56th GA, Company K
John Daniel PVT 10 May 1862 Franklin 56th GA. Company K
I.J. Daniel PVT 10 May 1862 Franklin 56th GA, Company K
M.J. Davis PVT 4 Aug 1863 Franklin 56th GA, Company K
W.O. Davis PVT 28 Feb 1864 Franklin 56th GA, Company K
Jackson Eady PVT 10 May 1862 Franklin 56th GA, Company K
L.H. Featherstone PVT 26 Nov 1862 Franklin 56th GA, Company K
F.D. Garrison PVT 5 Apr 1864 Franklin 56th GA, Company K
J.J. Hearn PVT 10 May 1862 Franklin 56th GA, Company K
J.B. Heard PVT 10 May 1862 Franklin 56th GA, Company K
G.W. Kelley PVT 10 May 1862 Franklin 56th GA, Company K
J.H. Lane PVT 10 May 1862 Franklin 56th GA, Company K
T.A. Lane PVT 11 Sep 1864 Franklin 56th GA, Company K
Obediah Owen PVT 18 Jan 1864 Franklin 56th GA, Company K
James Paschal PVT 10 May 1862 Franklin 56th GA, Company K
N.J. Reid PVT 10 May 1862 Franklin 56th GA, Company K
W.L. Strickland PVT 26 Oct 1863 Franklin 56th GA, Company K
Gabriel Spearman PVT 12 Apr 1864 Franklin 56th GA. Company K
W.M. Thomas PVT 4 Mar 1861
M.M. Wood PVT 10 May 1862 Franklin 56th GA, Company K

Company E – Captain Elijah H. Clarke (42 Men)
Name Rank Enlisted Where Original Command
E.H. Clark CPT 4 Mar 1862 Decatur 42nd GA, Company D
Ja. Moore 1LT 4 Mar 1862 Decatur 42nd GA, Company D
W.T. Smith 2LT 4 Mar 1862 Lawrenceville 42nd GA, Company B
S.A. Starr 1SGT 4 Mar 1862 Lawrenceville 42nd GA, Company B
J.W. Ozmer 2SGT 23 Jan 1862 Richmond 42nd GA, Company D
Angusth McCloud 3SGT 4 Mar 1862 Decatur 42nd GA, Company D
J.B. Mills 4SGT 12 Mar 1862 Atlanta 42nd GA, Company B
B.W.G. McCanlefe 1CPL 4 Mar 1862 Decatur 42nd GA, Company D
C.W. Argo PVT 4 Mar 1862 Decatur 42nd GA, Company D
Jepthah Atkins PVT 4 Mar 1862 Decatur 42nd GA, Company D
W.T. Arnold PVT 4 Mar 1862 Lawrenceville 42nd GA, Company B
J.S. Bryan PVT 4 Mar 1862 Lawrenceville 42nd GA, Company B
Ransom Brand PVT 4 Mar 1862 Lawrenceville 42nd GA, Company B
J.S. Bennett PVT 4 Mar 1862 Lawrenceville 42nd GA, Company B
J.W. Bennett PVT 12 May 1862 Atlanta 42nd GA, Company B
J.K.P. Cagle PVT 4 Mar 1862 Decatur 42nd GA, Company D
Cleveland Cagle PVT 13 May 1862 Atlanta 42nd GA, Company D
J.A. Clay PVT 13 May 1862 Atlanta 42nd GA, Company D
J.M.S. Clay PVT 17 Sep 1864 Decatur 42nd GA, Company D
J.H. Cobb PVT 4 Mar 1862 Decatur 42nd GA, Company D
Ezekiel Farris PVT 4 Mar 1862 Decatur 42nd GA, Company D

H.F. Freeman PVT 22 Mar 1864 Decatur 42nd GA, Company B
J.P. Freeman PVT 4 Mar 1862 Lawrenceville 42nd GA, Company B
H.C. Holcomb PVT 4 Mar 1862 Decatur 42nd GA, Company D
G.W. Haney PVT 4 Mar 1862 Lawrenceville 42nd GA, Company B
W.P. Hutchins PVT 4 Mar 1862 Lawrenceville 42nd GA, Company B
J.G. Johns PVT 4 Mar 1862 Decatur 42nd GA, Company D
T.J. Kenerley PVT 19 Jan 1864 Dalton 42nd GA, Company B
D.E. Jackson PVT 10 Oct 1862 Decatur 42nd GA, Company D
A.S. Lacy PVT 4 Mar 1862 Decatur 42nd GA, Company D
J.F. Miles PVT 4 Mar 1862 Decatur 42nd GA, Company D
J.M. McWilliams PVT 29 May 1861 Laredo, AR 42nd GA, Company D
J.F. Martin PVT 12 May 1862 Atlanta 42nd GA, Company B
D.P. Moore PVT 12 May 1862 Lawrenceville 42nd GA, Company B
J.B. Nash PVT 4 Mar 1862 Lawrenceville 42nd GA, Company B
G.S. Ozmer PVT 4 Mar 1862 Decatur 42nd GA, Company D
G.F. Smith PVT 20 Feb 1864 Dalton 42nd GA, Company B
D.C. Sheppard PVT 15 May 1862 Atlanta 42nd GA, Company D
B.T. Thomas PVT 22 Nov 1862 Decatur 42nd GA, Company D
J.W. Twilley PVT 4 Mar 1862 Decatur 42nd GA, Company D
Thomas Worthey PVT 4 Mar 1862 Lawrenceville 42nd GA, Company B
N.W. Wright PVT 4 Mar 1862 Decatur 42nd GA, Company D

Company F – Captain John S. Stubbs (41 Men)
Name Rank Enlisted Where Original Command
J.S. Stubbs CPT 42nd GA, Company I
W.L. Wilson 1LT 42nd GA, Company I
L.F. Stephenson 2LT 42nd GA, Company F
T.J. Peacock 1SGT 4 Mar 1862 Atlanta 42nd GA, Company I
A.M. Cox 2SGT 12 May 1862 Monroe 42nd GA, Company H
W.W. Tarrent 3SGT 4 Mar 1862 Atlanta 42nd GA, Company I
J.M. Huff 1CPL 4 Mar 1862 Monroe 42nd GA, Company H
J.P. Berry 2CPL 4 Mar 1862 Atlanta 42nd GA, Company I
B.J. Herndon 3CPL 4 Mar 1862 Atlanta 42nd GA, Company I
Robert Aycock PVT 22 Sep 1863 Decatur 42nd GA, Company H
H.H. Broach PVT 3 Oct 1863 Chickamauga 42nd GA, Company H
T.J. Boyd PVT 10 May 1864 Atlanta 42nd GA, Company F
E.J. Childress PVT 7 Feb 1864 Atlanta 42nd GA, Company F
T.M. Clegg PVT 4 Mar 1862 Monroe 42nd GA, Company H
G.W. Eason PVT 4 Mar 1862 Monroe 42nd GA, Company H
W.L. Eason PVT 4 Mar 1862 Monroe 42nd GA, Company H
J.B. Gordon PVT 4 Mar 1862 Atlanta 42nd GA, Company I
B.F. Grubbs PVT 4 Mar 1862 Atlanta 42nd GA, Company I
W.L. Grubbs PVT 4 Mar 1862 Atlanta 42nd GA, Company I
J.H. Green PVT 4 Mar 1862 Monroe 42nd GA, Company H
J.D. Griffith PVT 12 May 1862 Monroe 42nd GA, Company H
G.M. Herndon PVT 12 May 1862 Monroe 42nd GA, Company H

T.J. Holbrook PVT 4 Mar 1862 Atlanta 42nd GA, Company I
Allen Ivey PVT 22 Sep 1863 Monroe 42nd GA, Company H
S.A. Johnson PVT 12 May 1862 Monroe 42nd GA, Company H
J.B. Jett PVT 4 Mar 1862 Atlanta 42nd GA, Company I
J.T. Lowe PVT 19 Sep 1864 Palmetto 42nd GA, Company F
Nathan Lee PVT 4 Mar 1862 Atlanta 42nd GA, Company I
George Lyon PVT 4 Mar 1862 Atlanta 42nd GA, Company I
A.C. Mitchell PVT 4 Mar 1862 Monroe 42nd GA, Company H
W.T. Mayo PVT 12 May 1862 Monroe 42nd GA, Company H
Joe Milton PVT 4 Mar 1862 Atlanta 42nd GA, Company I
C. Owens PVT 4 Mar 1862 Atlanta 42nd GA, Company I
W.L. Payne PVT 4 Mar 1862 Atlanta 42nd GA, Company I
B.F. Roberts PVT 29 Feb 1864 Atlanta 42nd GA, Company I
J.T. Shierling PVT 4 Mar 1862 Atlanta 42nd GA, Company I
J.N. Simmons PVT 12 May 1862 Monroe 42nd GA, Company H
J.J. Towler PVT 12 Mar 1863 Decatur 42nd GA, Company H
H.C. Tucker PVT 29 Mar 1864 Monroe 42nd GA, Company H
D.J. Thompson PVT 23 Sep 1862 Decatur 42nd GA, Company H
J.D. Willis PVT 4 Mar 1862 Atlanta 42nd GA, Company I

Company G – Captain Strickland (39 Men)
Company roster is difficult to read.
Name Rank Enlisted Where Original Command
T.P. Loveless 1LT 29 Apr 1862 Campbellton 56th GA, Company C
W.A. Deweese 2LT 11 Mar 1862 Bartow 36th GA, Company D
J.P. Bradley 1SGT 2 Mar 1862 Dug Gap 36th GA, Company C
W.E. Lee 2SGT 5 Apr 1862 Atlanta 36th GA, Company E
A.J. Howell 3SGT Tunnel Hill 36th GA, Company G
E. Leatherwood 4SGT Tunnel Hill 36th GA, Company G
J.L. Holbrook 1CPL 5 Apr 1862 Atlanta 36th GA, Company E
N.B. Smith 2CPL 1 Sep 1862 Knoxville 36th GA, Company E
Joseph Phillips 3CPL Dalton
James Thomas 4CPL Dalton
M.M. Beavers PVT 4 Apr 1862 Atlanta 36th GA, Company E
E.H. Chadwick PVT 5 Apr 1862 Atlanta 36th GA, Company E
J.A. Dooly PVT 5 Apr 1862 Atlanta 36th GA, Company E
Joseph Davis PVT 15 Apr 1862 Dalton 36th GA, Company B
G.L. Deadman PVT 20 Jun 1864 Tunnel Hill 36th GA, Company G
W.G. Gentry PVT Tunnel Hill 36th GA, Company G
John Howell PVT 17 Dec 1864 Dalton 36th GA, Company G
C. Johnson PVT Dalton
W.F. Key PVT
Adam Kreischer PVT 2 Mar 1862 Dalton 36th GA, Company C
W.A. Keith PVT 2 Mar 1862 Dalton 36th GA, Company C
Joseph Nations PVT 20 Jan 1862 Tunnel Hill 36th GA, Company G
Peter Patrick PVT 1 Feb 1862 Dalton 36th GA, Company B

Joseph Patterson PVT 20 Jan 1862 Tunnel Hill 36th GA, Company G
Marion Pallijan PVT 2 Jan 1864 Dalton 36th GA, Company G
J.G. Pelfrey PVT 9 Feb 1864 Dalton 36th GA, Company G
J.M. Price PVT 20 Jan 1862 Tunnel Hill 36th GA, Company G
S.C. Pope PVT 3 Mar 1862 Dalton 36th GA, Company G
Thomas Rogers PVT Dalton 36th GA, Company G
E.C. Smith PVT 10 Apr 1862 Atlanta 36th GA, Company E
Henry Stoker PVT 20 Jan 1862 Tunnel Hill 36th GA, Company G
S.R. Sumner PVT 13 Apr 1862 Atlanta 36th GA, Company E
J.P. Worley PVT 21 Aug 1863 Atlanta
Isham Wood PVT 17 Apr 1864 Dalton 36th GA, Company B
J.C. Wilkey PVT 27 Apr 1862 Atlanta 36th GA, Company E
M.P. Woolbright PVT 20 Jan 1862 Tunnel Hill 36th GA, Company G
R.H. Ward PVT 20 Jan 1862 Tunnel Hill 36th GA, Company G
E.A. Ward PVT 20 Jan 1862 Tunnel Hill 36th GA, Company G
R.G. Williams PVT 15 Aug 1864 Macon

Company H – Captain James Moore (40 men)
Name Rank Enlisted Where Original Command
James Moore CPT 15 May 1862 Atlanta 42nd GA, Company D
J.O. Medlock 1LT 4 Mar 1862 Lawrenceville 42nd GA, Company A
H.H. Burgess 2LT 4 Mar 1862 Decatur 42nd GA, Company D
J.R. Noel 1SGT 4 Mar 1862 Lawrenceville 42nd GA, Company A
A.J. Furgurson 3SGT 14 Sep 1862 Tennessee 42nd GA, Company A
William Fitch 4SGT 4 Mar 1862 Alpharetta 42nd GA, Company C
J.A. Perkins 5SGT 4 Mar 1862 Alpharetta 42nd GA, Company C
R.P. Furgurson 1CPL 4 Mar 1862 Lawrenceville 42nd GA, Company A
W.A. Craig 2CPL 4 Mar 1862 Lawrenceville 42nd GA, Company A
J.H. Arnold PVT 6 Apr 1862 Camp McDonald 42nd GA, Company A
J.W. Arnold PVT 20 Sep 1862 Kentucky 42nd GA, Company A
J.W. Allen PVT 4 Mar 1862 Alpharetta 42nd GA, Company C
Sanford Allen PVT 31 Jan 1864 Dalton 42nd GA, Company C
D.J. Barnett PVT 4 Mar 1862 Lawrenceville 42nd GA, Company A
Fendal Bell PVT 4 Mar 1862 Lawrenceville 42nd GA, Company A
S.G. Brogden PVT 4 Mar 1862 Lawrenceville 42nd GA, Company A
Calvin Burton PVT 4 Mar 1862 Alpharetta 42nd GA, Company C
J.M. Barron PVT 4 Mar 1862 Alpharetta 42nd GA, Company C
C.C. Caldwell PVT 4 Mar 1862 Lawrenceville 42nd GA, Company A
Lindsay Conner PVT 4 Mar 1862 Alpharetta 42nd GA, Company C
William Davis PVT 4 Mar 1862 Lawrenceville 42nd GA, Company A
Isaac Dildy PVT 22 Jan 1863 Vicksburg 42nd GA, Company C
William Everett PVT 24 Mar 1862 Lawrenceville 42nd GA, Company A
J.M. Fowler PVT 21 May 1862 Atlanta 42nd GA, Company C
F.M. Ferguson PVT 12 Apr 1862 Camp McDonald 42nd GA, Company A
W.M. Garner PVT 1 Aug 1863 Atlanta 42nd GA, Company A
E.J. Hughes PVT 4 Mar 1862 Alpharetta 42nd GA, Company C

W.F. Johnson PVT 4 Mar 1862 Lawrenceville 42nd GA. Company A
George Jones PVT 4 Mar 1862 Lawrenceville 42nd GA, Company A
I.N. Lanier PVT 4 Mar 1862 Alpharetta 42nd GA, Company C
J.V. Mewborn PVT 4 Mar 1862 Lawrenceville 42nd GA, Company A
F.H. Nash PVT 4 Mar 1862 Alpharetta 42nd GA, Company C
Amos Parks PVT 4 Mar 1862 Lawrenceville 42nd GA, Company A
W.I. Pruett PVT 4 Mar 1862 Lawrenceville 42nd GA, Company A
S.C. Pittman PVT 4 Mar 1862 Lawrenceville 42nd GA, Company A
J.V. Roddy PVT 4 Mar 1862 Alpharetta 42nd GA, Company C
V.R. Stevenson PVT 4 Mar 1862 Lawrenceville 42nd GA, Company A
T.M. Tomlinson PVT 23 Dec 1862 Macon
M.M. Waits PVT 4 Mar 1862 Decatur 42nd GA, Company A
J.B. Whitworth PVT 4 Mar 1862 Lawrenceville 42nd GA, Company A

Company I – Captain Israel Hallman (43 Men)
Name Rank Enlisted Where Original Command
I. Hallman CPT 15 Jan 1862 Dalton 36th GA, Company A
J.R. Smith 1LT 10 Apr 1862 Atlanta 36th GA, Company F
J.B. Stewart 2LT 10 Apr 1862 Atlanta 36th GA, Company F
T.G. Stokes 1SGT 10 May 1862 Dalton 36th GA, Company I
W.N. Henderson 2SGT 1 Aug 1862 Decatur 36th GA, Company A
E.G. Nash 3SGT 13 May 1862 Atlanta 36th GA, Company K
J.R. Tarver 4SGT 10 Apr 1862 Dalton 36th GA, Company H
T.J. Grant 1CPL 15 Apr 1862 Dalton 36th GA, Company I
J.A. Mosley 2CPL 10 Apr 1862 Atlanta 36th GA, Company F
O.P. Ford 3CPL 13 May 1862 Atlanta 36th GA, Company K
M.C. Hooper 4CPL 10 Apr 1862 Dalton 36th GA, Company H
Thomas Alford PVT 11 Mar 1862 Dalton 36th GA, Company D
James Bagwell PVT 12 May 1862 Atlanta 36th GA, Company H
John Bagwell PVT 10 Apr 1862 Atlanta 36th GA, Company F
J.W. Cowart PVT 15 Apr 1862 Dalton 36th GA, Company I
J.B. Cochran PVT 11 Aug 1864 Stone Mtn
William Cristwell PVT 15 July 1863 Atlanta 36th GA, Company K
Richard Franklin PVT 12 May 1862 Atlanta 36th GA, Company K
R.F.M. Gittens PVT 1 Sep 1863 Atlanta 36th GA, Company F
Green Henderson PVT 12 May 1863 Vicksburg 36th GA, Company F
G.D. Hardman PVT 10 Apr 1862 Atlanta 36th GA, Company F
W.R. Hardman PVT 1 July 1864 Dalton 36th GA, Company F
Joseph Harris PVT 11 May 1862 Dalton 36th GA, Company D
W.D. Harris PVT 2 Jun 1864 Atlanta 36th GA, Company F
M.D. Hopkins PVT 12 May 1862 Atlanta 36th GA, Company K
J.F. Jones PVT 10 Apr 1862 Atlanta 36th GA, Company F
George Keller PVT 11 Mar 1862 Dalton 36th GA, Company D
W.A. Lietch PVT 12 May 1862 Atlanta 36th GA, Company K
W.C. Mangum PVT 1 Apr 1862 Dalton 36th GA, Company A
A. Miles PVT 1 Aug 1864 Atlanta

W.W. Mason PVT 12 May 1862 Atlanta 36th GA, Company K
W.E. McAlister PVT 25 Aug 1863 Atlanta 36th GA, Company F
J.N. McElroy PVT 1 Aug 1862 Tazewell 36th GA, Company F
Thomas McElrath PVT 15 Apr 1862 Dalton 36th GA, Company I
Thomas McKinney PVT 12 May 1862 Atlanta 36th GA, Company K
N.G. Medlock PVT 30 Jun 1864 Decatur
A.T. Payton PVT 15 July 1864 Atlanta 36th GA, Company F
J.P. Pinnell PVT 1 Jun 1864 Atlanta 36th GA, Company F
W.H. Townsend PVT 5 May 1862 Chattanooga 36th GA, Company I
A.H. Tucker PVT 10 Apr 1862 Atlanta 36th GA, Company F
T.P. Wells PVT 10 Apr 1862 Atlanta 36th GA, Company F
T.J. Willis PVT 15 Apr 1862 Dalton 36th GA, Company I
J.M. Wright PVT 11 Mar 1862 Dalton 36th GA, Company D

Company K – Captain Daniel F. Brewster (39 Men)
Name Rank Enlisted Where Original Command
D.F. Brouster CPT 2 Apr 1862 Campbellton 56th GA, Company A
E.C. Thomas 1LT 14 May 1863 Jackson, MS 56th GA, Company K
C.H. Mauldin 1SGT 15 May 1862 Dalton 36th GA, Company A
D.C. Coker 2SGT 28 Apr 1862 Atlanta 56th GA, Company E
F. Guep 3SGT 16 May 1862 Franklin
J.B. Wilkinson 4SGT 4 Mar 1862 Trenton 34th GA, Company F
R. Norrell 1CPL 16 May 1862 Dalton 36th GA, Company A
F. Burlitt 2CPL 16 May 1862 Trenton 34th GA, Company F
G. O'Neal 3CPL 16 May 1862 Trenton 34th GA, Company F
J. Spence 4CPL 16 May 1862 Cumming 56th GA, Company G
P.H. Brouster PVT 31 Mar 1864 Newnan 56th GA, Company A
J. Cole PVT 16 May 1862 Trenton 34th GA, Company F
Thomas Coin PVT 16 May 1862 Trenton 34th GA, Company F
G.W. Clark PVT 16 May 1862 Trenton 34th GA, Company F
S. Dodson PVT 16 May 1862 Trenton 34th GA, Company F
J.N. Fowler PVT 16 May 1862 Dalton 36th GA, Company A
J.M. Hamilton PVT 16 May 1862 Dalton 36th GA, Company A
W.C. Hawkins PVT 3 May 1862 Union Hill 56th GA, Company G
S.M. Hale PVT 16 May 1862 Trenton 34th GA, Company F
J.S. Hale PVT 16 May 1862 Trenton 34th GA, Company F
S.J. Hale PVT 1 Mar 1864 Dalton 34th GA, Company F
J.B. Larman PVT 15 May 1862 Dalton 36th GA, Company A
A.B. McKinsey PVT 15 May 1862 Dalton 36th GA, Company A
J.C. Nailor PVT 16 May 1862 Trenton 34th GA, Company F
T.C. Robinson PVT 28 Apr 1862 Atlanta 56th GA, Company E
C.C. Ross PVT 16 May 1862 Trenton 34th GA, Company F
T.J. Ragland PVT 28 April 1862 Atlanta 56th GA, Company E
L.H. Stephens PVT 16 May 1862 Dalton 36th GA, Company A
H.S. Strickland PVT 3 May 1862 Union Hill 56th GA, Company G
S.G. Seay PVT 1 July 1862 Trenton 34th GA, Company F

James Steel PVT 16 May 1862 Trenton 34th GA, Company F
Calvin Steel PVT 16 Aug 1863 Trenton 34th GA, Company F
C.M. Tatum PVT 16 May 1862 Trenton 34th GA, Company F
G.W. Tittle PVT 1 July 1864 Atlanta 34th GA, Company F
John Thomas PVT 16 May 1862 Trenton 34th GA, Company F
H.R. Wilkinson PVT 28 Apr 1862 Atlanta 56th GA, Company E
A.F. Williams PVT 4 Mar 1862 Calhoun
Thomas White PVT 1 Aug 1862
J.B. Walton PVT 6 May 1862 Newnan 56th GA, Company A

Appendix 5

Casualty Reports of C.S.A. Surgeon Gustavus Bankhead Thornton for A.P. Stewart's - later Henry Clayton's – Division for 1864
These reports were found by Timothy Burgess in loose leaf form at the Tennessee State Library and Archives. These casualty figures are higher than previously accepted figures and indicate a high level of skirmishing and combat in Atlanta campaign.
List of Casualties in A.P. Stewart's Division in the Battles of Rocky Face Ridge & Resaca, Georgia, May 8-15, 1864

Brigades	Killed	Wounded	Total
Stovall	24	195	219
Clayton	47	277	324
Gibson	17	75	92
Baker	22	141	163
Total	110	688	798

List of Casualties in A.P. Stewart's Division in the Battle of New Hope Church, Georgia, May 25, 1864 through June 9, 1864

Brigades	Killed	Wounded	Total
Stovall	18	122	140
Clayton	11	139	150
Gibson	14	75	89
Baker	14	73	87
Total	57	409	466

List of Casualties in A.P. Stewart's Division During Engagement Around Marietta, Georgia June 8, 1864 to July 6, 1864

Brigades	Killed	Wounded	Total
Stovall	9	66	75
Clayton	0	17	17
Gibson	0	5	5
Baker	4	24	28
Total	13	112	125

List of Casualties in Henry Clayton's Division During Engagements Around Atlanta, Georgia, July 6, 1864 to August 1, 1864

Brigades	**Killed**	**Wounded**	**Total**
Stovall	37	185	222
Holtzclaw	5	92	97
Gibson	53	361	414
Baker	33	175	208
Eldridge's Art	1	8	9
William's Art	6	20	26
Total	135	841	976

List of Casualties in Henry Clayton's Division During Engagements Around Atlanta & Jonesborough, Georgia, August 1864

Brigades	**Killed**	**Wounded**	**Total**
Stovall	3	113	116
Holtzclaw	11	122	133
Gibson	42	167	209
Eldridge's Art	18	18	
Total	56	420	476

Appendix 6

42nd Soldier Deaths in Federal POW Camps
Taken from *Confederate P.O.W.'s Buried in Northern Cemeteries* by Frances Ingmire and Carolyn Ericson, 1984. Original records were "compiled in the office of the Commissioner for Marking Graves of Confederate Dead, War Department, 1912."

Alton Prisoner of War Camp, Alton Illinois

Name	**Rank**	**Company**	**Death Date**	**North Alton Cemetery**
Josephus Deckert	PVT	C	8 Jul 1864	Confederate Section

Camp Chase, (near) Columbus, Ohio

Name	**Rank**	**Company**	**Death Date**	**Grave Marker**
William F. Barrett	CPL	I	16 Nov 1864	478
Leonard Bass	PVT	G	2 Nov 1864	390
William M. Carr	PVT	A	13 Oct 1864	300
Charles Green	PVT	B	20 Sep 1864	258
J.W. Huff	SGT	H	5 Feb 1865	1060
J.L. Johnson	PVT	H	23 Mar 1865	1737
Elisha Lake	PVT	G	14 Feb 1865	1239
James A. Miller	PVT	B	9 Jan 1865	727
James M. Rutledge	PVT	D	13 Feb 1865	1217
Noah B. Smith	PVT	I	9 Oct 1864	290
L. Victory	PVT	E	28 Feb 1865	1489

Camp Douglas, Chicago, Illinois

Name	Rank	Company	Death Date	Oak Woods Cemetery
Wilson C. Carter	PVT	I	29 Mar 1865	Confederate Mound
Daniel W. Chandler	PVT	G	11 Jan 1865	Confederate Mound
Cooper B. Chastine	PVT	C	4 Mar 1865	Confederate Mound
William Donehoo	PVT	I	19 Nov 1864	Confederate Mound
George W. Drake	PVT	I	14 Dec 1864	Confederate Mound
John T. Shepard	PVT	D	30 Jan 1865	Confederate Mound
William Sheppard	PVT	D	20 Jan 1865	Confederate Mound
John M. Singleton	PVT	A	25 Nov 1864	Confederate Mound
William S. Spain	PVT	D	17 Dec 1864	Confederate Mound
E.F. Turner	PVT	D	7 Mar 1865	Confederate Mound
H.F.M. Wilson	SGT	A	10 Dec 1864	Confederate Mound

Camp Morton, Indianapolis, Indiana

Name	Rank	Company	Death Date	Crown Hill Cemetery
Jarrett New	SGT	D	14 Jun 1864	Confederate Mound
John Martin Scott	PVT	A	16 Jan 1865	Confederate Mound

Point Lookout, Maryland

Name	Rank	Company	Death Date	Point Lookout Cemetery
Carlos Brown	PVT	D	16 May 1865	Confederate Section

Complete Roster of the Forty-Second Georgia
1,355 Men

Prepared by Cliff Roberts, Frank Clark & Joe Bailey

All soldiers enlisted on March 4, 1862, unless stated otherwise.

All soldiers are born in Georgia, unless stated otherwise.

The last complete roll of the 42nd Georgia that has survived is dated December 10, 1863.

Officers & Staff – 26 Men

Name Rank Birth Date

Benson Wesley Adams QM CPT 22 Apr 1825 2 Apr 1910

Married Lenora Lutner in Putnam County, Dec 23, 1856; Appointed Captain and Regimental Assistant Quartermaster from the Calhoun Guards (K), Nov 26, 1862; Captured at **Vicksburg,** MS July 4, 1863; On list of active quartermasters in Army of Tennessee, July 20, 1864; Surrendered at Washington, GA April 1865; Dry goods clerk in Eatonton; Founder of the Eatonton Methodist Church; Filed for Confederate pension in Putnam County in 1900; Buried at Pine Grove Cemetery, Putnam County.

Joseph Tarpley Baxter ORD SGT 4 Apr 1841 22 Apr 1904

Appointed Ordinance SGT from Gwinnett Beauregards(A); Captured at **Missionary Ridge,** TN Nov 25, 1863; Released at **Rock Island,** IL June 20, 1865; Farmer and Cotton Broker; Married Grace Woodward in Gwinnett County, Aug 26, 1886; Member of UCV Camp # 982 Lawrenceville, 1898; Cotton buyer in Gwinnett County in 1900; Buried at Level Creek Cemetery, Gwinnett County.

Edward Vance Branham QM SGT 18 Nov 1843 10 Aug 1902

Born in Putnam County; Enlisted as a private in Newton Rifles(F); Captured at **Vicksburg,** MS July 4, 1863; Appointed Regimental Quartermaster SGT Oct 1, 1863; Surrendered at **Greensboro,** NC, April 26, 1865; Became a physician after the war; Filed for Confederate pension in Newton County in 1900; Buried at Oxford Cemetery, Oxford, Newton County.

Henry Smith Cox SURG 6 Feb 1837 8 Mar 1894

Born in Tennessee; Studied at Jackson College in Columbia and took medical courses at University of Pennsylvania; Passed Medical Board at Holly Springs, MS Oct 27, 1862; Appointed Surgeon of the 42nd GA Jun 23, 1864; Formerly surgeon of the 26th TN Regiment; Surrendered at **Greensboro,** NC April 26, 1865; Married Rebecca Martin in Maury County, TN Dec 11, 1866; Physician in Columbia, TN after the war; Buried at Rose Hill Cemetery, Columbia, TN.

John A. Dunn ASST SUR 1840 After 1870

Born in Alabama; Appointed Assistant Surgeon Dec 19, 1862; Serving as Assistant Surgeon with Pioneer Corps Provost Guard Escort Company, Sep 28, 1864; Physician in Atlanta in 1870; Buried in Atlanta.

William Meigs Durham ADJ CPT 11 Oct 1845 21 Feb 1919

Enlisted as PVT in Calhoun Guards (K); Captured at **Vicksburg,** MS July 4, 1863; Present on Dec 1863 muster roll; Appointed Regiment Adjutant Dec 16, 1864; Surrendered at **Greensboro,** NC April 26, 1865; Graduated in 1880 from the Georgia College of Eclectic Medicine & Surgery in Atlanta; Became prominent Atlanta physician; Married Asta Bowen in Fulton County, Sep 21, 1888; Member of UCV Camp # 159 in Atlanta; Filed for Confederate pension in DeKalb County in 1914; Application endorsed by T.M. Armistead; Buried in Decatur Cemetery, DeKalb County.

James Henry Griffin SGT MAJ 18 Feb 1840 8 Feb 1911

Appointed Sergeant Major from Newton Rifles(F) in 1862; Captured at **Vicksburg**, MS July 4, 1863; Received Atlanta distribution hospital Feb 26, 1864; Recommended for promotion to regiment adjutant Jan 18, 1865; Married Susan Means in Newton County, Dec 26, 1865; Father-In-Law was Alexander Means, President of Emory College; Farmer in Newton County in 1880; Tax collector for the IRS in Atlanta in 1900; Buried in Oxford Cemetery, Oxford.

William S. R. Hardeman SURG 27 Jan 1831 Dec 1887

Older brother of Ebenezer & James Hardeman; Married Cynthia Wise in Oglethorpe County, Mar 18, 1858; 1LT in the Walton Tigers (H); Appointed Regimental Surgeon May 29, 1862; Resigned with disability Mar 24, 1863; Elected Captain of Company G, 10th GA State Guards Cavalry Aug 4, 1863; Mustered out Feb 2, 1864; Physician in Monroe, GA after the war; Buried at Rest Haven Cemetery, Walton County.

Robert Johnson Henderson COL 12 Nov 1822 3 Feb 1891

Born in Jasper County; Married Louisa Wood in Newton County, April 2, 1846; Representative to Georgia Assembly 1859; In the 1860 census his Covington plantation had real estate valued at $13,000 and personal property valued at $53,660; Captured at **Vicksburg**, MS July 4, 1863; Wounded at **Battle of Resaca**, GA May 15, 1864; Admitted to Fairgrounds Hospital #2 in Vineville, GA with kidney inflammation Aug 9, 1864; Returned to service Oct 19, 1864; Promoted to Brigadier General in 1865; Miller in Covington in 1880; Buried at Covington City Cemetery, Newton County.

James Henry CHAPLAIN

Aaron Green Hulsey SURG 4 Mar 1820 28 Jun 1880

Married Elizabeth Welch in Cobb County, Nov 26, 1857; Physician in Newton County 1860 census with real estate valued at $5,000 & personal property of $27,234; Appointed Assistant Surgeon from the Newton Rifles (F) Mar 22, 1862; Appointed Surgeon Jun 15, 1862; Farmer in Carroll County in 1880.

William Henry Hulsey LT COL 1 Oct 1838 17 May 1909

Nephew of Aaron Hulsey; Elected 1LT of Company F, 6th GA Infantry May 5, 1861; Elected Major of the 42nd Georgia Mar 20, 1862, Wounded at **Battle of Resaca**, GA May 15, 1864; Elected Lieutenant Colonel Jan 19, 1865; Married Marion J. Bateman in Houston County, April 26, 1865; Elected Mayor of Atlanta in 1869; Prominent Atlanta lawyer; Member of UCV Camp #159 in Atlanta; Judge of the Ordinary Court for Fulton County, 1900; Buried in Oakland Cemetery, Atlanta.

George Hoyle Jones QM SGT 23 Oct 1833 28 Mar 1915

Married Anna Elizabeth Martin in Gwinnett County, Dec 23, 1857; Appointed Quartermaster SGT from the Gwinnett Beauregards (A) in 1862; Detached for scouting duties by General Kirby Smith Aug 8, 1862; Assigned to the Confederate Quartermaster Department; Paroled at Kingston, Ga May 12, 1865; Filed for Confederate pension in Gwinnett County in 1912; Reported being assigned to the QM Department in Atlanta Sep 1862 until the end of the war; Farmer in Pinckneyville in 1880; Filed for Confederate pension in Gwinnett County in 1911; Buried at Norcross City Cemetery, Gwinnett County.

Silas Wesley Leak HOSP STE 1834 About 1870

Born in Newton County; Married Victoria Boring in Coweta County, Nov 13, 1859; Enlisted in Company C, 7th Georgia May 31, 1861 & discharged in Richmond, VA Dec 7, 1861; Enlisted in Newton Rifles (F) at Cumberland Gap, TN April 30, 1862; Appointed Hospital Stewart by Colonel Henderson May 12, 1862; Received two-month disability discharge in Knoxville, TN Sep 8, 1862.

Leonidas Felix Livingston QM SGT 3 Apr 1832 11 Feb 1912

Married Martha Griffin in Newton County, Dec 28, 1849; Transferred from Cobb's Legion to Newton Rifles (F) Mar 4, 1862; Appointed Assistant Quartermaster April 15, 1862; Transferred; Surrendered at Atlanta May 24, 1865; Vice President of the Georgia Agriculture Society; Served four years in the GA state legislature and twenty years as a U.S. Congressman from the 5th District of Georgia; Buried at Bethany Presbyterian, Covington.

Robert Flourney Maddox LT COL 3 Jan 1829 6 Jun 1899

Born in Putnam County; Married Nancy Reynolds in Newton County, Nov 27, 1860; Elected Lt. Colonel of the 42nd GA Mar 20, 1862; Appointed Colonel of 2nd GA Reserves May 7, 1864; Prominent Atlanta politician & banker after the war; Operated a tobacco & liquor store in the National Hotel from 1869 to 1879; In 1880 Maddox organized the Maddox-Rucker Banking Company which merged with the Atlanta National Bank in 1908; Member of UCV Camp # 159 in Atlanta; Buried at Oakland Cemetery, Atlanta.

Jonathan J. McClendon MAJ 24 Aug 1824 16 Dec 1913

Elected Major of the 42nd Georgia Consolidated Regiment on April 9, 1865; Surrendered at **Greensboro**, NC April 26, 1865; Previously Captain of Company K, 7th Regiment GA State Troops in 1861; Appointed 1LT & Adjutant of 32nd Georgia in 1862; Surrendered at Vicksburg, MS July 4, 1863; Owned plantations in South Georgia and Northern Alabama; Buried in Oakland Cemetery, Atlanta.

Egbert Bell Rosser SGT MAJ 7 Nov 1839 3 Nov 1912

Married Sarah Greer in Newton County, Mar 16, 1862; Appointed Sergeant Major of regiment from the Newton Rifles (F) Mar 22, 1862; Serving as Hospital Stewart Mar 4, 1863; Judge in Rockdale County; Cotton trader in Atlanta in partnership with Thomas Armistead of the Walton Tigers (H); Served eight years on the Fulton County Commission; Devoted member of the First Christian Church on Pryor Street; Died in his winter home in Deland, FL; Buried in Oakland Cemetery, Atlanta.

Joseph R. Simmons ACS CPT 1830 After 1890

Appointed Head of Regimental Commissary with rank of Captain, Mar 22, 1862; Captured at **Vicksburg**, MS July 4, 1863; Served as A.C.S. of Stovall's Brigade, Dec 7, 1864; Captured at Macon, GA April 20, 1865; Became wholesale grocer & commercial merchant in Atlanta in the 1870s; Lived at the Kimball House Hotel in 1877; Member of UCV Camp #159 in Atlanta.

Otto Spahr MUSICIAN 14 Jan 1830 9 Apr 1911

Born in Hesse-Darmstadt, Germany; Enlisted in Macon with the Campbell Sharpshooters (F) of the 30th Georgia June 10, 1864; Transferred to 42nd Georgia July 1, 1864; Served as regimental band leader for the Tennessee & Carolinas campaigns; Surrendered at **Greensboro**, NC April 26, 1865; Married Henrietta Patridge in Fulton County, Oct 17, 1872; Professor and teacher of vocal and instrumental music in Atlanta after the war; Founded Atlanta Conservatory of Music; Member of Masonic Order & Knights of Pythias; Member of UCV Camp #159 in Atlanta; Filed for Confederate pension in Fulton

County in 1910; Application endorsed by W.M. Durham; Buried at Oakland Cemetery in Atlanta.

Palasiah H. Stallings COM SGT 1836 After 1878

Elected 1st Sergeant of the Newton Rifles (F) Mar 4, 1862; Appointed Regiment Commissary Sergeant Feb 1863; Captured at **Vicksburg**, MS July 4, 1863; Present on Dec 31, 1864 muster roll; Surrendered at **Greensboro**, NC April 26, 1865; Farmer in Henry County, 1878.

John Smith Stubbs ADJ CPT Jun 1841 Dec 1912

Taught school in Cave Springs, GA before the war; Captured in Kentucky; Transported on US Steamer *Maria Denning* from Louisville, KY to Vicksburg, MS; Exchanged in Vicksburg Nov 15, 1862; Elected 2LT of Fulton Blues (I) April 2, 1863; Captured at Vicksburg, MS July 4, 1863; Elected 1LT Oct 14, 1863; Present on Feb 20, 1864 muster roll; Appointed Regiment Adjutant with rank of Captain, Jan 1865; Surrendered at **Greensboro,** NC April 26, 1865; Married Madge Simmons in Floyd County about 1867; Owned a general merchandise store in Cedartown after the war; Buried at Greenwood Cemetery, Cedartown.

Lovick Pierce Thomas LT COL 24 Mar 1835 2 May 1910

Born in Clarke County; Married Jennie Peeples in Gwinnett County, Jan 22, 1857; Ran a general store in Lawrenceville before the war; Elected CPT of the Gwinnett Beauregards(A) Mar 4, 1862; Captured at **Vicksburg**, MS July 4, 1863; Promoted to Major in Jan 1865, backdated to May 9, 1864; Promoted to Lt. Colonel of the 42nd Georgia Mar 1865; Surrendered at **Greensboro,** NC April 26, 1865; Elected Sheriff and Deputy Sheriff of Fulton County for multiple terms; Atlanta real estate investor; Married Jeannette Payne in Fulton County, Sep 7, 1886; President of the 42nd Georgia Survivors Association from 1884 until his death in 1910; Commander of UCV Camp # 159 in Atlanta; Buried in Oakland Cemetery, Atlanta.

John Stubbs Wilson ASST SUR 14 Feb 1831 1873

Older brother of William Wilson; Married Amaltha Gibert in Fulton County, Dec 15, 1857; Appointed Assistant Surgeon from the Fulton Blues(I) Jun 15, 1862; Wounded at **Battle of Resaca**, GA May 15, 1864; Admitted to Fairgrounds Hospital #1 in Vineville, GA Nov 3, 1864 with gunshot wound; Farmer & Physician in Adamsville area of Fulton County after the war. Buried in the Wilson Family Cemetery, Fulton County.

Robert Richardson Wood QM 3 Sep 1834 1 May 1903

Educated at Emory College in Oxford; Married Jackie Bates about 1855; Promoted to brigade quartermaster department & replaced by Covington merchant in the firm of Wood & Rogers; The Georgia Enterprise reported that Major Wood "was with President Davis' party on their retreat from Richmond until they arrived at Washington, Georgia and the organization was broken up." Farmer in Newton County in 1880; Buried at the Covington City Cemetery, Newton County.

Hugh M. Wylie ADJ 1812 16 Dec 1864

Owned 3 slaves in 1859 Fulton County tax list; 1LT in Atlanta militia company known as the Gate City Guard, Mar 1861; Appointed Adjutant of 42nd Georgia, April 7, 1862; Captured at **Vicksburg,** MS July 4, 1863; Admitted to Fairgrounds Hospital #1 in Atlanta April 28, 1864 with fever; Returned to duty May 22, 1864; Appointed A.A.A.G. July 24, 1864; Killed at the **Battle of Nashville**; Buried at Rest Haven Cemetery, Columbia, Maury County, TN.

Company A – Gwinnett Beauregards – 142 Men
Name Rank Birth Death

Lovick Pierce Thomas CPT 24 Mar 1835 24 May 1910
See regimental officers.

Andrew Marion Bramblett 1CPL 23 Jun 1830 2 Jan 1907
Born in South Carolina; Married Sallie Bradford in Gwinnett County, Aug 16, 1849; Elected 1CPL Mar 4, 1862; Promoted to Sergeant; Captured at **Vicksburg**, MS July 4, 1863; Present on Dec 10, 1863 muster roll; Filed for Confederate pension in Etowah County, AL in 1899; Reported that he left Confederate service in May 1865; Farmer in Howelton, Alabama in 1900; Buried in Walnut Grove Cemetery, Etowah County, AL.

Noah Richard Brogdon 2LT 1830 25 Jun 1862
Married Harriet Strickland in Gwinnett County, Mar 21, 1852; Elected 2LT Mar 4, 1862; Died of mumps at Cumberland Gap; Buried at Level Creek Cemetery, Suwanee, Gwinnett County.

William Green Davis 3SGT 1822 1887
Born in South Carolina; Married Margaret Stevenson in Gwinnett County, Jan 23, 1845; Elected 3SGT Mar 4, 1862; Sent to Vicksburg, MS hospital April 20, 1863; Surrendered at **Greensboro**, NC April 26, 1865; Farmer in Hog Mountain in 1880.

Hamilton G. Goza 4SGT 27 Mar 1838 25 Nov 1862
Married Sarah Hunnicutt in Gwinnett County, May 26, 1859; Elected 4SGT Mar 4, 1862; Died of disease at home, Nov 20, 1862; Buried at Mt. Carmel Methodist, Norcross.

Darling Peeples McDaniel 1LT 1 Mar 1825 24 Jan 1864
Married Mary Carroll in Gwinnett County, May 18, 1848; First Master of Norcross Lodge #228; Captured at **Vicksburg**, MS July 4, 1863; Killed by falling tree at Dalton, GA Jan 21, 1864; Memorial stone at First Baptist in Lilburn; Buried in Confederate Section at West Hill Cemetery in Dalton, Georgia.

John Oliver Medlock 2LT 1830 Jun 1911
Married Elizabeth Carroll in Gwinnett County, Dec 16, 1850; Captured at **Vicksburg**, MS July 4, 1863; Elected 1LT Aug 31, 1864; Surrendered at **Greensboro**, NC April 26, 1865; Farmer in Pinckneyville after the war; Buried at Norcross City Cemetery, Gwinnett County.

Thomas Wesley Mitchell 2SGT 20 May 1835 16 Jan 1864
Married Marceance Brooks in Gwinnett County, Jan 16, 1856; Detailed to Captain Calhoun's detachment in Knoxville, TN April 20, 1862 to guard political prisoners; Prison guard at Madison, GA Oct 31, 1862; Captured at **Vicksburg**, MS July 4, 1863; Died at Dalton, GA, Jan 16, 1864; Buried at Rock Springs Methodist, Buford, Gwinnett County.

William L. Pittard 5SGT 18 Apr 1842 28 Nov 1862
Brother of John Thomas Pittard; Died in Woodford County, Kentucky; Buried in McKinney Cemetery, Gwinnett County.

John M. Sexton 1SGT 1822 6 Nov 1862
Born in Spartanburg, SC; Married Mary Ann Blount in Gwinnett County, Aug 6, 1844; Died of typhoid fever at Bean Station, TN.

Cyrus T. Shelnutt 2CPL 1831 20 Jul 1863
Married Elizabeth Ann Martin in Walton County, Aug 23, 1849; Captured at **Vicksburg**, MS July 4, 1863; Died in Vicksburg hospital; Memorial stone at High Shoals Baptist in

Walton County.
Erwin Smith 3CPL 1838 15 May 1864
Appointed 2CPL Feb 1863; Appointed 4SGT Jun 20, 1863; Captured at **Vicksburg**, MS July 4, 1863; Killed at **Battle of Resaca**, GA.
Russell A. Whaley 4CPL 1 Mar 1838 17 Aug 1896
Married Rachel Amanda Johnson in 1858; Appointed 3CPL Feb 1863; Captured at **Vicksburg**, MS July 4, 1863; Present on Dec 10, 1863 muster roll; Deserted Nov 1864; Took US Oath of Allegiance in Chattanooga, TN Nov 18, 1864; Farmer in Berkshire in 1880; Buried in Whaley Cemetery, Gwinnett County.
John Quincy Allison PVT 12 Sep 1829 23 Sep 1906
Admitted to Fair Grounds Hospital #1 in Atlanta with rheumatism Nov 14, 1862; Admitted to Fairgrounds Hospital #2 in Atlanta with debility Feb 15, 1863; Returned to duty Mar 16, 1863; Captured at **Vicksburg**, MS July 4, 1863; Appointed **CPL**; Captured in the **Battle of Atlanta** July 22, 1864; Released at **Camp Douglas**, IL June 16, 1865; Farm hand in Vallejo, California in 1880; Married Maggie Cooper in 1885; Farmer in Douglas County in 1900; Filed for Confederate pension in Douglas County in 1915; Reported moving to Douglas County, in 1884; Buried at Buried at Rose Hill Cemetery in Austell.
Junius Hillier Armstrong PVT 1841 1862
Died at Bean Station, TN hospital.
Wiley F. Armstrong PVT 1839 9 Mar 1865
Older brother of Junius Armstrong; Captured at **Vicksburg**, MS July 4, 1863; Present on Dec 10, 1863 muster roll; Captured at **Battle of Resaca**, GA May 15,1864; POW at **Camp Morton**, IN; Transferred to Point Lookout, MD for exchange Feb 19, 1865; Died in Chimborazo Hospital #3 at Richmond, VA Mar 9, 1865.
James A. Arnold PVT 1842 30 July 1864
Enlisted in Lawrenceville Dec 30, 1862; Teamster in spring 1863; Captured at **Vicksburg**, MS July 4, 1863; Killed at **Battle of Knob Farm** near Kennesaw, GA.
James Harrison Arnold PVT 1842 1886
Sent to Vicksburg hospital April 5, 1863; Surrendered at **Greensboro**, NC April 26, 1865; Married Martha Evans in Madison County, AL Nov 25, 1866; House carpenter in Tupelo, MS in 1880.
John Wiley Arnold PVT 1831 After 1870
Older brother of James Harrison Arnold; Married Sarah Chamblee in Gwinnett County, Jan 4, 1852; Enlisted Sep 20, 1862 at Cumberland Ford, KY; Captured at **Vicksburg**, MS July 4, 1863; Surrendered at **Greensboro**, NC April 26, 1865; Farmer in Gwinnett County in 1870.
Larkin T. Bagwell PVT 1834 19 Mar 1863
Born in South Carolina; Married Sarah Miller in Gwinnett County, Mar 25, 1854; Wounded in the throat at **Battle of Chickasaw Bayou**, MS Dec 28, 1862; Died at Confederate hospital in Clinton, MS; Claim filed by Sarah Bagwell in Gwinnett County.
Charles G. Baker PVT
Deserted from his company near Vicksburg, MS Mar 4, 1863; Under arrest in Decatur Oct 31, 1863; Took U.S. Oath of Allegiance at Nashville, TN and released, Jun 30, 1864; Took Reconstruction Oath in Bartow County Jul 15, 1867.
John F. Baker PVT
Captured in Vicksburg rifle pit stormed by Federals under General Herron, June 23, 1863;

Paroled at **Vicksburg**, MS July 4, 1863; Deserted near Kingston, GA May 22, 1864; Took Loyalty Oath in Chattanooga, TN May 31, 1864, agreeing to stay north of the Ohio River; Took Reconstruction Oath in Bartow County Jul 15, 1867.

Daniel Jackson Barnett PVT 17 Mar 1837 Jan 1912
Older brother of John & Weldon Barnett; Married Narcissa Leach in Gwinnett County, May 27, 1858; Enlisted in Lawrenceville April 5, 1862; Detailed to Captain Calhoun's detachment in Knoxville, TN April 20, 1862 to guard political prisoners; Prison guard at Madison, GA Oct 31, 1862; Sent to Vicksburg, MS hospital Mar 10, 1863; Patient at Lauderdale Springs, MS hospital April 30, 1863; Absent sick on Oct 31, 1863 muster roll; Surrendered at **Greensboro**, NC April 26, 1865; Farmer in Crossville, Milton County in 1880; Farmer in Cobb County in 1900; Filed for Confederate pension in Cobb County in 1900; Application endorsed by J.O. Medlock; Buried at Lebanon Baptist in Roswell.

John Riley Barnett PVT Dec 1844 2 Nov 1862
Died in Lenoir, TN hospital.

William Weldon Barnett PVT 22 Apr 1847 19 Jul 1932
Enlisted in 1864 at age 17; Married Cynthia Thompson in DeKalb County, Dec 14, 1865; Worked in saw mill in Milton County in 1900; Filed for Confederate pension in Fulton County in 1907; Reported that he left the 42nd in MS on sick leave after the Tennessee campaign; Cut off from his command, he joined Company K of Carroll's Battalion, GA Militia in March 1865; Buried in New Hope Methodist, Fayette County.

Findell Bell PVT 1836 After 1880
Married Nancy Allen in Greene County, Jan 9, 1856; Captured at **Vicksburg**, MS July 4, 1863; Surrendered at **Greensboro**, NC April 26, 1865; Farmer in Cedar Shoals, Newton County in 1880.

Thomas J. Bennett PVT 1842 14 July 1863
Died in Vicksburg, MS hospital.

Simon Berry PVT 1830 22 May 1862
Married Sarah Blunt in Gwinnett County, Mar 2, 1854; Died in Tazewell, TN hospital; Sarah filed for Widow's pension in Gwinnett County in 1891; She reported "while in the service taken measles and relapsed and died at Tazewell. He never returned to his home."

Andrew Marshall Bolton PVT 1833 Before 1867
Born in Laurens County, South Carolina; Older brother of Sylvester Bolton; Married Frances Langley in Gwinnett County, Aug 29, 1858; Left sick in Knoxville, TN hospital April 23, 1862; Captured at **Vicksburg**, MS July 4, 1863; Present on Dec 10. 1863 muster roll; Widow living in Atlanta in 1892.

David Bradford Bolton PVT 7 May 1808 9 Jun 1877
Born in Laurens District, South Carolina; Married Margaret Ross in Gwinnett County, Nov 28, 1844; Listed in Gwinnett County 1864 Militia Census as 52-year-old farmer; Captured in Gwinnett County in 1864; Took US Oath of Allegiance in Louisville, KY Aug 6, 1864; Released to stay north of the Ohio River; Farmer in Pinckneyville in 1870; Buried at Mt. Carmel Methodist in Norcross.

David Sylvester Bolton PVT 19 Feb 1842 13 Nov 1902
Born in South Carolina; Captured at **Battle of Baker's Creek**, May 16, 1863; POW at **Camp Morton**, IN; Paroled at Fort Delaware July 3, 1863; Wounded at **Pumpkin Vine Creek**, June 5, 1864; Married Mattie Wiggins in Gwinnett County, Mar 18, 1872; Farmer in Pinckneyville after the war; Pension records report that he transferred Sept 1, 1864 to

Company H, 2nd Regiment GA Militia Cavalry; Buried in Norcross City Cemetery, Gwinnett County.

John R. Bostick PVT 1844 13 May 1862
Died in Gwinnett County home.

Alonzo Derastus Braziel PVT 14 Nov 1846 7 Feb 1909
Late enlistment; Married Ann Caroline McNeil in Cass County, TX, Feb 3, 1866; Farmer in Rains County, TX in 1900; Ann Braziel filed a widow pension in 1911 from Hardeman, TX and reported that her husband served two years in Company A of the 42nd GA under Captain Thomas; Buried in Cisco, Eastland County, TX.

Thomas Butler Braziel PVT 1841 6 Sep 1886
Older brother of Alonzo Braziel; Captured at the **Battle of Baker's Creek**, MS May 16, 1863; POW at **Camp Morton,** IN; Paroled at Fort Delaware DE July 3, 1863; Exchanged at Camp Lee, near Richmond, VA July 19, 1863; Present on Dec 10, 1863 muster roll; Married Elizabeth Goosby in Lincoln County, TN, Nov 20, 1867; Farmer in Emory, TX in 1880; Buried in Dougherty Cemetery, Rains County, TX; Wife filed for pension in Rains County in 1909.

Daniel Glenn Brogdon PVT 24 Aug 1841 15 Jun 1918
Captured at **Vicksburg,** MS July 4, 1863; Present on Dec 10, 1863 muster roll; Captured in Gwinnett County & sent to Louisville, KY where he took the U.S. Oath of Allegiance, July 8, 1864; Released to remain north of the Ohio River until war's end; Filed for Confederate pension in Gwinnett County in 1910; Reported being captured in May 1864; Patient at Georgia State Sanitarium in Milledgeville in 1900; Buried in Level Creek Cemetery, Suwanee.

Shadrack George Brogdon PVT 2 Sep 1839 26 Nov 1908
Older brother of Daniel Brogdon; Married Susannah Reese in Gwinnett County, Sep 23, 1860; Wounded at **Battle of Chickasaw Bayou,** Dec 28, 1862; Captured at **Vicksburg,** MS July 4, 1863; Wounded at **Battle of Resaca,** GA May 15, 1864; Surrendered at **Greensboro,** NC April 26, 1865; Farmer in Goodwin District in 1900; Buried in Level Creek Cemetery, Suwanee.

James H. Brooks PVT
Late enlistment; Admitted to Ocmulgee Hospital in Macon, Nov 7, 1864.

Felton Philo M. Brooks PVT 1842 25 May 1864
Married M.E. Burnett in Gwinnett County, Dec 27, 1861; Captured in Kentucky; Taken by steamboat from Louisville to Vicksburg, MS and exchanged on Nov 29, 1862; Sent to Vicksburg hospital April 1, 1863; Sent to general hospital in Lauderdale, MS; Killed at **Battle of New Hope Church,** GA May 25, 1864.

George W. Brown PVT 1842 29 Jun 1862
Detailed to Captain Calhoun's detachment in Knoxville, TN April 20, 1862 to guard political prisoners; Died in Knoxville, TN hospital; Left wife & one child.

George W. Burdett PVT 1842 Dec 1862
Born in South Carolina; Furloughed home from Bean Station, TN, Dec 12, 1862. Died in Bean Station, TN.

Stephen L. Burdett PVT 1815 2 Aug 1862
Born in Abbeville District, SC; Died of disease in Bean Station, TN.

William Thomas Burns PVT 5 Dec 1829 11 Mar 1871
Married Parthena Mason in Gwinnett County, Dec 15, 1853; Furloughed home from Atlanta, Dec 1862; Farmer in Hog Mountain in 1870; Buried at Liberty Baptist, Lilburn.
Samuel J. Burrell PVT 1829 Jan 1864
Born in South Carolina; Married Jane Suggle in Hall County, Feb 6, 1851; Captured at **Vicksburg**, MS July 4, 1863; Present on Dec 11, 1863 muster roll.
Curtis C. Caldwell PVT 1830 14 Feb 1897
Captured at **Vicksburg**, MS July 4, 1863; Wounded Nov 30, 1863; Surrendered at **Greensboro**, NC April 26, 1865; Farmer in Berkshire in 1880; Buried at Harmony Grove Methodist, Lilburn.
Allen W.K. Carr PVT 1828 17 July 1863
Married Nancy Ann McGinnis in Gwinnett County, Feb 16, 1853; Did prison time in 1858 for larceny; Confined to the Vicksburg barracks for leaving his company without leave Jan 7, 1863; Died in Vicksburg, MS hospital; Left wife and one child.
William M. Carr PVT 1842 12 Oct 1864
Married Caroline McGinnis in Gwinnett County, April 6, 1854; Captured at **Vicksburg**, MS July 4, 1863; Wounded at **Battle of Resaca**, Ga May 15, 1864; Captured near Atlanta July 19, 1864; Died of smallpox at **Camp Chase**, OH Oct 12, 1864; Buried in grave #300 at Camp Chase Confederate Cemetery.
Asa Ashmore Cole PVT 25 Apr 1845 25 Nov 1907
Younger brother of David Cole; Enlisted in Lawrenceville in July 1863; Captured near Franklin, TN in the retreat from Nashville, Dec 17, 1864; Released from **Camp Chase**, OH June 12, 1865; Married Lucinda Philips about 1867; Filed for Confederate pension in Gwinnett County in 1895; Application endorsed by R.P. Furgeson; Farmer in Cates in 1900; Buried at Camp Creek Primitive Baptist, Lilburn.
David C. Cole PVT 1842 30 Jun 1862
Died in Knoxville, TN hospital; Buried in Bethel Cemetery, Knoxville, TN.
William Henry Cole PVT 28 Jan 1843 29 Mar 1907
Detailed to Captain Calhoun's detachment in Knoxville, TN April 20, 1862 to guard political prisoners; Prison guard at Madison, GA Oct 31, 1862; Captured at **Vicksburg**, MS July 4, 1863; Present Dec 10, 1863; Married Emma Herrington in Gwinnett County, Jan 5, 1892; Farmer in Duluth in 1900; Buried in Duluth Methodist, Gwinnett County.
William A. Craig PVT Mar 1837 16 Nov 1907
Captured at **Vicksburg**, MS July 4, 1863; Blacksmith in Dalton winter camp; Surrendered at **Greensboro**, NC April 26, 1865; Married Martha Thomasine Carter in 1868; Filed for Confederate pension in DeKalb County, AL in 1899; Blacksmith in Collinsville, AL in 1900; Died in DeKalb County, AL.
James H. Culver PVT 29 Apr 1841 1909
Born in Hancock County; Captured at **Vicksburg**, MS July 4, 1863; Wounded at **Battle of New Hope Church** May 25, 1864; Filed for Confederate pension in Putnam County in 1898; Reported being present at Greensboro Surrender; Married Mary Sharpe in Putnam County, Dec 22, 1878; Farmer in Putnam County in 1880; Buried in Lester Cemetery, Athens, Clarke County.
Matthew T. Culver PVT 1843 20 Jun 1863
Died in Vicksburg, MS hospital.
Basil Wells Davis PVT 10 Apr 1843 19 Dec 1912

Captured at **Vicksburg**, MS July 4, 1863; Wounded at **Battle of Bentonville**, NC Mar 19, 1865; Sent to General Hospital #3 at Greensboro, NC; Married Mary Vanner in Gwinnett County, Aug 26, 1866; Farmer in Sugar Hill in 1900; Filed for Confederate pension in Gwinnett County in 1912; Application endorsed by J.B. Whitworth; Buried at Rock Springs Methodist, Buford.

George W. Davis PVT 20 Dec 1840 22 Apr 1862
Married Elizabeth Brown in Gwinnett County, Oct 6, 1861; Died of disease at home; Buried at Prospect Methodist, Lawrenceville.

Sanford Marion Davis PVT 22 May 1830 7 Sep 1863
Born in South Carolina; Married Jane Blount in Gwinnett County, Aug 30, 1850; Detailed to Captain Calhoun's detachment in Knoxville, TN April 20, 1862 to guard political prisoners; Prison guard at Madison, GA Oct 31, 1862; Captured at **Battle of Baker's Creek**, MS May 16, 1863; POW at **Camp Morton**, IN; Exchanged at Fort Delaware July 4, 1863; Died at home; Buried at Prospect Methodist, Lawrenceville.

William H. Davis PVT 1840 9 May 1862
Died in Knoxville hospital; Buried in Bethel Cemetery, Knoxville, TN.

Wilburn Zimri Defour PVT 19 Sep 1839 16 Aug 1896
Married Rachael Waldrop in Gwinnett County, Jan 19, 1860; Sent to Chattanooga hospital Nov 25, 1862; Absent in hospital on Dec 10, 1863 muster roll; Deserted at Resaca, GA May 1, 1864; Received as a Confederate deserter at Military Prison in Chattanooga Aug 16, 1864; Took U.S. Oath of Allegiance in Louisville, KY Sep 26, 1864, and released to remain north of the Ohio River; Farmer in Cherokee County after the war; Buried at Lebanon Methodist, Cherokee County.

James Jefferson Doss PVT 21 Oct 1829 20 Nov 1862
Married Rebecca Tinney in Gwinnett County, Oct 26, 1848; Married Jane Taylor in Gwinnett County, May 26, 1861; Died in Bean Station, TN hospital.

William Parker Doss PVT 1838 15 Nov 1862
Died of disease at Bean Station, TN; Buried at Bethel Cemetery, Knoxville, TN.

William Franklin Everett PVT 17 Nov 1831 25 May 1900
Married Serenna Braswell in Gwinnett County, Aug 9, 1855; Captured at **Vicksburg**, MS hospital July 4, 1863; Released from Federal authority in Vicksburg, MS Nov 24, 1863; Farmer in Jackson County, AL in 1880; Filed for Confederate pension in Jackson County, AL in 1899; Reports being honorable discharged in May 1865; Buried at Latham Cemetery #1 in Jackson County, AL.

James Marion Flowers PVT 17 Mar 1838 31 Jan 1896
Married Sarah McDaniel in Gwinnett County, Jan 31, 1857; Appointed **3CPL** April 1863; Captured at **Vicksburg**, MS July 4, 1863; Farmer in Pinckneyville in 1880; Buried at Forest Hill Cemetery, Birmingham, Jefferson County.

Andrew Jackson Furgerson PVT 1 Jan 1835 11 Oct 1920
Enlisted in Tennessee Aug 4, 1862; Captured in Vicksburg rifle pit stormed by General Herron's forces on the night of June 24, 1863; Paroled in **Vicksburg**, MS on July 4, 1863; Surrendered at **Greensboro**, NC April 26, 1865; Married Sarah Jones in Gwinnett County, April 4, 1886; Farmer in Gwinnett County in 1900; Filed for Confederate pension in Gwinnett County 1911; Application endorsed by H.F. Freeman; Buried at Pleasant Grove Methodist, Lilburn.

Francis Marion Furgerson PVT 6 Jun 1841 3 Aug 1896

Older brother to Andrew & Rufus Furgerson; Sent to Marion, MS hospital April 8, 1863; Surrendered at **Greensboro,** NC April 26, 1865; Married Mary Jane Weed in Gwinnett County, Dec 21, 1865; Farmer in Berkshire in 1880; Buried at Pleasant Grove Methodist, Lilburn.

Rufus P. Furgerson PVT 3 Nov 1843 6 Mar 1925

Captured at **Vicksburg,** MS July 4, 1863; Admitted to Madison, GA hospital May 18, 1864; Wounded at **Battle of Jonesborough,** GA Aug 31, 1864; Appointed **1CPL**; Surrendered at **Greensboro,** NC April 26, 1865; Farmer & grocery merchant in Rockbridge after the war; Buried at Stone Mountain Cemetery. DeKalb County.

Andrew James Garner PVT 1822 30 Jul 1863

Married Ellen Jane Cole in Gwinnett County, Aug 12, 1855; Captured at **Battle of Baker's Creek,** MS May 16, 1863; POW at **Camp Morton,** IN; Exchanged at Fort Delaware July 4, 1863; Died in Chimborazo Hospital #2 in Richmond, VA, July 30, 1863; Buried at Oakwood Cemetery, Richmond, VA.

David Thomas Garner PVT 20 Oct 1841 24 Feb 1917

Brother of James Franklin Garner; Captured at **Vicksburg,** MS July 4, 1863; Married Rhody Ann Carroll in Gwinnett County, Oct 5, 1865; Teamster in Boone, AR in 1900; Buried at Oak Hill Memorial, Booneville, AR.

James Franklin Garner PVT 8 Sep 1843 14 Mar 1883

Captured at **Vicksburg,** MS July 4, 1863; Pension record reports being present at Greensboro Surrender, April 26, 1865; Married Sarah Jane Garner July 16, 1865; Sheriff of Mason, TX in 1880; Buried at Old Gooch Cemetery, Mason County, Texas.

John Garner PVT 1838 8 Aug 1862

Died of disease in Knoxville, TN hospital; Buried at Bethel Cemetery, Knoxville.

William M. Garner PVT 1845

Enlisted in Atlanta Aug 1, 1863; Surrendered at **Greensboro,** NC April 26, 1865.

Benjamin F. Gholston PVT 1835 31 Aug 1864

Younger brother of Union Gholston; Appointed **1SGT**; Elected **2LT** July 1862; Killed at **Battle of Jonesborough**; Buried at Confederate Cemetery, Jonesboro, Clayton County.

Union F. Gholston PVT May 1833 13 Jan 1905

Married Martha Goza in Gwinnett County, Mar 7, 1852; Detailed to Captain Calhoun's detachment in Knoxville, TN April 20, 1862 to guard political prisoners; Prison guard at Madison, GA Oct 31, 1862; Captured at **Battle of Baker's Creek,** MS May 16, 1863; Prisoner at **Camp Morton,** IN; Exchanged from Camp Delaware, DE July 4, 1863; Captured near Atlanta July 27, 1864; Paroled at **Camp Chase,** OH and exchanged at Boulware & Cox's Wharves Mar 27, 1865; Married Margaret Nash in Gwinnett County, Oct 12, 1873; Atlanta dealer in coal in 1900; Buried at Gholston Cemetery, Gwinnett County.

Henry Jasper Ginn PVT 1833 1863

Married Frances Todd in Gwinnett County, Dec 1, 1853; Captured at Edwards Station, MS May 17, 1863; POW at **Camp Morton,** IN; Exchanged at City Point, VA July 6, 1863; Paid in Richmond, VA July 17, 1863; No further record.

Hamilton G. Goza PVT 27 Mar 1838 25 Nov 1862

Younger brother of John W. Goza; Married Sarah Hummeritt in Gwinnett County, May 26, 1859; Died at Lenoir Station, TN; Buried at Mt. Carmel Methodist, Norcross.

John W. Goza PVT 28 May 1835 20 Nov 1862

Married Mary Jahan in Gwinnett County, Jan 13, 1859; Died in Loudon, TN hospital

Nov 20, 1862; Buried at Stone Mountain Cemetery, DeKalb County.

William M. Grogan PVT 20 Nov 1835 11 May 1911

Married Rebecca Jackson about 1856; Captured at **Vicksburg**, MS July 4, 1863; Deserted regiment in McDonough Sep 16, 1864; Farmer in Pinckneyville after the war; Married Mary Frances Stancel about 1872; Filed for Confederate pension in Gwinnett County in 1903; Reported being detailed as a teamster in August 1864 until the end of the war; Buried at Mount Carmel Methodist, Norcross.

Harrison Miles Gunter PVT 4 Nov 1832 24 Feb 1924

Born in Walton County; Married Mary Corbin in Gwinnett County, Dec 22, 1859; Captured at **Vicksburg**, MS July 4, 1863; Present on Dec 10, 1863 muster roll; Farmer in Newtown, Milton County, after the war; Filed for Confederate pension in Milton County in 1907; Reported being home on sick furlough at close of war; Buried at Old Big Creek, Alpharetta.

William G. Haney PVT 10 Jul 1824 26 Feb 1901

Born in South Carolina; Discharged on account of rheumatism, Nov 4, 1862; Listed as 54-year-old miller in 1864 militia census; Married Amanda Steele in 1873; Farmer in Ball Flat, AL in 1880; Filed for pension in Cherokee County, AL in 1899; Buried in Gnatville Cemetery, Cherokee County, AL.

John T. Hanson PVT 1843 After 1885

Detailed to Captain Calhoun's detachment in Knoxville, TN April 20, 1862 to guard political prisoners; Prison guard at Madison, GA Oct 31, 1862; Captured at **Vicksburg**, MS July 4, 1863; Took the U.S. Oath of Allegiance in Louisville, KY after deserting on July 14, 1864, agreeing to remain north of the Ohio River until the close of the war; Farmer in Hog Mtn in 1885.

Elisha Hunter PVT 1833 12 Aug 1862

Married Sarah Adams in Gwinnett County, Jan 1, 1856; Died in Fair Grounds Hospital in Knoxville; Buried at Bethel Cemetery, Knoxville, TN.

William James Ingram PVT 3 Oct 1841 23 Jun 1920

Served in Ramsey's 1st Infantry in 1861; Captured at **Vicksburg**, MS July 4, 1863; Present on Dec 11, 1863 muster roll; Married Brunette Elizabeth **Haygood** in Monroe County, Oct 25, 1864; Farmer in Monroe County in 1870; Farmer in Upson County in 1880; Buried at Evergreen Cemetery in Perry, Georgia.

William F. Johnson PVT 28 Feb 1838 4 Apr 1905

Captured at **Vicksburg**, MS July 4, 1863; Surrendered at **Greensboro**, NC April 26, 1865; Married Mary E. Smith in 1883; Farmer in Jonesboro, AR in 1900; Buried at Locust Grove Baptist, Craighead County, Arkansas.

George Jones PVT 6 Jan 1840 28 Jun 1911

Captured at **Vicksburg**, MS July 4, 1863; Surrendered at **Greensboro**, NC April 26, 1865; Married Nancy Elizabeth Talton in 1870; Farmer in DeKalb County in 1900; Filed for Confederate pension in Fulton County in 1902; Application endorsed by Colonel L.P. Thomas; Buried at Rehoboth Cemetery, Tucker.

Robert Durgan Jones PVT 9 Dec 1831 15 May 1864

Married Martha Bates in Gwinnett County, Mar 12, 1854; Captured in Lexington, KY Nov 24, 1862; Taken by Steamboat *Mary Crane* from Louisville to Vicksburg, MS and exchanged on Nov 29, 1862; Captured at **Vicksburg**, MS July 4, 1863; Killed at **Battle of Resaca**.

Thompson H. Jones PVT 4 Apr 1834 2 Jul 1862
Married Martha Mills in Gwinnett County, Jan 6, 1858; Died in Strawberry Fields, TN hospital; Buried at Liberty Baptist, Lilburn.
Daniel David Keheley PVT 4 Oct 1835 13 Feb 1909
Married Frances Walker in Gwinnett County, Mar 17, 1859; Captured at **Vicksburg,** MS July 4, 1863; Captured in Gwinnett County in 1864 as a Rebel Deserter; Took the U.S. Oath of Allegiance in Louisville, KY on Sep 26, 1864, agreeing to remain north of the Ohio River until the close of the war; Filed for Confederate pension in Spalding County in 1898; Cripple living in Griffin, GA in 1900; Buried at Lebanon Primitive Baptist, Hampton, Henry County.
Amos Newton Kelley PVT 30 Sep 1836 23 Mar 1877
Married Susannah Matthews in Gwinnett County, Nov 6, 1856; Appointed **4CPL** April 17, 1863; Captured at **Vicksburg,** MS July 4, 1863; Farmer in Gwinnett County in 1870; Buried at Sweetwater Baptist, Duluth.
Andrew Jackson King PVT 13 Sep 1827 17 Sep 1904
Born in Jasper County; Married Nancy Lantern in Gwinnett County, Jun 14, 1849; Captured at **Vicksburg,** MS July 4, 1863; Paroled from General Hospital # 11 at Charlotte, NC May 8, 1865; Farmer near Lawrenceville, Ga after the war; Buried at Prospect Methodist, Lawrenceville.
John C. King PVT 7 Mar 1842 13 Jul 1919
Captured at **Vicksburg,** MS July 4, 1863; Captured in Gwinnett County in 1864 as a Rebel Deserter; Took the U.S. Oath of Allegiance in Louisville, KY on July 16, 1864, agreeing to remain north of the Ohio River until the close of the war; Filed for Confederate pension in Gwinnett County in 1910; Reported being captured at Murfreesboro, TN in Nov 1864; Married Sarah Baxter in Gwinnett County, Sep 30, 1873; Merchant in Suwanee in 1880; Farmer in Goodwin in 1900; Buried at Level Creek Cemetery, Suwanee.
Silas Kirby PVT 1845 May 1864
Born in South Carolina; Sent to Vicksburg, MS hospital April 12, 1863; Killed at **Battle of New Hope Church,** GA.
Augustus S. Mahaffey PVT 1841 20 Jun 1863
Appointed **4SGT** Feb 1863; Died of disease in Vicksburg, MS.
Absalom Parks Martin PVT 19 Feb 1844 4 Oct 1918
Younger brother of Moses Martin; Captured at **Vicksburg,** MS July 4, 1863; Taken to Louisville, KY as a Confederate deserter; Took U.S. Oath of Allegiance July 16, 1864, agreeing to remain north of the Ohio River until the close of the war; Married Elizabeth Sims in Jackson County, Aug 20, 1874; Farmer in Jackson County in 1880; Farmer in Rockwell County, TX in 1900; Buried at Sapulpa Cemetery, Creek County, OK.
Elijah Martin PVT 11 Aug 1843 21 May 1864
Captured at **Vicksburg,** MS July 4, 1863; Mortally wounded at **Battle of Resaca,** May 15, 1864; Buried at New Hope Methodist, Lawrenceville.
John Micah Martin PVT 6 Sep 1840 17 Aug 1919
Discharged for disability at Big Springs, TN Sep 7, 1862; Enlisted in Company I, 16th Georgia Cavalry in Jackson County in May 1864; Married Susannah Wood in Gwinnett County, Nov 16, 1865; Farmer near Lawrenceville in 1900; Buried at Martins Chapel, Lawrenceville.
Major John Martin PVT 1 Oct 1845 30 Mar 1938

Younger brother of Elijah Martin; Enlisted in Decatur Oct 18, 1863; Admitted to Madison, GA hospital July 11, 1864; Married Sarah Corbin in Gwinnett County, Sep 8, 1869; Farmer in Lawrenceville in 1900; Filed for Confederate pension in Cobb County in 1919; Reported being unable to rejoin command, and enlisted in CPT Bond's Company, Wofford's GA Militia Cavalry, Feb 1865; Buried at New Hope Methodist, Lawrenceville.

Moses Martin PVT 8 Dec 1831 19 Apr 1898

Married Hettie Elizabeth Pharr in Gwinnett County, Dec 13, 1855; Detailed to Captain Calhoun's detachment in Knoxville, TN April 20, 1862 to guard political prisoners; Prison guard at Madison, GA Oct 31, 1862; Appointed **5SGT** Feb 1863; Sent to Newton MS hospital Mar 18, 1863; Wounded in left leg at **Battle of Bentonville**, NC Mar 19, 1865; Leg amputated above the knee at General Hospital #3 at Greensboro, NC; Filed for Confederate pension in Gwinnett County in 1879; Application endorsed by Colonel L.P. Thomas; Tax collector of Gwinnett County for ten years; Farmer in Sugar Hill in 1880; Door Keeper for the Georgia General Assembly for 12 years; Buried at Old Suwanee Baptist, Gwinnett County.

Wash Lafayette McDaniel PVT 4 May 1832 5 Jul 1909

Married Elizabeth Carroll in Gwinnett County, Dec 4, 1854; Captured at **Vicksburg**, MS July 4, 1863; Admitted to Madison, GA hospital June 10, 1864; Returned to duty June 27, 1864; Farmer in Berkshire in 1900; Buried at Liberty Baptist, Lilburn.

Thomas Landrum Medlock PVT 22 Nov 1823 Feb 1864

Older brother of John Medlock; Married Rebecca Bellinger in DeKalb County, Aug 11, 1842; Enlisted in 42nd GA April 7, 1863 after previously serving as 1SGT of Company H, 16th Georgia. He had been wounded at Malvern Hill and released from Chimborazo Hospital #5 on Dec 14, 1862; Discharged with disability at Chickamauga, GA Oct 11, 1863; Died at home; Buried at Flint Hill Cemetery, Norcross.

Archibald Marion Mewborn PVT 23 May 1831 20 Oct 1862

Married Cynthia Noel in Gwinnett County, Aug 3, 1854; Died at Bean Station, TN hospital.

John Vestal Mewborn PVT 13 Jan 1838 22 Apr 1913

Brother of Archibald Mewborn; Captured at **Vicksburg**, MS July 4, 1865; Surrendered at **Greensboro**, NC April 26, 1865; Married Elizabeth Noel, sister of Cynthia, George & James Noel, in Gwinnett County, Aug 17, 1865; Farmer in Martins in 1900; Buried at Duluth Church Cemetery, Gwinnett County.

Joshua W. Morgan PVT 12 May 1826 15 Jul 1864

Married Elizabeth Harrison in Fayette County, Sep 24, 1850; Captured at **Vicksburg**, MS July 4, 1863; Captured at Dallas, GA June 3, 1864; Died of pneumonia at **Rock Island Prison**, Il Jul 15, 1864.

James B. Munday PVT 1 Jun 1831 17 Apr 1864

Married Georgia Ann Womack in DeKalb County, Apr 27, 1857; Died at home; Buried at Mount Carmel Methodist in Norcross.

Robert H. Munday PVT 1841 Oct 1905

Brother of James Munday; Wounded in right ankle at Vicksburg, MS Jun 1863; Captured at **Vicksburg**, MS July 4, 1863; Wounded above left ankle at **Battle of New Hope Church** May 25, 1864; Married Rosalie Hensolt in 1872; Filed for Confederate pension in Banks County in 1895; Application endorsed by M.M. Waits; Miller in Jackson County in 1900; Buried at Mount Carmel Methodist in Norcross.

George Washington Noel PVT 21 Oct 1845 25 Jun 1864
Younger brother of James Noel; Enlisted Dec 12, 1862 in Readersville, TN; Captured at **Vicksburg**, MS July 4, 1863; Died in hospital in 1864.
James Richard Noel PVT 25 Sep 1842 26 Dec 1925
Wounded at **Battle of Baker's Creek**, MS May 16, 1863; Appointed **1SGT**; Surrendered at **Greensboro**, NC April 26, 1865; Married Elizabeth Liddell in Gwinnett County, Feb 23, 1876; Worked in textile factory in Spaulding County in 1900; Carpenter in Bryan County, OK in 1900; Buried at Colbert Garden, Colbert, Bryan County, OK.
Sylvester Little Nunnally PVT 30 Dec 1840 28 Feb 1885
Detailed to Captain Calhoun's detachment in Knoxville, TN April 20, 1862 to guard political prisoners; Prison guard at Madison, GA Oct 31, 1862; Captured at **Vicksburg**, MS July 4, 1863; Captured near Atlanta July 23, 1864; Paroled at **Camp Chase**, OH and exchanged at Boulware & Cox's Wharves, VA Mar 12, 1865; Married Nancy Wright in Blount County, AL, Sep 28, 1879; Worked in a gristmill in Blount County in 1880; Buried at Wynnville Cemetery, Blount County, AL.
Amos Parks PVT 1846 Jun 1869
Captured at **Battle of Baker's Creek**, MS May 16, 1863; POW at **Camp Morton**, IN; Exchanged at Fort Delaware July 4, 1863; Surrendered at **Greensboro**, NC April 26, 1865; 1870 Gwinnett County mortality schedule shows his death from consumption at the age of 23.
Thomas Washington Phillips PVT Oct 1832 27 Aug 1917
Married Tabitha Bagwell in Gwinnett County, Dec 26, 1854; Captured in the advance rifle pits defending Vicksburg, MS June 23, 1863 and paroled at **Vicksburg**, MS July 4, 1863; Captured in retreat at **Franklin**, TN Dec 17, 1864; Released at **Camp Chase**, OH Jun 12, 1865; Farm hand in Jefferson County, AL in 1900; Buried at Shiloh Cemetery, Chapultepec, Blount County, AL.
John Thomas Pittard PVT 18 Apr 1844 18 Aug 1863
Captured at **Vicksburg**, MS July 4, 1863; Died at home in Gwinnett County, Aug 18, 1863; Buried at First Baptist in Lilburn.
John McGilvra Pittman PVT 12 Mar 1833 16 Apr 1915
Older brother of Simpson Pittman; Married Rebecca Waldrop in Gwinnett County, Mar 7, 1852; Captured at **Vicksburg**, MS July 4, 1863; Received at Atlanta distribution hospital May 18, 1864; Deserted at Dalton, GA; Took US Oath of Allegiance in Louisville, KY on Sep 26, 1864, and released to remain north of the Ohio River until the close of the war; Masonic Order; Farmer in Hog Mountain in 1900; Filed for Confederate pension in Gwinnett County in 1912; Reported leaving his command May 8, 1864 with rheumatism; Transferred to Atlanta then Greensboro hospitals; Captured in Norcross while on medical furlough; Application endorsed by J.B. Whitworth; Buried at Hog Mountain Baptist, Gwinnett County.
Simpson Clabus Pittman PVT 1837 After 1880
Married Adeline Hays in Gwinnett County, Jun 11, 1856; Captured at **Vicksburg**, MS July 4, 1863; Surrendered at **Greensboro**, NC April 26, 1865; Wagon dealer in Rockdale County in 1880.
Benson R. Prewitt PVT 1842 Jul 1888
Born in Cobb County; Married Matilda Sizemore in Gwinnett County, April 28, 1861;

Discharged Mar 2, 1862 with anchyloses in the elbow joint; Present on Dec 10, 1863 muster roll; 23-year-old farmer in 1864 Cobb County militia census; Day laborer in Floyd County in 1870.

William Isaac Prewitt PVT Aug 1846 Jul 1925

Born in South Carolina; Captured at **Vicksburg,** MS July 4, 1863; Present on Dec 10, 1863 muster roll; Surrendered at **Greensboro,** NC April 26, 1865; Married Nancy Ratliff in Blount County, AL, Nov 7, 1871; Farmer in Waterloo, AL in 1900; Buried in Wynnville Cemetery, Blount County, AL.

William D. Ramey PVT 11 Nov 1827 14 May 1863

Married Polly McDaniel in Gwinnett County, Oct 5, 1848; Assigned to hospital as nurse, Mar 10, 1863; Died in Brookhaven Hospital in Lauderdale Springs, MS.

James H. Reynolds PVT 1826

Enlisted Cumberland Gap, TN April 30, 1862; Cook in the Atlanta hospital April 1863; Admitted to Fairgrounds Hospital #1 in Atlanta Jan 5, 1863 with tuberculosis; Returned to duty Mar 16, 1863; Wounded at **Battle of Resaca,** May 15, 1864; Captured as deserter in Gwinnett County in Oct 1864; Took U.S. Oath of Allegiance at Louisville, KY on Oct 22, 1864; Released to remain north of the Ohio River until the close of the war.

James William Roberts PVT 3 Aug 1840 19 Mar 1922

Left sick at Knoxville, TN hospital April 22, 1862; Wounded in shoulder and both thighs at **Battle of Chickasaw Bayou** Dec 29, 1862; Captured at **Vicksburg,** MS July 4, 1863; Present on Dec 10, 1863 muster roll; Serving in Provost Guard in Atlanta Dec 1864; Married Elizabeth Kirby in Gwinnett County, Dec 21, 1865; Farmer in Goodwin in 1900; Member of UCV Camp # 982 in Lawrenceville; Filed for Confederate pension in Gwinnett County in 1913; Reported he "was on picket post and Cut off by the Yankees and made my way to Augusta, GA and couldn't get back to my command."; Died at Confederate Soldier's Home, Atlanta, Mar 19, 1922; Buried at Trinity Methodist, Duluth.

Joseph Wilson Roberts PVT 1832 Apr 1872

Sent to Knoxville hospital as nurse in 1862; Still on detached service in April 1863; Wounded at **Battle of Resaca,** May 15, 1864; Married Minerva Jackson in Gwinnett County, Mar 18, 1866; Cabinet maker in DeKalb County in 1870; Buried at Pauper's Cemetery, Duluth.

Larkin Harbrustar Roberts PVT 18 Jan 1831 18 Feb 1892

Married Deborah Goolsby in Gwinnett County, Jan 26, 1858; Enlisted Aug 11, 1862 in TN; Sent to hospital in Tazewell, TN, Sep 20, 1862; Farmer in Goodwins in 1880; Buried at First Baptist, Sugar Hill.

Sanford V. Roberts PVT 1837 10 Feb 1863

Born in Franklin County; Detailed to Captain Calhoun's detachment in Knoxville, TN April 20, 1862 to guard political prisoners; Prison guard at Madison, GA Oct 31, 1862; Died in Vicksburg hospital; Buried at Soldier's Rest, Cedar Hill Cemetery, Vicksburg, MS.

Cortes E. Ross PVT 1838 20 May 1863

Born in South Carolina; Sent to Vicksburg hospital, Jan 1863; Died in Enterprise, MS hospital.

Christopher A. Rowden PVT 1827 18 Apr 1863

Married Margaret Davis in Gwinnett County, Jun 13, 1859; Died in Brookhaven, MS hospital; Left wife and one child.

Joseph Scales PVT 1843 17 Apr 1863

Died at 1st Mississippi Hospital, Jackson, MS; Remarks: "Was very weak. Sat up too suddenly. Fainted and died."

John Martin Scott PVT 16 Jan 1865

Captured at **Buzzard's Roost**, GA May 13, 1864; Died at **Camp Morton**, IN; Buried Crown Hill Cemetery, Indianapolis, IN.

John M. Singleton PVT 1845 25 Nov 1864

Sent to Vicksburg hospital April 5, 1863; Captured by Chattahoochee River, GA July 31, 1864; Died at **Camp Douglas**, IL; Buried in Confederate Mound, Oak Woods Cemetery, Chicago, IL.

John C. Smith PVT 7 Apr 1834 3 Mar 1887

Left company without leave, Mar 10, 1863; Admitted to Lauderdale Springs, MS hospital April 22, 1863; Wounded below the elbow & admitted to Ocmulgee Hospital, Macon on June 6, 1864; Furloughed; Farmer in Hall County in 1880; Buried at Walnut Fork, Braselton, Jackson County.

Hosea Conine Stanley PVT 1828 12 Oct 1862

Died of disease at Gwinnett home. Buried at Stanley Cemetery, Gwinnett County.

George Marion Stevenson PVT 1833 8 May 1862

Brother of Reason, Van & William Stevenson; Died at Gwinnett home.

Reason D. Stevenson PVT 1828 22 May 1862

Born in South Carolina; Died of disease.

Van R. Stevenson PVT 3 Aug 1834 10 Jan 1867

Youngest brother of William, George and Reason Stevenson; Captured at **Vicksburg**, MS July 4, 1863; Received at Atlanta distribution hospital debilitated May 9, 1864; Surrendered at **Greensboro**, NC April 26, 1865; Married Salina Jackson in Gwinnett County, Jan 21, 1866; Named his only son Lovick Thomas Stevenson (1867-1882); Buried at Prospect Methodist, Lawrenceville.

William Thompson Stevenson PVT 1821 23 Nov 1863

Born in South Carolina; Married Sallie Sexton in Gwinnett County about 1843; Enlisted in Decatur Aug 23, 1863; Died of disease in LaGrange hospital on Nov 26, 1863 leaving a widow and 8 children; Buried at Stonewall Confederate, LaGrange, Troup County.

Emanuel Sudderth PVT 25 July 1829 3 Dec 1916

Married Sarah Brogdon (sister of Noah Brogdon) in Gwinnett County, Jan 15, 1852; Wounded in right knee at **Battle of Chickasaw Bayou**, MS Dec 29, 1862; Discharged by Board of Surgeons in Vicksburg, MS March 3, 1863 & approved by Board of Surgeons in Jackson, MS Mar 13, 1863; Applied for Confederate pension in 1889; Farmer in Sugar Hill in 1900; Buried at Buford City Cemetery, Gwinnett County.

Robert Hamilton Teague PVT 19 Dec 1843 25 Sep 1917

Captured at **Vicksburg**, MS July 4, 1863; Married Sarah Sexton in Gwinnett County about 1867; Farmer in Hog Mountain in 1880; Moved to Atlanta & employed as a carpenter; Filed for Confederate pension in Fulton County in 1907; Reported being wounded & in Cobb County hospital at end of war; Buried at Rock Springs Methodist, Buford.

Marion Jefferson Townley PVT 1842 5 Aug 1863

Married Susan Reid in Forsyth County, Dec 19, 1861; Enlisted at Bean Station, TN July 1, 1862; Transferred from Jackson, MS hospital to Brookhaven, MS hospital Feb 3, 1863; Sent to the Vicksburg hospital Mar 1, 1863; Captured at **Vicksburg**, MS hospital July 4, 1863; Died in Vicksburg hospital.

Wiley Simmons Townley PVT 5 Sep 1837 25 July 1925
Older brother of Marion Townley; Captured at **Vicksburg**, MS July 4, 1863; Married Catherine Wallis in Forsyth County, Dec 3, 1871; Farmer in Big Creek, Forsyth County, in 1900; Filed for Confederate pension in Dawson County in 1917; Reported being sick in Northern AL, Nov 1864 & granted furlough; Later joined General Wofford's Georgia militia until close of war; Buried at Concord Baptist, Cumming.
Maxwell Milton Waits PVT 27 July 1845 6 Mar 1931
Enlisted in Decatur Sep 4, 1863; Wounded in right hip at **Battle of Bentonville**, NC Mar 19, 1865; Surrendered at **Greensboro**, NC April 26, 1865; Married Sarah Lucinda Miller in Gwinnett County, Feb 18, 1866; Farmer near Duluth in 1900; Buried at First Baptist, Lilburn.
William Rhodes Waldrop PVT 3 Apr 1828 15 Mar 1864
Sent to Vicksburg hospital July 1, 1863; Killed at Dalton, GA March 1864.
Eli A. Whaley PVT 1832 26 Aug 1862
Married Lucy Thompson in Walton County, Dec 27, 1855; Died of disease in Morristown, TN hospital.
James Blackmon Whitworth PVT 15 Jul 1842 22 Dec 1926
Captured in the rifle pits in front of Vicksburg, MS siege lines on June 23, 1863; Paroled at **Vicksburg**, MS July 10, 1863; Surrendered at **Greensboro**, NC April 26, 1865; Married Caroline Lampkin in Gwinnett County about 1870; Farmer near Lawrenceville in 1900; Buried at Shadowlawn Cemetery, Lawrenceville.
I. Jasper Wiggins PVT 1826 5 Jun 1863
Born in South Carolina; Married Mary Ann Braswell in Gwinnett County, Sep 1, 1851; Killed during Siege of Vicksburg; Left wife & 4 minor children.
Hiram L. Wilson PVT Dec 1837 After 1900
Married Sarah Jane Grogan in Gwinnett County, Feb 2, 1862; Captured at **Battle of Baker's Creek**, May 16, 1863; Arrived as POW at **Fort Delaware**, DE June 9, 1863; Exchanged near Richmond, VA July 19, 1863; Captured in **Battle of Atlanta**, July 22, 1864; Paroled at Camp Chase, OH and exchanged at Boulware & Cox's Wharves, VA Mar 10, 1865; Farmer in Gadsden, AL in 1880; Laborer in Cordova, Walker County, AL in 1900.
Hugh Francis Marion Wilson PVT 1836 10 Jan 1865
Born in Anderson District, SC; Appointed **1SGT** Feb 1863; Captured at **Vicksburg**, MS July 4, 1863; Discharged in Dalton with approval of G.B. Thornton, Chief Surgeon, Stewart's Division, Feb 10, 1864; Captured as Sergeant, Company G of Hill's Georgia Dragoons near Warsaw, GA Sep 9, 1864; Died at **Camp Douglas**, IL; Buried at Confederate Mound at Oak Woods Cemetery, Chicago, IL.
William Bolton Worsham PVT 4 May 1843 5 Nov 1916
Captured at **Vicksburg**, MS July 4, 1863; Captured in Gwinnett County as deserter; Took U.S. Oath of Allegiance in Louisville, KY, Aug 10, 1864; Ordered to remain north of the Ohio River until the close of the war; Married Frances Salmon in 1869; Farmer in Piedmont, Calhoun County, AL in 1900; Filed pension application from Piedmont, AL in 1911; Buried at Goshen Valley Cemetery, Calhoun County, AL.

Company B – Independent Rebels (Gwinnett) – 146 Men
The Independent Rebels drew from the farming communities in Gwinnett County that were south and west of the county seat of Lawrenceville.

Name Rank Birth Death

Benjamin Putnam Weaver CPT 6 Jun 1831 1 Dec 1864

Born in North Carolina; Married Mary McGuire in Gwinnett County, Jan 31, 1851; Owned $2,000 in real estate & $2,425 in personal property in 1860 Gwinnett census; 30-day Certificate of Disability granted in Vicksburg, MS Feb 16, 1863; Mortally wounded at **Battle of Franklin**, TN Nov 30, 1864 & died Dec 1, 1864; Member of Masonic Order; Probably buried at Federal Cemetery at Stones River TN.

Jesse Sanford Bryan 1SGT 13 Jun 1819 12 Oct 1873

Born in South Carolina; Mexican War veteran; Married Sarah Powell in Gwinnett County, May 24, 1849; Captured at Camp Dick Robinson, KY Nov 10, 1862; Exchanged from the Steamboat *Mary Crane* in Vicksburg, MS Nov 29, 1862; Captured at **Vicksburg**, MS July 4, 1863; Surrendered at **Greensboro**, NC April 26, 1865; Buried at Bethany Baptist, Snellville.

John Clark Crow 2CPL 1828 8 Feb 1863

Married Lavina Harbin in Gwinnett County, Nov 28, 1850; Detailed to Captain Calhoun's detachment in Knoxville, TN April 20, 1862 to guard political prisoners; Prison guard at Madison, GA Oct 31, 1862; Died in Magnolia Hospital; Buried in Magnolia Cemetery, Pike County, MS.

William Perry Donaldson 2LT 25 Jun 1832 22 Dec 1887

Captured at **Vicksburg**, MS July 4, 1863; Married Synthia Ann Williams (sister of Newton & William Williams) in Gwinnett County, Dec 30, 1863; Wounded in the jaw on Atlanta trench line Aug 13, 1864; Present near Nashville, TN Dec 5, 1864; Farmer in Cates in 1880; Buried in Snellville Historic Cemetery, Gwinnett County.

Andrew John Ford 1LT 12 Sep 1824 21 Jan 1887

Born in Spartanburg, SC; Married Pamelia Hudson in Gwinnett County, Dec 15, 1850; Resigned at Cumberland Gap, TN July 8, 1862; Farmer in Cates in 1880; Buried in Ford Cemetery, Lilburn.

James E. Garner 2SGT 22 Nov 1820 20 Feb 1893

Born in Union County, SC; Married Margaret Lanier Garner about 1843; Served in April 1862; Farmer in Berkshire in 1880; Buried at Garner-Purcell Cemetery, Lilburn.

Joseph Andrew Hannah 4CPL 1 Apr 1841 23 Nov 1911

Married Garner in Gwinnett County, Sep 20, 1860; In Atlanta hospital Dec 31, 1862; Sent to Dalton hospital, Oct 1863; No further record; Farmer in Cates District after the war; Buried at Bethany Baptist, Snellville.

James Milton Henry 3CPL 1841 15 May 1864

Captured at **Vicksburg**, MS July 4, 1863; Killed at **Battle of Resaca.**

Thomas E. McCart 1CPL 30 Jun 1833 5 Jul 1909

Born in South Carolina; Married Susan Hannah in Gwinnett County, Oct 3, 1854; Admitted to Fairgrounds Hospital #1 in Atlanta Jan 5, 1863 with hepatitis; Returned to duty May 18, 1863; Patient at Stonewall Hospital, Montgomery, AL Nov 1864; Surrendered at **Greensboro**, NC April 26, 1865; Farmer near Rockbridge in 1900; Buried at Bethany Baptist, Snellville.

Samuel Sexton Paden CPL 12 Dec 1812 18 Jun 1890

Born in Laurens, South Carolina; Father of Samuel & William Paden; Married Margaret Massey in Gwinnett County, 22 Jan 1835; Captured at **Battle of Nashville**, Dec 16, 1864; Received at **Camp Chase**, Jan 4, 1865; Farmer in Martin's in 1880; Buried at Bethesda

Methodist, Gwinnett County.
William Wilson Russell 3SGT 8 Nov 1825 14 Nov 1863
Sent home from Jackson, MS hospital June 1863; Died in Gwinnett County home; Buried at Fairview Presbyterian, Lawrenceville.
William S. Starr 4SGT 28 Feb 1838 1887
Married Ann Switzer in Newton County, Mar 25, 1858; Captured at **Vicksburg** hospital, MS July 4, 1863; Transported with sick from Vicksburg to Mobile on US Steamer *St. Maurice*; Exchanged off Fort Morgan Aug 5, 1863; Sent to Quintard Hospital, Griffin, GA Nov 6, 1863; Miller near Social Circle, Walton County, in 1880.
William Henry Williams 2LT 18 May 1833 29 Apr 1903
Married Eliza Donaldson (sister of William Donaldson) in Gwinnett County, Dec 20, 1860; Promoted to **1LT** Aug 8, 1862; Captured at **Vicksburg**, MS July 4, 1863; Furloughed for 5 days near Lovejoy Station, GA, Sep 16, 1864; Present near Nashville, TN Dec 5, 1864; Member of Masonic Order; Farmer in Cates, Gwinnett County in 1900; Buried in William Pittman Williams Cemetery, Snellville.
Asa Duke Aderhold PVT 25 Mar 1832 22 Nov 1910
Married Nancy Ann Baggett in Gwinnett County, Sep 26, 1858; Sick furlough from Knoxville, TN April 22, 1862; Captured in the rifle pits outside of Vicksburg, MS on the night of June 24, 1863; Paroled at **Vicksburg**, MS July 4, 1863; Wounded in the foot at **Battle of New Hope Church**, May 25, 1864; Filed for Confederate pension in Douglas County in 1900; Reported being in Augusta hospital at close of war; Farmer in Fairplay, GA in 1900; Buried in Cold Springs Cemetery, Douglas County.
Isaac David Aderhold PVT 26 Jun 1834 26 May 1880
Born in Franklin County; Younger brother of Asa Aderhold; Married Elizabeth Jacobs in Gwinnett County, Sep 23, 1858; Captured at **Battle of Baker's Creek** May 16, 1863; POW at **Camp Morton**, IN; Paroled at Fort Delaware July 3, 1863; Captured at **Battle of Resaca**, May 15, 1864; Released from **Camp Morton**, IN May 18, 1865; Died of congestion in Montgomery County, Texas at the age of 45.
Elsberry B. Andrews PVT 1828 1880
Married Artamissy Davis in Gwinnett County, Sep 28, 1845; Captured at **Vicksburg**, MS July 4, 1863; Deserted & took the U.S. Oath of Allegiance in Chattanooga, TN Nov 18, 1864; Farm Laborer; Buried at Friendship Primitive Baptist, Snellville.
William Thomas Arnold PVT 9 Jul 1840 28 Oct 1925
Married Georgia Ann Shaw in Gwinnett County, Mar 17, 1859; Serving as teamster in 1862 & as teamster near Dalton Nov 1863; Surrendered at **Greensboro**, NC April 26, 1865; Farmer near Decatur, GA in 1900; Filed for Confederate pension in DeKalb County in 1908; Application endorsed by W.S. Lanier; Buried at Friendship Primitive Baptist, Snellville.
John Arch Austin PVT 9 Nov 1818 27 May 1889
Born in South Carolina; Married Eunice Hewitt in Gwinnett County, Aug 22, 1844; Father of William Monroe Austin; Captured at **Battle of Baker's Creek**, MS May 16, 1863; POW at **Camp Morton**, IN; Exchanged at Fort Delaware July 4, 1863; Absent, left home sick in Georgia, Dec 11, 1863; Farmer near Lawrenceville in 1880; Buried at Friendship Primitive Baptist, Snellville.
William Monroe Austin PVT 9 Jun 1844 10 Aug 1907
Enlisted March 1864 in Dalton; Married Clara Nash in Gwinnett County, Jan 19, 1865;

Farmer near Stone Mountain in 1900; Filed for Confederate pension in DeKalb County in 1906; Reported being wounded at Powder Springs, GA June 6, 1864; Discharged from Atlanta hospital as unfit for duty; Buried at Rock Chapel, Lithonia.

William Henry Bailey PVT 29 Oct 1845 Jan 1890

Admitted to Vicksburg hospital April 1863; Left sick in Mississippi on Dec 11. 1863 muster roll; Married Margaret Nash in Gwinnett County, July 9, 1868; Blacksmith in Cates, Gwinnett County in 1880; Buried at Friendship Primitive Baptist, Snellville.

George C. Baker PVT 1 Jan 1842 21 Nov 1914

Enlisted Nov 17, 1863 in Chickamauga, GA; Resident of Walker County; Deserted & took U.S. Oath of Allegiance in Nashville, TN Jan 4, 1864; Farmer in Walker County in 1900; Buried in Coulter Cemetery, Cassandra, Walker County.

Andrew P. Beaver PVT 1831 After 1880

Married Parmelia Langley in Gwinnett County, Aug 14, 1849; Married Lucinda Langley in Gwinnett County, Mar 30, 1859; Captured at Camp Dick Robinson, KY Nov 10, 1862; Exchanged at Vicksburg, MS from the Steamboat *Mary Crane* Nov 29, 1862; Captured at **Vicksburg**, MS July 4, 1863; Present on Dec 11, 1863 muster roll; Admitted to Madison, GA hospital May 18, 1864; Furloughed from Madison, Ga hospital with left hand fracture April 10, 1865; Farm labor in Floyd County in 1870; Farm laborer in Montgomery, AR in 1880.

Thomas Edward Beaver PVT 3 Feb 1835 14 Sep 1890

Married Frances Naylor in Forsyth County, 19 Mar 1854; Enlisted as conscript in Calhoun, GA July 7, 1862; AWOL on Dec 11, 1863 muster roll; Named a son General Sherman Beaver; Farmer in Conway County, Arkansas in 1870; Buried in Fairview Cemetery, Gainesville, Cooke County, Texas.

Isham Warren Bennett PVT 29 Aug 1840 20 Mar 1927

Captured at **Vicksburg**, MS July 4, 1863; Accidently wounded in left hand resulting in amputation of three fingers at Atlanta, July 27, 1864; Detailed for light duty at Salina Powder Works, Nov 1864; Part of a detachment that surrendered at Citronelle, AL May 4, 1865; Married Mary Jane Wiley in Gwinnett County, Aug 16, 1865; Filed for Confederate pension in Gwinnett County in 1889; Farmer in Bay Creek, Gwinnett County in 1900; Buried at Chestnut Grove Baptist, Grayson.

James W. Bennett PVT 3 Mar 1844 12 Jul 1930

Brother of Isham & Josiah Bennett; Captured at **Vicksburg**, MS July 4, 1863; Wounded in hand near Marietta, July 3, 1864; Deserted the regiment in Lovejoy Sep 20, 1864; Surrendered at **Greensboro**, NC April 26, 1865; Married Synthia Sutherland in Gwinnett County, Jul 19, 1866; Farmer near Lawrenceville in 1900; Filed for Confederate pension in Gwinnett County in 1901; Application endorsed by J.B. Mills; Buried at Chestnut Grove Baptist, Grayson.

John G. Bennett PVT 1842 7 Jan 1863

Wounded in lungs at **Battle of Chickasaw Bayou**, MS, Dec 28, 1862; Died of wounds, Jan 7, 1863 in Vicksburg, MS; Buried at Soldier's Rest, Cedar Hill Cemetery, Vicksburg.

Josiah S. Bennett PVT 9 Jun 1843 15 Jun 1923

Captured at **Vicksburg**, MS July 4, 1863; Received at Atlanta distribution hospital with pneumonia June 6, 1864; Admitted to Madison, GA hospital; Returned to duty June 25, 1864; Deserted the regiment in Lovejoy Sep 20, 1864; Surrendered at **Greensboro**, NC April 26, 1865; Married Melissa Langley in Gwinnett County, Feb 11, 1869; Farmer in

Goodwin, Gwinnett County in 1900; Filed for Confederate pension in Gwinnett County in 1903; Application endorsed by J.B. Mills; Buried at Old Suwanee Cemetery, Suwanee.

John A. Betts PVT 1834 20 July 1864

Older brother of Reddick Betts; Married Mary Wells in Gwinnett County, 28 Oct 1860; Sick in Vicksburg hospital March 1863; furloughed home by way of Atlanta; Died of gastritis in Griffin, GA hospital; Buried at Stonewall Cemetery, Griffin.

Reddick C. Betts PVT 1843 28 Jan 1863

Enlisted at Lenoir Station, TN Nov 1, 1862; Died of disease in Vicksburg, MS hospital; Buried at Soldier's Rest, Cedar Hill Cemetery, Vicksburg.

Henry Harrison Bracewell PVT 10 Dec 1840 25 Jan 1925

Married Elizabeth Paden (sister of Samuel & William Paden) in Gwinnett County, Jan 3, 1861; Captured at **Vicksburg**, MS July 4, 1863; Farmer in Goodwin, Gwinnett County in 1900; Filed for Confederate pension in Gwinnett County in 1920; Reported being captured near Decatur, AL in Nov 1864 & kept a prisoner in Chattanooga, TN until April 27, 1865.

Jeremiah Daniel Brand PVT 1833 25 Apr 1895

Captured at **Vicksburg**, MS July 4, 1863; Married Eliza Rawlins in Gwinnett County, April 22, 1864; Captured at **Battle of Nashville**, Dec 16, 1864; Discharged from **Camp Douglas**, IL June 2, 1865; Farmer in Gwinnett County after the war.

Ransom Brand PVT 14 Sep 1837 12 Mar 1925

Married Martha Summerlin in Gwinnett County, Feb 1, 1857; Detailed as waggoner by Colonel Robert Henderson April 19, 1862; Captured at **Vicksburg**, MS July 4, 1863; Deserted the regiment in Lovejoy, Sep 20, 1864; Surrendered at **Greensboro**, NC April 26, 1865; Farmer in Walton County after the war; Filed for Confederate pension in Walton County in 1897; Application endorsed by Silas Starr; Buried at Shiloh Primitive Baptist, Loganville, Walton County.

James W. Brewer PVT 1836 25 Oct 1862

Married Mary Thomason in Gwinnett County, Sep 20, 1855; Enlisted in Calhoun, GA July 7, 1862; Left sick on the march back from Kentucky; Died in Tazewell, TN.

John Rhodes Britt PVT 4 Dec 1829 17 Mar 1913

Married Sarah Ann Nash in Gwinnett County, Nov 10, 1850; Left sick in Kentucky in 1862; Captured in Versailles, KY & exchanged in Vicksburg, MS Dec 22, 1862; Present on Dec 11, 1863 muster roll; Farmer in Gwinnett County after the war; Buried at Friendship Primitive Baptist, Snellville.

John C. Brownlee PVT 10 Aug 1841 15 Jun 1897

Captured at **Vicksburg**, MS July 4, 1863; Left sick in Gwinnett County, Oct 31, 1863; Captured in Lawrenceville, GA July 25, 1864; Received at **Camp Chase**, OH Aug 6, 1864; Married Margaret Lockridge in Gwinnett County about 1866; Farmer in Gwinnett after the war; Buried at Bethesda Methodist, Lawrenceville.

Milton Samuel Brownlee PVT 19 Mar 1831 6 Mar 1909

Married Malissa Couch in Gwinnett County April 24, 1853; Working as nurse in Hospital #2 Vicksburg, MS April 1863; Captured at **Vicksburg**, MS July 4, 1863; Filed for Confederate disability pension in Gwinnett County in 1887; Reported being shot through the right wrist at the **Battle of Resaca**, May 15, 1864; Hand was permanently disabled. Farmer in Martins, Gwinnett County in 1900; Buried at Friendship Primitive Baptist, Snellville.

Thomas C. Brownlee PVT 1843 27 Mar 1863

Died of disease in Vicksburg, MS hospital; Buried at Cedar Hill Cemetery, Vicksburg.
Warren Dantal Brownlee PVT 2 July 1832 12 Jan 1894
Brother of John, Milton & Thomas Brownlee; Married Elizabeth Lanier in Gwinnett County, Mar 16, 1856; Captured at **Vicksburg**, MS July 4, 1863; Farmer in Cates in 1880; Buried at Friendship Primitive Baptist, Snellville.
Frances Marion Buchanan PVT 1833 26 Oct 1906
Married Nancy King in Gwinnett County, Nov 2, 1852; Captured at Washington Hospital in **Vicksburg**, MS July 4, 1863; Among sick & wounded transported on US Steamer *H. Chouteau* from Vicksburg to Mobile; Absent left sick at home on Dec 11, 1863 muster roll; Member Masonic Order; Farmer in Berkshire in 1880; Buried in Buchanan-King Cemetery, Lilburn.
John S. Cain PVT 1834 Before 1880
Married Sarah Brownlee (Sister of John, Milton, Thomas & Warren Brownlee) in Gwinnett County, Jan 15, 1854; Captured in Atlanta, July 24, 1864; Released from **Camp Chase**, OH May 15, 1865; Farmer near Stone Mountain in 1870.
Elias Cannon PVT 12 May 1820 2 Apr 1901
Born in Anderson District, SC; Granted discharge for chronic rheumatism & disease of the heart by medical board near Vicksburg, MS & approved by medical board in Jackson, MS Mar 24, 1863; Farmer in Mud Creek, Habersham County in 1870; Buried at Pleasant Grove Methodist, Habersham County.
Theophilus Cannon PVT 24 Nov 1824 29 Jul 1900
Married Sarah Jones in Gwinnett County, Feb 1, 1849; Discharged for disability, April 12, 1862; Enlisted in Company K, 64th Georgia. Filed for disability pension in Gwinnett County in 1893; Buried at Haynes Creek Primitive Baptist, Loganville.
Daniel Morgan Clower PVT 9 Aug 1846 28 Dec 1935
Filed pension application in 1919 from Gwinnett County; Reported joining the 42nd in Jan 1864; Fell sick in McDonough, GA in July 1864; After regaining health he joined Wofford's Battalion of Georgia Militia; Surrendered at Ellijay in April 1865; Married Sarah Williams in Gwinnett County, Dec 5, 1868; Farmer near Rockbridge in 1900; Elected to the Gwinnett County Commission; Buried at Zoar Methodist, Centerville, Gwinnett County.
John Wesley Clower PVT 14 Jan 1821 5 Mar 1906
Father of Daniel Clower; Married Sarah Martin in Gwinnett County, Aug 8, 1844; Captured at **Vicksburg**, MS July 4, 1863; Sick at home on Dec 11, 1863 muster roll; Farmer near Rockbridge; Buried at Zoar Methodist, Centerville, Gwinnett County.
William Caughtus Cofer PVT 15 Nov 1842 20 Feb 1926
Discharged with disability April 12, 1862: Buried at Camp Creek, Lilburn.
Elbert Collins PVT 1826 After 1890
Married Eliza Jane Prewitt in Gwinnett County, Mar 14, 1850; Captured at **Battle of Baker's Creek**, MS May 16, 1863; POW at **Camp Morton**, IN; Exchanged at Fort Delaware, DE July 4, 1863; Transferred to the Pioneer Corps according to Dec 11, 1863 muster roll; Farmer in Big Creek, Milton County in 1880; Filed for Confederate pension in Fulton County in 1890; Reported being shot in left thumb & dislocating left knee on July 1, 1864.
John Berry Collins PVT 1 Nov 1829 19 Nov 1862
Born in South Carolina; Married Amanda Steele in Gwinnett County, Jan 13, 1853; Died of disease in Atlanta hospital; Buried at First Baptist, Lilburn.

William Ervin Cruse PVT 25 Oct 1839 2 Nov 1862
Died at Bean Station hospital in Grainger County, TN.
John David Dameron PVT Dec 1845 19 Jun 1916
Born in Newton County; Late enlistment; Admitted to Fairgrounds Hospital #2 in Atlanta April 19, 1864; Furloughed April 27, 1864; Paroled in Charlotte, NC May 6, 1865; Married Susan Harwell in Fulton County, Nov 19, 1874; Dry goods salesman in Atlanta after the war; Buried at Oakland Cemetery in Atlanta.
George Washington Dutton PVT 15 Dec 1836 18 Jun 1922
Born in Calhoun County, Alabama; Discharged with disability after April 30, 1862; 28-year-old farmer in Gwinnett in 1864 Georgia Militia Census; Buried in Sadler Cemetery, Grayson County, Texas.
James Madison Flarity PVT 29 May 1826 8 Nov 1903
Married Rosanna Elizabeth Titshaw in Walton County, Dec 5, 1847; Captured at **Vicksburg**, MS July 4, 1863; Book agent in Banks County in 1900; Died in Confederate Soldiers Home in Atlanta.
John M. Franklin PVT 10 May 1841 20 Sep 1906
At Vicksburg, MS hospital April 1863; Deserted & took U.S. Oath of Allegiance in Louisville, KY Nov 10, 1864; Ordered to remain north of the Ohio River until the close of the war; Married Amanda Bauldran in DeKalb County in 1876; Farmer in DeKalb County in 1900; Buried at Pleasant Hill Baptist, Atlanta, DeKalb County.
George Washington Freeman PVT 1839 24 July 1863
Brother of James, Leroy, & Hamilton Freeman; Died in Montgomery, AL hospital.
Hamilton Franklin Freeman PVT 12 July 1847 27 Jan 1917
Enlisted Mar 22, 1864 in Decatur, GA at age of 17; Surrendered at **Greensboro**, NC April 26, 1865; Married Elizabeth Olds in Gwinnett County about 1870; Farmer near Rockbridge in 1880; Filed for Confederate pension in Gwinnett County in 1913; Application endorsed by S.A. Starr of Loganville; Buried at Zoar Methodist in Centerville.
James Freeman PVT 1833 19 Mar 1865
Brother of George, Hamilton & Leroy Freeman; Married Elizabeth Knight in Gwinnett County, Jan 13, 1859; Captured at **Vicksburg**, MS July 4, 1863; Present on Dec 11, 1863 muster roll; Wounded in the chest at **Battle of Bentonville**, NC Mar 19, 1865; Died in General Hospital #10, Salisbury, NC; Buried at Old Lutheran, Salisbury, Rowan County, North Carolina.
James Parks Freeman PVT 4 Jun 1840 16 Feb 1917
Married Serena Griswell in Gwinnett County, Dec 9, 1858; Appointed **2CPL** Jun 1, 1863; Captured at **Vicksburg**, MS July 4, 1863; Absent without leave Dec 11, 1863; Surrendered at **Greensboro**, NC April 26, 1865; Miller in Gwinnett County after the war; Filed for Confederate pension in Gwinnett County in 1910; Application endorsed by J.B. Whitworth; Buried at New Hope Methodist, Lawrenceville.
James R. Freeman PVT 12 Oct 1844 2 Aug 1929
Captured at **Battle of Nashville**, Dec 16, 1864; Released from **Camp Chase**, OH June 12, 1865; Married Mary Cato in Hall County, Dec 24, 1868; Farmer in Hall County in 1900; Filed for Confederate pension in Hall County in 1910; Reported transferring from Company I, 55th Georgia to the 42nd GA in winter of 1863; Buried at Harmony Hall Baptist, Gainesville.
Leroy E. Freeman PVT 9 Oct 1836 24 Oct 1904

Present at Atlanta Medical College Hospital, Nov 1862; Filed for Confederate pension in Gwinnett County in 1903; Reported being wounded in left arm at **Battle of New Hope Church**, May 25, 1864 & furloughed home; Joined Graham's Battalion of Georgia Militia as he was unable to rejoin command; Married Caroline Davis in Gwinnett County, Nov 9, 1865; Farmer near Rockbridge in 1900; Buried at Rockbridge Baptist in Gwinnett County.

Joseph F. Garner PVT May 1831 After 1910

Born in South Carolina; Married Elizabeth Hodgins in Gwinnett County, Mar 6, 1856; Captured at **Vicksburg**, MS July 4, 1863; Applied for Confederate pension in Fulton County in 1902; Reported being present at Greensboro Surrender, April 26, 1865; Colonel L.P. Thomas served as his witness; Peddler in Atlanta in 1900.

David J. Gattis PVT 1827 1887

Born in North Carolina; Discharged July 26, 1862 because of an injury in the limber region of the spine; Farmer in Rockbridge in 1880.

Charles Green PVT 1831 20 Sep 1864

Admitted to Fairgrounds Hospital #1 in Atlanta Nov 14, 1862 with pneumonia; Furloughed Nov 17, 1862; Captured in **Vicksburg**, MS July 4, 1863; Captured in **Battle of Atlanta** July 22, 1864; Died of smallpox at **Camp Chase**, OH Sep 20, 1864; Buried in Grave # 258.

William D. Gresham PVT 4 Oct 1839 22 May 1907

Captured at **Vicksburg**, MS July 4, 1863; AWOL on Dec 11, 1863 muster roll; Took U.S. Oath of Allegiance and released at Nashville, TN Jan 4, 1864; Married Mary Jackson in Gwinnett County, Dec 27, 1866; Farmer in Gwinnett County after the war; Filed for Confederate pension in Gwinnett County in 1906; Reported being captured at Battle of Missionary Ridge, Nov 25, 1863 & kept in Federal prison until the end of the war; Buried at Chestnut Grove Baptist, Grayson, Gwinnett County.

James Robert Hale PVT 28 Mar 1827 12 Feb 1863

Married Elizabeth Phillips in Gwinnett County, Oct 14, 1849; Enlisted in Murfreesboro, TN Dec 20, 1862; Died of pneumonia in Jackson, MS hospital; Buried at Greenwood Cemetery in Jackson, MS; Left behind a wife and five children.

Matthew M. Hale PVT 1834 After 1880

Brother of James Hale; Married Mary Ann Phillips in Gwinnett County, Dec 20, 1853; Married Rebecca Wright in Gwinnett County, Oct 23, 1856; Captured at **Vicksburg**, MS July 4, 1863; Deserted at Palmetto Station, GA Sep 20, 1864; Farmer in Magazine, Yell County, AR in 1880.

George Washington Haney PVT 16 Sep 1840 16 Apr 1898

Present on Dec 11, 1863 muster roll; Surrendered at **Greensboro**, NC April 26, 1865; Married Millie Rannals, Nov 29, 1866; Farmer near Rockbridge after the war; Buried at Zoar Methodist in Centerville.

John J. Haney PVT 1831 2 Sep 1863

Married Melissa Rawlins in Gwinnett County, Dec 29, 1859; Captured at **Vicksburg**, MS hospital July 4, 1863; Taken on U.S. Steamer *H. Chouteau* to Mobile; Died at home. Widow's pension filed in Gwinnett County in 1890; Reported her husband came home sick from Vicksburg and never recovered.

Stephen L. Haney PVT 1817 Jul 1880

Born in South Carolina; Married Elizabeth Dewes in Gwinnett County, Mar 16, 1845;

Present on Dec 11, 1863 muster roll; Farmer in Chulio District after the war; Died of meningitis; Buried at Wax Community Cemetery, Silver Creek, Floyd County.

Milton N. Harris PVT 1842 22 July 1863

Married Charlotte Juhan around 1861; Muster roll for Dec 11, 1863 reports Harris was left sick at Baker's Creek, MS & presumed to be dead; Captured at **Vicksburg**, MS hospital July 4, 1863; Died in Raymond, MS hospital.

John Washington Henry PVT 13 Apr 1839 20 Mar 1908

Married Sarah Donaldson in Gwinnett County, Jan 3, 1860; Wounded in the left knee at Bean Station, TN Dec 9, 1862; Absent on sick furlough Dec 1863; Filed for Confederate pension in Gwinnett County in 1888; Reported being disabled after gunshot wound Farmer near Rockbridge after the war; Buried at Sodom Cemetery, Pleasant Hill.

James G.R. Herringe PVT 10 Jan 1844 24 May 1864

Patient at French's Division Hospital, Enterprise, MS Jun 30, 1863; Absent without leave on Dec 11, 1863 muster roll; Died in Covington, GA; Buried at Chestnut Grove Baptist, Grayson.

Alexander Abner Hewatt PVT 1834 Apr 1921

Married Martha Brandon in Gwinnett County, Jun 1, 1857; Captured at **Vicksburg**, MS July 4, 1863; Sick in Dalton hospital Dec 11, 1863; Farmer in Cates District after the war; Filed for Confederate pension in Gwinnett County in 1913; Reported being placed on sick furlough in McDonough by Captain Weaver in Aug 1864; Home sick at the end of the war; Buried at Friendship Primitive Baptist, Snellville.

Berry Hollingsworth PVT 1827 15 May 1864

Born in South Carolina: Married Mary Betts in Gwinnett County, Dec 31, 1857; Captured at **Vicksburg**, MS July 4, 1863; Killed at **Battle of Resaca**, GA.

James M. Holt PVT 14 Aug 1826 After 1880

Married Harriot Puckett Juhan in Gwinnett County, Dec 2, 1847; Sent home sick from Camp McDonald, April 1862; 39-year-old farmer in Gwinnett County in 1864 Militia census; Farmer near Rockbridge after war; Buried at Holt-Juhan Cemetery, Centerville.

William P. Hutchins PVT 2 Dec 1846 9 Jul 1920

16-years-old at the time of his enlistment; Admitted to Fairgrounds Hospital #1 in Atlanta Nov 14, 1862; Returned to duty Dec 11, 1862; Captured at **Vicksburg**, MS July 4, 1863; Surrendered at **Greensboro**, NC April 26, 1865; Married Valinda Brand in Walton County, Sep 1, 1872; Married Julia Frances Perry in Walton County about 1890; Farmer in Vinegar Hill, Walton County, in 1900; Buried at Bay Creek Cemetery, Loganville.

William J. Jackson PVT 1836 8 Jul 1863

Captured at **Battle of Baker's Creek**, MS May 16, 1863; POW at **Camp Morton**, IN; Too sick to be exchanged from Fort Delaware hospital July 4, 1863; Died at Fort Delaware & buried at Fine Point Cemetery, NJ.

Allen Sampson Jenkins PVT 1839 25 May 1864

Married Narcissus Frances Jones about 1860; Killed at **Battle of New Hope Church**, GA.

Henry T. Johnson PVT 1845 6 Sep 1862

Died in Bean Station, TN hospital.

William Russell Jones PVT 1835 18 Sep 1862

Married Martha Bradbury in Gwinnett County, Oct 12, 1854; Wounded at **Battle of Tazewell**, TN Aug 7, 1862; Died at Bean Station, TN hospital.

John Everinaton Kennerly PVT 28 Sep 1842 8 Oct 1910

Captured at **Vicksburg**, MS July 4, 1863; Left in Vicksburg hospital; Married Sarah McConnell in Gwinnett County, Sep 6, 1865; Farmer & preacher in Cates District after the war; Buried at Chestnut Grove Baptist, Grayson.

Thomas Jefferson Kennerly PVT 12 Jun 1845 18 Aug 1913
Born in Newton County; Younger brother of John Kennerly; Enlisted in Dalton Jan 19, 1864; Surrendered at **Greensboro**, NC April 26, 1865; Married Lucy Morgan in Gwinnett County, Nov 5, 1868; Farmer in Hog Mountain in 1880; Farmer in Cullman County, AL in 1900; Buried at Corinth East Baptist, Cullman, Alabama.

Charles King PVT 1823 7 Apr 1863
Born in Orange County, North Carolina; Married Elizabeth Jane Mason in Gwinnett County, Sep 6, 1848; Died at Jackson, MS hospital; Buried at Soldier's Rest, Cedar Hill Cemetery, Vicksburg, MS.

William P. Kircus PVT 20 Sep 1837 2 Jan 1863
Married Margaret Clark in Gwinnett County, Dec 8, 1859; Sent home sick from Camp McDonald, April 1862; Buried at Harmony Grove Methodist, Auburn, Barrow County.

Wilburn F. Knight PVT 1 Jan 1823 7 Sep 1874
Enlisted in Dalton, Jan 25, 1864; Surrendered at **Greensboro**, NC April 26, 1865; Buried at Rawlings-Clark Cemetery, Snellville.

Jesse Lancaster PVT 1827 About 1869
Born in South Carolina; Married Amanda Stone in Union County, July 7, 1847; Enlisted in Calhoun, July 6, 1862; Captured at **Vicksburg**, MS July 4, 1863; AWOL on Dec 11, 1863 muster roll; Died in Big Creek, Forsyth County.

Miles Jefferson Langley PVT 30 Jun 1830 15 Mar 1863
Older brother of Thomas Langley; Married Elizabeth Ann Jackson in Gwinnett County, Sep 14, 1854; Detailed as wagoner April 19, 1862; Served as teamster; Died in Vicksburg, MS hospital; Buried at Soldier's Rest, Cedar Hill Cemetery, Vicksburg.

Thomas Curtis Langley PVT 3 Oct 1834 22 Apr 1895
Married Nancy Beavers in Forsyth County, April 3, 1858; Enlisted in Calhoun, July 6, 1862; Reported at Powell's Valley, East Tennessee Aug 16, 1862; Discharged at Lauderdale Springs, MS with Surgeon's Certificate for rheumatism Aug 6, 1863; Wagon Maker in Suwanee in 1880; Member of UCV Camp #159 in Atlanta; Buried at Suwanee Cemetery, Gwinnett County.

Francis M. Lanier PVT 1837 26 Jul 1863
Detached to wait on the sick at Tazewell, TN April 27, 1862; Captured at **Vicksburg**, MS July 4, 1863; Died in Montgomery, AL hospital; Left wife and two children.

Daniel Johnson Liddell PVT 1842 Jul 1890
Captured at **Vicksburg**, MS July 4, 1863; Appointed **2SGT** Oct 1, 1863 by order of Colonel Henderson; Captured in **Battle of Atlanta**, July 22, 1864; Paroled at **Camp Chase**, OH and exchanged at Boulware & Cox's Wharves, VA Mar 10, 1865; Married Isabella Liddell in Gwinnett County, Jan 1873; Farmer in Martins, Gwinnett County in 1880.

Daniel Meredith Liddell PVT 1844 10 Jun 1864
First cousin of Daniel Johnson Liddell; Captured at **Battle of Baker's Creek**, MS May 16, 1863; POW at **Camp Morton**, IN; Paroled at Fort Delaware July 3, 1863; Captured at **Battle of Missionary Ridge**, TN Nov 25, 1863; Died at **Rock Island Prison**, Grave #1217.

Jesse L. Long PVT 17 Oct 1830 18 Sep 1910
Married Mary Donaldson in Gwinnett County, Feb 25, 1858; Captured at **Vicksburg**, MS

July 4, 1863; Filed for Confederate disability pension in Gwinnett County in 1889; Reports being wounded in the left hand at Rocky Face Ridge, Feb 25, 1864; Farmer in Palmetto, GA in 1900; Buried at Ramah Baptist, Palmetto, Fulton County.

Daniel Marion Martin PVT 1833 20 Aug 1863

Married Elizabeth Maddox in Gwinnett County, Jan 6, 1856; Appointed **2SGT** Feb 1863; Captured in a **Vicksburg**, MS hospital July 4, 1863; Died of disease in St. Clair, AL hospital.

Joseph Franklin Martin PVT 12 Mar 1836 4 Nov 1895

Married Mary Matilda Mills in Gwinnett County, Dec 1, 1859; Enlisted as a private in Company H, 10th Georgia State Troops Oct 21, 1861; Detailed to Napier's Artillery Feb 11, 1862 & mustered out in May 1862; Enlisted in the 42nd Georgia at Bean Station, TN July 1, 1862; Captured at **Vicksburg**, MS July 4, 1863; Surrendered at **Greensboro**, NC April 26, 1865; Farmer in Smith County, TX in 1880; Buried at Bascom Cemetery, Smith County, Texas.

Samuel Cicero Martin PVT 14 Feb 1842 6 Feb 1910

Younger brother of Joseph Martin; Captured at **Vicksburg**, MS July 4; 1863; Promoted to **SGT**; Captured at **Battle of Wise's Fork**, NC Mar 10, 1865; Released at **Point Lookout**, MD Jun 29, 1865; Married Missouri Baker in Gwinnett County, Nov 29, 1889; Farmer in Pinckneyville in 1900; Buried at Mount Carmel Methodist, Norcross.

Samuel Newton Martin PVT 1830 30 Jun 1863

Married Georgia Gordon in Gwinnett County, Jun 18, 1858; Died at Vicksburg, MS hospital; Buried at Soldier's Rest, Cedar Hill Cemetery, Vicksburg.

Winship Stedman Massey PVT 24 Sep 1842 25 Apr 1911

Appointed **4CPL** Feb 1863; Captured at **Battle of Baker's Creek**, MS May 16, 1863; Sent to **Camp Morton**, IN & paroled from Fort Delaware, DE July 3, 1863; Captured near Greensboro, NC April 24, 1865; Married Mary Ann Garner about 1867; Teamster in Atlanta in 1900; Buried at Bethesda Methodist, Lawrenceville.

Eli J. Mathews PVT 1843 After 1867

Captured in Vicksburg, MS hospital July 4, 1863; Captured at **Battle of Nashville** Dec 16, 1864; Released from **Camp Chase**, OH Jun 12, 1865.

Daniel M. McDaniel PVT 28 Apr 1838 9 Jul 1862

Died of fever in Morristown, TN hospital; Memorial stone at Sweetwater Baptist, Duluth.

Charles Gordon McGuffey PVT 23 May 1823 3 Mar 1907

Married Elizabeth Caroline Sexton in Gwinnett County, Mar 17, 1853; At Atlanta hospital May 1863; Discharged with disability Aug 6, 1863; Listed as tanner in Gwinnett County in 1864 militia census; Farmer & preacher in DeKalb County, Alabama after the war; Buried at Henagar Baptist, DeKalb County, AL.

James L. McGuffey PVT 15 Feb 1831 After 1911

Younger brother of Charles McGuffey; Married Emeline Holt (Sister of James Holt) in Gwinnett County about 1863; Wounded in left wrist at **Battle of Wise's Fork**, NC Mar 10, 1865; Paroled from Charlotte, NC General Hospital #11 May 1865; Farmer near Rockbridge after the war; Filed for Confederate disability pension in Gwinnett County in 1890; Buried at Zoar Methodist, Centerville.

John A. Miers PVT 23 July 1828 7 Dec 1918

Married Sarah Humphries in Newton County, Jan 12, 1851; AWOL on Dec 11, 1863 muster roll; Surrendered at **Greensboro**, NC April 26, 1865; Married Lucretia Graham in

Newton County about 1865; Farmer near Rockdale after the war; Filed for Confederate pension in Rockdale County in 1911; Application endorsed by F.M. Chandler; Buried at Sodom Cemetery, Pleasant Hill.

James A. Miller PVT Feb 1833 9 Jan 1865

Married Martha Bagwell in Gwinnett County, Dec 29, 1854; Captured at **Vicksburg**, MS July 4, 1864; Absent sick in Gwinnett County Dec 11, 1863; Listed as shoemaker March 1864; Captured at **Battle of Atlanta** July 22, 1864; Died of pneumonia at Camp Chase, OH; Buried in grave #727.

Francis M. Mills PVT 12 Mar 1836 30 Mar 1933

Older brother of William Mills; Married Frances E. Martin in Gwinnett County, Jan 5, 1860; Captured at **Vicksburg**, MS July 4, 1863; Farmer in Pickneyville in 1900; Filed for Confederate Pension in Gwinnett County in 1911; Reports being present at the Greensboro Surrender; Application endorsed by James W. Bennett; Buried at Mount Carmel Methodist, Norcross. Lived 97 years.

John Thomas Mills PVT 31 Dec 1843 25 Mar 1925

Captured at **Battle of Nashville**, TN Dec 16, 1864; Released from **Camp Chase**, OH Jun 12, 1865; Married Emma Duncan in Gwinnett County about 1869; Prison guard in Atlanta in 1900; Buried at Mount Carmel Methodist, Norcross.

Joseph Berry Mills PVT 6 Apr 1837 14 Mar 1909

Older brother of John & Robert Mills; Married Mary Paden in Gwinnett County, Oct 22, 1858; Enlisted May 12, 1862 in Atlanta; Wounded at **Battle of Chickasaw Bayou**, MS Dec 28, 1862; Captured in a rifle pit at Vicksburg, MS on the night of June 24, 1863; Paroled in **Vicksburg** on July 4, 1863; Appointed **4SGT** in 1864; Surrendered at **Greensboro**, NC April 26, 1865; Filed for Confederate pension in 1899; Application endorsed by J.B. Whitworth; Farmer near Lawrenceville in 1900; Buried at Bethesda Methodist, Lawrenceville.

Robert Auston Mills PVT 1844 Before 1867

Brother of John and Joseph Mills; Captured at **Vicksburg**, MS July 4, 1863; Captured at **Battle of Nashville**, Dec 16, 1864; Released from **Camp Chase**, OH Jun 12, 1865; No further record.

William B. Mills PVT 22 Oct 1843 19 Feb 1929

Enlisted as a private in Company H of the 10th Georgia State Troops Oct 21, 1861; Mustered out in May 1862 & joined the 42nd Georgia in Atlanta May 12, 1862; Captured at **Vicksburg**, MS July 4, 1863; Present on Dec 11, 1863 muster roll; Married Rachel Jones in Gwinnett County, Aug 10, 1865; Farmer in Boone, AR in 1900; Buried at Callahan Cemetery, Magazine, Logan County, Arkansas.

David "Dock" Pinckney Moore PVT 12 Feb 1841 7 Feb 1904

Younger brother of William Moore; Married Tabitha Roquemore in Gwinnett County, Jan 11, 1860; Enlisted in Atlanta May 12, 1862; Captured at **Vicksburg**, MS July 4, 1863; Surrendered at **Greensboro**, NC April 26, 1865; Farmer in Rockbridge in 1900; Buried at Hillcrest Cemetery, Loganville.

William V. Moore PVT 1835 Jan 1864

Married Nancy Nichols in Gwinnett County, April 5, 1857; Admitted to Fairgrounds Hospital #1 in Atlanta Jan 5, 1863; Furloughed Jan 10, 1863; Captured at **Vicksburg**, MS July 4, 1863; Died of smallpox in Atlanta hospital.

James Berry Nash PVT 4 July 1842 23 May 1915

Born in Greenville County, SC: Absent sick on leave from Tazewell, TN May 19, 1862; Captured at **Vicksburg,** MS July 4, 1863; Received at Atlanta Distribution hospital with smallpox Feb 24, 1864; Admitted Madison, GA hospital Feb 26, 1864; Returned to duty Mar 19, 1864; Surrendered at **Greensboro,** NC April 26, 1865; Married Mary Ann Arnold in Gwinnett County in 1865; Farmer in Cates, Gwinnett County after the war; Buried at Friendship Primitive Baptist, Snellville.

John Thomas Nash PVT 6 Aug 1840 14 Aug 1864

Captured at **Vicksburg,** MS July 4, 1863; Transported with sick from Vicksburg to Mobile on US Steamer *St. Maurice*; Exchanged off Fort Morgan Aug 5, 1863; Died in Gwinnett County; Buried at Nash Cemetery, Lilburn.

Wilson D. Nichols PVT 2 Apr 1828 2 Jun 1912

Born in Walton County; Married Sarah Austin in Gwinnett County, Sep 10, 1846; Captured at **Vicksburg,** MS July 4, 1863; Detached to artillery Dec 1863; Married Sallie Burges in Gwinnett County, Aug 24, 1884; Farmer near Rockbridge after the war; Filed for Confederate pension in Gwinnett County in 1895; Reported serving for 20 months; Buried in Bethesda Methodist, Lawrenceville.

Francis Crowder O'Kelley PVT 16 Apr 1833 16 Apr 1882

Born in Oglethorpe County; Married Sarah Delilah O'Kelley Oct 15, 1855; Owned nine slaves in 1860; Nurse in Canton, MS hospital Jun 30, 1863; Druggist at Walker Division Hospital, Lauderdale, MS Aug 31, 1863; Presented to Medical Examining board under Surgeon E. Sea in Lauderdale, MS Sep 19, 1863, "We have carefully examined the said F.C. O'Kelley of Captain Weaver's Company and find him incapable of performing his duties of a soldier in Field service because of Ascites (liver disease) and therefore recommend that he be detailed as a Druggist, as he is a regular graduate of medicine & therefore competent to fill such a position."; Admitted to Fairgrounds Hospital #1 in Atlanta Nov 26, 1863 with hepatitis; Wounded at **Battle of Bentonville,** NC Mar 25, 1865; Taken to Greensboro Confederate Hospital #3; Physician in Cates, Gwinnett County in 1880; Buried at Eastview Cemetery, Conyers.

Samuel Winship Paden PVT 22 Sep 1843 19 May 1917

Son of Samuel Sexton Paden & brother of William S. Paden; Captured at **Vicksburg,** MS July 4, 1863; Promoted from ranks to **CPL** Oct 1, 1863 by order of Colonel Robert Henderson; Captured at **Battle of Nashville,** Dec 16, 1864; Released from **Camp Chase,** OH June 12, 1865; Married Jo Hannah Lockridge in Gwinnett County about 1866; Farmer in Gwinnett County; Buried at Bethesda Methodist, Lawrenceville.

William S. Paden PVT 6 Jan 1841 6 Oct 1862

Married Annie Jane Nash in Carroll County, Dec 25, 1859; Appointed **4CPL** Oct 1, 1862; Died of fever in Danville, KY Oct 6, 1862; Buried at Danville City Cemetery, Boyle County, Kentucky.

William Lenoir Parr PVT 11 Aug 1822 9 Dec 1862

Married Sarah Williams in Clarke County, Sep 5, 1839; Admitted to Fairgrounds Hospital #1 in Atlanta Nov 28, 1862 with pneumonia; Died in the hospital; Buried at Oakland Cemetery, Atlanta.

Daniel J. Peters PVT 1840 15 Jun 1864

Older brother of James Peters; Working as Teamster near Vicksburg, MS April 1863; Captured at **Vicksburg,** MS July 4, 1863; Sent to Chickamauga, GA hospital Oct 1863; Furloughed from Fairgrounds Hospital #1 in Atlanta Oct 31, 1863; Died in Gwinnett

County; Buried at Friendship Primitive Baptist, Snellville.

James M. Peters PVT 17 Oct 1845 9 May 1902

Appears in 1864 Militia Census as 18-year-old in Cates District, Gwinnett County; Enlisted in 1864; Severely wounded in right arm at **Battle of New Hope Church**, May 25, 1864; Granted medical furlough in Lauderdale, MS Jan 21, 1865; Married Lucinda Andrews in Gwinnett County in 1870; Filed for Georgia Disability Allowance in 1889; Farmer in Bay Creek, Gwinnett County in 1900; Buried at Friendship Primitive Baptist, Snellville.

Benjamin Franklin Phillips PVT 1831 27 Jun 1862

Married Mary Emeline Miller in Gwinnett County, Nov 15, 1850; Enlisted in Atlanta May 12, 1862; Died of measles at Knoxville, TN hospital; Buried at Bethel Cemetery, Knoxville, TN.

William Perry Phillips PVT 1839 5 Sep 1863

Younger brother of Benjamin Phillips; Captured at **Vicksburg**, MS July 4, 1863; Died at home in Gwinnett County.

John S. Rawlins PVT 1827 29 Dec 1862

Married Elizabeth Britt in Gwinnett County, May 22, 1859; Killed at **Battle of Chickasaw Bayou**, MS; Elizabeth filed for a widow's pension in Gwinnett County in 1891; S.H. Starr, W.H. Williams & J.B. Williams reported seeing him fall.

Oliver H.P. Rawlins PVT 1832 19 Jul 1863

Married Lucinda Green in Gwinnett County, Feb 4, 1855; Captured at **Vicksburg**, MS July 4, 1863; Died in Vicksburg, MS hospital.

William C. Richards PVT 1843 29 Dec 1862

Killed at **Battle of Chickasaw Bayou**; Buried at Soldier's Rest, Cedar Hills Cemetery, Vicksburg, MS; Claim filed by his father John Richards.

Henry William Robinson PVT 1829 12 Apr 1863

Born in North Carolina; Married Elizabeth Bridges in Catawba County, NC, Oct 27, 1850; Blacksmith living near Jefferson, GA in 1860; Died of smallpox in Vicksburg Hospital #2; Buried at Soldier's Rest, Cedar Hill Cemetery, Vicksburg.

George Clayton Smith PVT 30 Jul 1863

Married Jane Summerlin in Gwinnett County, Sep 2, 1855; Captured in **Vicksburg** hospital July 4, 1863; Died at Vicksburg, July 30, 1863; Left widow & three children; Buried at Soldier's Rest, Cedar Hill Cemetery, Vicksburg, MS.

George I. Smith PVT 1819 Before 1900

Captured at **Vicksburg**, MS July 4, 1863; Left sick at home on Dec 11, 1863 roll; Farmer in Goodwins District in 1880; Filed for Confederate pension in Gwinnett County in 1897; Reported being present at the 1865 Greensboro surrender.

Robert P. Smith PVT 1845 Before 1867

Captured at **Vicksburg**, MS hospital July 4, 1863; Transported with sick from Vicksburg to Mobile on US Steamer *St. Maurice*; Exchanged off Fort Morgan Aug 5, 1863; Sick at home on Dec 11, 1863 muster roll.

William T. Smith PVT 25 Feb 1842 19 Jan 1904

Born in Butts County; Joined Company H, 16th Georgia Aug 11, 1861 and was elected 2LT. Resigned in Virginia with sickness. Regained health in time to join the 42nd Georgia; Elected **2LT** Sep 1, 1862; Captured at **Vicksburg** July 4, 1863; Promoted to **CPT** in the closing days of the war; Surrendered at **Greensboro**, NC April 26, 1865; Married Mary J.

Jones in DeKalb County, Oct 27, 1865; Moved to Titus County TX & elected to the state legislature. Returned to Gwinnett County in 1882; Farmer and politician near Buford in 1900; Member of UCV Camp # 1300 in Buford; Member of the Baptist Church; A writer to the Atlanta Constitution noted that Smith was "Conservative in spirit, his actions were always the result of carefully considered measures and once convinced of their merit he threw the whole energy of his nature into the work of securing their adoption. As a debater there were few in the Senate who were his superiors."; Buried at Hillcrest Cemetery, Buford.

Jesse William Stapp PVT 12 Feb 1842 17 Jun 1932

Captured at **Vicksburg,** MS July 4, 1863; Wounded in the thigh & captured at **Battle of Resaca,** May 15, 1864; Released from **Camp Douglas,** IL Jun 16, 1865; Married Lucinda Cain in Gwinnett County, Nov 28, 1867; Farmer in Ringgold, GA in 1900; Filed for Confederate pension in Catoosa County in 1909; Application endorsed by J.B. Nash & F.M. Wills; Buried at Nathan Anderson Cemetery in Catoosa County.

John T. Stapp PVT 27 Jul 1845 After 1867

Younger brother of Jesse Stapp; Enlisted as a private in Company H of the 10th Georgia State Troops Oct 21, 1861; Mustered out & enlisted in the 42nd Georgia in Atlanta May 12, 1862; Wounded & captured at **Newton Station,** MS April 10, 1863 & paroled at Demopolis, AL Jun 5, 1863; Sent to hospital from Sweetwater, Gwinnett County Dec 11, 1863; Married Mary McCart in Gwinnett County, Aug 28, 1864; Signed Reconstruction Oath in Bartow County in 1867.

Benjamin Alexander Starr PVT 1846 28 Jun 1864

Brother of Silas & William Starr; Late recruit; Died of fever in Covington.

Silas Andrew Starr PVT 28 Apr 1844 14 Sep 1923

Captured at **Vicksburg,** MS July 4, 1863; Promoted from **1CPL** to **3SGT** Oct 1, 1863; Admitted to Fairgrounds Hospital #1 in Atlanta May 31, 1864 with dysentery; Furloughed June 13, 1864; Surrendered at **Greensboro,** NC April 26, 1865; Married Nancy Amanda Floyd in Newton County, Dec 13, 1866; Farmer in Walton County after the war; Filed for Confederate pension in Walton County in 1919; Buried at Loganville Memorial Gardens.

Dickson M. Steele PVT 1842 11 Aug 1863

Enlisted in company H of the 10th Georgia State Troops Oct 21, 1861; Mustered out & enlisted in the 42nd Georgia in Atlanta May 12, 1862; Captured in the Washington Hospital in **Vicksburg,** MS July 4, 1863; Taken by the Steamer *Chouteau* to New Orleans; Died in St. Louis Hospital, New Orleans, LA of typhoid fever Aug 11, 1863; Buried in Grave #503 at Cypress Grove #2, New Orleans.

James Andrew Tallent PVT

At Vicksburg, MS hospital April 1863; Transferred to Pioneer Corps before Dec 11, 1863.

Benjamin C. Thomason PVT 1839 Before 1867

Enlisted in Atlanta May 12, 1862; Wounded in the arm at **Battle of Baker's Creek,** MS May 16, 1863; Captured at **Vicksburg** Hospital #2, July 4, 1863; Transported with sick from Vicksburg to Mobile on US Steamer *S.S. Crescent*; Exchanged off Fort Morgan Aug 4, 1863; Left wounded in Gwinnett County Dec 11, 1863 muster roll.

Thomas M. Thomason PVT Feb 1840 After 1900

Brother of Benjamin & William Thomason; Enlisted in the 42nd Georgia May 12, 1862; Captured at **Vicksburg,** MS hospital July 4, 1863; Transported with sick from Vicksburg to Mobile on US Steamer *S.S. Crescent*; Exchanged off Fort Morgan Aug 4, 1863; Captured at **Battle of Nashville**, Dec 17, 1864; Released at **Camp Chase**, OH Jun 15, 1865; Filed for Confederate pension in Rockdale County in 1893; Farmer near Rockdale in 1900.

William P. Thomason PVT 1834 2 Jun 1863

Married Julia Britt in Gwinnett County, Dec 23, 1856; Enlisted in Atlanta May 12, 1862; Appointed **1CPL** Feb 1863; Died at **Vicksburg**, MS hospital; Left wife and one child.

Reuben Collins Weaver PVT 1841 About 1876

Enlisted at Woodson's Crossroads, TN Sep 1, 1862; Captured near Versailles, KY Oct 12, 1862; Exchanged at Vicksburg, MS from the Steamboat *Mary Crane* Nov 29, 1862; Detailed to guard prisoners in Atlanta Feb to May 1863; Captured at Macon, Ga April 1865; Married Georgia Napier in Dooly County, Dec 1, 1869; Engineer at mill in Sumter County in 1870.

James S. White PVT 1842 2 Jul 1863

Died in Vicksburg, MS hospital.

James T. Williams PVT 1836 After 1890

Born in North Carolina; Married Mary McGuffey in Gwinnett County, Nov 25, 1860; Captured at **Vicksburg**, MS July 4, 1863; Present on Dec 11, 1863 muster roll; Farmer near Rockbridge in 1880.

Newton T. Williams PVT 6 Oct 1843 16 Dec 1868

Brother of Lt. William Williams; Ancestor of William Williams who fought at Kings Mountain in 1780; Enlisted in Atlanta, May 12, 1862; Captured at **Vicksburg**, MS July 4, 1863; Sent to Dalton hospital Dec 5, 1863; Admitted to Fairgrounds Hospital #2 in Atlanta May 17, 1864; Buried at Williams-Pittman Cemetery, Snellville.

Columbus W. Wilson PVT 1842 1862

Died of camp fever.

George Willis Wilson PVT 1827 20 Aug 1863

Married Nancy Hamilton in Gwinnett County, Feb 14, 1850; Died in Lauderdale Springs, MS hospital; Left widowed wife & three children.

John William Wilson PVT 1838 15 Mar 1863

Brother of Thomas Wilson; Married Sarah McCune in DeKalb County, Dec 24, 1860; Died at Vicksburg hospital of fever; Buried at Soldier's Rest, Cedar Hill Cemetery, Vicksburg.

Thomas Oliver Wilson PVT 1845 21 Dec 1862

Died in Reidsville, TN hospital.

Hiram Alexander Worthy PVT 13 Sep 1836 12 Jan 1913

Younger brother of Thomas Worthy; Enlisted in Calhoun, GA July 28, 1862; Admitted to Fairgrounds Hospital #1 in Atlanta Nov 28, 1862 with fever; Furloughed Dec 10, 1862; Present on Dec 11, 1863 muster roll; Married Delilah Chaffin in Walton County, Dec 24, 1868; Farmer near Duluth after the war; Buried at Mt. Zion Baptist, Snellville.

Thomas Franklin Worthy PVT 19 Dec 1831 31 Jul 1922

Born in Morgan County; Married Martha Rawlins in Gwinnett County, July 10, 1859; Captured at **Vicksburg**, MS July 4, 1863; Surrendered at **Greensboro**, NC April 26, 1865; Farmer in Cates, Gwinnett County after the war; Filed for Confederate pension in Gwinnett County in 1908; Application endorsed by J.W. Bennett; Buried at Snellville Historical Cemetery.

Company C – Milton Tigers – 109 Men

Milton was a small rural county about 25 miles north of Atlanta. The company assembled in the county seat of Alpharetta. Several of the soldiers came from the southern portion of Forsyth County.

Name Rank Birth Death

Henry William Paris CPT 30 Sep 1818 23 May 1880

Stood 6'4"; Born in South Carolina; Married Martha Webb in Forsyth County, Jun 14, 1856; Sheriff of Milton County in 1860; Captured near Keene, KY; Exchanged at Vicksburg, MS from the Steamboat *Mary Crane*, Nov 29, 1862; Captured at **Vicksburg**, MS July 4, 1863; Resigned with disability Dec 21, 1863; Farmer in Milton County after the war; Buried at Old Big Creek Baptist, Fulton County.

John W. Allen 2CPL 27 Dec 1824 After 1900

Born in Morgan County; Older brother of Elisha Thomas Allen; Married Catherine Thomason in Walton County, Jan 10, 1844; Detailed as a nurse in Atlanta hospital Dec 10, 1862; Captured at **Vicksburg**, MS July 4, 1863; Surrendered at **Greensboro**, NC April 26, 1865; Farmer in Newton County in 1880; Filed for Confederate pension in Newton County in 1895.

Jonas Cain 2SGT 9 Dec 1824 27 Aug 1911

Married Mary Garner in Gwinnett County, April 24, 1845; Appears last on roll April 30, 1862; On 1864 Milton militia census as 41-year-old farmer exempt as member of GA State Guard; Farmer near Big Creek in 1880; Licensed preacher at Liberty Baptist in Gwinnett County; Filed for Confederate pension in Fulton County in 1895; Reports receiving medical discharge a few days before the Vicksburg surrender & later served in state militia; Member of UCV Camp # 1300 in Buford; Buried at Pleasant Hill Baptist, DeKalb County.

Francis Marion Dunbar 1SGT 30 Apr 1835 15 Sep 1889

Left sick in private home near Knoxville, TN by order of Colonel Robert Henderson, April 27, 1862; Elected **2LT** Jan 8, 1863; Captured at **Vicksburg**, MS July 4, 1863; Left sick at Mossy Creek, TN Oct 22, 1863; Sent to hospital Nov 25, 1863 by medical board; Married Mary Hubbard in Russell County, AL Oct 29, 1865; Wholesale sugar dealer in Opelika in 1880; Buried at Rosemere Cemetery, Opelika, Lee County, AL: "Captain F.M. Dunbar"on tombstone.

Russell Collins Emerson 1CPL 1830 10 Feb 1863

Died of disease in a Vicksburg, MS hospital; Buried at Soldier's Rest, Cedar Hill Cemetery, Vicksburg.

William Fitch 4CPL 1839 11 Mar 1881

Born in South Carolina; Appointed **SGT**; Captured at **Vicksburg**, MS July 4, 1863; Surrendered at **Greensboro**, NC April 26, 1865; Married Fannie Burnett in Fulton County, Sep 1, 1870; Railroad employee in Atlanta in 1880; Buried at Oakland Cemetery, Atlanta.

Russell Davis Garrison 2LT 5 Aug 1818 13 Oct 1862

Married Susan Key Camp, April 24, 1844; Died in Tennessee; Buried at Rest Haven, Alpharetta.

William Perry Jameson 5SGT 1837 4 Oct 1864

Born in Pickens County, SC; Married Nancy Phillips in Cherokee County, July 31, 1855; Appointed **4SGT** in 1862 & **2SGT** in Feb 1863; Patient at Enterprise, MS hospital April

30, 1863; Wounded & captured at **Battle of Resaca** May 15, 1864; Died of Neuralgia at prison hospital at **Alton**, Il; Buried at Confederate Cemetery, Alton, Madison, IL.

Singleton Allen Maxwell 1LT 18 Sep 1832 22 July 1864

Born in Elbert County; Married Mary Maxwell in Hart County, Aug 22, 1858; Captured at **Vicksburg**, MS July 4, 1863; Elected **CPT** Dec 21, 1863; Killed at **Battle of Atlanta**; Buried at Maxwell Cemetery, Alpharetta.

James M. McCollum 2LT 1834 16 Aug 1862

Died of disease at Bean Station, TN hospital; Buried at Rest Haven, Alpharetta. Replaced by M.H. Trammell.

Thomas A. Pannell 3SGT 18 Dec 1836 20 Sep 1912

Born in Forsyth County; Appointed **2SGT** in 1862 & **1SGT** Feb 1863; Detailed to Captain Calhoun's detachment in Knoxville, TN April 20, 1862 to guard political prisoners; Prison guard at Madison, GA Oct 31, 1862; Captured at **Vicksburg**, MS July 4, 1863; Admitted to Fairgrounds Hospital #2 in Atlanta with lung inflammation Jan 7, 1864; Captured at **Battle of Resaca** May 15, 1864; Sent to **Alton**, IL prison; Released at Fort McHenry, MD May 9, 1865; Married Mattie Mayne in Hall County, Oct 4, 1874; Millwright in Gainesville in 1880; Buried at Alta Vista Cemetery, Gainesville.

John B. Taylor 5CPL 1863

Appointed **3CPL** in place of Henry E. Jones (who was discharged) April 24, 1862; Admitted to Fairgrounds Hospital #1 in Atlanta Nov 28, 1862 with jaundice; Furloughed Dec 12, 1862; Wounded and captured at **Battle of Baker's Creek** May 16, 1863; Reduced in ranks by order of Colonel Henderson July 1, 1863. Died from his wounds.

Elisha Thomas Allen PVT 22 Sep 1834 31 May 1905

Born in Walton County; Married Mary Studdard in Walton County, July 10, 1856; Admitted to Fair Grounds Hospital #1 in Atlanta Oct 17, 1862; Returned to duty June 10, 1863; Captured at **Vicksburg**, MS July 4, 1863; Detached Oct 18, 1863 to guard prisoners at Chickamauga; Admitted to Fairgrounds Hospital #2 in Atlanta Nov 29, 1863 with fever; Deserted in Walker County July 16, 1864 and gave Oath of Allegiance in Chattanooga, TN July 18, 1864; Farmer in Walton County after the war; Filed for Confederate pension in Morgan County in 1897; Reported being captured in July 1864 and kept in prison in Lousiville, KY until the war ended; Buried at Studdard Cemetery, Rutledge, Morgan County.

Martin T. Allen PVT 2 Aug 1843 5 Jan 1926

First cousin of Elisha Allen; Enlisted in Mar 1863; Married Nancy Dick in Newton County, Mar 17, 1865; Farmer in Walton County after the war; Filed Confederate pension application in Walton County in 1902 & 1907; Reports being present at the Greensboro Surrender in April 1865; Filed for Confederate pension in Walton County in 1902; Buried at Alcovy Cemetery, Walton County.

Mathew L. Andrews PVT 2 Jan 1824 5 Jan 1894

Furloughed from Fair Grounds Hospital #1 in Atlanta Dec 12, 1862; Captured at Vicksburg, MS July 4, 1863; Dec 11, 1863 muster roll reports "sent to Dalton hospital, Oct 13, 1863"; Farmer in Milton County in 1880; Buried at Camp Creek Primitive Baptist, Lilburn.

John Enoch Bagwell PVT 11 Aug 1844 22 Nov 1909

Born in Edgefield County, SC; AWOL on Dec 11 1863 muster roll "since April 1, 1863"; Married Eliza Williams in Milton County in 1863; Farmer & preacher in Warrenton, AL in 1900; Buried at Brashiers Chapel, Arab, Marshall County, AL.

James Monroe Barron PVT 5 Dec 1821 18 Dec 1883
Born in Butts County; Detailed as wagoner for regiment April 14, 1862; Captured at **Vicksburg**, MS July 4, 1863; Surrendered at **Greensboro**, NC April 26, 1865; Married Rebecca Stark in Milton County about 1865; Farmer in Crossville in 1880; Buried at Union Primitive Baptist, Roswell.

John Thomas Barron PVT 15 Aug 1833 25 May 1864
Born in Butts County; Younger brother of James Barron; Captured at **Vicksburg**, MS July 4, 1863; Killed at **Battle of New Hope Church.**

George Washington Bennett PVT 1840
Enlisted in 1864; Captured at **Battle of Resaca**, May 15, 1864; Released at **Camp Morton**, IN May 18, 1865.

Daniel Washington Bennett PVT 1830 After 1875
Captured at **Vicksburg**, MS July 4, 1863; Dec 11, 1863 muster roll reports "absent sick in Milton County since Sep 18, 1863"; Farmer in Milton County in 1870.

John F. Bird PVT 1839 10 Sep 1862
Married Sarah Allen in Walton County, April 28, 1858; Left sick at Knox, TN April 28, 1862; Died in Fair Ground Hospital, Knoxville; Buried at Bethel Cemetery, Knoxville, TN.

Elbert G. Bowen PVT 25 Mar 1825 22 May 1863
Born in Anderson County, SC; Married Rachel Poole in Forsyth County, Jan 2, 1853; Sent to Knoxville, TN hospital Sep 25, 1862; Died in Atlanta hospital; Buried at Oakland Cemetery, Atlanta.

James Gibson Bowen PVT 30 Sep 1819 9 Apr 1899
Born in South Carolina; Older brother of Elbert Bowen; Married Frances Richards about 1844; Discharged in Knoxville, TN for chronic rheumatism Nov 4, 1862; On 1864 Militia Census for Milton County as a 44-year-old farmer; Farmer in Alpharetta in 1880; Buried at Morris Cemetery, Alpharetta.

David Daniel Brewer PVT 8 May 1836 Mar 1911
Born in Chatham, NC; Detailed to Captain Calhoun's detachment in Knoxville, TN April 20, 1862 to guard political prisoners; Prison guard at Madison, GA Oct 31, 1862; Captured at **Vicksburg**, MS July 4, 1863; Sick at home on Dec 11, 1863 muster roll; Married Mary Whitehead in Conway, AR Nov 23, 1879; Farmer in Washington County, AR in 1900; Buried at Kilgore Cemetery, Birdtown, Conway, AR.

William Henry Brooks PVT 18 Feb 1845 17 Feb 1911
Late enlistment; Filed for Confederate pension in Milton County in 1879; Reported being wounded in the arm at **Battle of Atlanta**, July 22, 1864; Arm amputated above the elbow; Married Mary Malinda Brooks in Milton County, Dec 19, 1869; Farmer in Big Creek in 1880; Farmer in Cross Keys in 1900; Member of UCV Camp #665 in Decatur; Buried at Johns Creek Baptist, Chamblee.

Daniel Bruce PVT 1836 26 Jul 1863
Married Nancy Fowler in Cherokee County, Nov 5, 1857; Captured at Washington Hospital in **Vicksburg**, MS July 4, 1863; Among Confederate sick taken by Steamer *H. Chouteau* to Mobile by way of New Orleans; Died in St. Louis Hospital in New Orleans; Buried at Cypress Grove #2, New Orleans.

Giles M. Burnett PVT Apr 1841 After 1910

Born in Union County, SC; Captured at **Vicksburg**, MS July 4, 1863; Married Margaret Bailey about 1867; Farmer in Duluth in 1900.

Spencer Washington Burnett PVT Dec 1841 4 May 1934
Born in Union County, SC: Older brother of Giles Burnett; Captured at **Vicksburg**, MS July 4, 1863; Married Martha Barnett in Milton County in 1865; Stated in 1907 & 1921 Alabama pension forms that he was wounded & disabled at **Battle of Resaca**, May 15, 1864; Farmer in Blount County, AL after the war; Buried at Wynnville Cemetery, Blount County, AL.

Calvin Burton PVT 1836 After 1883
Born in South Carolina; Married Martha Clark about 1858; Detailed regimental wagoner by Colonel Henderson Feb 9, 1863; Captured at **Vicksburg**, MS July 4, 1863; Surrendered at **Greensboro**, NC April 26, 1865; Farmer in Alpharetta in 1880.

Edmund Jackson Camp PVT 6 May 1828 14 May 1889
Married Nancy Thomason about 1850; Appointed Acting Quartermaster with the rank of **CPT** July 9, 1862; No further record; Miller in Walton County in 1864 Militia census; Living in Alpharetta in 1870 with $10,000 in real estate; Buried at Martin Cemetery, Alpharetta.

John Franklin Chaffin PVT 1834 19 Dec 1863
Detailed to Captain Calhoun's detachment in Knoxville, TN April 20, 1862 to guard political prisoners; Prison guard at Madison, GA Oct 31, 1862; Sent to Vicksburg, MS hospital April 23, 1863; Captured in Clinton, MS hospital May 15, 1863; Paroled by General Sherman at Demopolis, AL Jun 5, 1863; Died from sickness in Milton County Dec 19, 1863.

Cooper Bennett Chastain PVT 1824 4 Mar 1865
Born in South Carolina; Married Mary Jane Buchanan about 1848 in Anderson County, SC; Enlisted in 1864; Captured at the **Battle of Resaca**, May 15, 1864; Died of pneumonia as a prisoner at **Camp Douglas**, IL; Buried at Confederate Mound in Oak Woods Cemetery, Chicago.

Sterling Clark PVT 1830 1 Jun 1862
Detailed to Captain Calhoun's detachment in Knoxville, TN April 20, 1862 to guard political prisoners; Died in Gate City Hospital, Atlanta; Buried at Oakland Cemetery, Atlanta.

Lindsay E. Conner PVT 1831 Sep 1880
Married Sarah Gazaway in Milton County, July 1858; Captured at **Vicksburg**, MS July 4, 1863; Surrendered at **Greensboro**, NC April 26, 1865; Farmer in Newtown in 1880.

Thomas B. Cook PVT 1818 Jan 1885
Born in Union District, SC; Married Ann McConnell in Cherokee County, Nov 24, 1844; Appointed postmaster of Troy, Cherokee County in 1844; Paid at Bean Station, TN July 26, 1862; No further military record; Blacksmith in Milton County in 1864 militia census; Living in Marietta in 1880.

Smith F. Cottrell PVT 8 Apr 1827 Nov 1862
Born in South Carolina; Married Hannah Terry in Forsyth County, Mar 16, 1858; Sent to Versailles, KY hospital Oct 8, 1862; Died in Kentucky.

Josiah D. Davenport PVT 1830 20 Sep 1863
Married Winnie Taylor in Georgia, Aug 15, 1853; Captured at **Vicksburg**, MS July 4, 1863; Killed at **Battle of Chickamauga**, GA Sep 20, 1863; Winnie filed for widow pension

in Alpharetta in 1891; Reported that her husband transferred to another regiment shortly before the battle.

Augustus Clayton Dickert PVT 28 Apr 1836 22 Jan 1924
Admitted to Fairgrounds Hospital #1 in Atlanta Oct 17, 1862; Captured at **Vicksburg,** MS July 4, 1863; Captured at the Chattahoochee River July 5, 1864; Paroled at **Camp Morton,** IN Mar 15, 1865 & exchanged at Boulware & Cox's Wharves, James River, VA; Married Mary Owens in Milton County, Aug 21, 1867; Farmer in Sheltonville after the war; Member of UCV Camp # 159 in Atlanta; Filed for Confederate pension in Milton County in 1901; Application endorsed by M.H. Trammell; Buried at Sharon Baptist, Forsyth County.

Josephus Dickert PVT 1846 8 July 1864
Younger brother of Augustus Dickert; Late enlistment; Captured at **Battle of Resaca,** GA May 16, 1864; Died of fever as POW at **Alton** Prisoner of War Camp in Alton, Illinois; Buried at North Alton Confederate Cemetery.

Isaac Dildy PVT 1844
Living near Alpharetta with Claiborne & Elizabeth Vaughan in 1860; Enlisted in Vicksburg, MS Jan 22, 1863; Captured at **Vicksburg,** MS July 4, 1863; AWOL on Oct 31, 1863 muster roll; Surrendered at **Greensboro,** NC April 26, 1865; No record after the war.

Russell Jackson Fields PVT 1842 22 Jul 1862
Died in Bean Station, TN hospital.

Farr Fitch PVT 1824 2 Jul 1863
Born in South Carolina; Appointed **CPL** in Mar 1862; Reduced in rank to wagoner April 19, 1862; Killed at **Vicksburg,** MS; Claim filed by his mother Celia Fitch.

Joseph "Joel" M. Fowler PVT 1842 23 Oct 1862
Born in South Carolina; Died of disease in Knoxville hospital; Claim filed by his father Elias Fowler; Buried at Bethel Cemetery, Knoxville, TN.

John M. Fowler PVT 7 Nov 1827 25 Oct 1911
Married Elizabeth Carmichael in Cherokee County, Dec 19, 1848; Enlisted May 21, 1862; Captured at **Vicksburg,** MS July 4, 1863; Surrendered at **Greensboro,** NC April 26, 1865; Married Pora Hubbard in Cherokee County, Aug 27, 1893; Farmer in Cherokee County after the war; Filed for Confederate pension in Cherokee County in 1910; Application endorsed by M.H. Trammell; Buried at Pine Knott Cemetery, Lebanon, Cherokee County.

Washington Fowler PVT 1841 16 May 1863
Born in South Carolina; Killed at **Battle of Baker's Creek,** MS.

Clarke Gazeway PVT 1842 6 Jun 1863
Died in Vicksburg, MS hospital.

James M. Green PVT 1833 3 Dec 1886
Detailed to Captain Calhoun's detachment in Knoxville, TN April 20, 1862 to guard political prisoners; Prison guard at Madison, GA Oct 31, 1862; Captured at **Vicksburg,** MS July 4, 1863; Wounded at **Battle of Atlanta,** July 22, 1864; Farmer in Big Creek after the war; Buried at Shady Grove Baptist, Forsyth County.

Lewis M. Harmon PVT
Detailed to Captain Calhoun's detachment in Knoxville, TN April 20, 1862 to guard political prisoners; Prison guard at Madison, GA Oct 31, 1862; Captured at **Vicksburg,** MS July 4, 1863; Admitted to Fairgrounds Hospital #1 in Atlanta Oct 2, 1863 with debility.

M.C. Harmon PVT
Wounded at **Battle of Resaca**, May 15, 1864.
Alonzo Van Gordan Harris PVT 27 Dec 1829 1 Oct 1866
Married Winnie Phillips in Cherokee County, Jan 29, 1851; Sent to Vicksburg hospital Mar 21, 1863; Furloughed at Fairground Hospital #1 in Atlanta with debility Oct 12, 1863; Dec 11, 1863 roll shows him Absent, left sick in Atlanta; Buried in Phillips Cemetery, Alpharetta.
David F. Harris PVT 1 Dec 1830 8 Oct 1873
Brother of Van Harris; Married Elizabeth Wood in Cherokee County, Sep 19, 1854; Captured at **Vicksburg,** MS July 4, 1863; Absent left sick in Milton County on Dec 11, 1863 roll; Captured as Rebel Deserter in Forsyth County, July 1864; Gave Oath of Allegiance in Louisville, KY July 18, 1864, promising to remain north of the Ohio River until the end of the war; Buried at Midway Methodist, Forsyth County.
John W. Harris PVT 1839 Dec 1869
Enlisted in Atlanta, May 13, 1862; Captured at **Battle of Baker's Creek,** MS May 16, 1863; Taken to **Camp Morton,** IN & exchanged at Fort Delaware, July 3, 1864; Oct 31, 1863 muster roll reports Harris sick in Columbus since July 1863; Died of consumption in Cumming.
Evan D. Howell PVT 1822
Born in North Carolina; Married Margaret Knox in Forsyth County, Aug 19, 1845; Enlisted in Dalton, April 5, 1864; Surrendered at **Greensboro,** NC April 26, 1865.
Edward Jackson Hughes PVT 1843 After 1900
Born in South Carolina; Detailed to Captain Calhoun's detachment in Knoxville, TN April 20, 1862 to guard political prisoners; Prison guard at Madison, GA Oct 31, 1862; Sent to Vicksburg hospital April 23, 1863; Sent to French's Division hospital in Enterprise, MS; Left sick at Sweet Water, TN Oct 27, 1863; Surrendered at **Greensboro,** NC April 26, 1865; Married Louisa Erskine in Jasper County, TX, April 25, 1867; Farmer in Ellis County, TX in 1900.
Andrew Jackson Hunter PVT 1837 After 1900
Detailed to Captain Calhoun's detachment in Knoxville, TN April 20, 1862 to guard political prisoners; Prison guard at Madison, GA Oct 31, 1862; Appointed **5SGT** in April 1863; Detached Oct 18, 1863 to Chickamauga, GA to guard prisoners; Farmer in Craighead County, AR in 1900.
William Hope Jackson PVT May 1831 After 1903
Born in Gwinnett County; Married Louisa Nix about 1852; Discharged at Cumberland Gap by Captain Paris "by reason of having all the fingers of his left hand carried away by a gun that renders him unfit for military duty."; Married Mary McDaniel in Gwinnett County, Feb 6, 1882; Farmer in Forsyth County in 1900; Filed for Confederate pension in Forsyth County in 1889 and Cherokee County in 1903.
George Washington Jameson PVT 1844 13 Aug 1863
Detailed to Captain Calhoun's detachment in Knoxville, TN April 20, 1862 to guard political prisoners; Prison guard at Madison, GA Oct 31, 1862; Captured at Washington Hospital in **Vicksburg,** MS July 4, 1863; Placed with sick on U.S. Steamer *H. Chouteau* to Mobile by way of New Orleans; Died in New Orleans, LA Aug 13, 1863; Buried at Cypress Grove, New Orleans.
Henry Edward Jones PVT Sep 1835 10 Sep 1904

Born in South Carolina; Appointed **CPL** in April 1862; Discharged with disability April 24, 1862; Enlisted in Company F in 40th Georgia in 1864; Farmer in Fulton County after the war; Buried at Casey's Hill Cemetery, Atlanta.

John Washington Jones PVT Aug 1843 7 Jul 1923

Born in South Carolina; Detailed to Captain Calhoun's detachment in Knoxville, TN April 20, 1862 to guard political prisoners; Prison guard at Madison, GA Oct 31, 1862; Appointed **3CPL** July 1, 1863; Captured at **Vicksburg**, MS July 4, 1863; Wounded in **Battle at Rocky Face**, GA Feb 25, 1864; Wound necessitated amputation of left arm; Captured in Milton County July 21, 1864; Refused to take the Oath of Allegiance and taken to Camp Chase, OH; Paroled at Boulware & Cox's Wharves on James River, Mar 10, 1865; Married Nancy Lamberth in Milton County, July 30, 1872; Farmer in Milton County after the war; Filed for Confederate pension in Gwinnett County in 1907; Buried at Winters Chapel Methodist, Doraville.

William S. Jones PVT 22 Feb 1846 13 Jan 1924

Younger brother of John & Henry Jones; Enlisted in Atlanta Dec 20, 1862; Captured at **Battle of Baker's Creek**, MS May 16, 1863; Taken to **Camp Morton**, IN and paroled at Fort Delaware, July 4, 1863; Deserted in July 1864; Took Oath of Allegiance in Louisville, KY July 25, 1864 and agreed to stay north of the Ohio River until the end of the war; Enlisted in Company A, 145th Indiana; Farmer in German, Indiana in 1880; Buried at Garland Brook Cemetery, Columbus, Bartholomew County, Indiana.

Columbus Jackson Kent PVT 12 Nov 1834 20 Mar 1863

Married Mary Chafin in Cherokee County, Jan 4, 1857; Detailed to Captain Calhoun's detachment in Knoxville, TN April 20, 1862 to guard political prisoners; Prison guard at Madison, GA Oct 31, 1862; Died at Vicksburg, MS; Buried at Soldier's Rest, Cedar Hill, Vicksburg.

Francis Hopkins Kent PVT 12 Dec 1836 28 Jan 1904

Born in Clarke County; Younger brother of Columbus Kent; Married Susan Chafin in 1856; Detailed to escort political prisoners from Knoxville to Georgia, April 1862; Captured at **Vicksburg**, MS July 4, 1864; Captured at the Chattahoochee River in July 1864; Received in Louisville, KY as a Confederate deserter; Took the Oath of Allegiance July 27, 1864 and released to stay north of the Ohio River until the end of the war; Dairy farmer in Milton County after the war; Filed for Confederate pension in Fulton County in 1901; Buried at Union Primitive Baptist, Roswell.

Isaac T. Langley PVT 1845 After 1870

Married Nancy Brooks in Gwinnett County, Oct 20, 1861; Appointed **4SGT** July 1863; Captured at **Vicksburg**, MS July 4, 1863; Present on Dec 11, 1863 muster roll; Laborer in Milton County in 1870.

Jackson Langley PVT 1817 17 May 1863

Married Mary Ann Howell in Forsyth County, Mar 24, 1842; Wounded at **Battle of Baker's Creek**, May 16, 1863; Died from wounds at Lauderdale, MS hospital.

Isaac Newton Lanier PVT 5 Jan 1843 24 Aug 1915

Born in Clarke County; Older brother of Solomon Lanier; Captured at **Vicksburg**, MS July 4, 1863; Surrendered at **Greensboro**, NC April 26, 1865; Married Martha Jane Coleman in 1869; Famer in Milton County in 1880; Laborer in Atlanta in 1900; Filed for Confederate pension in Fulton County in 1900; Application endorsed by F.H. Kent; Buried at Old Roswell Cemetery, Roswell.

Solomon Jasper Lanier PVT 1843 28 Mar 1863
Died in Jackson, MS hospital.

John Ledford Lawson PVT July 1846 9 Jun 1887
Born in South Carolina; Captured at **Battle of Nashville** Dec 16, 1864; Released at **Camp Chase,** OH June 12, 1865; Married Lois Callie Wallace about 1866; Filed for a Confederate pension in Pike County, AR in 1887; Buried at Provencal Cemetery, Natchitoches, LA.

James Wright Lee PVT 1833 4 July 1876
Born in Anderson County, SC; Married Caroline Bagwell in Forsyth County about 1855; Captured at **Vicksburg,** MS July 4, 1863; AWOL on Dec 11, 1863 muster roll; Captured in Anderson, SC May 5, 1865; Farmer in Milton County in 1870.

Allen Brantley Mann PVT 8 Dec 1844 25 Dec 1926
Younger brother of Charles and George Mann; Appears on roll of 9th Artillery Battalion; No enlistment record in the 42nd Georgia survives; Filed for Confederate pension in Newton County in 1926; Claims he was a member of Company C of the 42nd Georgia and present at the final surrender; Also reports being captured in Atlanta in July 1864 and held for 16 days until exchanged; Married Mary Jane Martin in Cherokee County, Aug 13, 1868; Farmer in Walton County in 1900; Buried at Dover Church, Dialtown, Newton County.

Charles David Mann PVT 10 Jun 1839 16 Dec 1923
Captured at **Vicksburg,** MS July 4, 1863; Filed pension application in Marion County, AL; Reported being seriously wounded in the right knee at the Battle of Resaca May 15, 1864; Married Sarah Ann Parham in Milton County, Dec 27, 1864; Farmer in Fayette County, AL in 1880; Buried at Hebron Baptist, Montgomery County, MS.

George Washington Mann PVT 1843 26 May 1864
Born in Walton County; Grandfather born in Ireland; Younger brother of Charles Mann; Enlisted with his father in Decatur Sep 16, 1863, GA; Killed at **Battle of New Hope Church**; Large memorial stone to George Mann at the Confederate Cemetery in Marietta.

William Robert Bailey Mann PVT 1821 25 Jul 1892
Born in Abbeville County, SC; Father of Allen, Charles & George Mann; Married Sarah Elizabeth Harris in Walton County, Feb 11, 1841; Enlisted Sep 16, 1863 in Decatur, GA; Wounded at **Battle of New Hope Church** May 25, 1864; Blacksmith in Marion County, AL in 1880; Married Sarah Coons in Marion County, April 19, 1888; Buried at Hackleburg Cemetery, Marion, Alabama.

James Wesley Mansell PVT 18 Feb 1844 23 Dec 1913
Sent to Vicksburg, MS hospital Mar 16, 1863; Present on Oct 31, 1863; Married Gincy Willbanks in Cherokee County, Jul 9, 1865; Farmer in Milton County after the war; Buried at Providence Baptist, Alpharetta.

Oliver Newton Mattison PVT 26 Jan 1823 22 May 1901
Known as "Olly"; Born in Anderson County, SC; Married Emily Caroline Clement in Anderson County, SC, May 7, 1840; Admitted to Fairgrounds Hospital #1 in Atlanta Nov 28, 1862; Furloughed Dec 12, 1862; Detached to Chickamauga, GA to guard prisoners, Oct 18, 1863; Admitted to Bell Hospital in Rome, GA Nov 1, 1863; Assigned as shoemaker, Mar 2, 1864; Filed for Confederate pension in Milton County in 1898; Reported being deemed unfit for service in Nov 1863 & assigned to government work; Farmer in Milton County after the war; Buried at Sharon Baptist in Cumming.

Samuel Maulden PVT 1838 25 Aug 1862
Born in South Carolina; Detailed to Captain Calhoun's detachment in Knoxville, TN April 20, 1862 to guard political prisoners; Died in Bean Station, TN hospital; Claim filed by his father, James Maulden.
George Washington McFarland PVT 14 Feb 1827 22 May 1879
Married Elizabeth Edwards in Cherokee County, Dec 23, 1853; Captured in **Battle of Nashville**, TN Dec 16, 1864; Released from **Camp Chase**, OH June 12, 1865; Farmer in Big Creek in 1870; Died of brain fever, May 22, 1879; Buried at Jasmine Ridge Cemetery, Milledgeville.
Isaac Jackson Morris PVT 7 Apr 1847 22 Oct 1914
Late enlistment; Wounded in the hand and wrist at **Battle of Atlanta** July 22, 1864; Admitted to Direction Hospital in Griffin July 24, 1864; Right arm amputated above the elbow at Ocmulgee Hospital in Macon Sep 1864; Married Martha Ann Stephens in Cobb County, Aug 9, 1875; Filed for Disability Pension in Cobb County in 1879; Application supported by A.B. Vaughn, W.R. Morris, & A.J. Barker; Farmer in Cobb County after the war; Buried at Milford Cemetery, Marietta.
William Riley Morris PVT 28 Jul 1828 Aft 1905
Born in Gwinnett County; Married Temperance Shockley about 1854; Captured at **Vicksburg**, MS July 4, 1863; AWOL on Dec 11, 1863 muster roll since Sep 18, 1863; Captured in Milton County in July 1864; Took the Oath of Allegiance in Louisville, KY July 27, 1864, agreeing to remain north of the Ohio River until after the war; Farmer near Marietta in 1900; Filed for Confederate pension in 1897 in Cobb County.
Francis Hobson Nash PVT Feb 1828 4 Nov 1911
Married Francis Huley in Madison County, Jan 18, 1849; Detached to wait on Private F.M. Dunbar in private house April 27, 1862; Detailed regimental postmaster by order of Colonel Robert Henderson, Aug 23, 1862; Captured at **Vicksburg**, MS July 4, 1863; Surrendered at **Greensboro**, NC April 26, 1865; Farmer in Waco, TX in 1880; Living in the Choctaw Nation in 1900; Buried at Antlers City Cemetery, Pushmataha, Oklahoma.
Elias M. Norton PVT Oct 1841 After 1910
Brother of William Norton; Detailed to Captain Calhoun's detachment in Knoxville, TN April 20, 1862 to guard political prisoners; Prison guard at Madison, GA Oct 31, 1862; Appointed **1CPL** July 1, 1863; Captured at **Vicksburg**, MS July 4, 1863; Captured at **Battle of Nashville** Dec 16, 1864; Released from **Camp Chase**, OH Jun 12, 1865; Married Epsey Perkins in Milton County, Feb 7, 1867; Farmer in Hickory Flat, Cherokee County, GA in 1900; Living in East Kansas City, Mississippi in 1910.
William A. Norton PVT 26 May 1840 4 Dec 1908
Born in Whitfield County; Enlisted in Decatur Sep 16, 1863; Wounded at **Battle of Atlanta** July 22, 1863; Captured at **Penn's Bridge**, SC Feb 11, 1865; Released at **Point Lookout**, MD Jun 29, 1865; Married Lodemia Hillis in Van Buren County, TN Aug 19, 1867; Married Narcissa Naler in McLennan County, TX Nov 1, 1880; Widowed farmer in Bee County, TX in 1900; Buried at Clareville Cemetery, Bee County, Texas.
William M. Oliver PVT 1835 17 Aug 1863
Captured at **Vicksburg**, MS July 4, 1863; Died of disease at Raleigh, MS hospital.
John Albert Alexander Perkins PVT 22 Mar 1838 24 Dec 1906
Married Nancy Melinda Bailey in Milton County, Dec 18, 1859. Detailed to Captain Calhoun's detachment in Knoxville, TN April 20, 1862 to guard political prisoners; Prison

guard at Madison, GA Oct 31, 1862; Appointed **1CPL** April 1, 1863; Appointed **3SGT** July 1, 1863; Captured at **Vicksburg,** MS July 4, 1863; Surrendered at **Greensboro,** NC April 26, 1865; Blacksmith in Milton County in 1870; Watchman at a steel plant in Birmingham, AL in 1900; Died in a buggy accident; Nancy filed a widow's pension application in Ellis County, TX in 1913; Brother G.W. Perkins reported seeing his brother in camp in Dalton in the spring of 1864; Buried at Italy Cemetery, Ellis County, Texas.

William Jameson Perry PVT 20 Oct 1833 2 Jun 1864
Born in Pickens County, SC; Married Martha Nix in Cherokee County, Feb 17, 1853; Farmer near Alpharetta in 1860; Enlisted in Calhoun, GA Aug 15, 1862; Detailed as nurse in Knoxville hospital Oct 8, 1862; Mortally wounded at **Battle of Resaca** May 15, 1864; Died from gunshot wound in Fairgrounds Hospital #2 in Atlanta; Buried at Oakland Cemetery, Atlanta.

Albert Phillips PVT 27 Jun 1820 15 Dec 1902
Born in Burke County, NC; Married Malinda Jackson in Cherokee County, Sep 15, 1853; Sent to Vicksburg hospital April 25, 1863; On detached service to guard prisoners at Chickamauga, GA Oct 18, 1864; Member of Masonic Order; Farmer in Milton County after the war; Buried at Phillips Cemetery, Alpharetta.

Alman Phillips PVT 4 May 1827 17 Jul 1863
Brother of Albert & Ambrose Phillips; Detailed to Captain Calhoun's detachment in Knoxville, TN April 20, 1862 to guard political prisoners; Prison guard at Madison, GA Oct 31, 1862; Sent to Vicksburg hospital Feb 14, 1863; Died in Quitman, MS hospital; Brother Andrew Phillips served as executor of his estate.

Ambrose Phillips PVT 8 Mar 1832 8 Aug 1862
Married Mary Jane Jameson in Cherokee County, Dec 23, 1856; Died of camp fever at Bean Station, TN hospital.

John R. Owens PVT 1843 10 Nov 1862
Died at home, leaving a widowed wife.

Harvey Asberry Rainey PVT 27 Dec 1839 7 Jun 1925
Discharged with disability April 12, 1862; Re-enlisted in Decatur April 15, 1863; Captured at **Vicksburg,** MS July 4, 1863; Married Sarah Lamberth in Milton County about 1863; Captured in Milton County July 1864 as a Rebel deserter; Took Oath of Allegiance in Louisville, KY July 18, 1864, agreeing to stay north of the Ohio River until the end of the war; Farmer near Roswell in 1900; Filed for Confederate pension in Cobb County in 1910; Reported being captured near Marietta on July 19, 1864 and held at Camp Chase, OH until May 1865; Milton Trammell provided testimony to the application; Buried at Pleasant Hill Cemetery, Johns Creek.

Ira Sanford Rainey PVT 27 May 1841 12 Jan 1887
Younger brother of Harvey Rainey; Born in Oglethorpe County; Married Sarah Elizabeth Callahan in Milton County, Dec 1860; Detailed to Captain Calhoun's detachment in Knoxville, TN April 20, 1862 to guard political prisoners; Prison guard at Madison, GA Oct 31, 1862; Captured at **Vicksburg,** MS July 4, 1863; Captured near the Chattahoochee River July 5, 1864; Sent to **Camp Douglas,** IL July 18, 1864; Farmer in Newtown in 1880; Buried at Mount Pisgah Methodist, Alpharetta.

Pinckney Ferron Rainwater PVT 16 Oct 1811 Apr 1865
Born in Anderson County, SC; Married Sarah Eskew about 1841; Successful merchant in Alpharetta in 1860 census; Discharged with chronic rheumatism Nov 4, 1862.

Sydney Yancey Rainwater PVT 1819 5 Mar 1899
Born in South Carolina; Married Martha McGinnis in Forsyth County, October 10, 1843; Discharged with disability at Bean Station, TN Aug 2, 1865; Farmer in Milton County after the war.

Thomas Jefferson Parks Revis PVT Jan 1830 Apr 1903
Married Nancy Deverell in Cherokee County, Feb 5, 1854; Detailed to Captain Calhoun's detachment in Knoxville, TN April 20, 1862 to guard political prisoners; Prison guard at Madison, GA Oct 31, 1862; Later enlisted in Company F, 3rd Georgia Cavalry; Farmer in Coryell County, TX in 1880; Married Elizabeth Robbins around 1887 in Texas; Buried at Bee House Cemetery, Coryell County, Texas.

Alexander Erwin Roddy PVT 21 Jul 1843 4 May 1911
Detailed to Captain Calhoun's detachment in Knoxville, TN April 20, 1862 to guard political prisoners; Prison guard at Madison, GA Oct 31, 1862; Sent to Vicksburg hospital Mar 31, 1863; Present on Oct 31, 1863 muster roll; Filed for Confederate pension in Fulton County in 1893; Reported being shot "through the right thigh and privates" at **Battle of New Hope Church** May 25, 1864; Application endorsed by W.L. Calhoun; Married Francis Wright in Milton County, Dec 29, 1870; Watchman at a cotton mill in 1900; Buried in Hollywood Cemetery, Atlanta.

Franklin Crawford Roddy PVT Aug 1845 After 1911
One of four Roddy brothers in the Milton Tigers (C). No service record; Farmer near Alpharetta in 1900; Filed for Confederate pension in Milton County in 1911; Reported enlisting at Dalton in Jan 1864; Wounded at **Battle of Bentonville**, NC Mar 19, 1865; In Charlotte, NC hospital at close of war; Never married; Farmer in Alpharetta in 1900.

James Vernon Roddy PVT 1835 6 Aug 1886
Born in South Carolina; Detailed to Captain Calhoun's detachment in Knoxville, TN April 20, 1862 to guard political prisoners; Prison guard at Madison, GA Oct 31, 1862; Sent to Vicksburg hospital Jan 28, 1863; Transferred to First MS CSA Hospital in Jackson, MS Feb 6, 1863; Surrendered at **Greensboro**, NC April 26, 1865; Married Talitha Brown in Milton County, Feb 10, 1870; Farmer in Alpharetta in 1880.

William T. Roddy PVT 1841 Jan 1909
Detailed to Captain Calhoun's detachment in Knoxville, TN April 20, 1862 to guard political prisoners; Prison guard at Madison, GA Oct 31, 1862; Captured at **Vicksburg**, MS July 4, 1863; Captured at **Battle of Nashville**, Dec 16, 1864; Took Oath of Allegiance at Camp Chase, OH Mar 16, 1865; Enlisted in the 6th Regiment U.S. Infantry; Married Martha Bryant in Milton County, Feb 28, 1880; Farmer in Cherokee County, AL in 1900; Martha filed a widow petition in Cherokee County in 1909; Reports that he was injured in the right side at the Battle of Nashville.

Simeon Scoles PVT 1832 After 1874
Captured at **Vicksburg**, MS July 4, 1863; Sick at home on Oct 31, 1863 muster roll; Farmer in Newtown after the war.

Ransom Foster Sloan PVT 15 Nov 1837 2 Apr 1907
Born in South Carolina; Married Martha Pirkle in Hall County about 1860; Enlisted in Calhoun Aug 15, 1862; Left sick in Kentucky Oct 10, 1862; Farmer and Baptist preacher in Hall County after the war; Buried at Mulberry Baptist, Hoschton, Gwinnett County.

John T. Spence PVT 1838 26 Jul 1862
Born in North Carolina; Died in Bean Station, TN hospital.

Miles Waddy Spence PVT 22 Oct 1863
Returned to duty from Meridian, MS hospital April 20, 1863; Captured at **Vicksburg**, MS July 4, 1863; Died at home.
John Henry Thomas PVT 26 Mar 1844 3 Jul 1926
Feb 28, 1863 muster roll reports absent sick in Milton County since Sept 18, 1862; Captured at **Vicksburg**, MS July 4, 1863; Later served in 1st Battalion GA Militia; Married Martha Evans in Forsyth County in 1864; Married Nancy Jane Martin in Forsyth County Jan 5, 1879; Farmer in Coal Mountain after the war; Buried at Coal Mountain Cemetery, Forsyth County.
George S. Thomason PVT 1843 2 Oct 1862
Born in South Carolina; Detailed to Captain Calhoun's detachment in Knoxville, TN April 20, 1862 to guard political prisoners; Died of disease in Danville, Kentucky; Buried at Bellevue Cemetery, Danville, Kentucky.
Dr. Gideon B. Thomason PVT 19 May 1842 4 Jul 1918
Born in South Carolina; Profiled in *Memoirs of Georgia*, 1895; Commissioned orderly **SGT**; Captured at **Vicksburg**, MS July 4, 1863; Wounded at Smithfield, NC in 1865 and in a Raleigh hospital at the time of surrender; Married Jimmie Ann McCollum in Milton County, Nov 23, 1882; Graduated from the Atlanta Medical College in 1867; Physician in Alpharetta in 1900; Owned a 140-acre farm and 30 acres in the township of Alpharetta; Dr. Thomason was a Democrat, Methodist, and Master Mason; Buried at Rest Haven Cemetery, Alpharetta.
Joseph Macklin Thomason PVT 3 Feb 1840 28 Feb 1926
Born in Greenville, SC; Brother of Gideon & George Thomason; Married Kiziah Jane Allen in South Carolina, Feb 5, 1860; Admitted to Fairgrounds Hospital #1 in Atlanta Oct 17, 1862 with rheumatism; Sent to Vicksburg, MS hospital Mar 26, 1863 with pneumonia; Transferred to 1st CSA Hospital in Jackson, MS May 23, 1863; Served as teamster in MS through October 1863; Captured in Milton County as a Rebel deserter in July 1864; Took Oath of Allegiance in Louisville, KY July 27, 1864, agreeing to remain north of the Ohio River; Owned general store in Franklin County after the war; Filed for Confederate pension in Greene County in 1910; Reported being wounded and captured at Battle of Resaca, May 15, 1864 and kept in prison in Louisville, KY until end of war; Buried at Stone Mountain Cemetery, DeKalb County.
Milton Hampton Trammell PVT 14 Jun 1836 17 Feb 1915
Captured at **Vicksburg**, MS July 4, 1863; Elected Jr. **2LT** Oct 12, 1863; Left sick at Sweetwater, TN Oct 27, 1863; Captured near Cumming, GA Sep 7, 1864; Released from **Johnson Island**, OH Jun 16, 1865; Married Mary Elizabeth Weaver in 1865; Farmer in Alpharetta after the war; Buried at Old Big Creek Cemetery, Alpharetta.
Henry William Vaughn PVT 1841 1863
Captured at **Vicksburg**, MS July 4, 1863; Oct 31, 1863 muster roll reports Left sick in Milton County since Sep 18, 1863; Died in Milton County home in 1863; Buried at Shady Grove Baptist in Forsyth County.
Robert G. Vaughn PVT 1843 27 Oct 1862
Brother of Henry Vaughn; Detailed to escort political prisoners from Knoxville to Georgia, April 1862; Died in Bean Station, TN hospital.
Sinclair Sylvanus Wallis PVT 1824 After 1867
Born in Anderson District, SC; Sent to Vicksburg hospital Jan 28, 1863; Transferred to

1st MS CSA Hospital in Jackson, MS Feb 6, 1863 with "ascites"; Sick at home on Oct 31, 1863 muster roll since Sep 18, 1863; Appears in 1864 Militia Census for Milton County as 48-year-old farmer; On 1867 Loyalty Oath list for Milton County.

Newton Webb PVT 1844 Oct 1862

Brother-in-law of Captain Paris; Detailed to escort political prisoners from Knoxville to Georgia, April 1862; Left sick in Kentucky, Oct 17, 1862; Killed by "bushwhackers" in Kentucky.

Thomas J. Webb PVT 12 Mar 1837 16 Nov 1921

Born in Forsyth County; Married Margaret Matthews about 1860; Discharged from Cumberland Gap, TN with "phthisis pulmonatis" (consumption) on Sep 1, 1862; Farmer near Alpharetta in 1900; Buried at Old Big Creek Cemetery, Alpharetta.

William Andrew Wigley PVT 1807 After 1870

Born in Oglethorpe County; Married Susannah Dennis in Coweta County, Feb 4, 1830; Detailed to escort political prisoners from Knoxville to Georgia, April 1862; Discharged with a disability from Bean Station, TN Aug 2, 1862; Farmer in Franklin County, AL in 1870.

William Thomas Wills PVT 18 May 1841 12 Aug 1918

Younger brother of Andrew Wills; Appointed **2CPL** April 1863: Captured at **Vicksburg,** MS July 4, 1863; Captured at **Battle of Resaca** May 15, 1864; Released at **Camp Morton,** IN May 23, 1865; Married Martha Perry about 1870; Farmer in Milton County after the war; Buried at Midway Methodist, Forsyth County.

Company D – Fowler Guards – 122 Men

The Fowler Guards were the 7th Confederate company raised in DeKalb County. They were named after Captain John W. Fowler, who had raised the first company in DeKalb County in 1861 named the DeKalb Light Infantry.

Name Rank Birth Death

Nathan Clay CPT 1820 After 1867

Saddled with chronic diarrhea and bronchitis, Captain Clay applied for a disability certificate in Vicksburg, MS on February 4, 1863, and his resignation was accepted on Feb 23, 1863; Signed the Reconstruction Loyalty Oath in Panthersville in 1867.

Josiah Mint Fowler 2SGT 23 Oct 1844 28 Jan 1921

Appointed **5SGT** Oct 1863; Patient in Lauderdale Springs, MS hospital April 15, 1863; Present on Dec 11, 1863 muster roll; Married Emma Brantley Dec 20, 1870; Farmer in Evans District of DeKalb County after the war; Filed for Confederate pension application in DeKalb County in 1919; Reported being transferred in July 1864 to the 2nd Kentucky Cavalry & surrendering at war's end in Washington, Wilkes County; Farmer in DeKalb County after the war; Buried at Wesley Chapel Methodist, Decatur.

Robert Hollingsworth 5SGT 1838 6 Feb 1865

Born in South Carolina; Appointed **2SGT**; Detailed to Knoxville, TN hospital April 22, 1862 "to wait on the sick;" Served in Knoxville Hospital #2 as nurse until July 31, 1862; Captured in **Vicksburg,** MS July 4, 1863; Left sick in DeKalb County, Oct 2, 1863; Died Feb 6, 1865 in Army of Tennessee hospital of pneumonia.

Thomas J. Jackson 3SGT 27 Dec 1834 29 Jun 1900

Captured at **Vicksburg**, MS July 4, 1863; Described as "forage Master" near Dalton Feb 22, 1864; Married Margaret Tuggle in DeKalb County about 1888; Farmer in Panthersville after the war; Member of Masonic Order; Buried at Wesley Chapel Methodist, Decatur.

James Henry Jordan 4CPL 4 Oct 1833 20 Jan 1910

Married Sarah Fowler in DeKalb County, Nov 29, 1860; Captured in Harrodsburg, KY Oct 11, 1862; Sent to Vicksburg, MS for exchange by way of Cairo, Il on Steamboat *City of Madison* Dec 5, 1862; Appointed **3CPL** Oct 1863; Paroled as patient from General Hospital in Thomasville, NC May 1, 1865; Filed for Confederate pension in Clarke County in 1897; Night watchman in Athens in 1900; Buried at Oconee Hill, Athens.

Franklin W. Kendrick 1CPL 1832 24 Nov 1898

Married Caroline McWilliams, sister of Galen McWilliams, in DeKalb County, July 8, 1856; Detailed to Captain Calhoun's detachment in Knoxville, TN April 20, 1862 to guard political prisoners; Prison guard at Madison, GA Oct 31, 1862; Captured at **Vicksburg**, MS July 4, 1863; Admitted to Fairgrounds Hospital #1 in Atlanta April 27, 1864; Furloughed May 14, 1864; Paid in Atlanta by quartermaster Captain B.F. Baman June 2, 1864; Farmer in Wilkes County in 1880; Buried at Raytown Methodist, Taliaferro County.

David McWilliams CPL 1819 After 1880

Born in Lawrence District, SC; Married Emily Lyle in Henry County, July 2, 1843; Discharged with chronic rheumatism at Bean Station at Bean Station, TN Aug 2, 1862; Farmer in Panthersville in 1870.

Galen James McWilliams 4SGT 11 Dec 1843 29 Jul 1879

Captured at **Vicksburg**, MS July 4, 1863; Captured at **Battle of Bentonville**, NC Mar 19, 1865; Released at **Point Lookout**, MD June 29, 1865; Died of respiratory disease at the age of 37 in Panthersville July 29, 1879; Buried at Clifton Methodist, Atlanta.

Robert J. Parker 3CPL 8 Mar 1830 1 Aug 1908

Sent to Vicksburg, MS hospital Feb 18, 1863; Attached as nurse to Jackson's Cavalry Division Hospital in Old Marion, MS June 1, 1863; Married Elizabeth Stroud in DeKalb County in 1869; Filed for Confederate pension in DeKalb County in 1887; Reports being severely wounded in right shoulder and lung by a minie ball in Dalton on May 12, 1864; Right arm permanently disabled; Farmer in Panthersville in 1880; Buried at Parker Family Cemetery, Panthersville.

Nathan Turner 1SGT 1820 28 Jan 1863

Married Arminda Phillips in DeKalb County, Feb 17, 1842; Promoted to Brevet **2LT** Sep 8, 1862; Died of sickness in Vicksburg, MS; Left a wife and two children; Buried at Soldier's Rest, Cedar Hill, Vicksburg.

George M. Thomas 2CPL Mar 1831 24 Oct 1907

Born in Anderson County, SC; Married Ann Little in DeKalb County, Dec 1, 1853; Appointed **2CPL** Oct 1, 1863; Captured at **Vicksburg**, MS July 4, 1863; Captured at **Battle of Jonesborough** Aug 31, 1864; Paroled at **Camp Douglas**, Ill and exchanged at City Point, VA Mar 14, 1865; Farmer in Decatur in 1880; Filed for Confederate pension in DeKalb County in 1897; Buried in Old Fellowship Cemetery, Tucker.

William Coker White 1LT 12 Mar 1827 13 Dec 1912

Married Sarah Ann Batey in DeKalb County, Mar 20, 1849; Resigned with chronic rheumatism at Bean Station, TN July 20, 1862; Endorsed by Acting Surgeon W.K. Hardman

& Colonel Robert Henderson; Farmer in Haralson County after the war; Buried in Pleasant Grove Methodist, Haralson County.

Marshall Jefferson Wright 2LT 1835 10 Jun 1862
Married Judah Stowers in DeKalb County in 1856; Died of measles in Knoxville, TN hospital; Left widow & three children; Buried at Bethel Cemetery, Knoxville.

James H. Young 2LT 1833 29 Dec 1862
Married Mary Moore in DeKalb County, Feb 19, 1854; Elected **1LT** Aug 28, 1862; Killed at **Battle of Chickasaw Bayou**, MS; Left widow & three children.

Charles Willis Argo PVT Apr 1836 8 Sep 1919
Married Caroline Rachael Hooks in DeKalb County, Oct 10, 1854; Captured at **Vicksburg**, MS July 4, 1863; Surrendered at **Greensboro**, NC April 26, 1865; Reports being wounded at **Battle of Resaca** on his Alabama pension application; Married Louisa Moore in Jackson County, AL, Oct 18, 1894; Farmer in Jackson County, AL in 1900.

Benjamin Franklin Armstrong PVT 1842 4 Feb 1863
Born in Alabama; Married Kezziah Minerva Turner in Fulton County, May 22, 1861; Died of smallpox in Vicksburg, MS hospital.

James Isom Armstrong PVT 17 Dec 1843 13 Jun 1916
Captured at **Vicksburg**, MS July 4, 1863; Captured near Decatur with his brothers as deserters Sep 8, 1864; Discharged after taking the Oath of Allegiance from **Camp Douglas**, Il Oct 29, 1864; Married Amanda Hannah Hardman in DeKalb County, Aug 17, 1866; Farmer in Heard County in 1900; Filed for Confederate pension in Carroll County in 1905; Buried at Lee's Chapel, Carrollton.

John Thomas Armstrong PVT 12 Jul 1845 20 Dec 1920
Younger brother of Benjamin, James, & Lawson Armstrong; Wounded at **Battle of Resaca** May 15, 1864; Furloughed from Fair Grounds Hospital #1 in Atlanta May 21, 1864; Captured near Decatur as a deserter Sep 8, 1864; Discharged after taking the Oath of Allegiance from **Camp Douglas**, Il Oct 29, 1864; Married Emily Emma Lane in 1870; Merchant in Franklin, Heard County in 1880; Filed for Confederate pension in Heard County in 1889; Reported being seriously wounded in the left-shoulder at the Battle of Resaca May 15, 1864; Buried at Franklin City Cemetery, Heard County.

Lawson Moore Armstrong PVT 10 Oct 1837 28 Dec 1901
Born in Alabama; Married Lucy Ann Turner in DeKalb County, Mar 26, 1861; Captured at **Vicksburg**, MS July 4, 1863; Left sick in DeKalb County Oct 2, 1863; Captured near Decatur with his brothers as deserters Sep 8, 1864; Farmer & freight hand in McMinn County, TN after the war; Buried at New Zion Baptist, Athens, TN.

Benton Burford Ayers PVT 1842 16 May 1864
Older brother of Russell Ayers; Captured at **Vicksburg**, MS July 4, 1863; Wounded in the neck and captured at **Battle of Resaca**; Died in Federal hospital; Buried at Resaca Confederate Cemetery, Resaca.

Russell Thomas Ayers PVT 27 Nov 1844 After 1910
Went by "Rupel"; Enlisted in Decatur in Jan 13, 1863; On hospital muster roll in Marion, MS June 30, 1863; Wounded in the left thigh at **Battle of Resaca** May 15, 1864; Furloughed from Fair Grounds Hospital #1 in Atlanta May 20, 1864; Captured in Walton County July 23, 1864; Transferred from **Camp Chase**, OH and released from Point Lookout, MD June 4, 1865; Married Martha Perkins in DeKalb County, Nov 27, 1866; Minister living in Marietta in 1900.

Robert Elijah Bishop PVT 1831 9 Nov 1862
Married Emily Marbut in DeKalb County, Feb 1, 1855; Died in Bean Station, TN hospital.
John T. Bishop PVT 1824 26 Apr 1862
Older brother of Robert Bishop; Died at home.
John B. Blackstock PVT 1822 1863
Older brother of Terry Blackstock; Married Nancy Twilley in DeKalb County, Aug 9, 1860; Teamster in quartermaster department; Captured at **Vicksburg**, MS July 4, 1863; Died in service.
Terry T. Blackstock PVT 1835 12 Feb 1865
Married Martha Thomas in DeKalb County, Feb 20, 1859; Captured at **Vicksburg**, MS July 4, 1863; Detailed as hostler for Major Orme Dec 1863; Killed at **Orangeburg**, SC.
Carlos Brown PVT 16 May 1865
Late enlistment; Captured at **Edisto River**, SC Feb 10, 1865; Died as POW at **Point Lookout**, MD.
John Fannin Brown PVT Oct 1843 15 Mar 1926
Appointed regiment wagoner April 26, 1862; Appointed musician; Surrendered at **Greensboro**, NC April 26, 1865; Married Louisa Wilson in DeKalb County, Jan 19, 1868; Farmer in Carroll County in 1900; Filed for Confederate pension in Carroll County in 1919; Application endorsed by J.P. Cagle; Buried at Mount Zion Baptist, Douglas County.
William A. Brown PVT 1842 After 1880
Captured at **Vicksburg**, MS July 4, 1863; Admitted to Fairgrounds Hospital #1 in Atlanta June 6, 1864 with debility; Transferred to Vineville, GA hospital July 7, 1864; Widow pension application filed in Randolph County in 1916 reports that her husband was paroled in Greensboro, NC in 1865; Laborer in Russell County, AL in 1880.
Albert Clay Burgess PVT 13 Jul 1842 18 Feb 1904
Born in Greene County; Married Martha Vinson in DeKalb County, Dec 22, 1859; Captured at **Vicksburg**, MS July 4, 1863; Captured in Decatur, GA Sep 20, 1864; Released at **Camp Douglas**, Il June 30, 1865; Married Elizabeth Parker in DeKalb County, July 11, 1875; Farmer in Cumming, GA in 1900; Member of UCV Camp #665 in Decatur; Filed for Confederate pension in Forsyth County in 1900; Application endorsed by H.H. Burgess; Sent to Milledgeville Sanitarium in 1902; Buried at Wesley Chapel Methodist, Decatur.
Henry Harrison Burgess PVT 25 Jan 1839 25 Apr 1903
Older brother of Albert Burgess; Married Mary Elizabeth Wright in DeKalb County in 1858; Elected **2LT** Mar 10, 1863; Captured at **Vicksburg**, MS July 4, 1863; Furloughed Aug 16, 1864; Surrendered at **Greensboro**, NC April 26, 1865; Farmer in Panthersville in 1880; Elected clerk of the DeKalb County Superior Court in 1889; Served 14 years in that position until his death; Member of UCV Camp #665 in Decatur; Buried at Wesley Chapel Methodist, Decatur.
Robert Montgomery Cagle PVT 1843 17 Feb 1863
Born in North Carolina; Died at Vicksburg, MS Feb 17, 1863; Claim filed by his father, Robert Nelson Cagle; Buried at Rock Springs Primitive Baptist, Lithonia.
James Knox Polk Cagle PVT 17 Mar 1845 17 Apr 1928
Born in North Carolina; Younger brother of Robert Cagle; Captured at **Vicksburg**, MS July 4, 1863; Surrendered at **Greensboro**, NC April 26, 1865; Married Susan L. Miller in

DeKalb County, April 17, 1866; Farmer in Douglas County in 1880; Filed for Confederate pension in Carroll County in 1905; Application endorsed by L.P. Thomas; Insurance collector in Atlanta in 1910; Buried at Antioch Methodist, Fairburn.

Oliver Perry Campbell PVT 1834 4 July 1864

Younger brother of Shelton Campbell; On detached service to guard political prisoners in Madison, GA, 1862; Captured at **Vicksburg**, MS July 4, 1863; Sick at home on Dec 11, 1863 muster roll; Killed on skirmish line at Vining Station.

Shelton Harris Campbell PVT 22 Mar 1825 6 Feb 1891

Born in Virginia; Married Elizabeth Ann Chewning in DeKalb County, Dec 27, 1849; Captured at **Vicksburg**, MS July 4, 1863; Left sick in Sweetwater, TN Oct 26, 1863; On detached service in Atlanta "battoning shoes" March thru June 1864; General merchant in Stone Mountain in 1880; Buried at Rehoboth Cemetery, Tucker.

Benjamin Franklin Cardin PVT 14 Jan 1822 Mar 1910

Married Maneroz Kelly in Henry County, Oct 9, 1851; Captured at **Vicksburg**, MS July 4, 1863; Detached service as blacksmith in Dalton Dec 1863 thru March 1864; Returned to duty from St. Mary's Hospital, Union Springs, AL Oct 1, 1864; Married Permilia Philips in Henry County, Aug 19, 1866; Blacksmith in Loves, Henry County after the war; Filed for Confederate pension in Henry County in 1900; Reported being with his command in North Carolina at time of the surrender; Reported that he was currently an inmate of Henry County; Buried at Chafin Cemetery, McDonough.

David Chewning PVT 1823 18 Nov 1862

Born in North Carolina; Died in hospital; Claim filed by his father, Ambens Chewning; Buried at Rehoboth Cemetery, Tucker.

Rev. Elijah Henry Clarke PVT 13 Dec 1835 12 Jun 1898

Married Mary Burdette in Newton County, Feb 19, 1857; Valued his land at $4,000 in 1860 census and personal property at $10,900; Detailed at Knoxville, TN April 22, 1862 to wait on the sick; Elected 2LT Sep 1, 1862; Elected **1LT** Dec 29, 1862; Elected **CPT** May 1, 1863; Captured at **Vicksburg**, MS July 4, 1863; Admitted to Fairgrounds Hospital #1 in Atlanta Dec 2, 1863; Returned to duty Jan 15, 1864; Admitted to Fairgrounds Hospital #1 in Atlanta April 2, 1864 with kidney inflammation; Returned to duty May 27, 1864; Admitted to Fairgrounds Hospital #1 in Atlanta June 14, 1864 with debility; Returned to duty Aug 9, 1864; Surrendered at **Greensboro**, NC April 26, 1865; Married Martha America White in DeKalb County, Dec 17, 1871; Farmer in Lithonia in 1880; Minister of Ebenezer Methodist Church & later Wesley Chapel Methodist; Owned land and a sawmill in DeKalb County; Served term in Georgia House of Representatives; Served as a worshipful master of the Masons Pythagoras Lodge 41 in Decatur; Buried at Wesley Chapel Methodist, Decatur.

Cleveland A. Clay PVT 9 Jan 1836 28 Apr 1909

Younger brother of Jesse Clay; Married Nancy Jones in DeKalb County, Dec 12, 1858; Enlisted in Atlanta May 13, 1862; Wounded in the hand at **Battle of Chickasaw Bayou,** MS Dec 29, 1862; Captured at **Vicksburg**, MS July 4, 1863; Surrendered at **Greensboro,** NC April 26, 1865; Farmer near Decatur after the war; Filed for Confederate pension in DeKalb County in 1906; Application endorsed by H.C. Holcomb; Buried at Clay Cemetery, Atlanta.

James W. Clay PVT 9 Aug 1836 10 Aug 1900

Married Martha Collier in DeKalb County, Feb 2, 1860; Captured at **Vicksburg**, MS July

4, 1863; No further record; Farmer in Panthersville after the war; Buried at Parker Cemetery, Panthersville.

Jesse W. Clay PVT 1829 12 May 1863
Married Susannah Brown in DeKalb County, Oct 7, 1849; Enlisted May 15, 1862; Died of fever in Vicksburg City Hospital, MS; Survived by wife and six children; Buried at Soldier's Rest, Cedar Hill, Vicksburg.

Joseph A. Clay PVT 1838 Oct 1879
Enlisted in Atlanta May 15, 1862; Captured at **Vicksburg**, MS July 4, 1863; Surrendered at **Greensboro**, NC April 26, 1865; Married Green Partin in Henry County, Dec 6, 1865; Died of "flux" in Decatur.

David Felix Clotfelter PVT 3 Aug 1843 31 Mar 1910
Born in Henry County; Wounded in the foot at **Battle of Chickasaw Bayou**, Dec 29, 1862; Captured at **Vicksburg**, MS July 4, 1863; Surrendered at **Greensboro**, NC April 26, 1865; Married Mary J. Plunkett in 1866; Farmer living on Peek Street in Conyers in 1900; Filed for Confederate pension in Rockdale County in 1902; Reported being stricken with paralysis since June 1891; Buried at Smyrna Presbyterian Campground, Conyers.

James Humphrey Cobb PVT 24 Mar 1824 After 1900
Born in South Carolina; Detailed April 20, 1862 to guard political prisoners in Knoxville, TN; Captured at **Vicksburg**, MS July 4, 1863; Present on Dec 11, 1863 muster roll; Filed for Confederate pension in DeKalb County in 1898; Reported being present at High Point, NC surrender; Farmer in South Bend, Fulton County in 1900.

Joseph Moseley Collier PVT 1836 Sep 1864
Married Nancy Humphries in DeKalb County, Oct 14, 1858; Enlisted in Decatur Aug 15, 1863; Died in September 1864; Buried at Parker Cemetery, Panthersville.

William Thomas Collier PVT 5 Jul 1829 24 Apr 1913
Older brother of Joseph Collier; Married Eliza Jane Clay, sister of Cleveland Clay, in DeKalb County, Dec 29, 1853; Enlisted at Big Spring, TN Aug 16, 1862; Captured at **Battle of Baker's Creek**, MS May 16, 1863; Sent to **Camp Morton**, IN; Paroled at Fort Delaware, DE July 3, 1863; Sick in Federal hospital until Aug 1, 1863; Exchanged & taken to Episcopal Church Hospital in Williamsburg, VA; Present on Dec 11, 1863 roll; Farmer in Panthersville after the war; Filed for Confederate pension in DeKalb County in 1903; Reported being wounded and in the hospital at time of surrender; Filed 1910 pension application in DeKalb County; Reported being captured around Atlanta in July 1864 & taken to Fort Delaware, DE & returned home in May 1865; No Federal record of second capture; Buried at Parker Cemetery, Panthersville.

Joel Jackson Crockett PVT 13 Apr 1837 12 Dec 1923
Married Sarah Jordan in DeKalb County, Dec 2, 1857; Appointed **3SGT** June 1862; Captured at **Vicksburg**, MS July 4, 1863; Became PVT in Company E, 7th GA & wounded at the **Battle of the Wilderness**, May 6, 1864; Absent without leave from Nov 1, 1864 until end of war; Farmer & landlord in DeKalb County after the war; Buried at Wesley Chapel Methodist, Decatur.

James Wigmore Darby PVT 1840 25 Feb 1864
Married Georgia Vinson in Walton County, Oct 24, 1858; Served with brothers-in-laws Henry Clay Holcombe & Albert Burgess; Detailed to Captain Calhoun's detachment in Knoxville, TN April 20, 1862 to guard political prisoners; Prison guard at Madison, GA Oct 31, 1862; Captured at **Vicksburg**, MS July 4, 1863; Present on Dec 11, 1863 muster

roll; Killed at **Battle of Rocky Face**, GA; Buried in Confederate Section at West Hill Cemetery in Dalton, GA; Memorial stone at Vinson Farm Cemetery, DeKalb County.

Samuel T. Eades PVT 1842 20 Jul 1863

Captured at **Vicksburg**, MS July 4, 1863; Died in Vicksburg hospital.

William Ellison PVT Jan 1864

Captured at **Vicksburg**, MS July 4, 1863; Present on Dec 11, 1863 muster roll; Died at Newsome Hospital, Cassville, GA prior to April 7, 1864.

William C. Fargason PVT 1834 1863

Detailed to Captain Calhoun's detachment in Knoxville, TN April 20, 1862 to guard political prisoners; Prison guard at Madison, GA Oct 31, 1862; Admitted to CSA Post Hospital in Dalton Nov 21, 1862 with acute diarrhea; Furloughed for 30-days on Dec 10, 1862; Captured at **Vicksburg**, MS July 4, 1863; Left sick in DeKalb County Oct 2, 1863; Died at home.

Ezekiel F. Farris PVT 1825 1874

Born in South Carolina; Detailed to Captain Calhoun's detachment in Knoxville, TN April 20, 1862 to guard political prisoners; Prison guard at Madison, GA Oct 31, 1862; Captured at **Vicksburg**, MS July 4, 1863; Surrendered at **Greensboro**, NC April 26, 1865; In 1870, he was living in Panthersville and working as a rug cutter in a paper mill; Buried at Cool Springs Cemetery, DeKalb County.

Bennett T. Ford PVT 14 Feb 1823 7 Jun 1894

Married Mary Woodall in Henry County, Dec 16, 1847; Captured at **Vicksburg**, July 4, 1863; Left sick at Mossy Creek, TN Oct 20, 1863; Farmer in Panthersville after the war; Buried at Ford Cemetery, Stockbridge.

James T.W. Green PVT 20 Mar 1834 27 May 1899

Married Lucinda McClain in DeKalb County, Dec 9, 1858; Captured at **Vicksburg**, MS July 4, 1863; Captured near Decatur as a Rebel deserter Sep 8, 1863; Released at **Camp Douglas**, June 17, 1865; Farmer in Panthersville after the war; Member of the Masonic Order; Buried at Clifton Methodist, Atlanta.

Andrew Calvin Hamby PVT 1833 12 Feb 1884

Born in Morgan County; Married Nancy Eleanor McDuffee in DeKalb County, Oct 1, 1854; Admitted to Fairgrounds Hospital #1 in Atlanta Nov 20, 1862; Returned to duty Dec 26, 1862; Discharged from Fairgrounds Hospital, #1 in Atlanta, Oct 9, 1863, with "Irritatio Spinalis;" Carpenter in Atlanta after the war; Buried at Oakland Cemetery, Atlanta.

Elbert Franklin Harris PVT 4 Mar 1834 1 Sep 1899

Married Mahulda Toney in Fulton County, May 3, 1860; Captured at **Vicksburg**, MS July 4, 1863; Farmer in Panthersville in 1870; Buried at Sylvester Cemetery, Atlanta.

Cyrus Vance Henry PVT 6 Jul 1836 16 Dec 1915

Enlisted in Atlanta May 15, 1862; Appointed **CPL**; Nurse in City Hospital, Vicksburg, MS Feb thru May 1863; Captured at **Vicksburg**, MS July 4, 1863; Appointed **4SGT** Oct 1, 1863; Wounded at **Battle of New Hope Church**, GA May 25, 1864; Admitted to Fairgrounds Hospital #1 in Atlanta May 27, 1864 with gunshot wound; furloughed June 25, 1864; Wounded at the **Battle of Wise's Fork**, NC Mar 10, 1865; Finger on left hand amputated at Pettigrew Hospital #13, Raleigh, NC Mar 14, 1865; Married Henrietta Duncan in Gwinnett County, Feb 2, 1866; Farmer in Friendship, Hall County in 1900; Filed for Confederate pension in Hall County in 1899; Buried at Friendship Baptist, Buford.

Flavius J. Henry PVT 1841 31 Mar 1863
Younger brother of Cyrus Henry; Died of camp fever in Vicksburg, MS hospital; Claim filed by his father Joseph Tanner Henry of Stone Mountain, GA; Buried at Soldier's Rest, Cedar Hill, Vicksburg.
William W. Higginbotham PVT 1839 After 1896
Irish parents; Discharged with disability from Breckinridge's Division Hospital in Marion, AL Sep 18, 1863; Carpenter in Atlanta after the war.
Henry Clay Holcomb PVT 16 Dec 1842 22 Apr 1911
Detailed to Captain Calhoun's detachment in Knoxville, TN April 20, 1862 to guard political prisoners; Prison guard at Madison, GA Oct 31, 1862; Captured at **Vicksburg,** MS July 4, 1863; Surrendered at **Greensboro,** NC April 26, 1865; Married Millie Ann Vinson in DeKalb County, Dec 6, 1866; Farmer in Panthersville after the war; Filed for Confederate pension in DeKalb County in 1906; Application endorsed by Cleveland Clay; Buried at Indian Creek Cemetery, Scottsdale.
Jesse Marion Huey PVT 1838 30 Oct 1862
Married Mary Ann Clay in DeKalb County, Oct 20, 1859; Died in Tennessee hospital.
Daniel Evans Jackson PVT 13 Nov 1837 5 Feb 1891
Younger brother of James and John Jackson; Enlisted as 3SGT in Company E, 7th Georgia in Decatur May 29, 1861; Transferred to 42nd Georgia Oct 10, 1862; Captured at **Vicksburg,** MS July 4, 1863; Surrendered at **Greensboro,** NC April 26, 1865; Married Buena Florida Fowler in DeKalb County about 1867; Farmer near Decatur in 1880; Named his first son Lovick Pierce Jackson; Buried at Wesley Chapel Methodist, Decatur.
James W. Jackson PVT 14 Oct 1829 31 Jan 1906
Born in Greene County; Enlisted in Decatur Feb 27, 1863; Captured at **Vicksburg,** MS July 4, 1863; Transferred to Company E of the 66th Georgia, April 12, 1864; Farmer in Panthersville in 1870; Died in Atlanta's Confederate Soldier Home; Buried at Confederate Mound, Westview Cemetery, Atlanta.
John A.M. Jackson PVT 16 Sep 1832 27 Dec 1887
Captured at **Vicksburg,** MS July 4, 1863; Captured in Decatur Sep 8, 1864; Discharged from **Camp Douglas** after taking the Oath of Allegiance April 1, 1865; Wife Easter died in 1878; Farmer in Panthersville in 1880; Buried at Wesley Chapel Methodist, Atlanta.
Joseph "Joe" G. Johns PVT 5 Feb 1837 14 Dec 1903
Captured at **Vicksburg,** MS July 4, 1863; Among the sick shipped from Vicksburg on the Steamboat *St. Maurice* to Mobile by way of New Orleans, July 30, 1863; Surrendered at **Greensboro,** NC April 26, 1865; Married Mary Cooper in DeKalb County, Dec 7, 1865; Farmer in Cross Keys in 1880; Buried in Rehoboth Cemetery, Tucker.
James Green Berry Jones PVT 2 Aug 1821 15 Aug 1863
Married Elizabeth Hambrick in DeKalb County, Sep 5, 1850; Detailed to Captain Calhoun's detachment in Knoxville, TN April 20, 1862 to guard political prisoners; Prison guard at Madison, GA Oct 31, 1862; Captured at **Vicksburg,** MS July 4, 1863; Died in DeKalb County; Left a wife and six children; Buried at Indian Creek Cemetery, Scottsdale.
James Kelley PVT 1831 14 Mar 1863
Died at Vicksburg, MS hospital; Claim filed by his father Thomas Kelley from South River.
James Henry Kelley PVT 22 Sep 1830 30 Dec 1908
Older brother of Robert Kelley; Married Jane Hildebrand in DeKalb County, Nov 3, 1853;

Captured at **Vicksburg**, MS July 4, 1863; Among the sick shipped from Vicksburg on Steamboat *St. Maurice* for Mobile, by way of New Orleans, July 30, 1863; Left in DeKalb County, Oct 2, 1863; Wounded in the right foot at Binnicker's Bridge, SC Feb 29, 1865; Paroled from General Hospital #11 in Charlotte, NC May 7, 1865; Farmer in Panthersville after the war; Filed for Confederate Pension in DeKalb County in 1890; Buried at Clifton Methodist, Atlanta.

Robert Jackson Kelley PVT 17 Feb 1836 30 Dec 1908
Married Mary Talitha Stegar in Henry County, Dec 15, 1856; Captured at **Vicksburg**, MS July 4, 1863; Admitted to Fairgrounds Hospital #1 in Atlanta Dec 1, 1863; Returned to duty Jan 21, 1864; Teamster near Dalton, GA February 1864; Wounded through the right knee at **Battle of Bentonville**, NC Mar 19, 1865; Paroled at General Hospital, Thomasville, NC May 1, 1865; Farmer in Panthersville after the war; Buried at Oak Grove Baptist, Decatur.

Adiel Sherwood Lacy PVT 25 May 1835 14 Mar 1916
Married Cynthia Ann Moore in Taliaferro County, Jun 19, 1859; Captured at **Vicksburg**, MS July 4, 1863; Surrendered at **Greensboro**, NC April 26, 1865; Farmer in Panthersville after the war; Applied for Confederate pension in DeKalb County in 1911; Buried at South Bend Methodist, Atlanta.

Andrew Jackson Lee PVT 1833 1896
Born in South Carolina; Married Mary Elizabeth Morris in Campbell County about 1855, Enlisted in Atlanta May 15, 1862; Confined & court-martialed June 29, 1862; Teamster near Reedyville, TN Nov 1862; Captured at **Vicksburg**, MS July 4, 1863; Farmer in Campbell County in 1880; Buried at Red Oak Cemetery, Fulton County.

George Washington Lyon PVT 12 Aug 1838 8 Apr 1903
Married Nancy Williams in DeKalb County, Nov 1, 1860; Detailed as wagoner in Knoxville April 18, 1862; Transferred to Company H; Appointed regimental bandmaster; Captured at **Vicksburg**, MS July 4, 1863; Detailed Oct 16, 1863 to build a bridge; Surrendered at **Greensboro**, NC April 26, 1865; Music teacher in Atlanta after the war; Author of more than 15 sacred song books; Filed for Confederate pension in Fulton County in 1895; Application endorsed by W.H. Williams; Buried in Hollywood Cemetery, Atlanta.

Blanton W.G. McCanless PVT 3 Apr 1837 20 Mar 1917
Captured at **Vicksburg**, MS July 4, 1863; Appointed **4CPL** Oct 1863; 1CPL; Surrendered at **Greensboro**, NC April 26, 1865; Never married; Farmer & laborer in Cherokee County after the war; Member of UCV Camp # 1088 in Canton; Filed for Confederate pension in Cherokee County in 1910; Buried at Dogwood Hills, Waleska, Cherokee County.

William H. McClain PVT 3 May 1839 15 Feb 1888
Detailed to Captain Calhoun's detachment in Knoxville, TN April 20, 1862 to guard political prisoners; Prison guard at Madison, GA Oct 31, 1862; Serving with Quartermaster Department in Atlanta spring through fall 1863; Stock hand in Dalton Sep 1863; Married Laura Langford in Gwinnett County, Dec 5, 1872; Farmer in Gwinnett County in 1880; Buried at Harmony Grove Methodist in Lilburn.

Angus McLeod PVT 26 Dec 1826 18 Aug 1900
Married Martha Ann Ozmer in DeKalb County, Dec 27, 1855; Wounded in the forehead at **Battle of Chickasaw Bayou**, MS Dec 29, 1862; Captured at **Vicksburg**, MS July 4, 1863; Surrendered at **Greensboro**, NC April 26, 1865; Farm manager in Chattooga County in 1900; Buried in Lyerly City Cemetery, Chattooga County.

James L. McWilliams PVT 1845 After 1919
Brother of John W. McWilliams; Captured at **Vicksburg**, MS July, 4, 1863; In St, Mary's Hospital in West Point, MS with gunshot wound, Jan 9, 1865; Filed for Confederate pension in Campbell County in 1889; Reported being struck in the right shoulder by a bomb shell & breaking his collarbone at **Battle of Resaca**, May 15, 1864; Also suffered a gunshot wound to the left hand, "rendering the said hand useless to the extent that it is not worth to the applicant more than one half of its former value;" Former 2LT H.Y. Snow endorsed his application; Married in 1876; Farmer in Campbell County after the war.
James M. McWilliams PVT 9 May 1841 27 Jul 1868
Brother of Galen McWilliams; Enlisted in Company E, 7th GA State Infantry May 29, 1861; Left regiment without permission to enlist in an Arkansas Regiment July 21, 1861; Joined Company F, 19th Arkansas in Eldorado, AR Mar 1, 1862; Attached to Company D, 42nd Georgia Nov 1, 1863; Surrendered at **Greensboro**, NC April 26, 1865; Buried at Clifton Methodist, Atlanta.
John William McWilliams PVT 29 Aug 1843 10 Mar 1934
Brother of James L. McWilliams; Captured at **Vicksburg**, MS July 4, 1863; Admitted to Fairgrounds Hospital #2 in Atlanta May 24, 1864; Married Josephine Margaret Brown in DeKalb County, Oct 8, 1868; Successful grocery merchant in Edgewood District of DeKalb County; Operated a store and post office at the corner of Glenwood and Flat Shoals Road; Member of Masonic Order; Member of UCV Camp # 665 Decatur; Filed for Confederate pension in DeKalb County in 1919; Reported being seriously wounded in the retreat from Tennessee, December 1864 & being discharged somewhere near Corinth, MS; At home at time of surrender; Buried at Sylvester Cemetery in Atlanta.
Samuel J. McWilliams PVT 1838 Before 1867
Brother of Galen & James M. McWilliams; Married Harriet Wright in DeKalb County, Sep 4, 1860; Discharged by examining physician, April 12, 1862.
James B. Merritt PVT Oct 1840 19 Apr 1906
Born in Lincoln County; Discharged with tuberculosis from Jackson, MS hospital Mar 2, 1863; Married Margaret Bridwell in Fulton County, Jul 27, 1876; Retail grocer in Atlanta in 1880; Dairy Depot Manager for railroad in 1900; Filed for Confederate pension in Fulton County in 1902; Application endorsed by B.G. Mitchell; Buried at Oakland Cemetery, Atlanta.
James Franklin Miles PVT 1827 1892
Born in Tennessee; Married Margaret Bankston in Butts County, Jul 8, 1847; Detailed as carpenter at Charleston, TN Oct 24, 1863; Surrendered at **Greensboro**, NC April 26, 1865; Farmer in Cross Keys in 1880; Estate was probated in the Fulton County Ordinary Court of Lowndes Calhoun, July 4, 1892.
William A. Miller PVT 1826 After 1890
Born in South Carolina; Left sick in Vicksburg, MS Mar 1, 1863; Sent to Meridian, MS hospital; Farmer in Lithonia in 1880.
A.M. Mitchel PVT 4 Jan 1840 7 Dec 1913
Enlisted in Decatur in Nov 22, 1862; Present on Dec 11, 1863 muster roll; Buried in Rehoboth Cemetery, Tucker.
Benjamin Green Mitchell PVT 23 Nov 1845 1 Jan 1903
Captured at **Vicksburg**, MS July 4, 1863; Captured at **Battle of Nashville**, TN, Dec 16, 1864; Released from **Camp Chase**, OH Jun 12, 1865; Married Martha Johnson in DeKalb

County, Mar 25, 1859; Farmer in Lithonia in 1880; Grocer in Atlanta in 1900; Filed for Confederate pension in Fulton County in 1902; Application endorsed by J.B. Merritt; Buried at Wesley Chapel Methodist, Decatur.

Daniel Moore PVT 31 Dec 1839 14 Mar 1927

Born in Hall County; Married Martha Sheppard in DeKalb County, 6 Sep 1860; Enlisted in Decatur April 7, 1863; Captured at **Vicksburg**, MS July 4, 1863; Came into Federal lines as a deserter near Decatur Sep 26, 1864; Transferred from **Camp Douglas**, IL Mar 14, 1865 and paroled from Pt. Lookout, MD; Admitted to Jackson Hospital, Richmond, VA Mar 22, 1865; Farmer in Hog Mountain in 1880; Farmer in Muscadine, Cleburne County, AL in 1900; Appears in 1921 Confederate Census for Alabama living in Muscadine; Buried at Mountain View Baptist in Stone Mountain.

George W. Moore PVT 24 Oct 1834 23 Jul 1862

Married Susan Sheppard in DeKalb County, Dec 10, 1854; Died in Bean Station, TN hospital; Left wife and 5 children.

James Moore PVT 25 Dec 1829 25 Aug 1908

Born in Ireland; Married Jane Gallop Moore in DeKalb County, Dec 25, 1851; Appointed **1SGT** Sep 8, 1862; Captured at **Vicksburg**, MS July 4, 1863; Elected **1LT** Jan 7, 1864; Furloughed Aug 16, 1864 near Lovejoy Station; Appointed **CPT** late in the war; Surrendered at **Greensboro**, NC April 26, 1865; Farmer in Panthersville in 1880; Farmer in East Feliciana in 1900; Buried in Sewell Cemetery, Jackson, East Feliciana Parish, LA.

John Moore PVT 27 May 1837 Oct 1880

Brother-in-law to James Moore; Captured at **Vicksburg**, MS July 4, 1863; Married Clarissa Brown in DeKalb County, Sep 6, 1863; Farmer in Panthersville in 1880; Died in DeKalb County.

Alexander L. Morgan PVT 22 July 1863

Appointed **CPL**; Captured at **Vicksburg**, MS July 4, 1863; Died at Brandon Hospital in Rankin, MS

DeWitt Clinton Morgan PVT 1838 9 Aug 1863

Student in Decatur in 1860; Previously served as 2SGT in Company E, 7th GA; Discharged with disability at Camp Bartow, VA; Enlisted in 42nd GA; Appears last on roll Apr 30, 1862; Buried at Confederate Rest Cemetery, Point Clear, Baldwin County, AL; Marker says he was a member of Van Der Corput's Light Artillery.

Richard M. Morris PVT 9 Nov 1834 27 Jan 1928

Married Mary Park in Henry County in 1858; Enlisted at Camp McDonald Mar 22, 1862; Captured at **Vicksburg**, MS July 4, 1863; Elected **2LT** Jan 13, 1864; Placed on furlough by Medical Board near Atlanta Aug 19, 1864; Surrendered at **Greensboro**, NC April 26, 1865; Married Julia Cook in Henry County, Oct 5, 1865; Farmer in Stone Mountain in 1870; Married Caroline Mays in Henry County, Mar 18, 1875; Farmer in Anderson County, Texas; Filed for Confederate pension application in Palestine, TX in 1914; Buried in Little River Cemetery, Jones Prairie, Milam County, TX.

Joel Green New PVT Oct 1935 16 Dec 1923

Married Jane Cromwell in DeKalb County, Dec 15, 1851; Left sick in Knoxville, TN hospital April 23, 1862; Served in Company B, 4th U.S. Volunteer Regiment; Farmer in Woodstock, Cherokee County in 1900; Died in Acworth, GA.

Jarrett M. New PVT 1821 14 Jun 1864

Captured at **Edwards Station**, MS May 16, 1863; Paroled at Demopolis, AL June 5, 1863;

Left sick in DeKalb County, Oct 2, 1863; Captured in Tilton, GA May 14, 1864; Died of chronic diarrhea in **Camp Morton,** IN; Grave #910 at Green Lawn Cemetery, Crown Hill, Camp Morton, IN.

Samuel C. New PVT 1845 1864

Son of Jarrett New; Enlisted in Gwinnett County May 13, 1862; Captured at **Battle of Baker's Creek,** MS May 16, 1863; Paroled at Demopolis, AL Jun 5, 1863; Admitted to Fairgrounds Hospital #2 in Atlanta with bronchitis Dec 1, 1863; Present on Dec 11, 1863 roll; Believed to have been killed in service in Georgia.

George Smith Ozmer PVT 16 Dec 1843 19 May 1886

Younger brother of John Ozmer; Captured at **Vicksburg,** MS July 4, 1863; Surrendered at **Greensboro,** NC April 26, 1865; Married Ella Gunby in DeKalb County, Jan 13, 1870; Carpenter in Cherokee County in 1880; Buried at Wesley Chapel United Methodist, Decatur.

John Windsor Ozmer PVT 17 Nov 1836 28 Aug 1915

Married Elizabeth Mitchell in DeKalb County, Jan 3, 1856; Enlisted in Richmond, VA as private in Company E, 7th Georgia July 26, 1862; Promoted to 2SGT; Transferred to 42nd Georgia Nov 1863; Surrendered at **Greensboro,** NC April 26, 1865; Farmer in Lithonia in 1900; Filed for Confederate pension in DeKalb County in 1910; Application endorsed by H.C. Holcomb; Buried at Redan Community Cemetery, DeKalb County.

William Henry Pace PVT 7 Aug 1844 20 Sep 1916

Enlisted in Decatur Aug 10, 1863; Wounded at **Battle of Jonesborough,** Aug 31, 1864; Carpenter in Toccoa in 1900; Filed for Confederate pension in Stephens County in 1908; Reported being present at the Greensboro surrender; Buried at Toccoa City Cemetery, Stephens County.

Moses Richardson PVT 8 Jan 1830 15 Mar 1905

Married Mary Carroll in 1856; Physician in DeKalb County in 1860 census; Elected CPT of Company H, 16th Regiment GA Infantry, Aug 11, 1861; Unit mustered out one month later & Moses enlisted in the Fowler Guards (D), Mar 4, 1862; Serving as **Asst. Surgeon** of the 42nd Georgia in March 1864; Physician in Pinckneyville in 1880; Buried at Norcross City Cemetery.

James M. Rutledge PVT 7 Dec 1837 13 Feb 1865

Captured in Kentucky; Taken by steamboat from Louisville and exchanged in Vicksburg, MS on Nov 29, 1862; Captured at **Vicksburg,** MS July 4, 1863; Wounded at **Battle of Resaca,** GA May 15, 1864; Admitted to Fairgrounds Hospital #1 in Atlanta May 19, 1864 with gunshot wound; Deserted hospital June 4, 1864; Captured at **Battle of Nashville,** Dec 16, 1864; Died of pneumonia at **Camp Chase,** Ohio Feb 13, 1865; Buried in grave # 1217 in Camp Chase.

William H. Scarborough PVT 1834 Jul 1862

Born in Madison County; Married Nancy Ayers in Madison County, Jan 30, 1855; Died in TN hospital; Buried at Bethel Cemetery, Knoxville, TN.

David Crockett Sheppard PVT 15 Jun 1844 2 Jul 1917

Born in South Carolina; Enlisted in Bean Station, TN July 2, 1862; Captured at **Vicksburg,** MS July 4, 1863; Received at Atlanta distribution hospital May 18, 1864; Surrendered at **Greensboro,** NC April 26, 1865; Married Narcissa Smith in DeKalb County, Mar 7, 1867; Farmer in Evans after the war; Member of UCV Camp # 665 in Decatur; Buried at Mountain View Baptist in Stone Mountain.

John T. Sheppard PVT 1839 30 Jan 1865
Born in Greenville County, SC; Older brother of David and William Sheppard; Detailed to Captain Calhoun's detachment in Knoxville, TN April 20, 1862 to guard political prisoners; Captured at **Vicksburg,** MS July 4, 1863; Sick at home on Dec 11, 1863 muster roll; Deserted regiment in McDonough Sep 8, 1864; Captured as a deserter in Decatur, GA Sep 26, 1864; Received at **Camp Douglas,** Il Nov 1, 1864; Died as POW of pleurisy; Buried at Confederate Mound, Oak Woods Cemetery, Chicago.
William W. Sheppard PVT 1849 20 Jan 1865
Born in Greenville County, SC; Deserted regiment in McDonough Sep 8, 1864; Captured with his brother John as deserters in Decatur, Sep 26, 1864; Received at **Camp Douglas,** Il Nov 1, 1864; Died as POW of dysentery; Buried at Confederate Mound, Oak Woods Cemetery, Chicago.
Thomas E. Simpson PVT 1837 About 1878
Married Sarah Rowland in Coweta County, Sep 5, 1859; Present on Dec 11, 1863 roll; Vaccinated at St. Mary's Hospital in LaGrange, May 30, 1864; Farmer in Coweta County after the war.
William S. Spain PVT 1835 17 Dec 1864
Born in Virginia; Captured at **Vicksburg,** MS July 4, 1863; Captured as Deserter in Gwinnett County at South River, Sep 8, 1864; Received at **Camp Douglas,** Il Nov 1, 1864; Died as POW of smallpox; Buried at Confederate Mound, Oak Woods Cemetery, Chicago.
Rev. Martin P. Talton PVT 24 Oct 1825 12 Dec 1886
Born in Anson County, NC: Married Amanda Ford in DeKalb County, Sep 11, 1842; Late enlistment; Present on Dec 11, 1863 muster roll; Store clerk in Cross Keys in 1880; Baptist preacher; Buried at Rehoboth Cemetery, Tucker.
Sterling Tucker Tanner PVT 8 Apr 1844 18 Aug 1916
Went by "Tuck"; Captured at **Vicksburg,** MS July 4, 1863; Dec 11, 1863 roll reports Left sick in DeKalb County; Married Margaret "Willie" Whatley in DeKalb County, Mar 31, 1870; Carpenter in Atlanta in 1880; Farmer in Cullman County in 1900; Buried at Holly Pond Cemetery, Cullman, AL.
A.L. Thomas PVT
Captured at Vicksburg, MS July 4, 1863; No further record.
Benjamin T. Thomas PVT 9 Jul 1823 After 1880
Born in South Carolina; Enlisted in Decatur, Nov 22, 1862; Captured at **Vicksburg,** MS July 4, 1863; Wounded **Battle of Resaca,** GA May 15, 1864; Admitted to Fairgrounds Hospital #1 in Atlanta May 19, 1864 with contused wound; Surrendered at **Greensboro,** NC April 26, 1865; Farmer in DeKalb County in 1880.
John Hayden Thomas PVT 25 Jul 1818 15 Aug 1863
Born in South Carolina; Older brother of Benjamin & William Thomas; Captured at **Vicksburg,** MS July 4, 1863; Died in DeKalb County; Claim filed by his daughter Mary Jane Thomas.
William B. Thomas PVT 1826 30 Jan 1863
Born in South Carolina; Died in Jackson, MS hospital; Claim filed by his mother Martha Lester Thomas.
Lewis N. Thornton PVT 1827 1 Nov 1862
Married Sarah Ann Thomason in Jasper County, Jan 9, 1851; Died in TN; Left wife and 4 sons.

Elijah F. Turner PVT 1844 7 Mar 1865
Captured at **Vicksburg,** MS July 4, 1863; Captured as a deserter near Decatur, GA Sep 8, 1864; Received at **Camp Douglas,** Il Nov 1, 1864; Died as POW of ascites; Buried at Confederate Mound, Oak Woods Cemetery, Chicago.
Thomas C. Turner PVT 1826 13 Dec 1887
Married Nancy Collier in DeKalb County, Sep 10, 1846; Enlisted in Normandy, TN Nov 27, 1862; Captured at **Vicksburg,** MS July 4, 1863; Captured as a deserter near Decatur, GA Sep 8, 1864; Received at **Camp Douglas,** IL Nov 1, 1864; Discharged May 18, 1865; Worked in a paper mill in Panthersville after the war; Buried at Parker Cemetery, Panthersville.
Lewis Hiram Tweedell PVT 1842 8 Aug 1864
Left sick in Knoxville, TN hospital April 23, 1862; Captured at **Vicksburg,** MS July 4, 1863; Died in Foard Hospital, Forsyth, GA from wounds received in battle; Buried at Forsyth City Cemetery, Monroe County.
James Washington Twilley PVT 1835 After 1886
Enlisted in Atlanta, May 15, 1862; Captured at **Vicksburg,** MS July 4, 1863; Surrendered at **Greensboro,** NC April 26, 1865; Married Margaret Wheeler in DeKalb County, Oct 31, 1867; Farmer in Panthersville in 1880.
James J. Vaughn PVT 1835 19 Jan 1863
Born in South Carolina; Died of sickness in Vicksburg, MS; Buried at Soldier's Rest, Cedar Hill Cemetery, Vicksburg; Claim filed by his father, Griffin Vaughn.
Robert W. Waldrop PVT 1844 After 1880
Left at Montgomery, AL hospital Dec 23, 1862; Captured at **Battle of Nashville,** Dec 16, 1864; Received at **Camp Chase,** OH Jan 6, 1865; Released from Camp Chase June 12, 1865; Married Missouri Hooks about 1868; Farmer in Panthersville in 1880.
Jesse J. Warren PVT 1842 Before 1867
Older brother of John Warren; Given Surgeon certificate April 23, 1862.
John H. Warren PVT 1844 May 1864
Captured at **Vicksburg,** MS July 4, 1863; Severely wounded with gunshot wound to the thigh at **Battle of Resaca,** May 15, 1864; Leg amputated at upper third by Federal surgeon; Died shortly thereafter.
Thomas C. Webb PVT 1837 25 May 1864
Born in Anson County, NC; Older brother of William Webb; Married Emily Turner in DeKalb County, 19 July 1859; Captured at **Vicksburg,** MS July 4, 1863; Killed at **Battle of New Hope Church,** GA.
William Henry Webb PVT 9 Mar 1841 23 Oct 1912
Married Margaret Crowley in DeKalb County, Feb 21, 1861; Absent on sick leave April 11, 1862; No further record; Ran boarding house in Atlanta in 1900; Filed for Confederate pension in Fulton County in 1912; Reported serving in 42nd until receiving a medical discharge in Aug 1862; Served in the GA Militia from May 1863 to March 1865; Buried at Ousley Chapel Methodist, Decatur.
Nicholas Waters White PVT 22 Apr 1844 4 Apr 1908
Step-brother of Lt. William White; Captured at **Vicksburg,** MS July 4, 1863; Surrendered at **Greensboro,** NC April 26, 1865; Married Martha Ella in DeKalb County in 1873; Farmer in Rockmart in 1900; Buried at Rose Hill Cemetery, Rockmart.

Company E – Harper Guards – Newton County – 131 Men

Name Rank Birth Death

Thomas Jefferson Mercer CPT 13 May 1826 12 Oct 1886

Married Frances Caroline Livingston in 1848; At home on sick furlough March & April 1863; Admitted to Fairgrounds Hospital #1 in Atlanta Oct 2, 1863; Returned to duty Jan 5, 1864; Admitted to Fairgrounds Hospital #1 in Atlanta June 5, 1864; Furloughed June 30, 1864; Admitted Charlotte, NC hospital #11 Feb 24, 1865; Mill wright in Decatur in 1880; Buried at County Line Baptist, Covington.

John Almand 2LT 1833 18 Dec 1862

Elected 2LT Aug 15, 1862; Died in Conyers; Buried at Almand Cemetery, Conyers; Claim filed by his mother, Mary Peek.

Thomas Alexander Boyd 1CPL 25 Jul 1841 2 Jan 1929

Appointed color-bearer; Captured at **Vicksburg,** MS July 4, 1863; Surrendered at **Greensboro,** NC April 26, 1865; Married Mary Isabel Christian in Newton County, Sep 26, 1865; Farmer in Newton County in 1900; Filed for Confederate pension in Newton County in 1910; Reported he was on detail guarding a bridge in Danville, VA at time of surrender; Member of UCV Camp #305 in Covington in 1915; Buried at Hopewell Presbyterian, Newton County.

Drury Jefferson Brewer 5SGT 1840 After 1910

Captured at **Vicksburg,** MS July 4, 1863; AWOL on Dec 10, 1863 muster roll; Court martial trial in Dalton March 29, 1864; Death sentence remitted by army commander; Admitted to Madison, GA hospital May 18, 1864; Deserted near Cave Springs, GA; Taken into Federal custody near Decatur, AL Oct 31, 1864; Received at **Camp Douglas,** IL Nov 12, 1864; Released Mar 26, 1865; Enlisted in Company F, 6th US Infantry; Living in the Cherokee Nation, Indian Territory in 1900.

James Pinckney Cowan 3SGT 1840 18 Aug 1863

Born in North Carolina; Captured at **Vicksburg,** MS July 4, 1863; Died at home in Newton County; Claim filed by his father Alexander Cowan.

James J. Elliott 4CPL 1838 19 Mar 1865

Captured at **Vicksburg,** MS July 4, 1863; AWOL on Oct 31, 1863 muster roll; Killed at **Battle of Bentonville,** NC.

Joseph Heard 1LT 6 Dec 1839 11 Nov 1909

Married Cordelia Penick in Morgan County, Jan 17, 1861; Received 30-day medical leave in January 1863; Captured at **Vicksburg,** MS July 4, 1863; Surrendered at **Greensboro,** NC April 26, 1865; Farmer in Newton County in 1880; Carpenter in Polk County, FL in 1885; Buried at Wauchula Cemetery, Hardee, Florida.

Seth K. Huff 3CPL 1835 15 May 1864

Appointed color-bearer; Captured at **Vicksburg,** MS July 4, 1863; Killed at **Battle of Resaca.**

John W. Ivey 4SGT 29 Jul 1834 3 Mar 1905

Deserted army in March 1862; Married Nancy Grant in Newton County, Nov 28, 1865; Factory mechanic at Cedar Shoals in 1900; Buried at Liberty Cemetery, Porterdale.

James M. Mercer 2CPL 1 Jan 1836 16 Nov 1915

Captured at **Vicksburg,** MS July 4, 1863; Medical furlough in Atlanta, July 13, 1864; Admitted to Fairgrounds Hospital #1 in Atlanta with rubeola July 22, 1864; Admitted with dysentery Ocmulgee Hospital, Macon July 29, 1864; Married Sarah Middlebrook in Newton County, Jan 25, 1866; Farmer in Newton County in 1880; Filed for Confederate pension in Fulton County in 1899; Reported being discharged at Bentonville, NC at close of the war; Buried at Eastview Cemetery, Atlanta.

Zere Pendergrass Middlebrooks 2SGT 7 Feb 1834 3 Jul 1862
Married Penelope Walker in Newton County, Sep 29, 1857; Died of measles at Loudon, TN hospital; Buried at Middlebrook Cemetery, Porterdale.

William Henry H. Nolen 1SGT Apr 1841 1908
Married Georgia Anna Brewer about 1860; Sent home on sick furlough Mar 8, 1863; In Rome, GA hospital Dec 10, 1863; Wounded in **Battle of Atlanta,** July 22, 1864; Surrendered at **Greensboro,** NC April 26, 1865; Teacher in Buena Vista, AR in 1880; Clothing peddler in Harrison, AR in 1900; Buried at Columbia Cemetery, Waldo, Arkansas.

Lazarus Ozburn 2LT Nov 1825 4 Jan 1911
Married Teresa Brewer in Jasper County, Nov 9, 1847; Resigned with Surgeon's Certificate of Ordinary Disability Aug 10, 1862; Farmer in Newton County in 1870; Buried at Newburg Cemetery, Comanche, Texas.

John Clark Pope CPL 1831 1 Apr 1865
Married Emily Morgan in Jasper County, April 20, 1854; Enlisted in Atlanta May 13, 1862; Captured at **Vicksburg,** MS July 4, 1863; Admitted to Pettigrew Hospital #13, Raleigh, NC Mar 11, 1865; Died of wound to right hip; Reported being from Webbville in Jasper, County; Buried at Oakwood Cemetery, Raleigh, Wake County, NC.

James D. Walker 2LT 1834 29 Jun 1862
Died of fever.

William Jackson Allen PVT 25 Dec 1830 18 Apr 1908
Born in Edgefield District, SC; Enlisted at Cumberland Gap, TN in July 1862; Captured at **Vicksburg,** MS July 4, 1863; AWOL on Dec 10, 1863 muster roll; Married Sarah Elizabeth DeWitt in 1863; Filed for Confederate pension in Butts County in 1896; Shoemaker in Jackson, Butts County in 1900; Reported being too sick to continue service after Vicksburg capture; Buried at Oak Hill Cemetery, Griffin.

James Thomas Almand PVT 4 Feb 1828 26 May 1901
Married Cynthia Chaffin in Newton County, April 13, 1851; Enlisted in Atlanta May 13, 1862; Sent to Bean Station, TN hospital Aug 7, 1862; Sent to Vicksburg, MS hospital April 1, 1863; Discharged Sep 25, 1863 with a Certificate of Disability; Farmer in Lithonia in 1900; Buried in Almand Cemetery, Conyers.

Thomas Simeon Almand PVT 12 Aug 1838 15 Apr 1908
Married Mariah Wooley in Newton County, July 7, 1859; Captured at **Battle of Baker's Creek,** MS May 16, 1863; Prisoner at **Camp Morton,** IN; Exchanged at Fort Delaware, DE July 4, 1863; Admitted to Petersburg, VA hospital July 21, 1863; Filed for Confederate pension in Rockdale County in 1888; Reported being admitted to Knoxville, TN in Nov 1863 with typhoid fever; Lost sight in his right eye; Farmer in Rockdale in 1900; Buried at Almand Cemetery, Conyers.

Isham Meville Aycock PVT 14 Sep 1838 15 Apr 1908
Born in Virginia; Captured at **Vicksburg,** MS July 4, 1863; Wounded in the left leg in **Battle of Resaca,** May 15, 1864; Convalesced fighting gangrene in Camp Wright; Married Mary Giles in Clarke County, Dec 9, 1868; Farmer in Oconee County after the war; Married Nancy Burger in Oconee County, Mar 4, 1895; Member of UCV Camp #1124 in

Watkinsville; Buried in Hardigree-Aycock Cemetery, Watkinsville.
Rufus King Bennett PVT 2 Nov 1838 29 Sep 1920
Captured at **Vicksburg,** MS July 4, 1863; Surrendered at **Greensboro,** NC April 26, 1865; Married Indiana Frances Jones in Newton County, April 29, 1866; Father of 14 daughters & 2 sons; Farmer in Rockdale County in 1900; Filed for Confederate pension in Rockdale County in 1906; Reported suffering from typhoid fever as a soldier which has destroyed his eyesight; Buried at East View Cemetery in Conyers.
James Predaux Webb Biggers PVT 1 Mar 1844 20 May 1927
Younger brother of John Biggers; Enlisted in Covington Aug 19, 1862; Sent to Vicksburg, MS hospital April 1, 1863; Transferred to French's Hospital, Enterprise, MS June 30, 1863; Surrendered in Greensboro, NC April 26, 1865; Married Elizabeth Malone in Newton County, Dec 9, 1868; Married Victoria Burkhart in Newton County, Aug 12, 1879; Farmer in Gwinnett County in 1900; Filed for Confederate pension in Gwinnett County in 1907; Buried at Ebenezer Baptist, Dacula.
John Robert Ward Biggers PVT 18 Jul 1842 25 Jun 1862
Sent home sick from Camp McDonald; Buried at Biggers Cemetery, Newton County; Claim filed by his mother, Sarah Biggers.
Cyrus W. Bishop PVT 1843 1863
Enlisted in Covington Aug 19, 1862; Furloughed for 60 days from Confederate Post Hospital in Dalton, Dec 2, 1862; Captured at **Vicksburg,** MS hospital July 4, 1863; Among sick & wounded placed on US Steamer **St. Maria** for New Orleans & Mobile, July 30, 1863; AWOL on Dec 10, 1863 muster roll.
Henry James Boggus PVT 16 May 1843 19 Aug 1922
Appointed **3CPL** April 1863; Captured in **Vicksburg,** MS July 4, 1863; Appointed **1SGT** May 1864; Surrendered in **Greensboro,** NC April 26, 1865; Married Mary Stephenson in Newton County, Nov 8, 1866; Farmer in Gem Creek, Newton County after the war; Filed for Confederate pension in Newton County in 1919; Buried at Macedonia Baptist, Oxford.
Robert Woodson Boggus PVT 22 Aug 1835 3 Jul 1892
Older brother of Henry Boggus; Married Sarah Ann Hinton in Newton County, Nov 12, 1854; Enlisted in Atlanta May 13, 1862; Captured at **Vicksburg,** MS July 4, 1863; Surrendered at **Greensboro,** NC April 26, 1865; Mill worker in Newton Factory in Covington after the war; Buried at Resthaven Cemetery, Washington, Wilkes County.
James Thomas Boyd PVT 6 Dec 1832 26 Apr 1905
Married Susannah Thompson in Newton County, Dec 20, 1859; Enlisted as 3SGT in Company H, 6th Regiment, Georgia State Troops, Oct 21, 1861; Mustered out in Savannah, GA May 1862; Enlisted in the 42nd Georgia April 4, 1863; Appointed **SGT**; Surrendered at **Greensboro**, NC April 26, 1865; Married Zapora Thompson in Newton County, Oct 29, 1874; Farmer in Rocky Plains, Newton County after the war; Filed for Confederate pension in Newton County in 1903; Application endorsed by J.D. Ellis; Buried at Hopewell Presbyterian, Newton County.
William Perry Boyd PVT 1 May 1828 15 May 1864
Older brother of James Boyd; Enlisted in Atlanta May 13, 1862; Admitted to Fairgrounds Hospital #1 in Atlanta Nov 28, 1862; Returned to Duty Mar 27, 1863; Killed at **Battle of Resaca,** GA.
Elisha M. Brewer PVT 1842 10 Jun 1863
Older brother of William Brewer; Enlisted in Atlanta May 13, 1862; Absent Without

Leave since Dec 1, 1862.

William N. Brewer PVT 1844 After 1870

Enlisted in Covington May 12, 1862; Absent Without Leave since Dec 21, 1862; Present on Dec 11, 1863 muster roll; Surrendered at **Greensboro**, NC April 26, 1865; Farm laborer in Newton County in 1870.

James Monroe Bryan PVT 29 Mar 1822 15 Feb 1889

Married Mary Elizabeth Jenks in Newton County, Nov 21, 1846; Enlisted in Calhoun Aug 19, 1862; Captured at **Edwards Depot** in the Battle of Baker's Creek May 16, 1863; Sent to **Camp Morton**, IN & then paroled at Fort Delaware, DE July 4, 1863; Furloughed from Madison, GA hospital with gunshot wound Mar 6, 1865; Farmer in Bienville Parish, LA in 1880; Buried at Mount Olive Cemetery, Sparta, Bienville Parish, LA.

Joel C. Bryan PVT 1833 1863

Younger brother of James M. Bryan; Died at Vicksburg, MS in 1863.

Joel Madison Chafin PVT 1840 10 Jun 1863

Younger brother of William Chafin; Killed at Vicksburg, MS; Claim filed by his mother Martha Chafin.

William Thomas Chafin PVT 29 Oct 1829 27 Nov 1902

Married Mary Ann Almand (Sister of Thomas S. Almand) in Newton County, Nov 14, 1850; Sent to Vicksburg, MS hospital April 2, 1863; Present on Dec 11, 1863 muster roll; Farmer in Rockdale in 1880; Buried at Zion Baptist in Covington.

John Childers PVT 1827 1 Mar 1897

Born in McMinn County, TN; Given Certificate of Disability for Discharge in Knoxville, TN Nov 8, 1862; Married Nancy Fielder in Gwinnett County, April 7, 1867; Ditch builder in Lawrenceville in 1880; Filed for Indigent pension in Gwinnett County in 1897; Buried at Fairview Presbyterian, Lawrenceville.

Samuel Childers PVT 1836

Born in McMinn County, TN; Younger brother of John & William Childers; Sent to Knoxville, TN hospital Nov 1, 1862; AWOL on Dec 10, 1863 muster roll.

William Childers PVT 1833 July 1862

Present on June 28, 1862 roll; Assigned coffin & grave July 16, 1862; Buried at Bethel Cemetery, Knoxville, TN.

William Franklin Coady PVT 1820 After 1880

Born in Ireland; Married Rocella Kirkland in Walton County, April 25, 1847; Rock Mason in Henry County in 1860; At Knoxville, TN Convalescent Hospital Aug 31, 1862; Received Certificate of Disability & Discharge from Knoxville hospital Oct 15, 1862; Farmer in Butts County in 1870.

George J. Cook PVT 30 Apr 1833 23 Jun 1913

Married Sarah Sansom in Newton County, Oct 11, 1860; Enlisted in Atlanta May 13, 1862; Captured at **Vicksburg**, MS July 4, 1863; Surrendered at **Greensboro**, NC April 26, 1865; Farmer in Chambers County, AL in 1880; Landlord in Heard County in 1900; Buried at Prospect Cemetery, Franklin, Heard County.

James L. Cook PVT 1841 24 Apr 1862

Worked in the Newton Factory in Covington before the war.

Cornelius B. Cowen PVT 13 Mar 1841 18 May 1863

Younger brother of Harvey Cowen; Died in Enterprise, MS Confederate hospital; Buried at Enterprise Confederate Cemetery; Claim filed by his mother Nancy H. Cowen; Memorial at Hopewell Presbyterian, Newton County.

Harvey K. Cowen PVT 20 Dec 1838 27 Feb 1863

Enlisted in Atlanta, May 13, 1862; Died in Vicksburg, MS hospital #3; Claim filed by his mother Nancy H. Cowen; Buried at Cedar Hill, Vicksburg, MS; Memorial at Hopewell Presbyterian, Newton County.

James Calvin Cowen PVT 9 Aug 1843 7 Jan 1924

Born in Chester District, SC; Given a Certificate of Disability For Discharge at the University Hospital in Knoxville, TN Nov 25, 1862; Sick in Rome, GA hospital Dec 11, 1863; Wounded in left arm at **Battle of Wise's Forks**, NC Mar 10, 1865; Admitted to Pettigrew Hospital #13, Raleigh, NC Mar 10, 1865; Married J.L. Black in Newton County, Feb 14, 1867; Married Minnie Green in Newton County, Oct 28, 1883; Farmer in Newton County after the war; Buried at Hopewell Presbyterian, Newton County.

James F. Dick PVT 1841 2 Oct 1880

Born in Wilkes County; Father, James Dick, was Irish immigrant & War of 1812 veteran; Elected **3SGT** Mar 16, 1863; Captured in a rifle pit in front of the Vicksburg trenches by Federal troops under General Herron Jun 24, 1863; Paroled at **Vicksburg**, MS July 10, 1863; AWOL on Dec 11, 1863 muster roll; Surrendered at **Greensboro**, NC April 26, 1865; Married Laura Bryant in Newton County, May 18, 1876; Farmer in Newton County in 1880.

Henry H. Edwards PVT 12 Nov 1818 27 Dec 1862

Married Sutera Edwards Jun 16, 1840; Stonemason in Butts County in 1860; Admitted to Fairgrounds Hospital #1 in Atlanta Nov 28, 1862 with rheumatism; Died in hospital; Buried at Oakland Cemetery, Atlanta.

Thomas L. Edwards PVT 15 May 1864

Captured at **Vicksburg**, MS July 4, 1863; AWOL on Dec 11, 1863 muster roll; Killed at **Battle of Resaca**, GA.

George Robert Elliott PVT 1834 4 Nov 1864

Married Tabitha Crumbley in Henry County, Mar 22, 1859; Farmer in Snapping Shoals, Newton County in 1860; Captured at **Vicksburg**, MS July 4, 1863; Transferred to Company B, 53rd Georgia in exchange for Samuel L. McNair Nov 8, 1863; Wounded at Cedar Creek, VA Oct 19, 1864; Died from his wounds; Buried at Loudon Park Cemetery, Baltimore, MD.

William James Elliott PVT 1843 5 Apr 1863

Born in Henry County; Younger brother of George Elliott; Died at Big Shanty, Cobb County.

William B. Ellis PVT 13 Nov 1825 7 Sep 1898

Married Frances Hodges in Newton County, Jan 12, 1848; Enlisted in Covington, July 16, 1862; Detached as teamster in Vicksburg, MS April 1863; Captured at **Vicksburg**, MS July 4, 1863; Wounded in left leg at **Battle of Missionary Ridge**, TN Nov 25, 1864; Farmer in Cedar Shoals in 1880; Filed for Confederate pension in Newton County in 1895; Reported being present at the surrender in NC; Buried at Liberty Cemetery, Porterdale.

James Madison Fincher PVT 4 Jun 1841 18 Jan 1919

Captured at **Vicksburg**, MS July 4, 1863; Captured at **Battle of Nashville**, Dec 16, 1864; Received at **Camp Douglas**, IL, Dec 24, 1864; Released Jun 19, 1865; Married Frances Owens in Newton County, Nov 21, 1865; Merchant in Newton County in 1900; Moved store to Dodge County late in life; Buried at Woodlawn Cemetery, Eastman, Dodge County.

John Edward Fincher PVT 9 Feb 1845 19 Dec 1912
Younger brother of James Fincher; Married Sarah Warren in Butts County, May 15, 1864; Captured at **Battle of Nashville,** TN Dec 16, 1865; Released from **Camp Douglas,** IL Jun 19, 1865; Farmer in Jasper County after the war; Buried in family cemetery in Jasper County.

William H.H. Geiger PVT 15 Apr 1862

William T. Gilpin PVT
Appointed **2SGT** Mar 1863; Captured at **Vicksburg,** MS July 4, 1863; On detached duty in Macon, GA Arsenal Dec 10, 1863 Still at Arsenal August 1864.

Thomas D. Goodson PVT 1844 1862
Appointed **3CPL** & Color Bearer; Died in Bean Station, TN Hospital.

James M. Green PVT 1844 7 Mar 1863
Younger brother of William F. Green; Enlisted in Atlanta May 13, 1862; Wounded in throat at **Battle of Chickasaw Bayou,** Dec 29, 1862; Died in Vicksburg, MS; Buried at Cedar Hill, Vicksburg, MS; Claim filed by his father Levi Green.

William Frank Green PVT 1831 Before 1867
Married Cola Ann Fincher in Newton County, May 30. 1861; Captured at **Vicksburg,** MS July 4, 1863; AWOL on Dec 10, 1863 muster roll.

William Thomas Hailey PVT 29 May 1844 3 Jun 1916
Born in North Carolina; Appointed **2CPL;** AWOL on Dec 10, 1863 muster roll; Married Martha Jane Gerger in Newton County, July 12, 1868; Buggy maker in Conyers in 1900; Filed for Confederate pension in DeKalb County in 1912; Reported leaving his command near Jonesboro in Sep 1864 with sickness; Buried in Lithonia City Cemetery, DeKalb County.

Henry Gaither Harris PVT 24 Mar 1843 18 Apr 1930
Elected **2CPL** Mar 16, 1863; Captured at **Vicksburg,** MS July 4, 1863; Admitted to Fairgrounds Hospital #2 in Atlanta with fever Feb 25, 1864; Captured **Battle of Wise's Fork,** NC Mar 10, 1865; Released at **Point Lookout,** MD 27 Jun 1865; Married Amanda Phillips in Newton County, Oct 19, 1865; Miller in Tallapoosa County, AL in 1900; Buried in Macedonia Primitive Baptist, Clay County, AL.

James Thomas Harris PVT 7 Jun 1838 22 Apr 1862
Older brother of Henry Harris; Died of disease in Newton County; Claim filed by his father Hezekiah Harris.

Herbert S. Heard PVT 1844 16 Aug 1862
Died at Bean Station, TN hospital; Buried in Heard Family Cemetery, Newton County.

Hope Hull Hewell PVT 1835 May 1863
Enlisted as a private in Company H, 6th Regiment Georgia Troops Oct 21, 1861; Discharged with disability Jan 18, 1862; Sent to Vicksburg, MS hospital Apr 27, 1863; Sent to Jackson, MS hospital. Died in the hospital.

John Lewis Hinton PVT 15 Mar 1836 25 Nov 1900
Captured at **Vicksburg,** MS July 4, 1863; Farmer in Newton County after the war; Filed for Confederate pension in Newton County in 1889; Reported being wounded by shell at **Battle of Jonesborough** Aug 31, 1864; Buried at Liberty Cemetery, Porterdale, Newton County.

John W. Ingram PVT 1833 20 Jul 1863
Appointed color bearer; Captured at **Vicksburg**, MS July 4, 1863; Died at Vicksburg, MS hospital.
Benjamin Franklin Jones PVT 1834 Before 1900
Married Sarah Green in Newton County, Mar 7, 1854; Captured at **Battle of Baker's Creek**, MS May 16, 1863; POW at **Camp Morton**, IN until exchanged at Fort Delaware, DE July 3, 1863; Captured near Marietta July 4, 1864; Released at **Camp Douglas**, Il Jun 16, 1865; Farmer in Clay County, AL in 1880.
Wm. Christopher Columbus Jones PVT 12 Oct 1844 30 Apr 1924
Went by "Kit"; Enlisted in Covington Mar 20, 1864; Surrendered at **Greensboro**, NC April 26, 1865; Married Ann Edwards in Newton County, Jul 31, 1866; Farmer in Newton County after the war; Filed for Confederate pension in Newton County in 1909; Reported enlisting in Dalton in Jan 1864; Buried at Newton Factory Cemetery, Newton County.
James Martin King PVT 1838 29 Dec 1862
Older brother of John A. King; Married Mary Chaffin in Newton County, Dec 17, 1857; Enlisted in Atlanta May 13, 1862; Killed at **Battle of Chickasaw Bayou**, MS; Buried at Soldier's Rest, Cedar Hill Cemetery, Vicksburg, MS.
John Asberry King PVT 11 Dec 1839 12 Jun 1909
Born in Jasper County; Captured at **Vicksburg**, MS July 4, 1863; Present on Dec 10, 1863 roll; Married Annie Lowry in Newton County, Dec 19, 1872; Harness maker in Newton County in 1880; Filed for Confederate pension in Rockdale County in 1889; Reports being shot below the left knee at **Battle of Resaca**, May 15, 1864; Never fully recovered; Buried at Zion Baptist, Covington.
John B. King PVT 23 Jan 1844 19 Feb 1876
Son of Samuel King; Captured at **Vicksburg**, MS July 4, 1863; Appointed **2CPL**; Surrendered at **Greensboro**, NC April 26, 1865; Buried at Hopewell Presbyterian, Newton County.
Samuel Asbury King PVT 18 Oct 1820 29 Oct 1876
Born in Chester County, SC; Married Frances Ann Mercer in Newton County, Jan 17, 1843; Farmer in Newton County; Enlisted in Covington July 27, 1863; Present on Dec 10, 1863 roll.
Charles Kitchins PVT 1849 After 1910
Younger brother of John Kitchins; Enlisted in Covington Feb 1, 1865; Admitted to General Hospital #11 in Charlotte, NC Feb 25, 1865; Surrendered at **Greensboro**, NC April 26, 1865; Married Martha Paxton in Newton County, April 29, 1866; Farmer in Jasper County in 1880; Farmer in Newton County in 1910.
Henry Zachariah Kitchins PVT 11 Apr 1845 14 Apr 1902
Enlisted in Covington Feb 23, 1863; Captured at **Vicksburg**, MS July 4, 1863; Surrendered at **Greensboro**, NC April 26, 1865; Married Eliza Frances Nelms in 1867; Farmer in Heard County after the war; Buried at Prospect Methodist, Franklin, Heard County.
John Elbert Kitchins PVT 4 Apr 1847 4 Aug 1924
Private in 1864; Surrendered in Macon, GA April 30, 1865 as member of 1st Brigade, 2nd Cavalry Division; Married Emma Moore in Newton County, Dec 23, 1869; Farmer in Butts County after the war; Buried at Stark Methodist, Butts County.
Moses Peter Kitchins PVT 20 May 1829 14 Apr 1897
Married Permelia Pope in 1855; Enlisted May 13, 1862 in Atlanta; Captured at **Vicksburg**,

MS July 4, 1863; Admitted to Fairgrounds Hospital #2 in Atlanta with pneumonia Dec 11, 1863; Furloughed Dec 29, 1863; Surrendered at **Greensboro**, NC April 26, 1865; Farmer in Chambers County, AL after the war; Buried at Antioch Christian, Chambers County, AL.

Augustus James Lee PVT 3 Sep 1843 21 Mar 1865
Appointed **1SGT** Aug 31, 1862; Captured in Nicholasville, KY Oct 15, 1862; Transported to Cairo, IL on the steamboat *Belle Creole* Nov 18, 1862; Elected **2LT** Mar 15, 1863; Captured at **Vicksburg**, MS July 4, 1863; Wounded at **Orangeburg**, SC Feb 12, 1865; Died in General Hospital #1, Columbia, SC; Buried at Covington City Cemetery, Newton County.

John Wesley Leech PVT 8 Nov 1845 14 Jan 1898
Captured at **Vicksburg**, MS July 4, 1863; Surrendered at **Greensboro**, NC April 26, 1865; Married Emma Catherine Gilstrap in 1869; Farmer in Jasper County in 1870; Farmer in Morris County, TX in 1880; Buried at Holly Springs Cemetery, Cass County, Texas.

Joseph Elisha Lowery PVT 7 Dec 1842 4 Oct 1921
Appointed musician; Captured at **Vicksburg**, MS July 4, 1863; Surrendered at **Greensboro**, NC April 26, 1865; Married Georgia Ann Nix in Newton County, Nov 14, 1867; Carpenter in Newton County in 1880; Farmer in Marshall County, AL in 1900; Buried at Mt. Vernon Baptist, Marshall County, Alabama.

Dr. William Green Lowery PVT 9 Feb 1841 26 Sep 1891
Older brother of Joseph Lowery; Married Margaret Boyd in Newton County, Dec 6, 1860; Appointed musician; Captured at **Vicksburg**, MS July 4, 1863; Surrendered at **Greensboro**, NC April 26, 1865; Medical doctor in Newton County after the war.

Perry Oliver McCullough PVT 8 Sep 1836 20 Aug 1884
Born in South Carolina; Married Rebecca Underwood in Newton County, Aug 28, 1858; Captured at **Vicksburg**, MS July 4, 1863; Captured near Atlanta, Aug 23, 1864; Released at **Camp Chase**, OH June 11, 1865; Farmer in Jasper County in 1880.

Hugh Dixon McKitrick PVT 1842 16 Apr 1862
Born in South Carolina; Died from disease.

Samuel Lawson McNair PVT 8 Dec 1829 6 Dec 1913
Married Martha Jane Nisbit in Henry County, 6 Nov 1851; Enlisted in Company B, 53rd Georgia May 1, 1862; Transferred to 42nd Georgia Nov 8, 1863; Admitted to Fairgrounds Hospital #2 in Atlanta April 30, 1864; Wounded at **Battle of Wise's Forks** Mar 10, 1865; Admitted to Pettigrew Hospital #13 in Raleigh, NC with left thigh fracture; Captured in Raleigh, NC hospital April 13, 1865; Farmer in Clayton County after the war; Filed for Confederate pension in Fulton County in 1905; Buried at Philadelphia Presbyterian, Forest Park.

William Francis Mercer PVT 17 Feb 1828 12 Apr 1863
Brother of Captain Thomas Mercer & Lt. James Mercer; Married Francis Green in Jasper County, Dec 31, 1849; Died in Vicksburg, MS hospital.

William G. Morgan PVT 1842 About 1864
Enlisted in Covington Feb 23, 1863; Captured at **Vicksburg**, MS July 4, 1863; Left in hospital; Shipped from Vicksburg on US Steamer *St. Maurice* to Mobile, AL Aug 5, 1863; Present on Dec 10, 1863 roll.

Francis Marion Morris PVT 6 Apr 1842 1 Dec 1882
Married Delilah Tollison in Newton County, Dec 17, 1863; Enlisted in Covington Mar

23, 1864; Surrendered at **Greensboro,** April 26, 1865; Carpenter in Newton County after the war; Buried at Newton Factory Cemetery, Covington.

Isaac Newton Nolen PVT 4 Aug 1838 27 Mar 1900

Went by "Newton"; Married Elizabeth Tuggle in DeKalb County, Oct 26, 1859; Enlisted in Covington May 12, 1862; Captured in the Vicksburg rifle pits Jun 23, 1863; Paroled at **Vicksburg,** MS July 4, 1863; AWOL on Dec 10, 1863 roll; Admitted to St. Mary's Hospital, La Grange July 3, 1864 & returned to duty July 16, 1864; Surrendered at **Greensboro,** NC April 26, 1865; Buried at Wesley Chapel Methodist, Decatur.

James M. Nolen PVT 18 Jun 1846 6 Jan 1919

Enlisted in Covington April 13, 1864; Surrendered at **Greensboro,** NC April 26, 1865; Married Mollie Maddox in Newton County, Dec 24, 1873; Farmer in Buena Vista, AR in 1880; Farmer in Holland, TX in 1900; Buried in Simpson Cemetery, Eastland County, TX.

Stephen W. Nolen PVT 1844 26 Feb 1863

Younger brother of Isaac Nolen; His mother, Isabella McBride Nolen, had her last child, Robert Henderson Nolen, in 1861, when she was 47; Died in Vicksburg, MS hospital; Buried at Cedar Hill Cemetery, Vicksburg; Claim filed by his mother.

Aaron J. Parker PVT 1837 26 Mar 1863

Half-brother of John C. Parker; Died in Vicksburg, MS hospital.

James M. Parker PVT 6 Jul 1830 5 Feb 1914

Married Jemimah Catherine Polk in Jasper County, Dec 15, 1853; Enlisted in Covington May 12, 1862; Captured in the Vicksburg rifle pits Jun 23, 1863; Paroled at **Vicksburg,** MS July 4, 1863; Present on Dec 10, 1863 roll; Married Margaret Dick in Newton County, Dec 1, 1870; Farmer in Newton County after the war; Buried at High Point Cemetery, Newton County.

John C. Parker PVT 1845 1864

Enlisted in Covington Feb 23, 1863; Captured at **Vicksburg,** MS July 4, 1863; Absent without leave Dec 10, 1863 roll; Patient at Bragg Hospital, Newnan Jan 1, 1864; Admitted to Madison, GA hospital May 18, 1864; Died after May 10, 1864.

Samuel P. Pickett PVT 3 Apr 1837 28 Nov 1920

Enlisted in Covington Aug 1, 1862; Appointed musician; Sent to Vicksburg, MS hospital Aug 1, 1863; Present on Dec 10, 1863 roll; Buried in Pickett Cemetery, Newton County.

Wesley Piper PVT 1841 16 Jan 1863

Married Linna Warren in Newton County, Dec 8, 1859; Died of Variola in Atlanta hospital.

Charles Ellison Polk PVT 13 Sep 1828 21 May 1870

Married Susan Hodges in Butts County, Jun 17, 1852; Enlisted in Covington May 13, 1862; Captured in **Vicksburg,** MS July 4, 1863; Right thigh fractured in **Battle of Bentonville,** NC Mar 19, 1865; Admitted to Pettigrew Hospital #13 Raleigh, NC Mar 24, 1865; Captured at Raleigh, NC hospital April 13, 1865; Married Martha Glass in Newton County, July 22, 1866.

James Knox Polk PVT 3 Sep 1845 29 Dec 1906

Youngest brother of Charles, John, & Thomas Polk; Enlisted in Covington Aug 17, 1863; In Bell Hospital, Rome, GA Dec 10, 1863; Married Caroline Minter in Jasper County, Nov 2, 1867; Farmer in Jasper County after the war; "Mr. James Polk was killed by "Curb" Waldrop, a young man who had escorted the daughters of Polk to an entertainment. When upbraided by Polk for keeping them out until 2 a.m., Polk was shot by Waldrop." Buried at Williams-Payne Cemetery, Jasper County.

John Edmunds Polk PVT 19 Dec 1830 12 May 1863

Married Polly Ann Parker in Newton County, Dec 18, 1852; Enlisted in Covington May 12, 1862; Sent to Knoxville, TN hospital Nov 1, 1862; Died in Greenville, TN hospital.

Thomas William Obediah Polk PVT 4 Jan 1837 11 Mar 1914

Married Nancy Elizabeth Fisher in Henry County, Dec 11, 1860; Enlisted in Covington May 12, 1862; Appointed musician; Captured at **Vicksburg**, MS July 4, 1863; Farmer in Jasper County after the war; Filed for Confederate pension in Jasper County in 1889; Reported being wounded in the right thigh at Pine Mountain, Jun 24, 1864; Necessitated amputation of leg; Buried at West View Cemetery, Monticello.

Andrew Jackson Pope PVT 1840 25 Nov 1862

Died at Knoxville, TN hospital; Claim filed by his father William Pope.

John Clark Pope PVT 1831 1 Apr 1865

Enlisted in Covington May 12, 1862; Captured at **Vicksburg**, MS July 4, 1863; Admitted to Pettigrew Hospital #13 in Raleigh, NC Mar 11, 1865 with gunshot wound to right hip; Died in hospital; Buried in Confederate Section, Oakwood Cemetery, Raleigh, NC.

Henry B. Pope PVT 1842 5 Sep 1863

Enlisted in Covington May 12, 1862; Captured in Vicksburg rifle pits Jun 23, 1863; Paroled at **Vicksburg**, MS July 4, 1863; Died at home.

William M. Pope PVT 28 Sep 1833 29 Jul 1908

Older brother of Andrew & Henry Pope; Married Seletha Jane Maddox in Newton County, 11 Jan 1855; Enlisted in Covington May 12, 1862; Captured at **Vicksburg**, MS July 4, 1863; AWOL on Dec 10, 1863 roll; Wounded in the 1864 Tennessee Campaign; Admitted to Fairgrounds Hospital #2 in Corinth, MS with gunshot wound Dec 29, 1864; Furloughed from Way Hospital in Meridian, MS Jan 13, 1865; Farmer in Worthville in 1900; Buried at Stodghill Cemetery in Butts County.

Stephen Daniel Reagan PVT 1840 31 Aug 1912

Born in Anderson County, SC; Captured at **Vicksburg**, MS July 4, 1863; Present on Dec 10, 1863 roll; Married Rebecca Clementine Wallace in Newton County, Dec 21, 1865; Farmer in Newton County after the war; Filed for an indigent pension with epilepsy in Fulton County in 1897; Reported "becoming overheated from Marching while in the Service; I was taken with Typhoid Fever;" Buried at Old Concord, Stockbridge.

Thomas David Robinson PVT 1839 25 Jan 1863

Son of Rev. Luke Robinson; Married Martha Lindley in Newton County, Nov 7, 1858; Medical student in Atlanta in 1860; Died of variola in Atlanta, GA hospital.

David Lafayette Rogers PVT 29 Jun 1823 1903

Born in Jasper County; Married Sarah Almand in Newton County, Jan 2, 1845; Captured at **Vicksburg**, MS July 4, 1863; Present on Dec 10, 1863 muster roll; Married Martha Stewart in Newton County, Feb 9, 1865; Farmer in DeKalb County after the war; Filed for Indigent Pension in DeKalb County in 1899; Reported being present at Greensboro, NC surrender; Application supported by Lt. James M. Mercer who wrote, "I knew him to have made a good soldier."

John Robert Sams PVT 21 May 1830 15 May 1864

Born in Henry County; Married Margaret Lee in Newton County, Dec 13, 1854; Enlisted in Covington Aug 1, 1862; Elected **1SGT** Mar 16, 1863; Captured at **Vicksburg**, MS July

4, 1863; Admitted to Fairgrounds Hospital #1 in Atlanta Nov 19, 1863 with debility; Returned to duty Jan 15, 1863; Killed at **Battle of Resaca** May 15, 1864. His wife, Margaret Goodale, wrote an account of her experience in the war that was published in Macon Telegraph Dec 10, 1905. She wrote, "I never heard a word from him from May until July 27th, when he came riding up to the gate on a neighbor's mule. After he left Vicksburg he found the railroads all torn up, and he was ill and found but little to eat on his journey. He surely was feeble when he reached home."

Absalom Tire Smith PVT 6 Feb 1841 After 1901
Enlisted in Covington Feb 23, 1863; Captured at **Vicksburg,** MS July 4, 1863; Present on Dec 10, 1863 muster roll; Gave loyalty oath in Newton County in 1867; Filled Indigent Pension in Butts County in 1901; Reports being present at the surrender in Greensboro, NC; Application supported by William Pope.

James H. Smith PVT
Admitted to Fairgrounds Hospital #1 in Atlanta Nov 28, 1862 with rheumatism; Discharged from service April 13, 1863.

John O. Smith PVT
Enlisted in Covington Aug 1, 1862; Captured at **Vicksburg,** MS July 4, 1863; Surrendered at **Greensboro,** NC April 26, 1865.

John W. Smith PVT 1838 After 1880
Enlisted in Covington Aug 19, 1862; Captured at **Vicksburg,** MS July 4, 1863; Surrendered at **Greensboro,** NC April 26, 1865; Farmer in Newton County 1870; Farmer in DeKalb County in 1880.

Joseph A. Smith PVT 1838
Born in Monroe County; Sent to Vicksburg, MS hospital Feb 10, 1863; Transferred to Jackson, MS hospital; Captured at **Battle of Wise's Forks** Mar 10, 1865; Released from **Point Lookout,** MD Jun 19, 1865.

Madison M. Smith PVT 1840 12 Apr 1864
Married Mary Cameron in Newton County, Dec 7, 1860; Sent to Atlanta hospital Dec 1, 1862; Granted 15-day furlough from hospital Jan 10, 1863; AWOL on Dec 10, 1863 muster roll; Executed for desertion in Dalton, Georgia.

William Stockwell PVT 1822 1863
Married Mary Frances Shanks in Newton County, Mar 10, 1850; Sent to Vicksburg, MS hospital Feb 10, 1863; Died in Jackson, MS Soldiers Home hospital before Nov 1, 1863.

Charles W. Stone PVT 1865
Captured at **Vicksburg,** MS July 4, 1863; AWOL on Dec 10, 1863 muster roll; Killed near Smithfield, NC.

Fenton J. Taylor PVT 1820 Jun 1864
Born in Monroe County, TN; Married Rebecca Shanks in Newton County, April 5, 1849; Captured in **Battle of Baker's Creek,** MS May 16, 1863; Sent to **Camp Morton,** IN; Exchanged at Fort Delaware, DE July 4, 1863; Furloughed from Jackson Hospital, Richmond Aug 13, 1863; AWOL on Dec 10, 1863 roll; Received company punishment on April 14, 1864; Widows Pension filed in Newton County in 1891; Reported that in June 1864 Fenton Taylor was "shot near Kennesaw Mountain and died the next day."

Richard Taylor PVT
Captured in **Vicksburg,** MS July 4, 1863; Left in hospital sick; Shipped on US Steamboat

St. Maurice from Vicksburg to Mobile, AL, Aug 5, 1863; Admitted to Madison, GA hospital May 18, 1864; Surrendered at **Greensboro,** NC April 26, 1865.

Enoch Thomas PVT 1842 1 Jan 1863

Died at home Jan 1, 1863; Claim filed by his father Washington Thomas.

James W. Thomas PVT 1843 1865

Absent without leave since Mar 6, 1863; Buried at Oxford Cemetery, Oxford.

William B. Thompson PVT 1840 18 Jan 1864

Married Lucinda McCart in Newton County, Dec 12, 1862; Captured at **Battle of Baker's Creek,** MS May 16, 1863; Sent to **Camp Morton,** IN; Exchanged at Fort Delaware, DE July 4, 1863; AWOL on Dec 10, 1863 roll; Captured at **Battle of Missionary Ridge,** TN Nov 25, 1863; Sent to **Rock Island** Barracks, IL; Died of pneumonia as POW at Rock Island; Buried in grave # 218.

James Lamar Tucker PVT 12 Mar 1837 18 Oct 1917

Born in Bibb County; Married Annie Russell in Newton County, Nov 17, 1861; Enlisted in Covington Aug 1, 1862; AWOL since Dec 26, 1862; Present on Dec 10, 1863 roll; Wounded in the right leg at the Battle of Wise's Forks Mar 10, 1865; Admitted to Pettigrew Hospital #13 in Raleigh, NC; Farmer in Jasper County after the war; Buried at Providence Cemetery, Jasper County.

William E. Tucker PVT 10 Jun 1841 1 Jun 1892

Promoted to **4SGT**; Patient at St. Mary's Hospital in Dalton from Dec 4 to Dec 22, 1862; Captured at **Vicksburg,** MS July 4, 1863; Surrendered at **Greensboro,** NC April 26, 1865; Married Amanda Ivy in Newton County, Feb 5, 1879; OBIT reports, "He was in good financial circumstances and leaves some valuable real estate in Jackson and other places." Buried at Jackson City Cemetery, Butts County.

J.W. Thurman PVT 9 Mar 1864

Captured at **Battle of Baker's Creek** May 17, 1863; Died of inflammation of the lungs as POW at **Fort Delaware,** DE; Buried at Confederate Cemetery, Finn's Point, New Jersey.

John Tyler PVT 21 Apr 1862

Died at home.

Levi Victory PVT 28 Feb 1865

Late enlistment; Captured on the retreat near Franklin, TN Dec 17, 1864; Died at **Camp Chase,** OH; Buried in Confederate Cemetery grave #1489.

Elijah H. Vining PVT 22 Jun 1832 5 May 1887

Married Elizabeth Grant in Newton County, Sep 24, 1854; Enlisted as Conscript in Calhoun Aug 18, 1862; Discharged with rheumatism at Lenoir Station, TN Nov 22, 1862; Cotton Mill superintendent in Cedar Shoals in 1880; Buried at Liberty Cemetery, Porterdale.

Thomas J. Wiley PVT 1840 28 Apr 1863

Married Mary Harris in Walton County, Feb 10, 1859; Died at Vicksburg, MS hospital; Memorial stone at Jersey Methodist, Social Circle.

Thomas M. Wilkerson PVT

Captured in Vicksburg rifle pits night of June 24, 1863; Paroled in **Vicksburg,** MS July 4, 1863; Absent without leave Dec 10, 1863 roll.

William Wilson PVT 1834 13 Jun 1863

Enlisted in Covington May 12, 1862; Died of typhoid fever in Vicksburg, MS hospital.

Company F – Newton Rifles – 171 Men

This company drew men from western Newton County (which would become Rockdale County in 1870), northern Henry County, and the eastern fringe of DeKalb County. The Newton Rifles stood in the middle of the battle line for the 42nd and was designated the color company. J.M. Summers served as captain for the length of the conflict. 42 men were enlisted in Atlanta by Lt. Elias Woolley on May 13, 1862.

Name Rank Birth Date

James Madison Summers CPT 24 Dec 1830 27 Feb 1915
Married Lucinda Watts in Newton County, Aug 11, 1852; Successful miller near Conyers at the start of the war; Captured at **Vicksburg,** MS July 4, 1863; Present near Nashville, TN Dec 5, 1864; Wounded in Tennessee Campaign; Admitted to Fairgrounds Hospital #2 in Corinth, MS with gunshot wound Jan 5, 1865; Married Flora Jane Stephenson (Sister to the 3 Stephenson brothers) in Newton County in 1865; Farmer near Covington 1880; Butcher in south Fulton County 1900; Filed for Confederate pension in Fulton County in 1906; Reported that General Beauregard had ordered him to Atlanta a few weeks before the final surrender to "establish a camp of direction for absent soldiers"; Buried at Confederate Knoll, Westview Cemetery, Atlanta.

Charles Andrew Bailey 5SGT 13 Nov 1840 30 Mar 1921
Born in Jasper County; Promoted to **4SGT**; Furloughed from Fair Grounds Hospital #1 in Atlanta Oct 22, 1862; Sick at home Dec 9, 1863; Married Emma Turner in Randolph County, April 26, 1865; Farmer in Newborn after the war; Filed for Confederate pension in Jasper County in 1897; Reported getting typhoid pneumonia in Bean Station, TN in 1862, and assigned to part time duty in Covington on April 26 1864; Buried at Newborn Cemetery, Newton County.

William Bradley Christian 4SGT 14 May 1835 Jan 1905
Captured at **Vicksburg,** MS July 4, 1863; Wounded at **Battle of Resaca** May 15, 1864; Admitted to Fairgrounds Hospital #2 in Atlanta May 16, 1864; Married Susan Robinson in Newton County, 1864; Farmer in Newton County in 1880; Moved to Cullman County, AL about 1894; Buried at Fairview Cemetery, Cullman County, Alabama.

Reuben Carter Cook 2LT 26 Sep 1813 26 Mar 1899
Born in Elbert County; Married Martha Lackey in Newton County, Jan 26, 1840; Wealthy farmer near Covington in 1860 with net worth of $25,085; Resigned "unconditionally" at Bean Station, TN July 8, 1862; Buried at Cook Family Cemetery, Newton County.

Eugene Summerfield Dickson 2CPL 9 Sep 1843 16 Nov 1862
Died in hospital in Nicholasville, Kentucky. Buried at Maple Grove Cemetery, Nicholasville, Kentucky; Claim filed by his father David Dickson.

Levi T. Gay 1CPL Jul 1843 1876
Absent April 1862; Married Mary Elizabeth Richardson; Signed 1867 Oath Book in Newton County; Died in Corinth, Van Zandt County, Texas.

William Newton Hicks 1LT 1840 6 Aug 1862
Died in Bean Station, TN hospital; "I was not in any of this (Kentucky) marching or fighting, however, having been left behind to stay with the first Lieutenant, who was sick in a private house with camp fever, and who by the way was one of the most exemplary young men I ever saw,"Aug 24, 1862 letter from Dr. Ira Lawson Gunter; Buried in Hicks

Family Cemetery in Newton County; Claim filed by his mother Hester Hicks.
Benjamin A. Leak 3CPL 1838 12 Jul 1862
Died in Morristown, TN hospital; Claim filed by his grandfather Silas H. Starr.
John Thomas Ogletree 3SGT 9 Feb 1838 16 Oct 1921
Born in Fayette County; Married Mary Ann Elizabeth McCollum in Newton County, Mar 27, 1862; Captured at **Vicksburg**, MS July 4, 1863; Surrendered at **Greensboro**, NC April 26, 1865; Farmer in Newton County in 1880; Moved to Cullman County, Alabama; Buried at Pleasant Grove Methodist in Hulaco, Morgan County, Alabama.
Palasiah H. Stallings 1SGT 1827 19 Aug 1882
Appointed Commissary Sergeant for regiment Feb 1863; Captured at **Vicksburg**, MS July 4, 1863; Surrendered at **Greensboro**, NC April 26, 1865; Farmer in Henry County in 1880.
Leonidas Franklin Stevenson 2SGT 25 May 1840 31 Mar 1906
Younger brother of John Stevenson; Married Janet Thompson in Newton County, Oct 17, 1861; Promoted to **1SGT** Oct 1862; Captured at **Vicksburg**, MS July 4, 1863; Elected **2LT** Dec 26, 1863; Surrendered at **Greensboro**, NC April 26, 1865; Farmer in Rocky Plains after the war; Served 20 years on the Newton County Board of Education. Buried at Hopewell Presbyterian in Newton County. Newspaper obit reports that he was captain of the Provost Guard.
William Booker Tindall 4CPL 23 Jul 1841 20 Nov 1890
Sent to Knoxville, TN hospital Oct 1862; Captured at **Vicksburg**, MS July 4, 1863; Listed as **Commissary SGT**; Patient in Atlanta, GA hospital Dec 1863; Furloughed from Madison, GA hospital with chronic diarrhea April 10, 1865; Married Suzanne Tooke in Madison County, Florida, Dec 11, 1870; Farm laborer in Madison County, FL after the war.
Elias Woolley 2LT 4 Aug 1824 11 Sep 1893
Born in Barnwell, SC; Older brother of Isaiah, Jordan, & Vardeman Woolley; Married Susan Rhodes in Newton County, Sep 17, 1846; Resigned with health disability at Bean Station, TN July 9, 1862; Admitted to Fairgrounds Hospital #1 in Atlanta Nov 23, 1862 with bronchitis; Furloughed Nov 27, 1862; Reenlisted as a private in Decatur, GA Sep 17, 1863; In Atlanta hospital Dec 1863; Admitted Madison, GA hospital May 18, 1864; Farmer near Rockdale after the war; Buried at Almand Cemetery in Conyers.
Reuben Alexander Aaron PVT 4 Sep 1843 12 Jan 1923
Born in Jasper County; Present Sep 24, 1863; Married Emma Elizabeth Welborn about 1877; Farmer in Stansells District after the war; Filed for Confederate pension in Newton County in 1889; Reported being shot in left hand at **Battle of Resaca**, May 15, 1864. Two fingers were amputated, and vision was lost in his left eye; Buried at Carmel Cemetery in Newton County.
Henry Pryor Almand PVT 7 Jan 1821 13 Jan 1903
Born in Elbert County; Married Celia Alice Wilson in Newton County, Nov 25, 1838; Married Zella Farrell in DeKalb County, Dec 27, 1855; Filed for Confederate pension in Rockdale County in 1891; Reported working as a commissary teamster for Company F of the 42nd Georgia; Involved in a wagon accident in Alabama in November 1864 where he lost three fingers; Father of 15 children; Senior partner in the firm H.P. & D.M. Almand of Conyers; Buried at Old Conyers Cemetery in Newton County.
Suprey Anderson PVT 1812 1864

Born in South Carolina; Captured in Harrodsburg, KY; Taken to Vicksburg, MS for exchange on US Steamer *Metropolitan* Dec 4, 1862; Left in hospital in Chattanooga, TN Feb 17, 1863; Patient at Gilmer Hospital in Atlanta, June 30, 1864.

Edmond Jasper Argo PVT 17 Nov 1844 9 Sep 1919
Born in DeKalb County; Sent to Marion, MS hospital from Vicksburg April 1863; Sick at home on Dec 9, 1863 muster roll; Married Mary Ann Hollingsworth in Rockdale County, Oct 12, 1871; Railroad engineer in 1880; Buried at Eastview Cemetery in Conyers.

Francis Marion Argo PVT 14 Feb 1829 17 Apr 1908
First cousin of Edmund Argo; Married Susan Adaline Hooks in DeKalb County, Sep 6, 1853; Enlisted May 12, 1862 in Atlanta; Captured at **Vicksburg,** MS July 4, 1863; Present on Dec 9, 1863 roll; Farmer in Cobb County after the war; Buried at Collins Springs Primitive Baptist in Smyrna.

Elisha Warren Atha PVT 18 Dec 1846 13 Dec 1901
Enlisted in Dalton Jan 23, 1864; Admitted to Pettigrew Hospital in Raleigh, NC on Feb 17, 1865 with shell wound to the right side of his head; Surrendered at **Greensboro,** NC April 26, 1865; Married Roxie Ann Swords in Walton County, Mar 2, 1869; Farmer in Gwinnett County after the war; Filed for Confederate pension in both Gwinnett & Walton counties in 1898; Reported receiving head wound from exploding shell while building breastworks in Nashville. Shell fragment broke his skull; Letters of support for his application filed by J.M. Gresham, T.F. Atha, T.E. Brand, & E.T. Moon; Buried in Bay Creek Cemetery in Loganville.

Thomas F. Atha PVT 23 Oct 1843 26 Nov 1915
Born in Walton County; Uncle of Elisha Atha; Present on Dec 9,1863 roll; Wounded at **Edisto River** Feb 1865; Surrendered at **Greensboro,** NC April 26, 1865; Married Martha Frances Moon in Walton County in 1871; Worked in cotton mill in Monroe in 1900; Filed for Confederate pension in Walton County in 1910; Buried at Sharon Baptist in Loganville.

Jesse A. Bailey PVT 1831 14 Jan 1863
Born in Lexington County, SC; Enlisted in Atlanta May 13, 1862; Buried at Soldier's Rest, Cedar Hill Cemetery in Vicksburg, MS; Claim filed by his mother Mary A. Bailey in Gwinnett County.

John B. Beard PVT Sep 1845 Dec 1903
Enlisted in Dalton Feb 9, 1864; Surrendered at **Greensboro,** NC April 26, 1865; Married Mary Jane Estes in Newton County, Nov 16, 1871; Farmer near Rockdale in 1900; Buried at Prospect Methodist in Newton County.

Bertram Jefferson Bellah PVT 1844 11 Aug 1863
Enlisted in Atlanta May 13, 1862; Captured at **Vicksburg,** MS while sick in hospital, July 4, 1863; Died at home; Buried at Presbyterian Campground in Conyers; Claim filed by his father Thomas W. Bellah.

Andrew Jackson Berry PVT 1829 3 Sep 1864
Married Harriet Cole in Newton County, 20 Jan 1856; Lived next to Colonel Robert Henderson in Covington in 1860 census; Enlisted in Atlanta May 13, 1862; Admitted to Federal hospital in Chattanooga May 20, 1864 with leg wound received at **Battle of Resaca,** May 15, 1864; Died in hospital.

James Franklin Berry PVT Feb 1844 2 Jul 1915
Married Martha Taylor in Henry County, Mar 4, 1869; Farmer in Newton County after

the war; Filed Confederate pension in 1909 in Newton County; Reported joining Company F in Dec 1863; Reported being wounded in the leg at **Battle of Resaca,** May 15, 1864; Spent three weeks in Atlanta where gangrene set in. Sent home by a board of doctors; Buried at County Line Baptist in Covington.

James E. Berry PVT 1810 Before 1867

Born in South Carolina; Father of James & William Berry; Miller in Henry County in 1860; Enlisted in Madison May 25, 1863; Attached as nurse in Loring Hospital in Lauderdale, MS Jun 28, 1863; Present on Dec 9, 1863 muster roll; Shoemaker Jan 1864; Received at St. Mary's Hospital in LaGrange Aug 11, 1864.

William A. Berry PVT 19 Mar 1836 1909

Born in Henry County; Son of James E. Berry; Captured at **Vicksburg,** July 4, 1863; Present Dec 9, 1863 roll; Farmer in Henry County after the war; Filed Confederate pension in Henry County in 1895; Reported being home on sick furlough at time of surrender.

Henry T. Black PVT 26 Feb 1842

Younger brother of Thomas Black. Never returned from Col. Henderson's general furlough for all sick recruits in April 1862; Enlisted as 2LT in Company F, Georgia 2nd Reserves in Atlanta on May 7, 1864.

Thomas Shelly Black PVT 1840 1878

Appointed 1CPL of Company H of the 6th Regiment Georgia State Troops Oct 21, 1861; Mustered out in Savannah April 1862; Enlisted in Company F in Atlanta May 12, 1862; Wounded at **Battle of Chickasaw Bayou** Dec 29, 1862; Muster roll for Dec 9, 1863 lists him as "absent, at home wounded;" Married Eliza Lorene Livingston in Newton County, Jan 12, 1869; Elected elder of the Bethany Church in 1871; Moved to Covington in 1877 & became elder of Covington Presbyterian; Buried in Covington City Cemetery.

T.J. Boyd PVT

Enlisted in Atlanta May 10, 1864; Surrendered at **Greensboro,** NC April 26, 1865.

John William Bradshaw PVT 1839 6 May 1864

Married Mary Ann Edwards in Newton County, 29 Jun 1862; Enlisted in Woodson Roads, TN Aug 20, 1862; Captured at **Vicksburg,** MS July 4, 1863; Transferred to Company E of the 53rd Georgia in exchange for Frank M. Knight on Nov 1, 1863; Killed at the **Battle of the Wilderness,** Virginia.

Edward Vance Branham PVT 18 Nov 1843 10 Aug 1902

Born in Putnam County; Son of Methodist minister Walter Branham; Captured at **Vicksburg,** MS July 4, 1863; Appointed 2nd **Master SGT** 20 Sep 1863; Surrendered at **Greensboro,** NC April 26, 1865; Medical physician after the war; Married Julia Billups in Clarke County, Oct 7, 1880; Filed for Confederate pension in Newton County in 1900 as a patient in the state sanitarium; Buried at Oxford Historical Cemetery in Newton County.

John Wesley Brantley PVT 18 Feb 1844 29 Apr 1933

Enlisted in Atlanta, May 13, 1862; Captured in Washington Hospital, **Vicksburg** MS July 4, 1863; Sent with sick on US Steamer *Chouteau* to Mobile, AL; Admitted to Fairgrounds Hospital #1 in Atlanta Oct 16, 1863; Deserted from hospital Dec 21, 1863; Captured in Marietta, GA July 3, 1864 and sent to **Camp Morton,** Indiana; Exchanged at Camp Lookout, MD Mar 23, 1865; Farmer in DeKalb County after the war; Filed for Confederate pension in DeKalb County in 1919; Reported being present at surrender in Bentonville, NC; Buried at Redan Community Cemetery in DeKalb County.

Warren J.H. Brantley PVT 19 Mar 1842 9 Sep 1870

Older brother of John Brantley; Married Mary Zackens in Henry County about 1865; Died of tuberculosis contracted while a soldier; Buried at Brantley Family Cemetery in Henry County.

Bennett W. Brightwell PVT 1844 18 Jan 1863

Enlisted Aug 20, 1862 in Newton County; Died of disease in an Atlanta hospital; Claim filed by his father Moses Brightwell.

Elisha Nail Brown PVT 20 Nov 1844 22 Apr 1910

Enlisted in Atlanta May 12, 1862; Captured at **Vicksburg,** MS July 4, 1863; Present on Dec 9, 1863 muster roll; Married Sudie Elizabeth Hale in Newton County, Dec 6, 1866; Miller in Rockdale in 1880: Atlanta jeweler in 1900; Filed Confederate pension in Fulton County in 1899; Reported being shot through the hand at the Battle of New Hope Church, May 25, 1864 resulting in amputation of finger.

William Asa Brown PVT 15 Dec 1846 20 Oct 1921

Enlisted in Covington April 19, 1864; Surrendered at **Greensboro,** NC April 26, 1865; Married Mary Frances Duke in Winston County, AL Dec 21, 1871; Farmer in Winston County after the war; Buried at Ashridge Methodist in Winston County, Alabama.

Rhesa Gilmore Butler PVT 5 Jan 1832 12 Jul 1863

Married Permelia Plunkett in Newton County, Jan 12, 1854; Enlisted in Atlanta May 13, 1862; Sent to hospital in Vicksburg, MS; Died in Confederate hospital in Meridian, MS; Claim filed by his wife Permelia Butler.

Young Drewry Butler PVT 15 Aug 1842 17 Mar 1926

Younger brother of Rhesa Butler; Enlisted in Atlanta May 13, 1862; Discharged from Knoxville, TN hospital with Certificate of Discharge Nov 12, 1862; Appointed 1CPL of Company F, 2nd Georgia Reserves on May 7, 1864; Married Nancy Adaline Rogers in Douglas County about 1865; Farmer in Douglas County after the war; Buried at Prays Mill Baptist in Douglasville.

John M. Cason PVT 9 Jun 1833 25 Mar 1921

Born in Walton County; Married Martha Patterson in Newton County, Dec 5, 1858; Captured at **Vicksburg,** MS July 4, 1863; Sick at home on Dec 9, 1863 muster roll; Surrendered at **Greensboro,** NC April 26, 1865; Farmer in Newton County after the war; Buried at Sardis Cemetery in Newton County.

Francis Marion Chandler PVT 2 Sep 1842 5 Dec 1921

Born in Cherokee County, AL; Wounded in face at **Battle of Chickasaw Bayou,** MS Dec 29, 1862; Captured at **Battle of Baker's Creek,** May 16, 1863; Sent to **Camp Morton,** IN; Exchanged at Camp Lee near Richmond, VA July 4 1863; Surrendered at **Greensboro,** NC April 26, 1865; Married Lucy Granger in Newton County, May 5, 1870; Farmer near Rockdale after the war; Filed for Confederate pension in Rockdale County in 1892; Reported being shot in the right cheek at Battle of Chickasaw Bayou on Dec 29, 1862; Ball lodged in the back of his skull and caused him issues the rest of his life; Buried at Philadelphia Methodist in Rockdale.

George Washington Chandler PVT May 1846 19 Aug 1911

Born in Cherokee County, AL; Younger brother of Frank Chandler; Enlisted in Dalton Oct 15, 1864; Surrendered at **Greensboro,** NC April 26, 1865; Married Sophronia Caroline Everett in DeKalb County, July 14, 1867; Farmer near Rockdale after the war; Filed for Confederate pension in Rockdale County in 1906; Application endorsed by E.L. Turner & W.P. Plunkett; Buried at Underwood Memorial in Conyers.

Ezekiel Jesse Childress PVT 16 Dec 1846 2 Feb 1902
Enlisted in Atlanta Feb 7, 1864; Surrendered at **Greensboro**, NC April 26, 1865; Married Parthena Willis in Fulton County, Dec 16, 1868; Farmer at Bryants Crossroads in Fulton County after the war; Buried at Utoy Primitive Baptist in Atlanta.
Elijah Willis Christian PVT 18 Dec 1830 28 Oct 1862
Younger brother of William B. Christian. Died in Frankfurt, Kentucky hospital; File made by his father E.W. Christian.
Reuben M. Christian PVT 1834 25 Dec 1863
Captured in Frankfurt, KY Oct 16, 1862; Taken to Vicksburg, MS aboard US Steamboat *Cairo* and exchanged on Nov 29, 1862; Captured at **Vicksburg**, MS July 4, 1863; Died in Institute Hospital, Atlanta; Buried at Oakland Cemetery, Atlanta, Georgia.
James Drayton Cleland PVT 1838 After 1895
Enlisted in Atlanta May 12, 1862; Captured at **Vicksburg**, MS July 4, 1863; Present on Dec 9, 1863 muster roll; Married Elizabeth Saye in Fulton County, April 22, 1880; Night watchman in Atlanta in 1880; City fireman in 1885.
John Crowell PVT 13 Nov 1828 1902
Born in North Carolina; Wagon maker in 1860; Enlisted in Atlanta May 13, 1862; Present on Dec 9, 1863 muster roll; Farmer & wheel wright in Rockdale County after the war; Filed for Confederate pension in Rockdale County in 1897.
John David Dameron PVT Dec 1845 19 Jan 1916
Late enlistment; Surrendered at **Greensboro**, NC April 26, 1865; Married Susan Ann Harwell in Fulton County, Nov 19, 1874; Dry goods salesman in Atlanta after the war; Buried at Oakland Cemetery in Atlanta.
James R. Daniel PVT 26 Apr 1842 7 Jan 1911
Born in Randolph County, AL; Enlisted in Company H, 6th Georgia State Troops Oct 21, 1861; Mustered out in Savannah; Enlisted in Company F in Atlanta May 12, 1862; Appointed **2CPL**; Captured at **Vicksburg**, MS July 4, 1863; Present on Dec 9, 1863 muster roll; Captured in fight for **Edisto Bridge** near Orangeburg, SC Feb 11, 1865; Released at **Point Lookout**, MD June 26, 1865; Married Sarah Hawkins in Newton County, Jan 13, 1870; Farmer in Rockdale after the war; Filed for Confederate pension in Rockdale County in 1903; Reported getting home from prison on July 5, 1865; Buried at Honey Creek Baptist in Conyers.
Benjamin Franklin Day PVT 1840 28 Aug 1862
Married Harriet Catherine Kirkpatrick in Newton County, Oct 12, 1857; Took in training at Camp McDonald; Died at home in Newton County.
John Michael Day PVT 12 Sep 1846 8 Jun 1924
Enlisted at Jones Station, TN Aug 8, 1862; Captured at **Vicksburg**, MS July 4, 1863; Sick at home on Dec 9, 1863 muster roll; Married Elizabeth Parnell in Newton County, 1 Jan 1867; Farmer in Rockdale County after the war; Filed for Confederate pension in Rockdale County in 1919; Reported leaving command after the Battle of Bentonville, NC; Buried at Underwood Methodist in Conyers.
Thomas Jefferson Day PVT 23 Oct 1823 9 Apr 1862
Older brother of Benjamin Day and father of John Day; Married Mary Elizabeth Phillips in Newton County, Dec 23, 1844; Died of measles during training; Buried at Salem Baptist in Conyers.
Wiley Floyd Dennard PVT 17 Aug 1847 6 Mar 1931

Enlisted July 15, 1864; Surrendered at **Greensboro**, NC April 26, 1865; Married Mary Ann Winburn in Newton County, Dec 12, 1867; Married Mary Ogle in Fulton County, May 8, 1886; Worked for the Sanitary Department of Atlanta after the war and as a **night watchman for the city; Filed for Confederate pension in Fulton County in 1901; Application endorsed by Colonel L.P.** Thomas who described Dennard as "a brave soldier who was never absent from his assignment"; Buried at Hollywood Cemetery in Atlanta.

David Richmond Dial PVT 8 May 1834 30 Oct 1917

Born in Walton County; Married Sarah Shanks in Newton County Dec 8, 1858; Captured at **Vicksburg**, MS July 4, 1863; Present on Dec 9, 1863 muster roll; Married Mary Ann Victoria Starr in Rockdale County, May 12, 1874; Farmer in Stansells in 1880; Filed for Confederate pension in Newton County in 1899; Reported being discharged at Bentonville "at the surrender;" Member of UCV Camp #305 in Covington; Buried at Carmel Cemetery in Newton County.

Edmund Wiley Runnels Dobbs PVT 8 Dec 1833 7 May 1910

Born in Franklin County; Married Mary Kirkpatrick in Newton County, Jan 30, 1862; Captured at Washington Hospital in **Vicksburg**, MS July 4, 1863; Among sick delivered by US Steamer *Chouteau* to Fort Morgan, AL Aug 6, 1863; Surrendered at **Greensboro**, NC April 26, 1865; Present on Dec 9, 1863 muster roll; Farmer in Newton County after the war; Filed for Confederate pension in Newton County in 1898; Reported being wounded at **Battle of Resaca** May 15, 1864 and **Battle of Nashville**, Dec 16, 1864; Buried at Lane Farm Cemetery in Newton County.

Nathan Judson Dobbs PVT 1832 6 Aug 1863

Married Mary Amanda Shaw in Bartow County, Dec 28, 1858; Enlisted in Atlanta May 13, 1862; Captured at **Vicksburg**, MS hospital July 4, 1863; Died en route on Federal steamer to New Orleans; Claim filed by his widow Mary A. Dobbs in Newton County.

Doctor H. B. Doolittle PVT 3 Nov 1843 10 Apr 1862

Died of measles in Tennessee; Buried at Oxford Historical Cemetery in Newton County; Inscription reads, "He died a soldier's death far from home in Confederate service."

John Simmons Downs PVT 5 Dec 1842 17 Jul 1900

Married Sarah Freeman in Newton County, April 14, 1860; Present on Dec 9, 1863 roll; Farmer near Lawrenceville in 1900; Filed for Confederate pension in Gwinnett County in 1899; Reported being wounded defending Mill Creek Gap on May 10, 1864, and wounded in the arm below the left shoulder during skirmish near Orangeburg, SC Feb 17, 1865; Still an open sore in 1900; Buried at Chestnut Grove Baptist in Grayson.

Jesse C. Edwards PVT 1838 8 Dec 1862

Married Zillah Haney in Newton County, June 22, 1862; Enlisted July 20, 1862; Died of camp fever in University Hospital in Knoxville; Buried in Bethel Confederate Cemetery in Knoxville, TN.

William Franklin Edwards PVT 22 Feb 1844 21 Mar 1926

Captured at **Vicksburg**, MS July 4, 1863; Present on Dec 9, 1863 muster roll; Made **3SGT** & regimental color bearer after **Battle of Resaca**, May 15, 1864; Surrendered at **Greensboro**, NC April 26, 1865; Married Mary Jane Gore in Newton County, Nov 25, 1866; Member of UCV Camp #305 in Covington; Brought the regimental battle flag to all regimental reunions and UCV events; Filed for Confederate pension in Newton County in 1919; Farmer in Stansells District after the war; Buried at Hominy Creek Baptist in Rockdale.

Joshua Elisha Ellis PVT 1810 9 Jun 1886
Born in Hancock County; Married Sarah Ann Hughes in Walton County, Sep 5, 1833; Father of John Ellis; Late enlistment; Wounded at **Binniker's Bridge**, SC Feb 9, 1865; Farmer in Newton County.
John Daniel Ellis PVT 29 Jun 1844 31 Aug 1925
Married Sarah Ann Ewing in Newton County, Nov 4, 1860; Enlisted Aug 8, 1862 at Jones Station, TN; Captured & exchanged in the Kentucky campaign; Captured at **Vicksburg,** MS July 4, 1863; Present on Dec 9, 1863 muster roll; Surrendered at **Greensboro**, NC April 26, 1865; Farmer at Gum Creek after the war; Member of UCV Camp #305 in Covington; Filed for Confederate pension in Newton County in 1910; Application endorsed by W.F. Edwards; Buried at Sardis Cemetery in Newton County.
William F. Findley PVT 22 Aug 1862
Died of camp fever in Tazewell, TN; Claim filed by his wife Marian Findley in Newton County.
William W. Freeman PVT 1837 Before 1886
Married Darcus Parker in DeKalb County, Mar 4, 1860; Enlisted in Atlanta May 13, 1862; Captured at **Vicksburg,** MS July 4, 1863; Surrendered at **Greensboro,** NC April 26, 1865; Farmer in Newton County in 1880; Widowed wife married Luke Howell in DeKalb County, Sep 3, 1886.
John Burton Graham PVT 5 Oct 1826 1 Jul 1914
Married Susan Hughey in Newton County, Mar 4, 1862; Captured at **Vicksburg,** MS July 4, 1863; Present on Dec 9, 1863 muster roll; Captured along Atlanta siege lines Aug 7, 1864; Prisoner at **Camp Chase,** OH; Sent to Point Lookout, MD and exchanged on Mar 27, 1865. Farmer in Rockdale County after the war; Member of UCV Camp #1684 in Conyers; Buried at Eastview Cemetery in Conyers.
Lindsey S. Grainger PVT 22 Mar 1832 1911
Married Sarah Elizabeth Smith in Newton County, Dec 7 1855; Sent to GA prison March 2, 1859 for "stabbing;" Appointed waggoner by order of Colonel Henderson April 1862; Captured at **Vicksburg,** MS July 4, 1863; Present on Dec 9, 1863 muster roll; Filed for Confederate pension in Rockdale County in 1901; Reported being detailed to guard the wagon train from Mississippi to South Carolina. He was with the wagon train in Griffin, GA at the time of the surrender; Carpenter and farmer in Rockdale after the war; Buried in Philadelphia Church in Conyers.
William Jackson Green PVT 25 Jul 1844 1919
Born in South Carolina; Enlisted in Atlanta May 13, 1862; Captured at **Vicksburg,** MS hospital July 4, 1863; Present on Dec 9, 1863 muster roll; Wounded and captured at **Battle of Resaca** May 15, 1864; Right leg amputated at US hospital in Nashville,, TN; Prisoner at **Camp Douglas,** IL; Exchanged at Point Lookout, MD Mar 18, 1865; Married Nancy Johnson in Newton County, Aug 4, 1869; School teacher in Ashland, AL in 1880; Buried at Liberty Cemetery in Clay County, Alabama.
James Henry Griffin PVT
See Regimental officers
Winston Gunn PVT 1818
Born in Jones County; Married Mildred Smith in Newton County, April 22, 1850; Accidently wounded in back and hips in Kentucky Oct 1, 1862; Discharged for disability, Feb 20, 1863; Farm laborer in Newton County in 1870.

Ira Lawson Gunter PVT 20 Dec 1830 30 Oct 1904
Married Ann Elizabeth Williamson in Henry County, Sep 30, 1852; A practicing physician in 1860; Enlisted in Atlanta May 13, 1862; Served as nurse in Bean Station, TN; Discharged from company to serve in Confederate hospitals in Atlanta; Passed his Confederate Medical Boards Mar 12, 1863; Physician and preacher in Henry County after the war; Elected to Georgia Assembly from Henry County in 1889; Buried at Ozias Primitive Baptist in Stockbridge.

Jeremiah B. Hamby PVT 1833 18 Jul 1833
Son of Rev. Isaac Hamby; Sent to Vicksburg, MS hospital; Captured in **Battle of Baker's Creek,** MS, May 16, 1863; POW at **Camp Morton,** IN; Died in hospital at Fort Delaware, DE; Buried at Finn's Point National Cemetery, Salem, New Jersey; Claim filed by his father Isaac Hamby.

George L. Hamilton PVT 1832 After 1880
Married Martha Hardin in Newton County, Mar 17, 1853; Enlisted in Atlanta May 13, 1862; Promoted to **2SGT**; Captured at **Vicksburg,** MS July 4, 1863; Muster roll for Dec 9, 1863 shows "absent sick at home;" Farmer in Rockdale after the war.

Alexander F. Hammock PVT 12 Mar 1825 5 Apr 1865
Older brother of Mansel Hammock; Elected **2CPL** Oct 5, 1862; Died in Canton, MS hospital; Buried at Canton City Cemetery, Mississippi.

Mansel Washington Hammock PVT 27 Apr 1830 8 Jul 1863
Enlisted in Atlanta May 13, 1862; Sent to Vicksburg, MS hospital April 1863; Captured in Mississippi hospital & sent to Fort Delaware, DE; Buried at Finn's Point National Cemetery, Salem, New Jersey; Claim filed by his widow Mary E. Hammock in Newton County.

Robert J. Hardin PVT 1845 17 May 1864
Born in Henry County; Died from wounds received at **Battle of Resaca,** May 15, 1864 in 20th Corps army hospital.

Samuel H. Hardin PVT 1837 19 May 1864
Married Lucy Ann Seamans in Newton County, Dec 5, 1861; Enlisted in Atlanta May 13, 1862; Captured at **Vicksburg,** MS July 4, 1863; Present on Dec 9, 1864 muster roll; Died from wounds received at **Battle of Resaca,** May 15, 1864.

Thomas H. Hardin PVT 3 May 1830 25 Dec 1905
Older brother of Robert & Samuel Hardin; Married Mary McCullum in Newton County, Feb 8, 1855; Enlisted in Atlanta May 13, 1862; Captured at **Vicksburg,** MS July 4, 1864; Absent in Atlanta hospital Dec 9, 1864 muster roll; Farmer in Rockdale after the war; Buried at Ebenezer Methodist in Conyers.

John A.H. Harper PVT 1811 After 1870
Born in South Carolina; Married Susan Oliver in Elbert County, Sep 25, 1832; Hotelkeeper in Covington in 1860; Enlisted at Tiner's Station, TN Nov 11, 1863; Captured at **Battle of Bentonville** Mar 21, 1865; Shipped from New Bern, NC to Hart's Island, New York Harbor & released June 15, 1865.

Addison P. Harper PVT 1827 After 1880
Born in North Carolina; Present on Dec 9, 1863 roll; Wounded at **Battle of Nashville,** TN Dec 14, 1864; Admitted to Fairgrounds Hospital #2 in Corinth, MS with gunshot wound Dec 28, 1864; Farmer near Covington in 1880.

Uel M. Harper PVT 1824 Before 1882

Born in South Carolina; Married Roxie Dixon about 1856; Merchant in Conyers in 1860; Captured at **Vicksburg**, MS July 4, 1863; Present on Dec 9, 1863 muster roll; Liquor dealer in Conyers in 1870; Store clerk in 1880; Will probated in Rockdale County in 1882.

George Floyd Hays PVT 19 Feb 1829 30 Aug 1901

Married Elizabeth Frances Marks in Newton County, Feb 3, 1857; Admitted to Fairgrounds Hospital #1 in Atlanta Oct 17, 1862; Returned to duty Jan 15, 1863; Admitted to Fairgrounds Hospital #1 in Atlanta Feb 5, 1863; Returned to duty Mar 16, 1863; Captured at **Vicksburg**, MS July 4, 1863; Present on Dec 9, 1863 muster roll; Captured near Atlanta Aug 7 1864; Prisoner at **Camp Chase**, OH; Released from Point Lookout, MD May 18, 1865; Farmer in Morgan County in 1880; Farmer in Newton County 1900.

Henry G. Hays PVT 1843 14 Feb 1863

Nephew of George Hays; Died of fever at home; Buried at Martin-Hays Cemetery in Newton County; Claim filed by his father Martin H. Hays.

Joseph William Hays PVT 20 Jul 1843 30 Oct 1863

Half-brother of George Hays; Detailed to wait on the sick at Maynardville, TN by Colonel Henderson April 25, 1862; Captured at **Vicksburg**, MS July 4, 1863; Died at home in Newton County; Claim filed by his father Robert L. Hays.

William David Heath PVT 6 Jan 1842 31 Oct 1922

Discharged with disability April 12, 1862; Married Mary Williams in Newton County, Oct 23, 1862; Enlisted in Company H, 1st Regiment, Confederate Engineer Troops Oct 22, 1863; Became a Methodist preacher after the war; Buried at Oak Hill Cemetery in Prattville, Alabama.

Thomas Moreland Hicks PVT 30 Jun 1829 21 Jan 1904

Born in South Carolina; Brother of Warren & William Hicks; Married Elizabeth Martha Downs in Newton County, July 14, 1854; Enlisted in Atlanta May 13, 1862; Captured in Nicholasville, Kentucky Oct 15, 1862; Taken on US Steamer *Metropolitan* to Vicksburg, MS and exchanged on Dec 4, 1862; Surrendered at **Greensboro**, NC April 26, 1865; Farmer in Newton County after the war; Father of 11 children; Filed for Confederate pension in Newton County in 1905; Application endorsed by J.W. Stephenson; Buried at Prospect Methodist in Covington.

Warren R. Hicks PVT 1827 2 Dec 1862

Born in South Carolina; Married Tinsey Downs in Newton County, July 29, 1849; Enlisted in Atlanta with his brother Thomas Hicks May 13, 1862; Captured in Nicholasville, KY Oct 15, 1862; Taken on US Steamer *Metropolitan* to Vicksburg, MS for exchange, but died en route.

Robert S. Hilley PVT 22 Jun 1822 25 Sep 1897

Born in Shelby County, Alabama; Married Louisiana O'Flarity; Detailed as waggoner; Injured and discharged after a mule fell & broke his hip Aug 31, 1862; Stock dealer in Atlanta in 1880; Buried at Westview Cemetery in Atlanta.

Christopher Columbus Hudson PVT 29 Mar 1831 22 May 1921

Married Sarah Mann in Newton County, Aug 18, 1859; Discharged with disability in Knoxville, TN Sep 1, 1862; Exempt as blacksmith in 1864 Newton County militia census; Farmer in Rockdale in 1880; Farmer in Bosque County, TX in 1900; Filed for Confederate pension in Bosque County in 1905; Reported serving in the 42nd until "J.E. Johnston surrendered;" Buried at Riverside Cemetery in Iredell, Texas.

William Henry C. Hudson PVT 1832 27 Jun 1863

Buried at Soldier's Rest, Cedar Hill Cemetery in Vicksburg, MS.
Aaron Green Hulsey PVT 4 Mar 1820 28 Jun 1880
See regimental officers.
William James PVT 15 May 1864
Late enlistment; Killed at **Battle of Resaca**, May 15, 1864.
John N. Jones PVT 19 May 1845 27 May 1915
Enlisted in Dalton Jan 29, 1864; Surrendered at **Greensboro**, NC April 26, 1865; Farmer in Walton County after the war; Filed for Confederate pension in Walton County in 1895; Reported that, while building breastworks at Resaca, GA, he fell on his right shoulder while carrying a heavy log. Spent six months recovering in a Confederate hospital; Buried at Bay Creek Baptist in Walton County.
George M.D. King PVT 1844 After 1880
Admitted to St. Mary's Hospital in Dalton Nov 6, 1862; Captured at **Vicksburg**, MS July 4, 1863; Present on Dec 9, 1863 muster roll; Married Mary Lou Veal in Newton County, July 6, 1865; Farmer in Stansells District after the war.
John Marion King PVT 29 May 1847 6 Jul 1909
Younger brother of George King; Born in Alabama; Married Cynthia Woolley in Newton County, Jan 11, 1866; Farmer in Henry County in 1900; Filed for Confederate pension in Henry County in 1891; Reported being shot in the chest with ball lodged by his spine at **Battle of Resaca** May 15, 1864; Buried at Philadelphia Methodist in Conyers.
John Franklin Knight PVT 1834 3 Mar 1890
Married Mary Ewing in Newton County, Mar 15, 1858; Enlisted in Dalton Aug 20, 1862; Surrendered at **Greensboro**, NC April 26, 1865; Farmer in Gwinnett County in 1880.
Franklin Means Laird PVT 7 Jun 1832 27 Dec 1863
Married Minerva Wells in Newton County, Sep 14, 1854; Enlisted at Woodson's Crossroads, TN Aug 20, 1862; Patient in Meridian, MS hospital Mar 21, 1863; Hospital cook in July 1863; Furloughed from Fairgrounds Hospital #2 in Atlanta Dec 11, 1863; Died at home; Buried at Philadelphia Methodist in Conyers; Claim filed by his wife Minerva Lard.
Alfred Benjamin Lane PVT 14 Mar 1831 31 Dec 1911
Born in Clarke County; Married Mary Margaret Johnson in Newton County, Dec 13, 1850; Captured at **Vicksburg**, MS July 4, 1863; Admitted to Fairgrounds Hospital #1 in Atlanta Oct 2, 1863 with fever; Present on Dec 9, 1863 muster roll; Teamster near Dalton Feb 1864; Ran a mill near McDonough in 1880; Member of Masonic Order; Filed for Confederate pension in Henry County in 1897; Reported being present until the surrender of General Johnston; Buried at Bethany Baptist in McDonough.
Matthew Phillip Lane PVT 1823 12 Feb 1863
Married Mary Hurst in Newton County, Dec 12, 1850; Enlisted in Atlanta May 13, 1862; Will probated in Newton County Sep 6, 1869.
Silas Wesley Leak PVT 1834 1870
Married Victoria Boring in Coweta County, Nov 13, 1859; Physician in Butler County, Alabama in 1860; Enlisted at Cumberland Gap, TN April 30, 1862; Detailed as hospital steward by Colonel Henderson May 12, 1862; Provided medical discharge by Medical Director G.B. Thornton Sep 8, 1862; Died in Panola County, Texas.
Leonidas Felix Livingston PVT 3 Apr 1832 11 Feb 1912
See Regimental officers.
Rev. J.T. Lowe PVT 1844

Enlisted in Palmetto Sep 19, 1864; Surrendered at **Greensboro**, NC April 26, 1865; Student at Emory College in 1870; Methodist minister after the war.

Richard M.J. Lunsford PVT 5 Dec 1843 16 Jun 1910

Captured at **Vicksburg**, MS July 4, 1863; Present on Dec 9, 1863 muster roll; Patient at Catoosa Hospital in Griffin July 27, 1864; Furloughed from Charlotte, NC hospital Jan 24, 1865; Farmer in Newton County after the war; Filed for Confederate pension in Newton County in 1888; Reported being hit by a shell in the right thigh at **Battle of Atlanta**, July 22, 1864, causing a compound-fracture & shortening his leg by four inches; Application endorsed by Lt. T.A. Starr; Buried at Carmel Cemetery in Newton County.

William Joel Lunsford PVT 16 Jul 1842 22 Mar 1912

Wounded at Cumberland Gap, TN June 18, 1862; Captured at **Vicksburg**, MS July 4, 1863; In Atlanta hospital on Dec 9, 1863 muster roll; Surrendered at **Greensboro**, NC April 26, 1865; Married Mary Frances Adams in 1867; Farmer in Newton County after the war; Filed for Confederate pension in Monroe County in 1907; Application endorsed by Silas Morgan; Buried at Maynard Baptist in Newton County.

Joel Patterson Marks PVT 5 Jul 1837 10 Oct 1906

Married Mary Elizabeth Almand in Newton County, Feb 23, 1864; Captured at **Vicksburg**, MS July 4, 1863; Present on Dec 9, 1863 muster roll; Farmer in Rockdale in 1900; Buried at Mt. Zion Primitive Baptist in Conyers.

John Calvin Marks PVT 11 Nov 1823 30 Jun 1863

Older brother of Joel Marks; Married Martha Grenada in Newton County, Oct 4, 1849; Married Susannah Grenada in Newton County, May 14, 1854; Sent to Vicksburg, MS hospital; Died of influenza in Lauderdale Springs, MS hospital.

William Reese McClung PVT 1829 16 Dec 1862

Born in South Carolina; Married Mary Ann Stewart about 1852; Died at home; Buried at McClung Cemetery in Conyers; Claim filed by his widow Mary A. McClung.

James B. McCord PVT 1841 About 1895

Wounded at **Battle of Baker's Creek**, MS May 16, 1863; Captured at **Vicksburg**, MS July 4, 1863; Absent at Atlanta hospital on Dec 9, 1863 muster roll; Married Sarah Killingsworth in Clay County, Jan 27, 1870; Farmer in Clay County after the war.

Ferdinand M. McCullough PVT Dec 1828 1906

Born in Jasper County; Married Lourania Slagle in Newton County, Jan 13, 1853; Captured at **Vicksburg**, MS July 4, 1863; Present on Dec 9, 1863 muster roll; Surrendered at **Greensboro**, NC April 26, 1865; Farmer in Newton County after the war; Filed for Confederate pension in Newton County in 1905; Application endorsed by J.W. Stephenson; Buried at Tabernacle Cemetery in Newton County.

Jefferson A. McDonald PVT 1846 After 1910

Younger brother of Marcus & Oscar McDonald; Late enlistment; Married Rebecca A. Stanton in Newton County, Sep 26, 1867; Married Ann Hays in Newton County, Jan 6, 1876; Farmer in Newton County 1910.

Marcus A. McDonald PVT 23 Jan 1842 27 Feb 1907

Captured at **Vicksburg**, MS July 4, 1863; Present on the Dec 9, 1863 muster roll; Married Harriet Stowers in Newton County, Dec 30, 1863; Captured in Covington Aug 21, 1864; Released from **Camp Chase**, OH June 11, 1865; Farmer in Rockdale in 1880; Farmer in Cranford, AL in 1900; Buried at Methodist Church Cemetery in Arab, Marshall County, Alabama.

Oscar Virgil McDonald PVT 7 Sep 1835 After 1900
Married Martha Ann Powell in 1856; Enlisted in Atlanta, May 13, 1862; Captured at **Vicksburg,** MS July 4, 1863; Present on Dec 9, 1863 muster roll; Surrendered at **Greensboro,** NC April 26, 1865; Carpenter in Atlanta after the war; Filed for Confederate pension in Fulton County in 1896; Application endorsed by J.T. Stewart.
Ethan A. Melton PVT 11 Nov 1823 27 Sep 1863
Married Elizabeth Reid in Newton County, Jan 6, 1846; Enlisted in Decatur, GA April 30, 1863; Captured at **Vicksburg,** MS July 4, 1863; Buried at Salem Methodist in Covington; Claim filed by his wife Elizabeth M. Melton.
Thomas D. Miers PVT 1828 13 Feb 1863
Married Sarah Snead in Gwinnett County, Aug 3, 1853; Enlisted in Atlanta May 13, 1863; Sent to Vicksburg, MS hospital where he died; Claim filed by his wife Sarah Miers.
Henry P. Mitchell PVT 1831 1863
Married Pitney Crenshaw in Newton County, May 22, 1856; Enlisted in Atlanta, May 13, 1862; Captured at **Vicksburg,** MS July 4, 1863; Listed on Dec 9, 1863 muster roll as "absent, left in Miss."
John W. Moore PVT Feb 1837 After 1910
Enlisted at Woodson's Crossroads, TN Aug 20, 1862; Captured at **Battle of Baker's Creek,** MS May 16, 1863; Sent to **Camp Morton,** IN; Exchanged at Fort Delaware, DE July 4, 1863; Sick at home on Dec 9, 1863 muster roll; Received clothing at Marshall Hospital in Columbus, GA July 22, 1864; Married Mary Jeffers in Newton County, Dec 18, 1865; Farmer in Rockdale County after the war; Filed for Confederate Pension in Rockdale County in 1895; Reported being with his company until the Greensboro, NC surrender.
Silas Morgan PVT 20 Apr 1846 23 Nov 1922
Enlisted in Covington April 23, 1864; Surrendered at **Greensboro,** NC April 26, 1865; Married Mary Hackney in Taliaferro County, Nov 17, 1869; Farmer in Newton County after the war; Filed for Confederate pension in Newton County in 1910; Application endorsed by J.W. Stephenson; Buried in Starrsville Cemetery in Newton County.
James David Murdock PVT 15 Dec 1826 21 Aug 1863
Born in Newberry County, SC; Married Elizabeth Ann Plunkett in Newton County, Sep 9, 1847; Farmer in Haralson County in 1860; Enlisted in Vicksburg, MS Feb 16, 1863; Captured at **Battle of Baker's Creek** May 16, 1863; Sent to **Camp Morton,** IN; Exchanged at Camp Delaware, DE July 4, 1863; Died in Polk Confederate hospital in Rome, GA; Buried at Myrtle Hill Cemetery in Floyd County.
John P. Murdock PVT 1829 After 1880
Born in South Carolina; Younger brother of James Murdock; Married Elizabeth Plunkett in Newton County, Dec 29, 1853; Enlisted in Atlanta May 13, 1862; Captured at **Vicksburg,** MS hospital July 4, 1863; Present on Dec 9, 1863 muster roll; Surrendered at **Greensboro,** NC April 26, 1865; Farmer in Whiteville, Dorsey County, AR in 1880.
Green K.P. Odom PVT 1845
Younger brother of John Odom; Enlisted in Vicksburg, MS Mar 5, 1863; Captured at **Vicksburg,** MS July 4, 1863; Present on Dec 9, 1863 muster roll.
John W. Odom PVT 24 Jan 1831 22 Apr 1890
Married Rebecca Tucker in Newton County, Mar 5, 1856; Enlisted in Atlanta May 13, 1862; Captured in Kentucky; Taken by steamboat from Louisville to Vicksburg, MS and exchanged on Nov 29, 1862; Captured at **Vicksburg,** MS July 4, 1863; Present on Dec 9,

1863 muster roll; Railroad employee killed in accident at Whitehall Street crossing in Atlanta; Buried at Oakland Cemetery in Atlanta.

Joel A. Owens PVT 20 Oct 1862

Claim filed by father William Owens.

William T. Owens PVT 8 Apr 1843 26 Apr 1862

Buried at Owens Cemetery in Newton County; Claim filed by his father Anderson Owens.

James Aaron Parker PVT 25 Dec 1834 25 Dec 1902

Married Maza Veal in DeKalb County, Nov 27, 1856; Enlisted in Atlanta May 13, 1862; Captured at Washington Hospital in **Vicksburg,** MS July 4, 1863; Transported with sick on U.S. Steamer *Suffolk* from Vicksburg to Mobile & exchanged off Fort Morgan July 22, 1863; Absent sick at home on Dec 9, 1863 muster roll; Farmer in Rockdale County after the war; Buried at Parker Cemetery in Conyers.

John Wesley Persall PVT 20 Aug 1845 13 Feb 1920

Enlisted in Decatur Sep 29, 1863; Surrendered at Garrett House Hospital in Greensboro, NC April 29, 1865; Married Rebecca Baker in Newton County, July 15, 1877; Farmer in Rockdale County after the war; Filed for Confederate pension in Rockdale County in 1895; Reported losing his two middle fingers of his right hand at **Battle of Bentonville** Mar 19, 1865; Application supported by Captain J.M. Summers; Buried at Prospect Methodist in Covington.

William A. Persall PVT 11 Apr 1836 31 Jul 1905

Older brother of John Persall; Married Susan Elmira Moore Aug 30, 1860; Captured at **Vicksburg,** MS July 4, 1863; Present on Dec 9, 1863 muster roll; Deserted regiment in McDonough Sep 6, 1864; Laborer in Winston County, AL in 1900; Filed for Confederate pension in DeKalb County, AL in 1897; Died at Confederate Soldier's Home in Atlanta; Buried at Confederate Knoll at Westview Cemetery in Atlanta.

Richard F. Phillips PVT

Married Nancy Dobbs in Newton County, Dec 26, 1858; Captured at **Vicksburg,** MS July 4, 1863; Present on Dec 9, 1863 muster roll; Wounded at **Battle of Resaca** May 15, 1864.

Noel Marcellus Pitts PVT 1841 1 Jun 1863

Elected **1LT** Sep 1, 1862; Died of smallpox at Vicksburg, MS hospital.

Charles Wesley Plunkett PVT 1820 1864

Born in South Carolina; Married Malinda Mangrum about 1841; Enlisted in Decatur Sep 5, 1863; Absent in Atlanta Hospital Dec 9, 1863 muster roll.

Minor Hezekiah Plunkett PVT 1841

Married Martha Jane Hamilton in Newton County, Aug 19, 1860; Captured at **Vicksburg,** MS July 4, 1863; Present Dec 9, 1863 muster roll.

James Wiley Plunkett PVT 1821 1893

Born in South Carolina; Enlisted in Decatur Sep 5, 1863; Present on Dec 9, 1863 muster roll; Admitted to Ocmulgee Hospital in Macon July 22, 1864; Married Nancy Underwood in Gwinnett County, April 5, 1874; Farmer in Gwinnett County after the war.

William Pinckney Plunkett PVT 17 Dec 1844 8 Dec 1931

Enlisted in Covington Mar 1, 1864; Furloughed from Fairground Hospital #1 May 14, 1864; Detailed as bridge builder on Memphis & Chattanooga Railroad Dec 1864; Surrendered at **Greensboro,** NC April 26, 1865; Married Seleta Ann Hamilton in Newton County, Dec 16, 1866; Farmer in Rockdale County after the war; Member of UCV Camp # 1684 in Conyers; Filed for Confederate pension in Rockdale County in 1919; Buried in

Old Conyers Cemetery.

Isom Berry Pritchett PVT 15 Oct 1826 4 Jul 1907

Married Mary Ann Henderson in Newton County, Jan 23, 1849; Enlisted in Atlanta, May 13, 1862; Captured at **Vicksburg**, MS July 4, 1863; Wounded at **Battle of Missionary Ridge**, TN Nov 25, 1863; Filed for Confederate Pension in Henry County in 1897; Reported being wounded & permanently disabled at **Battle of Resaca**, May 15, 1864; Farmer in Clayton County in 1880; Buried at Burks Cemetery in Stockbridge.

Pleasant L. Puckett PVT 19 Jul 1828 28 Jul 1894

Married Winnie Rice in Newton County, May 18, 1856; Captured at **Vicksburg**, MS July 4, 1863; Wounded at **Battle of Resaca**, GA May 15, 1864; Admitted to Fairgrounds Hospital #1 in Atlanta May 16, 1864 with gunshot wound; Furloughed June 4, 1864; Surrendered at **Greensboro**, NC April 26, 1865; Farmer in Rockdale County after the war; Buried at Salem Baptist in Conyers.

William Simmons Puckett PVT 24 Jun 1842 25 Dec 1920

Younger brother of Pleasant Puckett; Captured at **Vicksburg**, MS July 4, 1863; Surrendered at **Greensboro**, NC April 26, 1865; Married Elizabeth Lancaster in Newton County, Mar 8, 1866; Miller in Rockdale County after the war; Filed for Confederate pension in Rockdale County in 1909; Buried at Salem Baptist in Conyers.

William Eugene Reagan PVT 24 Jul 1832 10 Jan 1907

Older brother of Stephen Reagan in Company E; Born in North Carolina; Married Susan Green in Newton County, Sep 28, 1856; Enlisted in Jones Station, TN July 20, 1862; Detailed as shoemaker Jan 1864; Married Helen Lard in Newton County, July 13, 1865; Farmer in Rockdale County in 1880; Farmer in Clay County, AL in 1900; Buried at Lystra Cemetery in Ashland, Alabama.

James Joseph Reagin PVT 5 Nov 1837 25 Jul 1914

Born in DeKalb County; Married Lydia Narcissus New in DeKalb County, Dec 25, 1856; Enlisted in Atlanta, May 13, 1862; Captured at **Vicksburg**, MS July 4, 1863; Present on Dec 9, 1863 muster roll; Granite cutter near Stone Mountain in 1900; Filed for Confederate pension in DeKalb County in 1902; Captain J.M. Summers reports in the application that Reagin was transferred in the fall of 1863 to Stephens Light Artillery "with my consent;" Buried at Lithonia City Cemetery in DeKalb County.

Samuel R. Rodgers PVT 1843 27 Jul 1863

Enlisted in Atlanta May 13, 1862; Captured at **Vicksburg**, MS July 4, 1863; Died in Enterprise, MS hospital; Claim filed by his father George Rogers.

Egbert Bell Rosser PVT 7 Nov 1838 3 Nov 1912

See Regimental officers.

William M. Rowden PVT 1828 28 May 1910

Married Martha Center in Newton County, OCT 19, 1854; Enlisted in Atlanta May 13, 1862; Appointed **4CPL** in Oct 1862; Captured at **Vicksburg**, MS July 4, 1863; Absent sick at home on Dec 9, 1863 muster roll; Detailed as bridge builder on Memphis & Chattanooga Railroad Dec 1864; Promoted to **4SGT**; Surrendered in **Greensboro**, NC April 26, 1865; Farmer in DeKalb County after the war; Member of UCV Camp #665 in Decatur.

James M. St. Johns PVT 1847 8 Jan 1865

Enlisted in Company C, 66th Georgia Aug 1, 1863; Deserted Aug 17, 1863; Late enlistment in 42nd Georgia; Wounded in 1864 Tennessee Campaign; Admitted to Fairgrounds

Hospital #2 with gunshot wound in Corinth, MS; Died in hospital.

George Washington Shaw PVT 21 Apr 1832 15 Mar 1920

Married Mary Jane Glenn in Morgan County, Feb 28, 1854; Sent to Knoxville, TN hospital; Absent sick at home Dec 9, 1863 muster roll; Admitted to Fairgrounds Hospital #1 in Atlanta Dec 12, 1863 with rheumatism; Farmer in DeKalb County in 1880; Living in Upshur County, TX in 1882; Farmer in Chilton County, AL in 1900; Buried at Providence Baptist in Jemison, Alabama.

James P. Shaw PVT 1832

Discharged April 12, 1862 with "weak eyes;" Re-enlisted in Company K.

Judge W. Shaw PVT 1830 20 Jan 1863

Born in Morgan County; Married Mary Ann Watson in Newton County, Jan 1, 1854; Enlisted in Atlanta May 13, 1862; Wounded in **Battle of Chickasaw Bayou**, MS Dec 29, 1862; Arm amputated; Died from his wounds in Vicksburg, MS hospital; Buried at Soldier's Rest, Cedar Hill Cemetery in Vicksburg, MS; Claim filed by his widow Mary Shaw.

Sidney J. Shaw PVT 21 Sep 1825 19 Sep 1909

Older brother of George Shaw; Married Victoria Watson in Henry County, Dec 21, 1848; Married Susan Russell in Newton County, Oct 13, 1859; Captured at **Vicksburg**, MS July 4, 1863; Detailed as bridge builder on Memphis & Chattanooga Railroad Dec 1864; Surrendered at **Greensboro**, NC April 26, 1865; Farmer in DeKalb County in 1880; Farmer in Cullman County, AL in 1900.

John H. Smith PVT 1833 After 1874

Appointed wagon master on April 29, 1862; Patient in Breckenridge Hospital, Marion, MS June 30, 1863 thru Oct 31, 1863; Absent in Atlanta hospital Dec 9, 1863 muster roll; Detailed as bridge builder on Memphis & Chattanooga Railroad Dec 1864; Married Caroline Ivy in Newton County, Nov 9, 1873.

Marcus D. Smith PVT 6 Sep 1842 After 1900

Sent to Vicksburg, MS hospital; Promoted to **CPL**; Wounded at **Battle of Resaca** May 15, 1864; Surrendered at **Greensboro,** NC April 26, 1865; Married Sarah Moss in Newton County, Nov 21, 1865; Farmer & miller in Henry County after the war; Filed for Confederate pension in Henry County in 1898; Buried at Timberridge Presbyterian in Henry County.

William J. Smith PVT 18 Apr 1862

Died from camp illness April 18, 1862.

Theophilus A. "Fish" Stansell PVT 9 Oct 1841 21 Apr 1892

Discharged with kidney disease on April 12, 1862; Enlisted in Rome, GA as CPL in Company H, Georgia 3rd Cavalry on May 13, 1862; Farmer in Newton County; Buried at Cannon Cemetery in Walton County.

Tyre Adolphus Starr PVT 19 Dec 1840 19 Aug 1901

Went by "Dolph"; Elected **2LT** Sep 1, 1862; Captured in **Vicksburg**, MS hospital July 4, 1863; Promoted to **1LT** Oct 14, 1863; Admitted to Pettigrew Hospital, Raleigh, NC with wound to left thigh, Mar 11, 1865; Ran general store and post office in Starrsville after the war; Buried at Starr Family Cemetery in Starrsville.

John James Stephenson PVT 9 Nov 1837 11 Feb 1880

Older brother of Leonidas & Jack Stephenson; Enlisted in Atlanta May 13, 1862; Elected Brevet **2LT** Sep 1, 1862; Captured at **Vicksburg**, MS July 4, 1863; Promoted to **2LT** Oct

14, 1863; Married Keren James in Newton County, Feb 4, 1864; Surrendered at **Greensboro**, NC April 26, 1865; Named son after Lovick Thomas; Owned the Stephenson House Hotel by the city square in Covington; Buried at Covington City Cemetery.

John Williamson Stephenson PVT 7 Mar 1844 26 Aug 1913

Went by "Jack"; Younger of John & Leonidas Stephenson; Enlisted Mar 20, 1864 in Covington; Surrendered at **Greensboro**, NC April 26, 1865; Married Mary Willingham in Newton County, Aug 29, 1872; Farmer in Stansells after the war; Served as Newton County Treasurer; Buried at Southview Cemetery in Covington.

John Thomas Stewart PVT 3 Mar 1831 18 Jan 1919

Born in South Carolina; Married Eliza A. Lindsey in Newton County, Aug 28, 1856; Captured at **Vicksburg**, MS July 4, 1863; Admitted to Fairgrounds Hospital #1 in Atlanta Nov 2, 1863 with typhoid fever; Wounded at **Battle of Rocky Face** May 8, 1864; Promoted to **1SGT**; Surrendered at **Greensboro**, NC April 26, 1865; Carpenter in Atlanta after the war; Married Mary Veal in DeKalb County, Nov 29, 1877; Filed for Confederate pension in Fulton County in 1917; Application endorsed by T.M. Armistead; Buried at Sylvester Cemetery in DeKalb County.

Augustus Swan PVT 10 May 1844 15 Nov 1862

Enlisted at Jones Station, TN Aug 8, 1862; Died of disease in Loudon, TN; Claim filed by his father Thomas D. Swan.

William Hamilton Taylor PVT 1837 8 Jul 1863

Married Rachel Mann in Gwinnett County, Feb 7, 1856; Died of fever in Vicksburg, MS hospital.

Manson Grice Townsend PVT 9 Oct 1821 9 Sep 1887

Married Elizabeth Jenkins in Henry County, Jan 6, 1847; Captured at **Vicksburg**, MS July 4, 1863; Surrendered at **Greensboro**, NC April 26, 1865; Farmer in Rockdale County in 1880; Buried at Old Conyers Cemetery in Rockdale County.

George J. Trimble PVT 1845 15 Sep 1862

Claim filed by his mother Martha Trimble.

Henry M. Veal PVT 1844

Appointed **3CPL** Feb 1863; Captured at **Vicksburg**, MS July 4, 1863; Present on Dec 9, 1863 muster roll; Wounded at **Battle of Rocky Face**, GA Feb 25, 1864; Appointed **SGT**; Admitted to Fairgrounds Hospital #2 in Atlanta with gunshot wound Feb 26, 1864; Furloughed Mar 11, 1864; Admitted to Asylum Hospital, Madison, GA July 10, 1864; Returned to duty July 16, 1864.

Minton G. Veal PVT 1829

Married Mary Christian in Newton County, July 3, 1848; Enlisted in Atlanta May 13, 1862; Patient at Marion, MS hospital June 30, 1863; Admitted to Fairgrounds Hospital #1 in Atlanta Oct 2, 1863 with gunshot wound; Present on Dec 9, 1863 muster roll; Admitted to Fairgrounds Hospital #2 in Atlanta Feb 25, 1864; Took 1867 Oath of Allegiance in Newton County;

Augustus H. Wallis PVT 1839 7 Apr 1864

Sent to Tazewell, TN hospital; Absent sick at home on Dec 9, 1863 muster roll; Died in Kingston, GA hospital; Claim filed by his father Moses Wallis.

William H. Wallis PVT 1836 16 Nov 1862

Married Malinda Estes in Newton County, Dec 11, 1856; Buried at Maple Grove Cemetery, Nicholasville, Kentucky.

William Jasper Wallis PVT 1842 Before 1905
Enlisted May 13, 1862 in Atlanta; Captured at **Vicksburg,** MS hospital July 4, 1863; Absent sick at home on Dec 9, 1863 muster roll; Captured at **Pen's Bridge,** SC Feb 11, 1865; Released at **Camp Lookout,** MD June 22, 1865; Married Anna Chandler in 1866; Farmer in DeKalb County in 1900.

James Washington Watson PVT 10 Jul 1830 Apr 1912
Born in Surry County, NC; Married Matilda Jane Shaw about 1856; Enlisted in Atlanta May 13, 1862; Captured at **Vicksburg,** MS July 4, 1863; Present on Dec 9, 1863 muster roll; Farmer in Henry County in 1880; Farmer near Jemison, AL in 1900; Buried at Providence Baptist in Chilton County, Alabama.

John Watson PVT 1832 30 Dec 1863
Born in North Carolina; Younger brother of James Watson; Married Aletha Hooton in Newton County, Nov 15, 1855; Enlisted in Company H, 6th Georgia State Troops, Oct 21, 1861; Discharged with disability April 14, 1862; Enlisted in Atlanta May 13, 1862; Captured at **Vicksburg,** MS July 4, 1863; Received on U.S. Steamer *Maurice* in Mobile Bay July 30, 1863; Died in Henry County home.

William Jackson Watson PVT 25 Aug 1836 After 1930
Enlisted in Company H, 6th Georgia State Troops, Oct 21, 1861; Mustered out in Savannah May 1862; Enlisted at Woodson's Crossroads, TN Aug 25, 1862; Captured at **Vicksburg,** MS July 4, 1863; Absent sick at home on Dec 9, 1863 muster roll; Married Elizabeth Persall in 1866; Farmer in Henry County after the war; Filed for Confederate pension in Henry County in 1906; Reported being sent home from Atlanta hospital on medical furlough in fall 1864; Application supported by J.R. Daniel; Buried at Zion Baptist in Covington.

Iverson Vestal Wells PVT 11 Sep 1843 1895
Enlisted at Woodson's Crossroads, TN Aug 20, 1862; Admitted Dalton hospital Dec 24, 1862; Present on Dec 9, 1863 muster roll; Surrendered at **Greensboro,** NC April 26, 1865; Married Sarah Winfield in Newton County, Mar 8, 1866; Farmer in Newton County after the war; Buried at Underwood Memorial in Conyers.

Francis M. Whitehead PVT Mar 1830 2 Dec 1900
Married Nancy Bostwick in Newton County, Dec 25, 1851; Patient in Knoxville, TN hospital Oct 1862; Sick in Atlanta hospital on Dec 9, 1863 muster roll; Farmer in Morgan County in 1880; Postmaster in community of Winston in from 1893 to 1900.

John Henry Wilson PVT 1822 6 Oct 1899
Married Nancy Whitehead in Newton County, Dec 27, 1842; Sent to Vicksburg, MS hospital; Discharged May 13, 1863; Farmer in Rockdale County after the war; Buried at Sodom Cemetery in Gwinnett County.

William Francis Wilson PVT 24 Sep 1843 18 Apr 1923
Born in Morgan County; Son of John Wilson; Enlisted in Decatur Mar 11, 1863; Sent to French's Division Hospital in Enterprise, MS spring 1863; Admitted to Fairgrounds Hospital #1 in Vineville, GA Aug 11, 1864 with dysentery; Returned to duty Sep 16, 1864; Surrendered at **Greensboro,** NC April 26, 1865; Married Sarah Ann Smith about 1882; Filed for Confederate pension in Rockdale County in 1895; Farmer in Rockbridge, Gwinnett County in 1900; Buried at Sodom Cemetery in Gwinnett County.

Isaiah Woolley PVT 1837 Feb 1837
Married Sarah Ann Lynn in Newton County, Aug 14, 1860; Enlisted in Atlanta May 13,

1862; Cook at City Hospital in Vicksburg Feb 1863; Captured at City Hospital in **Vicksburg,** MS July 4, 1863; Among sick exchanged in Mobile, AL July 30, 1863; Present on Dec 9, 1863 muster roll; Farmer in Lithonia in 1870; Buried at Old Conyers Cemetery in Rockdale County.

Jordan Madison Woolley PVT 1838 17 Nov 1862

Married Julia Ann Hudson in Newton, County, Nov 1, 1857; Enlisted in Atlanta May 13, 1862; Died in Knoxville, TN hospital; Buried at Bethel Cemetery in Knoxville, TN.

Vardiman Woolley PVT 1833 12 Dec 1862

Born in South Carolina; Younger brother of Elias, Isiah, & Jordan Woolley; Married Elizabeth Overton in Newton County, Oct 19, 1853; Enlisted in Atlanta May 13, 1862; Captured in Kentucky; Sent from Louisville on US Steamer *Maria* for exchange in Vicksburg, MS Nov 15, 1862; Died in hospital; Claim filed by widow Elizabeth Woolley.

Company G – Walton Blues – 151 Men

This company was drawn principally from the northern half of Walton County.

Name Rank Birth Date

Enoch Edwards McCollum CPT 30 Aug 1831 7 Jul 1862

Born in Habersham County; Married Krista Breedlove in Walton County, Oct 24, 1854; Master Mason; Died of measles at Bean Station, TN; Buried at Breedlove Family Cemetery in Walton County; Claim filed by wife Kalizta McCollum.

Thomas W. Davis CPT 18 Jul 1863

Enlisted as 2LT in Company K; Promoted to **1LT** Mar 20; 1862; Elected **CPT** of Company G in March 1863; Died of disease in Vicksburg, MS; Claim filed by his wife Ada H. Davis in Edgefield County, SC in April 1864.

Alvin H. Beall 5SGT 1843 13 Nov 1862

Born in Walton County; Died of measles in Atlanta; Buried at Oakland Cemetery in Atlanta.

John Foster Crawford 2LT Jan 1831 22 Oct 1902

Married Marth Bonds in Gwinnett County, Aug 18, 1853; Married Lucy Roberts Culbreath in Walton County, Jan 24, 1860; Elected 2LT Sep 1, 1862; Admitted to Fairgrounds Hospital #1 in Atlanta Nov 28, 1862; Furloughed Dec 1, 1862; Resigned with disability Feb 24, 1863; Farmer in Waco, Texas after the war; Died in Alameda County, CA.

Hiram Hugh Edgar 3SGT 1830 Sep 1862

Oldest of six Edgar brothers serving in the Walton Blues (G); Married Mary Matthews in Walton County in 1848; Discharged with disability at Woodson Crossroads, TN Sep 17, 1862; Died shortly thereafter.

Henry Harden Green 2LT 27 Aug 1829 31 May 1906

Married Frances America Malcom in Walton County, Nov 19, 1850; Wounded in the leg at Battle of Chickasaw Bayou, MS Dec 29, 1862; Elected **2LT** of Company G Oct 14, 1862; Captured at **Vicksburg,** MS July 4, 1863; Appeared before medical board near Atlanta Aug 19, 1864; Farmer in Good Hope after the war; UCV member; Buried at Green Cemetery, Walton County; Widows pension filed in Walton County in 1912; Reported that her husband was seriously wounded and disabled at the Battle of Resaca, May 15,

1864.

William David Malcom 2LT 4 Feb 1833 9 Feb 1863

Married cousin Rebecca Malcom in Morgan County, Dec 15, 1853; Elected **CAP** Sep 1, 1862; Severely wounded in the leg at **Battle of Chickasaw Bayou,** MS Dec 29, 1862; Died from his wound; Buried at Green Cemetery in Walton County.

Eli Mayfield 1SGT 27 Jul 1822 28 Aug 1864

Married Amanda Ross Burson in Walton County, Sep 5, 1841; Captured at **Vicksburg,** MS July 4, 1863; Died of disease; Buried at Burson Farm Cemetery in Walton County.

David Burson McCord 1CPL 1842 29 Dec 1862

Killed at **Battle of Chickasaw Bayou,** MS.

George Washington McDonald 4CPL 1829 29 Jul 1863

Married Martha Ann Nelson in Walton County, Jan 17, 1847; Appointed **1CPL** Dec 1862; Captured at **Vicksburg,** MS July 4, 1863.

Henry Britton Mikel 1LT 2 Jun 1835 8 Jan 1920

Captured at **Vicksburg,** MS July 4, 1863; Surrendered at **Greensboro,** NC April 26, 1865; Married Jemima Seale in 1866; Deputy sheriff of Milton County in 1871; Farmer in Newtown in 1880; Filed for Confederate pension in Crisp County in 1918; Buried at Mount Pisgah Methodist in Johns Creek.

John W. Sims 4SGT 7 Mar 1834 1864

Born in Clarke County; Married Elizabeth Gillespie in Walton County, Jan 16, 1855; Appointed 4CPL of Company C, 9th Regiment Georgia State Troops Jun 13, 1861; Discharged with disability from Richmond, VA hospital Sep 18, 1861; Sick in Knoxville, TN hospital April 22, 1862; Medical discharge Mar 26, 1863.

James H. Smith 3CPL 5 Apr 1832 1918

Married Sarah Adair in Walton County, Feb 3, 1854; Captured at **Battle of Atlanta,** July 22, 1864; Buried in McGregor Cemetery in McLennan County, Texas.

Cicero Washington Terrell 2SGT 1 Jan 1834 19 Dec 1828

Enlisted as PVT in Company F of 16th Georgia State Troops July 19, 1861; Discharged with disability at Yorktown, VA Nov 4, 1861; Captured at **Vicksburg,** MS July 4, 1863; Married Margaret Jane Brooks about 1866; Telegraph lineman living in Chattanooga in 1880; Filed for Confederate pension application in Cobb County in 1889; Reported being shot in the chest & hip June 27, 1864 near Marietta; Application supported by H.B. Mikel; Died in Chattanooga, TN; Buried at Forest Hills Cemetery in Chattanooga.

John Benjamin Adams PVT 1837 6 Jul 1888

Patient at Lauderdale Springs, MS hospital April 22, 1863; Admitted to 1st Mississippi hospital in Jackson, MS June 18, 1863; Present on Dec 11, 1863 muster roll; Attached to Company H of the 24th South Carolina August 1863; Returned to 42nd Georgia in October 1863; Admitted to Ocmulgee Hospital in Macon August 14, 1864 with shell wound to right forearm; furloughed Sep 28, 1864; Married Minerva Clack about 1876; Farmer in Walton County in 1880; Filed for Confederate pension in Walton County in 1889; Reported being wounded in right elbow in July 1864; Buried at David Smith Family Cemetery in Barrow County.

James Gower Adams PVT 4 Jan 1827 19 Jul 1863

Born in Clarke County; Married Martha Whitehead in Clarke County, June 28, 1847; Captured at **Vicksburg,** MS hospital July 4, 1863; Died in Vicksburg hospital.

Richard Sion Adams PVT 22 Nov 1845 26 Jan 1896

Late enlistment; Wounded in the foot at **Battle of New Hope Church** May 25, 1864; Took 1867 Loyalty oath in Walton County; Married Victoria Stone about 1870; Physician in Duluth in 1880.

James R. Aiken PVT 1827 14 Jul 1862

Older brother of William Aiken; Died in his Walton County home; Claim filed by his father James S. Aiken.

William H. Aiken PVT 1830 9 Apr 1862

Born in Clarke County; Died in his Walton County home; Claim filed by his father James S. Aiken.

James B. Austin PVT 19 Aug 1846 25 Dec 1932

Enlisted Feb 17, 1864; Captured at **Battle of Nashville**, Dec 16, 1864; Discharged from **Camp Douglas**, IL June 18, 1865; Married Harriett Thomas in Blount County, AL, Oct 1, 1879; Filed for Confederate pension in Walton County in 1907; Application endorsed by J.M. Odom; Buried at Carter Hill Christian Church in Barrow County.

James L. Austin PVT 1844 9 Jul 1864

Enlisted Nov 26, 1863; Died of disease.

John E. Austin PVT 25 Mar 1840 30 Dec 1920

Brother of James, Middleton, & William Austin; Enlisted as a private in Company B, 3rd Battalion of Georgia State Troops; Mustered out & enlisted in 42nd Georgia in Monroe May 13, 1862; Captured at **Battle of Baker's Creek**, MS May 16, 1863; Sent to **Camp Morton**, IN; Exchanged at Fort Delaware, MD July 4, 1863; Surrendered at **Greensboro**, NC April 26, 1865; Married Juda Adams in Walton County, Jan 1, 1868; Filed for Confederate pension in Walton County in 1898; Worked in fruit industry in south Georgia; Buried in Cedar Hill Cemetery in Bleckley County.

John W. Austin PVT 1843 13 Jun 1862

Older brother of James B. Austin; First cousin of James, John, Middleton, & William Austin; Enlisted as a private in Company B, 3rd Battalion of Georgia State Troops; Mustered out & enlisted in 42nd Georgia in Monroe May 1, 1862; Died in Knoxville, TN hospital; Claim filed by his father William Austin.

Middleton Austin PVT 1841 Before 1867

Captured at **Vicksburg**, MS July 4, 1863.

William W. Austin PVT 1833 14 Jul 1862

Married Lucy Patrick in Walton County, Nov 11, 1852; PVT in the Joe Brown Rough & Ready Militia from Cut Off, GA June 18, 1861; Enlisted as a private in Company B, 3rd Battalion of Georgia State Troops; Mustered out & enlisted in 42nd Georgia in Monroe May 13, 1862; Died of disease in Knoxville, TN hospital; Claim filed by his widow Lucy Austin.

Theophilus Jackson Barnes PVT

Captured at **Vicksburg**, MS July 4, 1863; Transferred to Captain Ritter's battery of Georgia light artillery Nov 10, 1863; Captured in Columbia, TN Dec 24, 1864; Released from **Camp Chase**, OH June 13, 1865.

Cicero Newton Baxter PVT 12 Jan 1842 22 Feb 1911

Captured at **Vicksburg**, MS July 4, 1863; Admitted to Fairgrounds Hospital #1 in Atlanta Oct 2, 1863 with debility; Surrendered at **Greensboro**, NC April 26, 1865; Married Mary Ann in Walton County, Nov 26, 1865; Farmer in Walton County after the war; Filed for Confederate pension in Walton County in 1902; Application endorsed by W.L. Johnston;

Buried at High Shoals Methodist in Walton County.
John Jasper Baxter PVT 7 Mar 1844 18 Mar 1865
Younger brother of Cicero Baxter; Captured at **Vicksburg**, MS July 4, 1863; Present on Dec 11, 1863 muster roll; Killed at **Battle of Bentonville**, NC.
Benjamin C. Blair PVT 18 Jul 1839 20 Feb 1863
Born in Habersham County; Entered CSA Post Hospital in Dalton debilitated Nov 21, 1862; Died in Jackson, MS hospital of pneumonia.
James Russell Boman PVT 1839 Before 1867
Married Sarah Jane Stanley in Walton County, May 27, 1858; Enlisted May 12, 1862 in Monroe; Admitted to 1st MS Hospital in Jackson MS with gastric discomfort Mar 25, 1863; Transferred to Meridian, MS hospital May 14, 1863; Absent without leave on Dec 11, 1863 muster roll; Captured at **Battle of Jonesborough** Sep 1, 1864; Exchanged at Rough & Ready Sep 19, 1864.
Joseph Boman PVT 1837 Before 1867
Older brother of James Bowman; Married Sarah Terrell about 1857; Enlisted in Monroe May 12, 1862; Captured at **Vicksburg**, MS July 4, 1863; Estate probated in Walton County Nov 4, 1872. Named his son William Robert Henderson Boman.
William Posey Bone PVT 1825 Before 1900
Captured at **Vicksburg**, MS July 4, 1863; AWOL on Dec 11, 1863 muster roll; Wounded at **Battle of Resaca**, May 15, 1864; Surrendered at **Greensboro**, NC April 26, 1865; Married Martha Turner in Walton County, Dec 2, 1870; Farmer in Newton County in 1880.
Leonard Boss PVT 1839 2 Nov 1864
Half-brother of Andrew & Henry Boss from Company K; Captured at **Vicksburg**, MS July 4, 1863; Absent without leave on Dec 11, 1863 muster roll; Appears on 1864 Georgia militia census in Walton County as "Vicksburg prisoner;" Captured after the **Battle of Atlanta** July 23, 1864; Died of smallpox in **Camp Chase**, OH; Buried in Confederate Cemetery, grave # 390.
Isaac Monroe Brantley PVT 4 May 1841 1 Sep 1929
Captured at **Vicksburg**, MS July 4, 1863; AWOL on Dec 11, 1863 muster roll; Married Letty Braswell in Walton County, Oct 19, 1865; Filed for Confederate pension in Chilton County, AL in 1907; Reported being paroled in Atlanta in May 1865; Buried at Mars Hill Baptist in Chilton County, AL.
George A. Braswell PVT Apr 1862
Married Mary Weatherford in Walton County, Oct 4, 1857; Died at Camp McDonald.
Lewis Edward Braswell PVT 4 Aug 1835 14 Nov 1915
Married Nancy Miller in Walton County, Jan 12, 1860; Patient in Knoxville, TN hospital August 1862; Captured at **Vicksburg**, MS July 4, 1863; Present on Dec 11, 1863 muster roll; Farmer in Walton County in 1880; Farmer in Smith County, Texas in 1900; Filed for Confederate pension in Marion County, AL in 1907; Buried at Musgrove Cemetery in Fayette County, AL.
James Madison Brown PVT 6 Jan 1836 2 Jan 1915
Married Mary Jane Wilson in Walton County in 1855; Enlisted in Monroe May 12, 1862; Furloughed from Fairgrounds Hospital #1 in Atlanta Dec 10, 1862; Captured at **Vicksburg**, MS July 4, 1863; Farmer in Walton County after the war; Filed for Confederate pension in Walton County in 1900; Reported being "honorable discharged at the surrender of Johnston's army;" Buried at Hill Baptist Church in Walton County.

William Brown PVT 15 Oct 1862
Died in Knoxville, TN hospital.
John Thomas Bulloch PVT 1838 1863
Older brother of Reuben Bulloch; Married Arsenia McElhannon in Walton County, Jan 27, 1861; Enlisted in Monroe May 12, 1862; Granted Certificate of Disability with typhoid fever in Jackson, MS hospital 13 Mar 1863.
Reuben Thomas Bulloch PVT 1839 21 Jul 1863
Married Cynthia Hughes in Walton County, Dec 15, 1859; Captured in Washington Hospital, **Vicksburg,** MS July 4, 1863; Sent with sick on US Steamer *Chouteau* for exchange in Mobile via New Orleans, LA; Admitted to St. Louis Hospital in New Orleans with acute diarrhea July 20, 1863; Buried at Cypress Grove Cemetery #2, New Orleans.
Jackson Burns PVT 16 May 1864
Late enlistment; Wounded at **Battle of Resaca** May 15, 1864; Died of his wounds.
Eldridge M. Burton PVT 1834 16 Apr 1875
Enlisted as private in Company C of the 9th Regiment of Georgia Infantry; Company was formerly the "Walton Guards," a local militia unit under Captain George Hillyer; Discharged with disability in Richmond, VA Aug 10, 1861; Captured at **Vicksburg,** MS July 4, 1863; Captured at **Battle of Nashville,** TN Dec 16, 1864; Released from **Camp Douglas,** IL June 18, 1865; Married his brother's widow, Mary Rusk Burton in Walton County, Sep 6, 1865; Farmer in Walton County in 1870.
John Clarke Burton PVT 1831 5 Jan 1863
Older brother of Eldridge Burton; Married Mary Catherine Rusk in Walton County, Nov 6, 1851; Died of smallpox in 1863.
James Patrick Butler PVT 1 Aug 1837 24 Apr 1883
Married Elander Austin in Walton County, Jan 2, 1859; Deserted May 26, 1863; Present on Dec 11, 1863 muster roll; Farmer in Walton County after the war; Buried at Rays Methodist Chapel in Oconee County.
Daniel Wallace Chandler PVT 5 Feb 1835 10 Jan 1865
Married Winnie White in Gwinnett County, Jan 13, 1858; Present on Dec 11, 1863 muster roll; Captured at **Battle of Resaca** May 15, 1863; Died a prisoner of war of erysipelas at **Camp Douglas,** IL.
George Washington Clack PVT 1836 4 Oct 1863
Married Rhoda Harrison in Walton County, Feb 27, 1855; Captured in Kentucky Oct 1862; Taken by steamboat from Louisville to Vicksburg, MS and exchanged on Nov 29, 1862; Captured in Washington Hospital at **Vicksburg,** MS July 4, 1863; Transported with sick on US Steamer *Suffolk* from Vicksburg to Mobile; Received off Fort Morgan on July 22, 1863; Claim filed by wife Rhoda Clack.
Seaborn "Seb" Jones Clack PVT 22 Apr 1830 4 Nov 1909
Brother of George Clack; Married Drucilla Holloway in Walton County, Sep 14, 1855; Captured in Versailles, KY Oct 10, 1862; Paroled in Harrodsburg, KY Dec 28, 1862; Sent from Louisville to Vicksburg via Cairo by steamboat for exchange; Jan 11, 1863; Captured in Benton, TN near Chattanooga Nov 30, 1863; Entered Rock Island, Il prison Dec 13, 1863; Exchanged at Boulware & Cox's Wharves, James River, VA Mar 5, 1865; Sick paroled prisoner admitted to Camp Winder General Hospital, Richmond, VA Mar 7, 1865; Farmer in Gwinnett County after the war; Buried in Seaborn Clack Cemetery in Gwinnett County; Marker dedicated to his service in the 42nd Georgia dedicated April 26, 2001.

James Russell Coker PVT 10 Dec 1841 1 Apr 1936
Captured at **Vicksburg**, MS July 4, 1863; Absent sick at home on Dec 11, 1863 muster roll; Detailed as shoemaker in Dalton, Jan 31, 1864; Married Melinda Wells in Walton County, Feb 6, 1868; Married Mildred Wills in Walton County, Jan 11, 1872; Stone cutter in Jackson County in 1900; Filed for Confederate pension in Jackson County in 1906; Reported losing sight in one eye from exposure in 1863; Reported surrendering with unit at Bentonville in 1865; Buried at Rose Hill Cemetery in Winder.
Andrew Jackson Cooper PVT 2 Feb 1836 25 May 1907
Younger brother of John & William Cooper; Married Maryann Sarah Culbreath in Chambers County, AL Nov 1, 1859; Enlisted in Monroe May 12, 1862; Captured at **Vicksburg**, MS July 4, 1863; Received at Atlanta distribution hospital with diarrhea May 8, 1864; Furloughed from Stout Hospital, Madison, GA with chronic diarrhea July 5, 1864; Grant leave to Cusseta, Chambers County, AL; Grocer in Waco, TX in 1880; Buried in Hico Cemetery, Hamilton County, Texas.
John Z. Cooper PVT 21 Feb 1829 1909
Older brother of Andrew and William Cooper; Married Eliza Ann Jackson in Clarke County, Dec 18, 1851; Enlisted in Jones Station, TN Aug 17, 1862; Captured at **Vicksburg**, MS July 4, 1863; Prison guard under Captain Hurtel in military prison in Atlanta June 30, 1864; Appointed **5SGT** in 1865; Keeping livery stable in Athens in 1870; Buried at Grey Hill Cemetery in Commerce.
William Allen Cooper PVT 15 Aug 1834 8 Dec 1906
Born in Clarke County; Admitted to 1st MS Hospital in Jackson, MS Jan 18, 1863; Furloughed Mar 18, 1863; Present on Dec 11, 1863 muster roll; Surrendered at **Greensboro**, NC April 26, 1865; Married Mary Davis in Clarke County, Dec 18, 1866; Farmer in Jackson County after the war; Buried in Center Cemetery in Commerce.
William Augustus Craft PVT 1826 1864
Married Margaret Barber in Walton County, Sep 19, 1849; Admitted to Fairgrounds Hospital #1 in Atlanta Nov 28, 1862; Deserted hospital Dec 3, 1862; Captured at **Vicksburg**, MS July 4, 1863; AWOL on Dec 11, 1863 muster roll.
Jabel Calvin Crow PVT Dec 1832 After 1900
Older brother of Stephen Crow; Married Sarah Ann Reynolds in Clarke County, Mar 14, 1855; Making coffins Mar 18, 1863; Factory worker in Walton County in 1870.
Stephen E. Crow PVT 1839
Absent sick in in Knoxville, TN hospital April 22, 1862; Making coffins Mar 18, 1863.
James C. Davis PVT 12 Nov 1830 2 Oct 1900
Married Sarah Davis in Walton County, May 15, 1853; Captured at **Vicksburg**, MS July 4, 1863; Admitted to Fairgrounds Hospital #1 in Atlanta Oct 2, 1863; Returned to duty Oct 5, 1862; Captured at **Battle of Nashville**, Dec 16, 1864; Released from **Camp Chase**, OH June 12, 1865; Farmer in Walton County after the war; Filed for Confederate pension in Walton County in 1899; Application endorsed by J.W. Odom; Buried at New Hope Methodist in Between, Walton County.
James Solomon Davis PVT 28 Dec 1844 4 Nov 1940
Son of William Davis; Enlisted in Dalton Jan 25, 1864; Surrendered at **Greensboro**, NC April 26, 1865; Married Mary McDaniel in Walton County, Dec 29, 1870; Farmer in Walton County after the war; Filed for Confederate pension in Gwinnett County in 1897; Buried at Davis-McDaniel Family Cemetery in Winder.

John William Davis PVT 1842
Son of William Davis; PVT in the Joe Brown Rough & Ready Militia from Cut Off, GA June 18, 1861; Left sick in Knoxville, TN hospital April 22, 1862; Patient at 1st MS Hospital in Jackson, MS Mar 25, 1863; Patient at Marion, MS hospital Jun 30, 1863; AWOL on Dec 11, 1863 muster roll; Surrendered at **Greensboro**, NC April 26, 1865.
William Green Davis PVT 1820 14 Jun 1863
Father of James & John Davis; Married Elizabeth Baccus in Walton County, Jan 30, 1842; Detached to wait on sick at Knoxville, TN hospital April 22, 1862.
Absalom Edgar PVT 1844 11 Jan 1863
Enlisted at Woodson's Crossroads, TN Sep 13, 1862; Granted 30-day furlough from Atlanta hospital Jan 10, 1863; Died three days later in Atlanta hospital; Claim filed by his father Hugh Edgar.
Elisha Edgar PVT 22 Dec 1837 18 Apr 1914
Married Eliza Jane Thompson (sister of Alfred Thompson) in Walton County, Oct 20, 1859; Left sick in Knoxville, TN hospital April 22, 1862; Captured at **Vicksburg**, MS July 4, 1863; Left sick at home Dec 11, 1863 muster roll; Surrendered at **Greensboro**, NC April 26, 1865; Farmer in Cut Off after the war; Married Camelia McCord in Walton County, Dec 14, 1880; Filed for Confederate pension in Walton County in 1905; Buried at Edgar Family Cemetery, Bethlehem, Barrow County.
John Edgar PVT 24 Aug 1833 8 Oct 1907
Married Mahulda Ann Arthur in Walton County, Dec 19, 1854; Enlisted in Monroe May 12, 1862; Captured at **Vicksburg**, MS July 4, 1863; Present on Dec 11, 1863 muster roll; Farmer in Jackson County after the war; Buried in Stratham City Cemetery in Barrow County; Widow's pension filed in Jackson County in 1904; Reports her husband was present at the Greensboro surrender.
John Henry Edgar PVT 26 Sep 1846 6 Feb 1937
Youngest of six Edgar brothers; Enlisted in Dalton in April 1864; Admitted to Madison, GA hospital May 18, 1864; Furloughed May 20, 1864; Married Eliza Adams in Walton County, Jan 30, 1868; Farmer in Cut Off after the war; Filed for Confederate pension in Jackson County in 1906; Reported being declared disabled by regimental physician in Shelby Springs, AL hospital Dec 1864; Application endorsed by J.R. Coker who wrote that Henry Edgar left the command at Florence, AL in Dec 1864; Buried at Chapel Christian in Winder.
Matthew Edgar PVT 1841 7 Dec 1889
Admitted to 1st MS Hospital in Jackson, MS with typhoid fever Feb 3, 1863; Present on Dec 11, 1863 muster roll; Married Mary Adams in Walton County, Dec 21, 1865.
Samuel A. Edmonds PVT 7 Jan 1844 29 Feb 1932
Appointed **1CPL** in 1863; Captured at **Vicksburg**, MS July 4, 1863; Married Emily Austin in Walton County, Dec 20, 1867; Farmer in Dacula after the war; Filled for a Confederate pension in Gwinnett County in 1912; Reported being wounded in the neck at **Battle of Jonesborough** Aug 31, 1864; On medical furlough at time of surrender; Buried at Hebron Baptist in Dacula, Gwinnett County.
John W. Edwards PVT 12 May 1826 1 May 1905
Older brother of Wiley Edwards; Married Frances Kilgore in Walton County, July 8, 1853; Appointed **5SGT** in Mar 1863; At home sick on Dec 11, 1863 muster roll; Surrendered at **Greensboro**, NC April 26, 1865; Farmer in Walton County after the war; Filed for

Confederate pension in Walton County in 1895; Application endorsed by Ransom Brand; Buried at Bay Creek Cemetery in Loganville.

Wiley W. Edwards PVT 12 Mar 1836 1 Jul 1896

Married Mahala Tucker in Walton County, Dec 6, 1860; Captured at **Vicksburg**, MS July 4, 1863; Captured at Salisbury, NC April 12, 1865; Released from **Camp Chase**, OH June 13, 1865; Farmer in Walton County in 1880; Buried at Morris-Edwards Family Cemetery in Bethlehem, Bartow County.

Henry H. Evans PVT 1837 After 1894

Married Martha Wiggins in Walton County, Nov 9, 1856; Enlisted in Monroe May 12, 1862; Captured at **Vicksburg**, MS July 4, 1863; Present on Dec 11, 1863 muster roll; Farmer in Walton County in 1880; Filed for Confederate pension in Walton County in 1894; Reported being run over by a horse in Sweetwater, TN breaking his ribs.

William D. Evans PVT 1836 Mar 1862

Placed on indefinite leave by Cpt McCollum April 20, 1862; Died of disease shortly thereafter.

James Chapel Forrester PVT 10 May 1823 19 Dec 1883

Married Martha Mayfield in Walton County, Dec 15, 1844; Married Mary Ann Odum in Walton County, April 1, 1858; Furloughed from Fair Grounds Hospital #1 in Atlanta Dec 11, 1862; Captured at **Vicksburg**, MS July 4, 1863; Present on Dec 11, 1863 muster roll; Wounded in hospital Aug 25, 1864; Farmer in Flowery Branch in 1880; Buried at Alcovy Cemetery in Walton County.

Samuel William Forrester PVT 19 Nov 1832 15 Apr 1903

Married Sarah Ann Harris in 1852; Captured at **Vicksburg**, MS July 4, 1863; Surrendered at **Greensboro**, NC April 26, 1865; Farmer in Broken Arrow District after the war; Married Nancy Henderson in Walton County, Mar 4, 1886; Buried at Forrester Family Cemetery in Walnut Grove, Walton County.

John G. Fuller PVT 1825 15 Aug 1898

Married Nancy Bowman in Gwinnett County, April 26, 1849; Captured at **Vicksburg**, MS July 4, 1863; Present on Dec 11, 1863 muster roll; Farmer in Cut Off after the war; Filed for Confederate pension in Walton County in 1889; Reported being wounded in left hand and having three fingers amputated in Vicksburg, MS May 23, 1863; Buried at Smith's Chapel in Winder.

John Franklin Greeson PVT 12 Jan 1846 23 Dec 1921

Filed for Confederate pension in Gwinnett County in 1913; Reported joining Walton Blues(G) in Dalton in Jan 1863; Reported being in Montgomery, AL hospital in Feb 1865 sick with chills and fever; Married Nancy Ann Sells about 1867; Farmer in Gwinnett County after the war; Buried at Nazareth Methodist in Winder, Barrow County.

James John Gunter PVT 9 Dec 1839 5 Dec 1926

Captured at **Vicksburg**, MS July 4, 1863; Captured at **Battle of Jonesborough** Sep 1, 1864; Exchanged at Rough & Ready, GA Sep 19, 1864; Surrendered at **Greensboro**, NC April 26, 1865; Married Missouri Cook in 1867; Farmer in Oconee County after the war; Filed for Confederate pension in Oconee County in 1895; Member of UCV Camp # 1124 in Watkinsville; ZBuried at Bethabara Baptist in Oconee County.

William Henry Gunter PVT 13 Jul 1843 23 Apr 1900

Younger brother of James Gunter; PVT in the Joe Brown Rough & Ready Militia from Cut Off, GA June 18, 1861; Captured at **Vicksburg**, MS July 4, 1863; Absent sick at home

Dec 11, 1863 muster roll; Filed for Confederate Pension in Walton County in 1888; Reports arm was disabled from gunshot wound at **Battle of Resaca** May 15, 1864; Married Mattie Barrett in Walton County, Jan 13, 1875; Farmer in Broken Arrow in 1880; Appointed US Postmaster of Social Circle in 1897; Buried at Circle View Cemetery in Social Circle.

William G. Hailes PVT 11 Oct 1837 11 Jul 1862
Married Sarah Judith Breedlove in Walton County, Nov 15, 1856; Died of measles at home; Buried at Old Breedlove Cemetery in Walton County.

William D. Hamilton PVT 1839 1862
Died of disease; Claim filed by his father E.J. Hamilton in Feb 1863.

John Harper PVT 1825 After 1880
Dec 11, 1863 muster roll reports him absent with General Joseph Johnston's Mississippi command; Farmer in Oconee County in 1880.

James H. Harrison PVT 1840 16 Jun 1864
PVT in the Joe Brown Rough & Ready Militia from Cut Off, GA June 18, 1861; Captured at **Vicksburg**, MS July 4, 1863; Present on Dec 11, 1863 muster roll; Mortally wounded at **Battle of Resaca** May 15, 1864.

John D. Harrison PVT 3 Dec 1843 5 May 1906
Captured at **Vicksburg**, MS July 4, 1863; Present on Dec 11, 1863 muster roll; Took Oath of Allegiance in Chattanooga July 1, 1864; Married America Treadwell in Walton County, Oct 11, 1866; Farmer in Cut Off after the war; Buried at Bethlehem Methodist in Barrow County.

Gideon Hendrick PVT 30 Dec 1826 20 Jul 1889
Married Mary Thompson in Walton County, Jan 6, 1848; Captured at **Vicksburg**, MS July 4, 1863; Absent sick at home on Dec 11, 1863 muster roll; Farmer in Blount County, AL in 1880; Buried in Hood Cemetery in Royal, AL.

Nathan S. Hendrick PVT 1832 Before 1871
Wounded in arm at **Battle of Chickasaw Bayou**, MS Dec 29, 1862; Admitted to 1st Mississippi Hospital in Jackson, MS Jan 18, 1863; Returned to duty Mar 29, 1863; Captured at **Vicksburg**, MS July 4, 1863; Will probated in Walton County April 3, 1871.

Rufus Philip Hughes PVT 22 Dec 1837 29 July 1863
Married Martha Allen in Walton County, Jan 3, 1859; Captured at **Vicksburg**, MS July 4, 1863; Buried at Hughes Family Cemetery in Loganville.

William Joseph Hughes PVT 5 Jul 1836 18 Oct 1909
Older brother of Rufus Hughes; Married Sarah Odum in Walton County, Oct 4, 1860; Captured at **Vicksburg**, MS July 4, 1863; Absent sick at home Dec 11, 1863 muster roll; Farmer in Walton County in 1880; Farmer in Blount County, AL in 1900; Buried at Mount Moriah in Blount County, Alabama.

George A. Ivie PVT 1835 7 Apr 1873
Married Sarah Maulden in Gwinnett County, Oct 3, 1854; Boot maker in Stone Mountain in 1860; Granted Certificate of Disability for consumption from Knoxville, TN hospital Jan 24, 1863; Died in Fairland County, Indiana.

James Aubrey Johnston PVT 10 Dec 1838 14 Dec 1862
Older brother of William Johnston; Enlisted in Calhoun Aug 23, 1862; Granted Certificate of Disability for "imbecility attended with epileptic convulsion," Dec 13, 1862;" Died at home; Claim filed by his father Robert A. Johnston in Walton County.

William Lafayette Johnston PVT 20 Jan 1843 28 Oct 1918
Enlisted in Calhoun Aug 23, 1862; Captured at **Vicksburg**, MS July 4, 1863; Present on Dec 11, 1863 muster roll; Appointed **1SGT** July 1864; Surrendered at **Greensboro**, NC April 26, 1865; Married Mary Taylor LaBoon in Walton County, Mar 28, 1867; Farmer in Walton County after the war; Married Susanna Smith in Walton County, Sep 10, 1891; Buried at Mount Vernon Christian in Monroe.
David Kilgore PVT Nov 1837 26 Apr 1913
Married Seaney Austin in Walton County, June 26, 1859; PVT in the Joe Brown Rough & Ready Militia from Cut Off, GA June 18, 1861; Received Certificate of Discharge from Vicksburg, MS Feb 20, 1863; Suffered from severe smoke inhalation for "about eighteen months caused by heat & smoke from his burning dwelling;" Farmer in Jackson County in 1880; Farmer in Tarrant County, TX in 1900; Died in Arlington, Texas.
Frances Marion Kilgore PVT 1843 20 Apr 1862
Brother of David & William Kilgore; Accepted commission as ensign of Militia District 250 in Walton County Feb 19, 1862; Died from disease during training.
Thomas J. Kilgore PVT 1841 9 Jul 1862
Died in Social Circle, GA.
Elisha Lake PVT 1828 14 Feb 1865
Married Mary Ridgeway in Walton County, Jan 28, 1849; Captured at **Battle of Nashville**, TN Dec 16, 1864; Died of smallpox as POW at **Camp Chase**, OH; Buried in Grave #1239 in Camp Chase Confederate Cemetery.
Robert Alexander Landress PVT 15 Jun 1843 5 Jul 1911
Sick in Knoxville, TN hospital April 22, 1862; Captured at **Vicksburg**, MS July 4, 1863; Present on Dec 11, 1862 muster roll; Surrendered at **Greensboro**, NC April 26, 1865; Married Martha Ann Edgar in Walton County, Dec 8, 1866; Farmer in Chattooga County in 1900; Applied for Confederate pension in Floyd County in 1909; Application endorsed by A.G. Thompson; Buried at Fellowship Baptist in Rome, Georgia.
Matthew Landress PVT 16 Dec 1838 14 Apr 1919
Older brother of Robert Landress; Married Martha Sims in Walton County, Aug 7, 1856; Admitted to 1st Mississippi Hospital in Jackson, MS July 7, 1863; Present on Dec 11, 1863 muster roll; Admitted to Madison, GA hospital in debilitated condition Mar 30, 1864; Returned to duty April 2, 1864; Surrendered at **Greensboro**, NC April 26, 1865; Farmer in Gwinnett County after the war. Died in Dacula.
Francis Marion Maughon PVT 28 Nov 1825 22 Jun 1905
Married Saleta Barrett in Walton County, Jun 16, 1857; Enlisted in Company H, 11th Georgia Mar 23, 1862; Discharged with disability in Richmond, VA Nov 16, 1862; Enlisted in Walton Blues(G) Sep 23, 1863; Patient in 2nd Georgia Hospital in Augusta Oct 31, 1864; Furloughed at home on Feb 28, 1865 roll; Farmer in Morgan County after the war; Buried at Carlton Family Cemetery in Good Hope, Walton County.
James Shannon Mayfield PVT 1845 16 Aug 1864
Enlisted in Manchester, TN Nov 15, 1862; Attached to Loring Hospital in Lauderdale, MS as nurse July 7, 1863; Furloughed Sep 24, 1863; At home sick on Dec 11, 1863 muster roll.
William Radford Mayfield PVT 2 Oct 1828 29 Dec 1862
Son of Rev. Robert Mayfield; Married Martha Jane Hamilton in Walton County, Oct 31, 1850; Promoted to **2LT**; Killed at the **Battle of Chickasaw Bayou**, MS Dec 29, 1862.

Jesse David McCord PVT 1838 29 Dec 1862
Older brother of David Burson McCord; Killed at **Battle of Chickasaw Bayou**, MS.
James M. McCree PVT 22 Jul 1863
Captured in Vicksburg, MS hospital July 4, 1863.
Irvin H. McDaniel PVT 1834 1864
Half-brother of Green McDaniel; Married Dertha Boleman in 1858; Enlisted in Monroe May 12, 1862; Captured at **Vicksburg,** MS July 4, 1863; AWOL on Dec 11, 1863 muster roll since Dec 9, 1862.
Green Jackson McDaniel PVT 16 May 1829 4 May 1911
Captured at **Vicksburg,** MS July 4, 1863; Arrested by Lt. J.H. Camp of 16th Georgia as a deserter and taken to the Camp of Instruction for the 42nd Georgia in Decatur, GA Nov 1863; AWOL on Dec 11, 1863 muster roll; Married Mary Culberson Stewart; Farmer in Cut Off, Walton County after the war; Buried at Davis-McDaniel Family Cemetery in Winder.
John Franklin McGaughey PVT 29 Apr 1841 11 Jun 1912
Captured at **Battle of Baker's Creek,** MS May 16, 1863; POW at **Camp Morton,** IN; Paroled at Fort Delaware, DE July 3, 1863; Absent sick at home on Dec 11, 1863 muster roll; Married Susan Hailes in Walton County, Sep 25, 1866; Physician in Oconee County in 1900; Filed for Confederate pension in Oconee County in 1907; Reported that he was captured at Battle of New Hope Church, GA May 25, 1864.
Napoleon Monroe McGaughey PVT 12 Apr 1843 10 May 1921
Younger brother of John McGaughey; Transferred to the Walton Tigers (H); Captured at **Vicksburg,** MS July 4, 1863; Absent sick at home on Dec 11, 1863 muster roll; Married Nancy Hale in Walton County July 16, 1864; Married Martha West in Greene County, April 10, 1877; Filed for Confederate pension in Greene County in 1910; Reported being with his command at the Greensboro surrender; Buried at Penfield Cemetery in Greene County.
John William Micheal PVT 24 Mar 1836 19 Nov 1881
Older brother of William Michael; Captured at **Vicksburg,** MS July 4, 1863; Present on Dec 11, 1863 muster roll; Captured at **Battle of Resaca,** May 15, 1864; Released from **Alton,** IL military prison May 23, 1865; Married Nancy Clark Johnston in Walton County, Oct 1, 1868; Carriage maker in Good Hope after the war; Buried at Mount Carmel Church in Walton County.
William David Michael PVT 30 Nov 1841 28 Nov 1887
Captured at **Vicksburg,** MS July 4, 1863; Left sick at home Dec 11, 1863 muster roll; Captured at **Missionary Ridge,** TN Nov 25, 1863; Prisoner at **Rock Island,** IL; Enlisted in US Army for frontier service Oct 6, 1864; Married Mary Catherine McGaughey in Walton County, Nov 15, 1869; Farmer in Walton County after the war; Buried at McGaughey Cemetery in Walton County.
Robert William Miller PVT 28 Apr 1836 11 Apr 1916
Married Sarah Edmonds in Walton County, Aug 25, 1857; Captured at **Vicksburg,** MS July 4, 1863; Among the sick taken by US Steamer *St. Maurice* from Vicksburg to Mobile, AL Aug 5, 1863; Absent sick at home Dec 11, 1863 muster roll; Farmer in Cut Off after the war; Buried at Bethlehem Methodist in Barrow County.
Benjamin Franklin Moore PVT 8 Jul 1843 24 Feb 1900
Captured at **Vicksburg,** MS July 4, 1863; Present on Dec 11, 1863 muster roll; Appointed

regimental musician; Surrendered at **Greensboro,** NC April 26, 1865; Married Mary Frances Thompson in Walton County, Nov 8, 1866; Miller in Walton County after the war; Buried in Thompson Family Cemetery in Walton County.

Edwin Moore PVT 1839 30 Apr 1862

Sick in Knoxville, TN hospital April 22, 1862; Buried at Bethel Confederate Cemetery in Knoxville, TN; Claim filed by his mother, Ally E. Morris.

Addison W. Nowell PVT 1843 16 Sep 1863

Captured at **Vicksburg,** July 4, 1863; Patient in Point Clear, Baldwin County, AL hospital August 31, 1863.

James William Nowell PVT 26 Jan 1841 16 Dec 1928

Married Parmelia Thomas in 1861; Enlisted in Monroe May 12, 1862; Absent with Johnston's command on Dec 11, 1863 muster roll; Surrendered at **Greensboro,** NC April 26, 1865; Retail clerk in Commerce in 1870; Filed for Confederate pension in Jackson County in 1914; Buried at Rose Hill Cemetery in Barrow County.

James W. Oats PVT 1839 22 May 1862

Absent sick in Maynardville, TN hospital; Claim filed by his mother Sarah A. Oats.

James L. Odum PVT 13 Jan 1835 3 Aug 1919

Married Elizabeth Kilgore in Walton County, Nov 22, 1857; Absent sick in Knoxville, TN hospital April 22, 1862; Captured at **Vicksburg,** MS July 4, 1863; Absent sick at home on Dec 11, 1863 muster roll; Farmer in Cut Off after the war; Buried at Kilgore Family Cemetery in Walton County.

John Wesley Odum PVT 12 Oct 1838 6 Feb 1915

Married Elizabeth Mayfield in Walton County, Dec 7, 1856; Captured at **Vicksburg,** MS July 4, 1863; Present on Dec 11, 1863 muster roll; Filed for Confederate pension in Walton County in 1898; Reported being taken with measles in Knoxville, TN in May 1862 and typhoid fever & pneumonia in 1863; Farmer in Cut Off after the war; Buried at Bethlehem Methodist in Barrow County.

William Elijah Odum PVT 11 Aug 1840 27 Apr 1917

Younger brother of James & John Odum; Married Ellender Thompson in Walton County, Nov 6, 1859; Present on Dec 11, 1863 muster roll; Captured at **Battle of Resaca** May 15, 1864; POW at **Camp Morton,** IN; Released at Point Lookout, MD Mar 15, 1865; Farmer in Cullman County, AL in 1880; Buried at Cluster Springs Cemetery in Jefferson County, Alabama.

Calvin C. Patrick PVT 13 Jul 1834 9 Dec 1905

Brother of Robert Patrick; Captured at **Battle of Baker's Creek** May 16, 1863; Prisoner at **Camp Morton,** IN; Exchanged at Fort Delaware, DE July 4, 1863; Captured at **Battle of Nashville** Dec 16, 1864; Released from **Camp Chase,** OH June 12, 1865; Married Sarah Judith Breedlove in Walton County, Nov 22, 1868; Farmer in Walton County after the war; Buried at Center Hill Baptist in Walton County.

Luke David Patrick PVT 1818 12 Jul 1862

Married Emeline Cynthia Davis in Walton County, Nov 5, 1845; Mechanic in Cut Off in 1860.

Robert R. Patrick PVT 1837 10 Aug 1863

Captured at **Vicksburg,** MS hospital July 4, 1863; Left on US Steamer *St. Maurice* on July 30, 1863 from Vicksburg to Mobile, AL; Died of disease in Atlanta hospital; Buried at Oakland Cemetery in Atlanta.

James Benjamin Peeler PVT 1842 1886
Younger brother of William Peeler; Born in Elbert County; Captured at **Vicksburg** July 4, 1863; Sick at home on Dec 11, 1863 muster roll; Surrendered at **Greensboro**, NC April 26, 1865; Married Sarah Dickson in Walton County, Feb 2, 1868; Married Huldy Dobbs in Oconee County, Jan 3, 1878; Factory hand in Good Hope, Walton County in 1880; Buried at Old High Shoals Baptist in Walton County.
William C. Peeler PVT Jun 1840 4 May 1910
Captured at **Vicksburg**, MS July 4, 1863; Surrendered at **Greensboro**, NC April 26, 1865; Married Rachel LeCroy in Walton County, Dec 24, 1865; Worked in cotton factory in Madison County, AL after the war; Died in Madison County, Alabama.
Jasper Robert Reeves PVT 5 Jan 1834 29 Dec 1862
Killed at **Battle of Chickasaw Bayou** Dec 29, 1862.
William W. Ridgeway PVT 27 Jan 1833 20 Apr 1884
Born in South Carolina; Married Elizabeth Thompson in 1851; Captured at **Vicksburg**, MS July 4, 1863; Absent without leave on Dec 11, 1863 muster roll; Farmer in Cut Off in 1870; Farmer in Blount County, AL in 1880; Buried at Wynnville Cemetery in Blount County, Alabama.
Edward S. Rutledge PVT 17 Oct 1863
Went by "Dock"; Captured at Vicksburg, MS July 4, 1863.
James A. Sharpton PVT 1833 1863
Enlisted in Calhoun Aug 23, 1862; Deserted on Dec 22, 1862; Killed by Sergeant H. Boyd while resisting arrest on Oct 17, 1863.
Hezekiah Thomas Sheppard PVT 4 Feb 1840 7 Sep 1909
Admitted to Fairgrounds Hospital #1 in Atlanta Jan 14, 1863; Listed with General Johnston's Mississippi command on Dec 11, 1863 muster roll; Teamster with supply train of Army of Tennessee Feb 28, 1865; Married Octavia Patman Nov 28, 1867; Farmer and miller in Walton County after the war; Buried in Sparks City Cemetery in Cook County, Georgia.
Jefferson Virgil Sheppard PVT 1845 17 Jul 1863
Younger brother of Hezekiah Sheppard; Captured at **Vicksburg**, MS July 4, 1863.
Charles Holt Smith PVT 18 Jun 1844 22 Dec 1914
Enlisted at Woodson's Crossroads, TN Sep 13, 1862; Captured at **Vicksburg**, MS July 4, 1863; Surrendered at **Greensboro**, NC April 26, 1865; Married Serena Smith in Walton County, Dec 22, 1875; Farmer in Cut Off after the war; Buried at David Smith Family Cemetery in Barrow County.
Jesse H. Smith PVT 1839 10 Dec 1862
Married Martha Richardson in Walton County, Oct 8, 1861; Died in Bean Station, TN.
William Job Smith PVT 16 Jan 1846 3 Oct 1899
Younger brother of Charles Smith; Enlisted in Dalton Feb 18, 1864; Wounded at **Battle of Resaca**, GA May 15, 1864; Farmer in Cut Off in 1880; Buried at David Smith Family Cemetery in Barrow County.
Sampson Still PVT 10 Jul 1822 8 Sep 1863
Born in Edgefield County, SC; Married Jemima Morgan in Walton County, Dec 1, 1842; Died in Marion, AL hospital.
Thomas Virgil Stone PVT 1841 3 Apr 1862
Died in first month of enlistment.

George Franklin Sturdivant PVT 9 Mar 1841 13 Feb 1905
Married Sarah Elizabeth Griffith in Walton County, Feb 24, 1859; Captured at **Battle of Baker's Creek** May 16, 1863; Sent to **Camp Morton,** IN; Exchanged at Fort Delaware, DE July 4, 1863; Sick at home on Dec 11, 1863 muster roll; Farmer in Walton County after the war; Filed for Confederate Pension in Walton County in 1897; Reported being wounded at the Battle of New Hope Church May 25, 1864; Buried at Bethlehem Methodist in Barrow County.
William H. Terrell PVT 12 Aug 1836 3 Oct 1898
PVT in the Joe Brown Rough & Ready Militia from Cut Off, GA June 18, 1861; Enlisted in Monroe May 12, 1862; Captured in **Vicksburg,** MS hospital July 4, 1863; Taken with sick by US Steamer *St. Maurice* from Vicksburg to Mobile, AL; Exchanged at Fort Morgan Aug 5, 1863; Married Buena Vista Ballew about 1866; Farmer near Marietta after the war; Filed for Confederate pension in Cobb County in 1889; Reported being shot in the right breast and permanently disabled at Battle of Kennesaw Jun 27, 1864; Buried at Sardis Cemetery in Cobb County.
Truman Fletcher Thomas PVT 18 Mar 1831 29 Dec 1862
Enlisted in Monroe May 12, 1862; Married Mary Caroline Thompson in Walton County, Dec 8, 1855; Killed at **Battle of Chickasaw Bayou,** MS; Claim filed by his wife Mary C. Thomas in Walton County.
Alfred G. Thompson PVT 31 Mar 1837 5 Dec 1918
Married Nancy Chandler about 1860; Enlisted in Monroe May 12, 1862; Admitted to Fairgrounds Hospital #1 in Atlanta Nov 28, 1862 with debility; Deserted from hospital Dec 3, 1862; Captured at **Vicksburg,** MS July 4, 1863; Sick at home on Dec 11, 1863 muster roll; Surrendered at **Greensboro,** NC April 26, 1865; Farmer in Cut Off after the war; Filed for Confederate pension in 1910; Buried at Bethlehem Methodist in Barrow County.
David H.C. Thompson PVT 9 Jan 1846 17 Mar 1932
Younger brother of Alfred & Thomas Thompson; Enlisted in Atlanta Feb 18, 1864; Surrendered at **Greensboro,** NC April 26, 1865; Farmer in Cut Off after the war; Filed for Confederate pension in Barrow County in 1919; Application endorsed by J.B. Austin; Buried at Bethlehem Methodist in Barrow County.
Hilliard Thompson PVT 20 May 1838 1 Aug 1863
Married Lydia Tabitha Harrison in Walton County, Jan 29, 1860; Captured in **Vicksburg,** MS hospital July 4, 1863.
Thomas C. Thompson PVT 1839 15 Sep 1877
Captured at **Vicksburg,** MS July 4, 1863; Present on Dec 11, 1863 muster roll; Surrendered at **Greensboro,** NC April 26, 1865; Farmer in Cut Off in 1870; Estate probated in Walton County in 1882.
Franklin J. Thurmon PVT 1841 7 Jan 1922
Admitted to Fairgrounds Hospital #1 in Atlanta Nov 28, 1862 with pneumonia; Captured at **Vicksburg,** MS July 4, 1863; Present on Dec 11, 1863 muster roll; Captured at **Battle of Nashville** Dec 16, 1864; Discharged from **Camp Douglas,** IL June 20, 1865; Farmer in Walton County after the war; Filed for Confederate pension in Walton County in 1920.
Thomas Thurman PVT 29 Dec 1862
Enlisted Aug 23, 1862; Killed at **Battle of Chickasaw Bayou,** MS.
Benjamin Franklin Tuck PVT 16 Aug 1832 8 Apr 1862

Brother of James Tuck; One of five Tuck brothers to die in the war; Married Lucinda Boss in Walton County, Nov 10, 1853; Died in Camp McDonald, GA.

James William Tuck PVT 29 Aug 1829 1864

Married Mary Frances Upshaw in Walton County, Oct 19, 1851; Captured at **Vicksburg,** MS hospital July 4, 1863; Shipped with sick on US Steamer *St. Maurice* from Vicksburg to Mobile July 30, 1863; Exchanged at Fort Morgan Aug 5, 1863; Sick at home on Dec 11, 1863 muster roll; Died while in service in Dalton.

John L. Tucker PVT 1839 21 Apr 1862

Died of disease.

Silas Washington Vickers PVT 6 Jul 1837 9 Aug 1915

Born in Clarke County; Captured at **Vicksburg,** MS July 4, 1863; Present on Dec 11, 1863 muster roll; Filed for Confederate pension in Cherokee County in 1895; Application endorsed by H.B. Mickel; Died in Cedartown, Polk County.

Abraham Cowan Warren PVT 17 Sep 1831 24 Jun 1894

Brother of Jacob Warren; Married Margaret Peeler in Clarke County, May 15, 1852; Farmer in Good Hope in 1880; Filed for Confederate pension in Oconee County in 1894; Reported being afflicted with measles in Knoxville in 1862; Condition "settled in his lungs & heart" and he was granted a medical discharge on July 25, 1864.

Jacob L. Warren PVT 20 Sep 1839 19 Feb 1909

Captured at **Vicksburg,** MS July 4, 1863; Present on Dec 11, 1863 muster roll; Admitted to Way Hospital in Meridian, MS Jan 15, 1865; Surrendered at **Greensboro**, NC April 26, 1865; Married Rebecca Peeler in Walton County, Dec 23, 1866; Farmer in Morgan County in 1880; Member of UCV Camp #1124 in Watkinsville; Farmer in Dark Corner, Oconee County in 1900; Filed for Confederate pension in Oconee County in 1904; Buried at Rays Chapel Methodist Church in Oconee County.

Frances Marion Weatherford PVT 1828 1888

Married Louisa Sims in Walton County, Sep 1, 1850; Enlisted in Monroe May 12, 1862; Captured at **Vicksburg,** MS July 4, 1863; Present on Dec 11, 1863 muster roll; Detailed as Division Provost by order of General Henry Clayton Mar 8, 1864; Surrendered at **Greensboro,** NC April 26, 1865; Farmer in Rockdale County in 1880; Buried at Weatherford Family Cemetery in McDonough, Henry County.

John Chester Weatherford PVT 17 Jun 1835 26 Apr 1917

Appointed **3SGT** in March 1863; Captured at **Vicksburg,** MS July 4, 1863; Present on Dec 11, 1863 muster roll; Surrendered at **Greensboro,** NC April 26, 1865; Cotton mill operator in Morgan County in 1870; Married Nancy Pridgeon in Oconee County, Oct 6, 1881; Cotton mill machinist in Oconee County in 1900; Buried at High Shoals Holiness Cemetery in Oconee County.

John William Weatherford PVT 14 Apr 1840 6 Oct 1910

Younger brother of F.M. Weatherford; Married Elizabeth O'Kelley in Walton County, April 14, 1861; Captured at **Vicksburg,** MS hospital July 4, 1863; Shipped with sick by US Steamer *St. Maurice* from Vicksburg to Mobile July 30, 1863; Exchanged at Fort Morgan Aug 5, 1863; Sick at home on Dec 11, 1863 muster roll; Furloughed from Madison, GA hospital, May 20, 1864; Admitted to Madison, GA hospital July 15, 1864; Farmer in Walton County after the war; Buried at Shoal Creek Baptist in Walton County.

John W.S. Whitehead PVT Nov 1832 7 Feb 1907

Born in Clarke County; Captured at **Battle of Baker's Creek** May 16, 1863; POW in

Camp Morton, IN; Exchanged at Fort Delaware, DE July 4, 1863; Admitted 1st Mississippi Hospital in Jackson, Mississippi Jan 18, 1863; Sent to Rome hospital Oct 18, 1863; Married Harriet Davis in Walton County, May 23, 1867; Farmer in Gwinnett County after the war; Buried at Bethesda Methodist in Lawrenceville.

William S. Whitehead PVT 5 Nov 1831 28 Jan 1863

Born in Jackson County; Married Mary Betts in Jackson County, Jan 4, 1855; Married Eleanor Jane Betts in Jackson County, June 5, 1862; Died of disease at City Hospital in Vicksburg, MS; Claim filed by his wife Eleanor Betts.

James T. Wiggins PVT 25 Dec 1841 21 Mar 1915

Captured at **Vicksburg**, MS July 4, 1863; Present on Dec 11, 1863 muster roll; Surrendered at **Greensboro**, NC April 26, 1865; Married Sarah Thornton in Walton County, Jan 5, 1870; Farmer in Walton County after the war; Buried at Campton Methodist in Monroe.

Templeton Williams PVT 1843 22 Apr 1862

Died in first month of enlistment.

Billington A. Wilson PVT 1837 19 Mar 1865

Born in Clarke County; Patient in Knoxville, TN hospital April 22, 1862; Appointed **4CPL**; Captured at **Vicksburg**, MS July 4, 1863; Present on Dec 11, 1863 muster roll; Killed at **Battle of Bentonville**, NC.

Company H – Walton Tigers – 156 Men

This company drew men from south and east of the county seat of Monroe & a handful from Greene County.

Name Rank Birth Date

Jett Thomas Mitchell CPT 22 Apr 1822 21 Oct 1898

Resigned July 21, 1862 with a disability of the spinal column "which prevents him from performing his duties as a soldier;" Lawyer & farmer in Broken Arrow; Buried at Mitchell Cemetery in Walnut Grove.

William James Akridge 2LT 21 Jun 1827 1 Jan 1888

Born in DeKalb County; Married Mary Baker in DeKalb County, June 30, 1853; Farming in Calhoun County, AL in 1860; Sent to Walton County April 16, 1862 by Colonel Henderson to boost recruiting; Resigned at Lenoir Station, TN on Nov 1, 1862 after receiving a medical discharge for diabetes; Farmer in Walton County after the war; Buried at Circle View Cemetery in Social Circle.

Edmond Bryan Gresham 3SGT 1839 Before 1876

Older brother of James Gresham; Married Elizabeth Crawley in Walton County, Oct 9, 1859; Captured at **Vicksburg**, MS July 4, 1863; Present on Dec 11, 1863 muster roll; Farmer in Walton County in 1870.

James Monroe Gresham 2LT 24 Dec 1840 5 Jul 1913

Married Martha Atkinson in Walton County, Nov 14, 1865; Captured at **Vicksburg**, MS July 4, 1863; Present on Dec 11, 1863 muster roll; Captured at **Battle of Nashville**, TN Dec 16, 1864; Released from **Johnson's Island** Prison June 16, 1865; Farmer, school teacher, and state legislator for Walton County after the war; Commander of R.E. Lee UCV Camp 1055 in Monroe; Buried at Gresham Cemetery in Walton County.

William Samuel R. Hardeman 1LT 27 Jan 1831 23 Dec 1887

See Regimental Officers.

Jonathan Wilson Huff 1SGT 1827 5 Feb 1865

Married Elizabeth Smith about 1852; Captured at **Vicksburg**, MS July 4, 1863; Captured at **Battle of Nashville**, TN Dec 16, 1865; Died of smallpox at **Camp Chase**, OH; Buried at Confederate Cemetery, Camp Chase.
William H. Lloyd 4CPL 23 Sep 1862
Died at Bean Station, TN.
William H. Marston 2CPL 1832 11 Feb 1902
Born in Clarke County; Married Mary Elizabeth Andrew in Walton County, July 5, 1860; Shoemaker at time of enlistment; Received Certificate of Disability for pulmonary disorder Mar 11, 1863; Painter in Conyers in 1880.
Edward D. Queen 4SGT 1829 After 1880
Born in North Carolina; Married Susan Ann Harrelson in Newton County, Dec 31, 1851; Transferred to the Harper Guards (E) on Sep 30, 1862; Captured at **Vicksburg**, MS July 4, 1863; Absent without leave on Dec 10, 1863 muster roll; Listed on May 20, 1865 Roll of Prisoners surrendered at Augusta; Married Angeline Sealey in Maury County, TN Aug 23, 1868; Laborer in Spring Hill, TN in 1880.
William F. Smith 1CPL 5 Mar 1842 After 1914
Married Anna Baccus in Walton County, Oct 18, 1860; Reduced to ranks May 17, 1863; Placed on detached service by Colonel Henderson Oct 20, 1863; Captured at Washington Hospital in **Vicksburg**, MS July 4, 1863; Placed with sick on US Steamer *Chouteau* from Vicksburg to Mobile; Exchanged at Fort Morgan July 22, 1863; Machinist in Social Circle in 1900; Filed for Confederate pension in Walton County in 1914; Reported being present at Greensboro surrender April 26, 1865.
William Joseph Williams 2SGT 19 Sep 1831 8 Sep 1891
Captured at **Vicksburg**, MS July 4, 1863; Present on Dec 10, 1863 muster roll; Elected 2LT in Dalton **2LT** Feb 25, 1864; Married Anna Blasingame in Walton County, Dec 30, 1867; Farmer in Walton County after the war; Buried at Blasingame Cemetery in Walton County.
Seaborn "Seb" O. Adcock PVT 10 Jul 1815 24 Dec 1887
Married Sarah Peters in Walton County, Oct 25, 1843; Left sick in Knoxville, TN hospital April 22, 1862; Detailed to Division Ordinance Train as teamster Oct 12, 1862; Admitted to Fair Grounds Hospital #1 in Atlanta Nov 28, 1862; Captured at **Vicksburg**, MS July 4, 1863; Captured with wagons on Missionary Ridge, TN Nov 25, 1863; Transferred for exchange from **Rock Island**, IL prison Mar 5, 1865; Farmer in Walton County after the war; Buried at Blasingame Cemetery in Walton County.
Joseph Elbert Andrews PVT 17 May 1837 15 Feb 1923
Captured at **Vicksburg**, MS July 4, 1863; Wounded at **Battle of New Hope Church** May 25, 1864; Captured at **Battle of Atlanta**, July 22, 1864; Released from **Camp Douglas**, IL May 16, 1865; Married Louise Frances Lowery in Walton County, April 1, 1869; Carpenter & farmer in Newton County after the war; Buried in Eastview Cemetery in Conyers.
William M. Argo PVT 4 Mar 1818 18 May 1880
Born in South Carolina; Married Elizabeth Whitley about 1838; Father of Edmond Jasper Argo from Newton Rifles (F); Appointed **4CPL** in Feb 1863; Attached as nurse to General Hospital in Enterprise, MS May 27, 1863; Captured at **Vicksburg**, MS July 4, 1863; Wounded at **Missionary Ridge**, Nov 25, 1864; Appointed **1CPL** Dec 1, 1863; Admitted

to Ocmulgee Hospital in Macon with paralysis from lightning strike Aug 23, 1864 & furloughed Sep 6, 1864; Farm laborer in Newton County in 1870; Buried at McClung Cemetery in Conyers.

James M. Armistead PVT 20 Aug 1833 10 Feb 1909
Married Martha Everett in Gwinnett County, Dec 29, 1857; Appointed **5SGT** Feb 1863; Captured at **Vicksburg,** MS July 4, 1863; Furloughed from Fair Grounds Hospital #1 in Atlanta Nov 24, 1863; Patient at Empire Hotel Hospital in Atlanta April 30, 1864; Farmer in Gwinnett County after the war; Buried at Cedar Creek Baptist in Winder.

Thomas Macon Armistead PVT 21 Sep 1843 18 Mar 1922
Appointed **3SGT** Feb 1863; Captured at **Vicksburg,** MS hospital July 4, 1863; Shipped with sick from US Steamer *St. Maurice* from Vicksburg to Mobile; Exchanged at Fort Morgan Aug 5, 1863; Present on Dec 11, 1863 muster roll; Tax receiver for Fulton County for 34 years; Married Annie Hudson in Fulton County, May 16, 1881; Filed for Confederate Pension in Fulton County in 1890; Reported being severely wounded in the left hip at the Battle of Resaca May 16, 1864, rendering his left leg useless; Application endorsed by Captain W.L. Calhoun; Member of UCV Camp # 159 in Atlanta; Buried at Westview Cemetery in Atlanta.

Zachariah Jabez Armistead PVT 9 Jun 1847 24 Jul 1877
Younger brother of Thomas Armistead; Admitted to General Hospital in Lauderdale Springs, MS April 2, 1863; Appointed **4CPL** Oct 14, 1863; Returned to duty from Fair Grounds Hospital #1 in Atlanta May 24, 1864; Captured near Atlanta Aug 7, 1864; Prisoner at **Camp Chase,** OH; Transferred to Point Lookout, MD for release Mar 18, 1865; Employee of the Georgia Railroad; Married Clara Perdue about 1872; Killed in a RR accident near Union Point in Greene County; Buried at Armistead Family Cemetery in Newton County.

Robert Aycock PVT 13 Aug 1818 15 Sep 1896
Claimed $5,000 in wealth in 1860; Enlisted in Decatur Sep 15, 1863; Surrendered at **Greensboro,** NC April 26, 1865; Farmer in Walton County after the war; Buried at Rest Haven Cemetery in Monroe.

William M. Baccus PVT 1844 1862
Died at home.

Nicholas R. Baker PVT 1830 2 Dec 1862
Married Rebecca Etheridge in Gwinnett County, Jan 27, 1850; Enlisted in Monroe May 12, 1862; Died in Atlanta hospital; Claim filed by his wife Rebecca in Newton County.

Robert E. Bird PVT 1845 1901
Married Sarah Ivy in Walton County, Dec 16, 1862; Enlisted July 15, 1863; Shoemaker in Walton County in 1880; Farmer in 1900.

James Andrew Blankenship PVT 21 Mar 1846 26 Dec 1913
Born in Cherokee County; Captured at **Vicksburg,** MS hospital July 4, 1863; Transported with sick from US Steamer *St. Maurice* from Vicksburg to Mobile; Exchanged at Fort Morgan Aug 5, 1863; Sick at home on Dec 11, 1863 muster roll; Married Nancy Plunkett in Newton County, Feb 16, 1868; Farmer in DeKalb County in 1880; Filed for Confederate pension in Fulton County in 1900; Reported being wounded in the left hand at **Battle of New Hope Church** May 25, 1864; Furloughed with blood poisoning; Application supported by T.M. Armistead and L.P. Thomas; Member of UCV Camp #159 in Atlanta; Buried at Union Grove Baptist in Henry County.

Beverly Frederick Blasingame PVT 2 Feb 1842 12 Apr 1911
PVT in Guards of the Sunny South Militia from Social Circle, GA Sep 18, 1861; Detailed to Captain Calhoun's detachment in Knoxville, TN April 20, 1862 to guard political prisoners; Prison guard at Madison, GA Oct 31, 1862; Captured at **Vicksburg,** MS July 4, 1863; Married Mary Brockwell in Morgan County, Dec 19, 1865; Farmer in Blasingame after the war; Filed for Confederate pension in Walton County in 1890; Reported being struck in left leg breaking the tibia at **Battle of Bentonville,** NC Mar 19, 1865; Reported being shot in the hand at **Battle of Nashville** Dec 16, 1864; Buried at Blasingame Cemetery in Walton County.

James Powell Blasingame PVT 1845 2 May 1864
Younger brother of William Blasingame; Enlisted in Monroe Sep 20, 1863; Died of disease at Gilmer Hospital in Marietta.

John Franklin Blasingame PVT 7 Feb 1840 24 Dec 1889
Older brother of Beverly Blasingame; PVT in Guards of the Sunny South Militia from Social Circle, GA Sep 18, 1861; Captured at **Vicksburg,** MS July 4, 1863; Placed on detached service Oct 20, 1863; Married Lena Marett in Marshall County, MS, Sep 12, 1865; Carpenter in Hood County, TX in 1880; Buried at Granbury Cemetery, Hood County, Texas.

William Thomas Blasingame PVT 4 Jul 1830 30 May 1863
Married Elizabeth Brown in Walton County, Nov 6, 1850; Married Mary Barrett in Morgan County, Jan 19, 1855; Claimed $5,500 in wealth in 1860; Appointed **2CPL** May 11, 1863; Died in Mississippi.

Thomas Haynes Bonner PVT 1822 28 Nov 1863
Enlisted May 12, 1862 as substitute for George Malcom; Admitted to Fairgrounds Hospital #1 in Atlanta Nov 14, 1862; Deserted hospital Dec 4, 1862; Captured in **Vicksburg,** MS hospital July 4, 1863; Paroled at Vicksburg July 16, 1863; Died in Vicksburg; Memorial stone at Memory Hill Cemetery in Milledgeville.

Harrison Henderson Broach PVT 26 Mar 1820 1 Jun 1909
Born in Clarke County; Married Elizabeth Peters in Walton County, Oct 25, 1843; Enlisted at Chickamauga Oct 4, 1863; Surrendered at Greensboro, NC April 26, 1865; Farmer in Blasingame after the war; Filed for Confederate pension in Walton County in 1895; Buried at Peters Cemetery in Monroe.

William H. Brooks PVT 6 Jan 1814 25 July 1863
Married Adaline Adams in Walton County, Oct 15, 1832; Day laborer in Monroe in 1860; Enlisted in Monroe May 12, 1862 as a substitute for P.S. Rogers; Captured at **Vicksburg,** MS July 4, 1863; Died in Vicksburg.

Alexander G. Bulloch PVT 4 May 1836 1862
Died in Knoxville, TN hospital.

Hamilton C. Buse PVT 1832 9 Sep 1862
Married Sarah Arnold in Newton County, Dec 30, 1852; Died at Bean Station, TN hospital; Claim filed by his widow Sarah Buse.

Benjamin Franklin Callahan PVT 1838 15 Nov 1862
Younger brother of Jacob Callahan; PVT in Guards of the Sunny South Militia from Social Circle, GA Sep 18, 1861; Enlisted as PVT in Company B, 3rd Battalion, Georgia State Troops Oct 30, 1861; Mustered out May 1862; Enlisted in Company H in Monroe May 12, 1862; Died in Knoxville, TN hospital.

Jacob Jocephus Callahan PVT 17 Mar 1827 6 Dec 1892
Born in Clarke County; Married Sarah Herndon in Walton County, Nov 18, 1845; Received Certificate of Discharge at Woodson's Crossroads, TN on Aug 24, 1862 as soldier "has no duty since regiment left camp of instruction;" Miller in Walton County; Buried at Herndon Family Cemetery in Social Circle.
Charles L. Carter PVT 1838 1864
Married Mary Eberhart in Oglethorpe County, Nov 27, 1856; Received Certificate of Discharge in Vicksburg, MS on Feb 8, 1863 as "unhealthy and weakly all his life;" Listed as disabled and exempt on 1864 Walton County militia census.
George Washington Cawley PVT 20 Jul 1864
PVT in Guards of the Sunny South Militia from Social Circle, GA Sep 18, 1861; Captured at **Vicksburg**, MS July 4, 1863; Killed along skirmish line in **Battle of Peachtree Creek**, GA,
Oliver Hazzard Perry Chick PVT 20 Sep 1831 26 Jul 1862
Married Sarah Herrin in Walton County, Dec 7, 1854; Discharged with Certificate of Disability Mar 29, 1863 from Vicksburg, MS; Farmer in Blasingame after the war; Buried at Chick Family Cemetery in Monroe.
Jacob Holloway Clegg PVT 1828 1864
Married Martha Elizabeth Greer in 1855; Overseer in 1860; Enlisted in Monroe May 12, 1862; Discharged April 1863; Listed on 1864 Walton County militia census with surgeon's certificate; Confederate widow pension applied for in Arkansas in 1892.
Joel Boone Clegg PVT 1841 1864
PVT in Guards of the Sunny South Militia from Social Circle, GA Sep 18, 1861; Enlisted in Company B, 3rd Battalion, GA State Troops Oct 30, 1861; Mustered out May 1862; Enlisted in Company H May 12, 1862; Captured at **Vicksburg**, MS hospital July 4, 1863; Shipped with sick from Vicksburg to Mobile on US Steamer *St. Maurice* July 30, 1863; Exchanged at Fort Morgan Aug 5, 1863; Present on Dec 11, 1863 muster roll.
Joseph R. H. Clegg PVT
Captured at **Vicksburg**, MS July 4, 1863; Admitted to Ocmulgee Hospital in Macon with electric shock from lightning strike Aug 23, 1864; Furloughed from May Hospital, Madison, GA with partial paralysis Sep 17, 1864.
Joseph S.M. Clegg PVT 1834 21 Nov 1862
PVT in Guards of the Sunny South Militia from Social Circle, GA Sep 18, 1861; Died in Fair Grounds Hospital in Knoxville, TN; Claim filed by his father James Clegg in Walton County.
Thomas Marion Clegg PVT 1 Mar 1837 1 Aug 1898
Captured at **Vicksburg**, MS hospital July 4, 1863; Shipped with sick from Vicksburg to Mobile on US Steamer *St. Maurice* July 30, 1863; Exchanged at Fort Morgan Aug 5, 1863; Sick at home on Dec 11, 1863 muster roll; Detailed to provost guard in Macon Dec 1864; Surrendered at **Greensboro**, NC April 26, 1865; Married Sarah Akridge in Walton County, Dec 2, 1866; Farmer in Blasingame after the war; Buried at Clegg Family Cemetery in Walton County.
William T. Clegg PVT 1828 14 Sep 1862
Older brother of Thomas & Joel Clegg; Married Darcus Bulloch in Walton County, Dec 16, 1847; Died in hospital near Cumberland Gap, TN; Claim filed by his widow Darcus F. Clegg in Walton County.

Henry D. Coleman PVT 6 Nov 1827 11 Apr 1862
Married Martha Robinson in Walton County May 14, 1854; Died in Atlanta during the first week of training.
George Washington Coleston PVT 1842 25 Oct 1862
Left at Knoxville, KY sick April 26, 1862.
Abraham Moses Colton PVT 29 Feb 1820 10 Jan 1894
Born in Connecticut; Married Nancy Crawley in Morgan County, Jun 1, 1845; Merchant in Social Circle in 1860; Enlisted Sep 4, 1863; Shoemaker at Dalton camp, Jan 31, 1864; Admitted to Fairgrounds Hospital #1 in Vineville, GA Aug 11, 1864; Furloughed Sep 8, 1864; Granted medical disability at Forest Hospital, Lauderdale, MS Mar 29, 1865; Paroled at Meridian, MS hospital May 12, 1865; Appointed postmaster of Social Circle in 1865; Dry goods merchant after the war; Buried at Circle View Cemetery in Social Circle.
Alexander M. Cox PVT 19 Jul 1838 31 Dec 1905
Older brother of John & Charles Cox; Married Adeline McGaughey in Morgan County, Dec 22, 1859; Enlisted in Monroe May 12, 1862; Employed in Vicksburg, MS hospital May 1863; Captured in **Vicksburg**, MS hospital July 4, 1863; Shipped for exchange to Mobile Bay, AL Aug 4, 1863; Present on Dec 11, 1863 muster roll; Surrendered at **Greensboro**, NC April 26, 1865; Married Mary Catherine Berry in 1868; Farmer in Newton County after the war; Buried at Cox Family Cemetery in Walton County.
Charles A. Cox PVT 7 Dec 1843 7 Jun 1911
PVT in Guards of the Sunny South Militia from Social Circle, GA Sep 18, 1861; Enlisted as 1CPL of Company B, 3rd Battalion, Georgia State Troops Oct 30, 1861; Mustered out May 1862; Enlisted in the 42nd Georgia Nov 15, 1862; Captured at **Vicksburg**, MS July 4, 1863; Captured along Atlanta siege lines Aug 7, 1864; Prisoner at **Camp Chase**, OH; Exchanged at Point Lookout, MD Mar 18, 1865; Filed for Confederate pension in Walton County in 1895; Reported catching typhoid pneumonia on Vicksburg siege lines.
John James Cox PVT 3 Apr 1840 30 Apr 1923
Ensign in Guards of the Sunny South Militia from Social Circle, GA Sep 18, 1861; Elected 2LT of Company B, 3rd Battalion, Georgia State Troops Oct 30, 1861; Mustered out May 1862; Enlisted in 42nd Georgia May 15, 1862; Elected **2LT** Aug 28, 1862; Captured at **Vicksburg**, MS July 4, 1863; Wounded and disabled at **Battle of Missionary Ridge**, TN Nov 25, 1864; Furloughed from Madison, GA hospital with gunshot wound Mar 6, 1865; Married Susan Hester in Walton County, Feb 2, 1865; Farmer in Blasingame after the war; Buried at Cox Family Cemetery in Walton County.
James Robert Darby PVT 1842 2 Nov 1862
Married Mary Henrietta Broach; Granted Certificate of Disability Nov 1, 1862; Died the next day at Summit Hospital in Knoxville, TN; Buried at Bethel Cemetery, Knoxville, TN; Claim filed by his widow Mary Darby.
Joseph J. Davis PVT 11 Sep 1862
Died at Bean Station, TN hospital.
John James Daws PVT 12 Jul 1828 2 Dec 1904
Born in Virginia; Married Permelia Adcock in Morgan County Jan 25, 1849; Captured at **Vicksburg**, MS July 4, 1863; Present on Dec 11, 1863 muster roll; Wounded & disabled in left arm at **Battle of New Hope Church**, May 25, 1864; Paroled at Augusta May 20, 1865; Farmer in Walton County after the war; Buried at Ivey Cemetery in Walton County.
John N. Day PVT 1836 1862

Married Sophronia Sludin in Walton County, Nov 29, 1860; PVT in Guards of the Sunny South Militia from Social Circle, GA Sep 18, 1861; Died in Morristown, TN hospital; Will probated in Walton County by Sophronia Day Oct 20, 1862.

William Durden PVT 1835 1880

Older brother of Wilson Durden; Married Mary Darnell in Walton County, April 11, 1853; Enlisted in Monroe Sept 20, 1863; Present on Dec 11, 1863 muster roll; Died in collision; Buried at Studdard Cemetery in Morgan County.

Wilson Lumpkin Durden PVT 5 Feb 1839 10 Jul 1895

Married Sarah Malcom in Walton County, June 12, 1859; PVT in Guards of the Sunny South Militia from Social Circle, GA Sep 18, 1861; Enlisted in Company B, 3rd Battalion, Georgia State Troops Dec 13, 1861; Mustered out May 1862; Enlisted in Monroe May 12, 1862; Captured at **Vicksburg**, MS July 4, 1863; Sent to hospital Oct 30, 1863; Farmer in Broken Arrow in 1880.

George Washington Eason PVT 1834 After 1870

Appointed **3CPL** Feb 1863; Captured at **Vicksburg**, MS July 4, 1863; Present on Dec 10, 1863 muster roll; Surrendered at **Greensboro**, NC April 26, 1865; Farmer in Walton County in 1870. Never married.

Josiah Sydney Eason PVT 1826 1891

Older brother of George, James & William Eason; Enlisted in Monroe Sep 20, 1863; Wounded at **Battle of Atlanta** July 22, 1864; Deserted from May Hospital in Madison April 16, 1865; Farmer in Walton County after the war. Never married.

James Madison Eason PVT 9 Jun 1833 1909

Married Bertie Mosely about 1855; Enlisted in Atlanta May 12, 1862; Transferred to Company C in 1863; Captured at **Vicksburg**, MS July 4, 1863; Captured as a deserter in Milton County July 1864; Took Oath of Allegiance in Louisville, KY July 18, 1864 agreeing to stay north of the Ohio River; Farmer in Milton County after the war; Filed for Confederate pension in Tift County in 1909; Reported transferring to Company C ten days after enlistment; Reported being captured by the Chattahoochee River about July 1, 1864.

William Lampkin Eason PVT 27 Oct 1838 7 Jan 1872

Married Edena Simmons in Walton County, Dec 4, 1859; Captured at **Vicksburg**, MS July 4, 1863; Detailed Oct 9, 1863 to attend to Colonel R.F. Maddox's horse; Surrendered at **Greensboro**, NC April 26, 1865; Farmer in Walton County in 1870.

Thomas Sanford Edmonds PVT 29 Dec 1840 27 May 1923

Brother of Samuel Edmonds in Company G; Present on Dec 10, 1863 muster roll; Furloughed from Madison, GA hospital with left leg wound Mar 13, 1865; Married Martha McGaughey in Walton County, Oct 19, 1873; Filed for Confederate pension in Walton County in 1898; Reported being wounded through the left leg just below the knee at the Battle of Nashville, Dec 16, 1864; Application endorsed by Lt. J.M. Gresham; House painter in Monroe in 1900; Buried at Rest Haven Cemetery in Monroe.

Edmond Bolton Edmondson PVT 15 Sep 1837 19 Nov 1884

Married Mary Frances Dally about 1859; Captured at **Vicksburg**, MS July 4, 1863; Present on Dec 10, 1863 muster roll; Captured at Social Circle July 23, 1864; Prisoner at **Camp Chase**, OH; Transferred to Fort Delaware, DE for exchange Mar 1864; Patient at Jackson Hospital, Richmond, VA Mar 16, 1865; Farmer in Blasingames in 1880.

William M. Everett PVT 4 Oct 1842 After 1910

Enlisted in Calhoun Aug 22, 1862; Dec 10, 1863 muster roll report Everett sick at home since Dec 1862; Married Susan Sorrells in Walton County, Jan 10, 1866; Farmer in Walton County after the war; Filed for Confederate pension in Walton County in 1897; Reported contracting measles at Woodson's Crossroads, TN; Never able to perform duties again and sent home; Formally discharged April 1864; Buried at Everett Family Cemetery in Walton County.

David M. Fowler PVT 1834 After 1883

Admitted to Fairgrounds Hospital #1 in Atlanta Nov 28, 1862; Furloughed Dec 1, 1862; Admitted to Fairgrounds Hospital #1 in Atlanta Dec 2, 1863; Returned to duty Jan 15, 1864; Patient at Ocmulgee Hospital in Macon with complications from thigh wound April 3, 1865; Furloughed April 11, 1865; Farmer in 1870; Owned 100 acres in Blasingame in 1883 tax roll.

Nathaniel Lumpkin Galloway PVT 1 Jun 1832 1 Sep 1907

Enlisted in Decatur Feb 23, 1863; Promoted to Assistant Surgeon, CSA May 26, 1863; Promoted to Full Surgeon of 43rd Georgia Feb 8, 1864; Surrendered at **Greensboro**, NC April 26, 1865; Married Lucy Hughes in Walton County, Nov 26, 1868; Farmer and "allopath" physician in Monroe after the war; Buried at Rest Haven Cemetery in Monroe.

Elisha Garrett PVT 1825 1864

Enlisted in Decatur Mar 12, 1863; Appointed **3CPL** in 1863; Captured at **Vicksburg**, MS July 4, 1863; Discharged from Cannon Hospital in Dalton Jun 29, 1864 with "extreme emaciation & debility;" Never married.

Amos H. Green PVT 4 Jan 1846 1913

Captured at **Vicksburg**, MS July 4, 1863; Sick at home on Dec 10, 1863 muster roll; Furloughed from disabled camp in Lauderdale, MS with gunshot wound to left hip Jan 21, 1865; Farm laborer in Broken Arrow after the war; Never married; Filed for Confederate pension in Walton County in 1888; Reported being hit by shell near Atlanta Aug 8, 1864 breaking his left hip and three ribs.

John H. Green PVT 8 Jun 1843 3 Jul 1920

Older brother of Amos Greene; Married Susan Johnson in Walton County, Jun 21, 1860; Captured at **Vicksburg**, MS July 4, 1863; Sick at home on Dec 10, 1863 muster roll; Surrendered at **Greensboro**, NC April 26, 1865; Farmer in Walton County after the war; Filed for Confederate pension in Walton County in 1897; Buried at Corinth Memorial Gardens in Loganville.

Caleb T. Greer PVT 4 Nov 1833 28 Nov 1902

Enlisted Mar 22, 1862; Issued Certificate of Disability for Discharge Oct 16, 1862 from Knoxville, TN hospital; Married Amanda Edmondson in Morgan County, Dec 18, 1864; Blacksmith in Walton County in 1870; Member of Masonic Order; Buried at Hurricane Cemetery, Claiborne Parish, LA.

John Henry Greer PVT 19 Sep 1845 16 Jan 1930

Captured at **Vicksburg**, MS July 4, 1863; Present on Dec 10, 1863 muster roll; Married Eliza Frances Kelly in Walton County, Feb 1869; Farmer in Walton County in 1880; Farmer in Claiborne Parish in 1900; Buried at Hurricane Cemetery, Claiborne Parish, LA.

Charles Hezekiah Griffith PVT 27 Aug 1833 7 Jun 1902

Married Matilda Sturdivant in Walton County, Dec 18, 1856; Enlisted in Monroe May 12, 1862; Captured at **Vicksburg**, MS July 4, 1863; Sick at home on Dec 10, 1863 muster roll; Captured at Covington July 22, 1864; Prisoner at **Camp Chase**, OH; Transferred to

City Point, VA for exchange Mar 4, 1865; Farmer in Walton County after the war; Filed for Confederate pension in Walton County in 1901; Reported being captured in Oxford, GA in July 1864; Buried at Harper-McRae Cemetery in Oconee County.

John David Griffith PVT 4 Dec 1828 5 May 1910
Older brother of Charles Griffith; Married Louisa Collier about 1845; Enlisted in Monroe May 12, 1862; Captured at **Vicksburg**, MS July 4, 1863; Sick at home on Dec 10, 1863 muster roll; Surrendered at **Greensboro**, NC April 26, 1865; Married Elizabeth Rhodes in Clarke County, Nov 4, 1866; Filed for a Confederate pension in Jackson County in 1899; Reported being struck in the back by shell at Battle of Atlanta July 22, 1864; Working in cotton mill in Athens in 1900.

Ebenezer Ben Hardeman PVT 25 Jan 1837 19 Feb 1897
Born in Oglethorpe County; Married Iantha Lynch in Putnam County, Dec 13, 1859; Enlisted in Monroe May 12, 1862; Received Certificate of Disability for Discharge from Knoxville, TN hospital Nov 4, 1862; Farmer near Monroe in 1880; Buried at Salem Methodist in Covington.

James Harrison Hardeman PVT 27 Dec 1840 22 Aug 1862
Younger brother of Samuel and Ebenezer Hardeman; Died at home; Buried at Old Baptist Cemetery in Monroe; Claim filed by his father Bisber Hardman.

James W. Harris PVT 7 May 1828 After 1880
Married Emily St. Johns in Newton County, Nov 11, 1845; Enlisted in Company A, 3rd Battalion, Georgia State Troops Nov 18, 1861; Discharged Dec 12, 1861; Left sick in Knoxville, TN hospital April 22, 1862; Captured at **Vicksburg**, MS July 4, 1863; Present on Dec 10, 1863 muster roll; Farmer in Walton County in 1880.

Ganaway Hawk PVT 1832 July 1862
Married Ann Abercrombie in Walton County, July 15, 1852; Wealthy Walton farmer in 1860; Died at home.

Jesse Musser Haynie PVT 21 Jun 1837 26 May 1901
Attached as nurse to Cleveland, TN hospital Mar 26, 1863; Dec 10, 1863 muster roll reports Hayne on detached duty in MS since May 15 1863; Captured as a deserter in Walton County Oct 27, 1864; Took Oath of Allegiance in Louisville, KY Nov 10, 1864; Agreed to remain north of the Ohio River; Granite quarryman at Stone Mountain in 1900; Buried at Sone Mountain Cemetery in DeKalb County.

U. Posey Haynie PVT 24 Jan 1863
Married Sarah Sturdivant in Walton County, Jan 5, 1859.

Asariah Head PVT 18 Jan 1847 6 Aug 1897
Late enlistment; Wounded at **Battle of New Hope Church**, May 25, 1864; Admitted to Fairgrounds Hospital #1 in Atlanta June 5, 1864 with gunshot wound; Furloughed June 17, 1864; Married Mary Ann Partee in Morgan County, Oct 10, 1871; Buried at Studdard Cemetery in Rutledge, Morgan County.

Egbert Monroe Herndon PVT 2 Jun 1832 1864
Older brother of Elisha, George, & John Herndon; Captured at **Vicksburg**, MS July 4, 1863; Present on Dec 10, 1863 muster roll; Transferred to Ritter's Company, Georgia Light Artillery.

Elisha Madison Herndon PVT 27 Nov 1842 28 Jul 1864
PVT in Guards of the Sunny South Militia from Social Circle, GA Sep 18, 1861; Enlisted in Company B, 3rd Battalion, Georgia State Troops Oct 30, 1861; Mustered out April

1862; Enlisted in 42nd Georgia in Calhoun Aug 22, 1862; Captured at **Vicksburg,** MS July 4, 1863; Present on Dec 10, 1863 muster roll; Transferred to Company H of the 11th Georgia; Killed near Petersburg, VA.

George Mosley Herndon PVT 2 Jan 1835 4 Oct 1920

Married Marissa Whatley in Clarke County, Jan 25, 1858; Enlisted in Company B, 3rd Battalion, Georgia State Troops Oct 30, 1861; Mustered out May 1862; Enlisted in 42nd Georgia in Monroe May 12, 1862; Captured in **Vicksburg,** MS hospital July 4, 1863; Sick at home on Dec 10, 1863 muster roll; Surrendered at **Greensboro,** NC April 26, 1865; Married Icey Conner in Walton County, Dec 13, 1883; Farmer in Blasingame after the war; Filed for Confederate pension in Walton County in 1897; Application endorsed by J.M. Gresham; Died in Confederate Soldiers Home in Atlanta; Buried on Confederate Crest in Westview Cemetery in Atlanta.

John McDonald Herndon PVT 1839 12 Jun 1862

Married Mary Dickinson Gregg in Walton County, Feb 1, 1859; Accidentally killed on night picket duty by member of his company at Cumberland Gap, TN; Buried at Peters Cemetery in Monroe; Claim filed by his widow Mary Herndon.

John D. Herring PVT 1839 10 Aug 1863

PVT in Guards of the Sunny South Militia from Social Circle, GA Sep 18, 1861; Enlisted in Company B, 3rd Battalion, Georgia State Troops Oct 30, 1861; Mustered out April 1862; Enlisted in 42nd Georgia May 12, 1862; Captured in **Vicksburg,** MS hospital July 4, 1863; Taken with sick by US Steamer *St. Maurice* from Vicksburg to Mobile; Exchanged off Fort Morgan Aug 5, 1863; Died at St. Clair near Mobile; Claim filed by his father John Herring.

David M. Hogan PVT 1826 8 Aug 1863

Married Nancy Vickery in Walton County, Sep 28, 1848; Captured at Washington Hospital in **Vicksburg,** MS July 4, 1863; Taken with sick by US Steamer *Suffolk* from Vicksburg to Mobile; Exchanged off Fort Morgan July 22, 1863; Died in Mobile hospital.

James Madison Huff PVT 9 Sep 1838 11 Apr 1900

Younger brother of Sgt. Jonathan Huff; Captured at **Vicksburg,** MS July 4, 1863; Present on Dec 11, 1863 muster roll; Surrendered at **Greensboro,** NC April 26, 1865; Married Ora Ann Peters in Walton County, Jan 25, 1866; Farmer in Newton County in 1870; Buried at Old Conyers Cemetery.

Francis Marion Huston PVT 8 Oct 1840 After 1910

Younger brother of Young Huston; Captured at Vicksburg, MS hospital July 4, 1863; Taken with sick by US Steamer *St. Maurice* from Vicksburg to Mobile; Exchanged off Fort Morgan Aug 5, 1863; Died Sick at home on Dec 10, 1863 muster roll; Married Ellen Booth in Walton County, Mar 29, 1877; Farmer in Cut Off after the war; Filed for Confederate pension in Walton County in 1897.

Young Alexander Huston PVT 31 Mar 1832 25 Dec 1907

PVT in Guards of the Sunny South Militia from Social Circle, GA Sep 18, 1861; Captured at **Vicksburg,** MS hospital July 4, 1863; Taken by US Steamer *St. Maurice* from Vicksburg to Mobile; Exchanged off Fort Morgan Aug 5, 1863; Captured at **Battle of Nashville** Dec 16, 1864; Prisoner at **Camp Chase,** OH; Released June 12, 1865; Married Margaret Odom in Walton County, Jan 7, 1868; Farmer near Lawrenceville in 1900; Filed for a Confederate pension in Gwinnett County in 1901; Buried at Friendship Primitive Baptist in Snellville.

Allen Ivey PVT 30 Oct 1820 28 Oct 1898
Married Elizabeth Jane Towler in Morgan County, April 24, 1845; Overseer in Walton County in 1860; Transferred from Company K, 66th Georgia on Sep 20, 1863; Dec 10, 1863 muster roll lists Ivey as sick at Frank Ramsey Hospital in Cassville; Surrendered at **Greensboro,** NC April 26, 1865: Farmer in Blasingame in 1880; Buried at Towler Family Cemetery in Social Circle.
Brazil G.H. Ivey PVT 1842 After 1883
Left sick at Knoxville, TN hospital April 22, 1862; Captured at **Vicksburg,** MS July 4, 1863; Sick at home on Dec 10, 1863 muster roll; Married Fannie Bennett in Clarke County, Jan 4, 1872; Working on Walton farm in 1880.
Edmond M. Jackson PVT 25 Dec 1839 After 1909
PVT in Guards of the Sunny South Militia from Social Circle, GA Sep 18, 1861; Enlisted in Company B, 3rd Battalion, Georgia State Troops Oct 30, 1861; Mustered out April 1862; Enlisted in 42nd Georgia May 12, 1863; Captured at **Vicksburg,** MS July 4, 1863; Present on Dec 10, 1863 muster roll; Wounded in the right thigh at **Battle of Wise's Forks,** NC; Admitted Pettigrew General Hospital in Raleigh, NC Mar 11, 1865; Sent home July 1865; Married Martha Crawley in Morgan County, Nov 14, 1865; Farmer in Morgan County in 1870; Liveryman in Social Circle in 1900; Filed for Confederate pension in Chattooga County in 1909; Reported being on crutches for six months and paroled from Raleigh, NC hospital three months after the war ended.
John T. Jackson PVT 1842 1864
Captured at **Vicksburg,** MS July 4, 1863; Sick at home on Dec 10, 1863 muster roll; Patient at Oxford, GA hospital Jan 8, 1864; Admitted to Fairground Hospital #2 in Atlanta Jan 29, 1864 with kidney inflammation.
William M. Jackson PVT 10 Oct 1837 20 Nov 1910
Born in Hancock County; Captured at **Vicksburg,** MS July 4, 1863; Present on Dec 10, 1863 muster roll; Captured in retreat from Franklin, TN Dec 17, 1864; Prisoner at **Camp Chase,** OH; Paroled & sent to New Orleans for exchange May 2, 1865; Never married; Filed for Confederate pension in Walton County in 1899; Buried at Mountain Creek Baptist in Monroe.
James L. Johnson PVT 1836 23 Mar 1865
Married Nancy Stephens in Walton County, Dec 22, 1859; Dec 10, 1862 muster roll reports Johnson sick at home since June 1862; Captured at **Battle of Nashville,** TN Dec 16, 1865; Died of debility as prisoner at **Camp Chase,** OH; Buried in Confederate Cemetery at Camp Chase, grave # 1737.
John William Johnson PVT 9 Oct 1838 After 1910
Born in Jackson County; Left sick at Priceville, TN hospital April 26, 1862; Captured at Perryville, KY; Sent from Louisville to Vicksburg, MS for exchange on US Steamer *Maria Denning* Nov 15, 1862; Sick at home on Dec 10, 1862 muster roll; Wounded near Dalton May 15, 1864; Captured at **Battle of Nashville,** TN Dec 16, 1864; Prisoner at **Camp Chase,** OH; Paroled May 15, 1865; Married Clementine Allgood about 1869; Farmer in Walton County after the war; Filed for Confederate pension in Walton County in 1897.
Scott A. Johnson PVT 8 Feb 1829 16 Feb 1902
Married Jane Zuber in Oglethorpe County, June 2, 1850; Enlisted in Monroe May 12, 1862; Captured at **Vicksburg,** MS July 4, 1863; Present on Dec 10, 1863 muster roll; Surrendered in **Greensboro,** NC April 26, 1865; Took 1867 Loyalty Oath in Walton County;

Married Elizabeth Fambrough in Spring Hill, AR Mar 29, 1896; Buried at Huckabee Cemetery in Homestead County, AR.

James M. Johnston PVT 1823 Before 1870

Born in Smith County, TN; Enlisted in Monroe May 12, 1862; Issued Certificate of Disability for Discharge from Atlanta hospital Aug 19, 1863 with heart disease.

George Anderson Knight PVT 22 Nov 1832 30 Sep 1912

Born in Washington County; Married Julia Hawk in Walton County, Aug 14, 1859; Appointed **SGT**; Captured at Washington Hospital in **Vicksburg,** MS July 4, 1863; Taken with sick by US Steamer *H. Chouteau* from Vicksburg to Mobile for exchange; Struck by lightning and seriously injured on picket duty in front of Atlanta trench line Aug 19, 1864; Captured at **Battle of Wise's Forks,** NC Mar 10, 1865; Released at **Point Lookout,** MD June 28, 1865; Married Mahela Bacchus in Walton County, Sep 19, 1865; Farmer in Blasingame after the war; Filed for Confederate pension in Walton County in 1891; Dr. N.L. Galloway reported Knight's nervous system never recovered from lightning strike. He was now mute and blind in one eye; Buried at Thomas & Knight Cemetery in Walton County.

Lewis T. Knight PVT 1827 27 Jun 1864

Born in South Carolina; Married Martha Vickery in Walton County, May 29, 1849; Captured at **Vicksburg,** MS July 4, 1863; Present on December 10, 1863 muster roll; Killed near Kennesaw Mountain, GA.

John W. Lemmons PVT 1815 After 1871

Married Nancy Stewart in Morgan County, Aug 8, 1833; Married Rebecca Stewart in Walton County, Oct 13, 1841; Enlisted in Monroe May 12, 1862 as substitute for O.L. Stroud; Captured at **Vicksburg,** MS July 4, 1863; Present on Dec 10, 1863 muster roll; Buried at Day Family Cemetery in Social Circle.

James M.T. Livingston PVT 1827 24 Jan 1863

Died in Smith Hospital in Mossy Creek, TN; Marker at Bethel Baptist in Walton County.

William M. Lyon PVT 22 Dec 1844 20 Dec 1908

Younger brother of regimental band leader G.W. Lyon; Enlisted in Decatur Mar 14, 1863; Appointed regimental musician Dec 31, 1863; Surrendered at **Greensboro,** NC April 26, 1865; Mechanic in Atlanta after the war; Buried at Oakland Cemetery in Atlanta.

David H. Malcom PVT 1823 25 May 1864

Married Harriot Towler in Morgan County, April 24, 1845; Enlisted in Monroe Sep 20, 1863; Present on Dec 10, 1863 muster roll; Killed at **Battle of New Hope Church** May 25, 1864; Buried at Green Cemetery in Walton County.

George Washington Malcom PVT 16 Jul 1830 25 Jun 1891

Uncle of James Malcom; Discharged in May 1862 after furnishing Thomas Bonner as his substitute; Served in Company H of Georgia State Troops; Buried at Preston Cemetery in Walton County.

James David Malcom PVT 11 Nov 1843 5 Jan 1822

PVT in Guards of the Sunny South Militia from Social Circle, GA Sep 18, 1861; Enlisted in Monroe May 12, 1862; Captured in **Vicksburg,** MS hospital July 4, 1863; Shipped with sick from US Steamer *St. Maurice* from Vicksburg to Mobile; Exchanged at Fort Morgan Aug 5, 1863; Sick at home on Dec 10, 1863 muster roll; Captured at **Battle of Nashville,** TN Dec 16, 1864; Prisoner at **Camp Chase,** OH; Released June 13, 1865; Married Rebecca Malcom in Walton County, Sep 10, 1865; Farmer in Walton County after the war;

Filed for Confederate pension in Walton County in 1919; Application endorsed by T.S. Edmonds; Buried at Green Cemetery in Walton County.

Sidney Elisha Marston PVT 1834 7 Dec 1862

Married Sarah Andrews in Newton County, Oct 1, 1857; Died at Fairgrounds Hospital in Knoxville, TN; Widow's pension filed in Henry County in 1893; Reported that she was present in Knoxville when her husband died of typhoid pneumonia. She brought his body home and buried his remains in Newton County.

William Thomas Mayo PVT 16 Jun 1831 15 Sep 1899

Married Barbara Calendar O'Kelly in Walton County, July 23, 1857; Enlisted in Monroe May 12, 1862; Captured at **Vicksburg**, MS July 4, 1863; AWOL on Dec 10, 1863 muster roll; Surrendered at **Greensboro**, NC April 26, 1865; Farmer in Blasingame in 1880; Buried at Herndon Cemetery in Social Circle.

Henry John "Jake" McAdams PVT 10 Mar 1832 28 Mar 1910

Born in Oglethorpe County; Married Amanda Wilkins in Walton County, Jan 16, 1853; Captured in **Vicksburg**, MS hospital July 4, 1863; Taken with sick by US Steamer *St. Maurice* to Mobile & exchanged at Fort Morgan Aug 5, 1863; Surrendered at **Greensboro**, NC April 26, 1865; Farmer in Gwinnett County after the war; Filed for Confederate pension in Gwinnett County in 1901; Buried in Chapel Christian Church in Barrow County.

James M. McAfee PVT

Appointed **4SGT** Feb 1863; Admitted to Fairgrounds Hospital #1 in Atlanta Nov 14, 1862 with rheumatism; Returned to duty Dec 19, 1862; Captured at **Vicksburg**, MS July 4, 1863; Present on Dec 10, 1863 muster roll; Wounded at **Battle of Resaca**, GA May 16, 1864.

Charles W. McAllister PVT 16 Jun 1862

Claim filed by his mother Mary Ann McAlister in Walton County.

Francis Jerome McDowell PVT 1840 28 Jul 1863

Older brother of Joseph McDowell; PVT in Guards of the Sunny South Militia from Social Circle, GA Sep 18, 1861; Captured in **Vicksburg**, MS hospital July 4, 1863; Died in Vicksburg.

Joseph D. McDowell PVT Dec 1841 23 Aug 1915

Captured in **Vicksburg**, MS hospital July 4, 1863; Taken with sick by US Steamer *St. Maurice* from Vicksburg to Mobile & exchanged at Fort Morgan Aug 5, 1863; Sick at home on Dec 10, 1863 muster roll; Surrendered at **Greensboro**, NC April 26, 1865; Married Jeanette Williams in Walton County, Dec 14, 1876; House carpenter in Social Circle after the war; Buried in Circle View Cemetery in Social Circle.

Napoleon Monroe McGaughey PVT 12 Apr 1843 10 May 1921

Sick at home on Dec 10, 1863 muster roll; Captured in **Vicksburg**, MS hospital July 4, 1863; Taken with sick by US Steamer *St. Maurice* from Vicksburg to Mobile & exchanged at Fort Morgan Aug 5, 1863; Married Nancy Hale in Walton County, July 16, 1864; Transferred to Walton Blues (G); Married Martha West in Greene County, April 10, 1877; Farmer in Greene County after the war; Filed for Confederate pension in Greene County in 1911; Reported being present at the Greensboro surrender; Buried at Penfield Cemetery in Greene County.

David Isaac McIntosh PVT Apr 1846 After 1910

17-year-old farmer in 1864 militia census for Walton County; Enlisted in Monroe April 19, 1864; Married Jersey Chosewood in Walton County, Jan 28, 1866; Married Sarah

Maughon in Walton County, May 21, 1876; Farmer in Broken Arrow after the war; Filed for Confederate pension in Walton County in 1901; Reported being sent to hospital July 22, 1864 and furloughed home; McIntosh was in Augusta returning to his command when the war ended.

Asa Columbus Mitchell PVT 1832 1878

Married Susan Tucker in Walton County, Oct 6, 1859; Captured at **Vicksburg**, MS July 4, 1863; Present on Dec 10, 1863 muster roll; Wounded at **Battle of Resaca**, GA May 16, 1865; Admitted to Madison, GA hospital May 19, 1864; Wounded at **Battle of Wise's Forks**, NC Mar 19, 1865; Surrendered at **Greensboro**, NC April 26, 1865; Married Sarah E. Tucker in Walton County, May 3, 1866.

Joshua Marion Mitchell PVT 13 Oct 1826 10 Mar 1911

1LT in Guards of the Sunny South Militia from Social Circle, GA Sep 18, 1861; Elected 1LT of Company B of the 3rd Battalion of Georgia State Troops Oct 30, 1861; Mustered out in Savannah, GA May 1861; Enlisted in Company G May 12, 1862; Elected **CPT** July 22, 1862; Captured at **Vicksburg**, MS July 4, 1863; Reprimanded by Court Martial Court Jan 19, 1864; Wounded at **Battle of Rocky Face** Feb 25, 1864; Returned to duty from Madison, GA hospital May 23, 1864; Captured at **Battle of Nashville**, TN Dec 16, 1864; Released from **Johnson Island**, OH prison June 17, 1865; Buried at Indian Creek Cemetery in Scottsdale, DeKalb County.

James N. Myears PVT 20 Sep 1833 6 Jul 1904

Married Sarah Ann Baker in Walton County, Nov 16, 1852; Enlisted in Monroe May 12, 1862; Promoted to **Chaplain** Nov 2, 1862; Granted 30-day furlough from Atlanta hospital Jan 6, 1863; Captured at **Vicksburg**, MS July 4, 1863; Regimental Chaplain commission granted Aug 4, 1863; Present on Dec 10, 1863 muster roll; Appears on brigade inspection report near Atlanta Aug 19, 1864; Minister in Villa Rica in 1870; Preacher in Walker County in 1900; Buried in West Hill Cemetery in Dalton.

John Pendleton O'Kelley PVT 14 Nov 1830 25 Jan 1897

Married Adeline Ford in Walton County, Aug 31, 1857; Assigned to wait on sick at Knoxville, TN hospital April 22, 1862; Admitted to Atlanta hospital Oct 2, 1863; Patient at Cannon Hospital in Union Springs, AL July 15, 1864; Filed for Confederate pension in Walton County in 1891; Reported suffering from rheumatism in Knoxville, TN & sent to Atlanta hospital and then furloughed for six months; Buried at Willow Springs Church in Social Circle.

Erby B. Osburn PVT 15 Apr 1830 30 Apr 1901

Sick at home on Dec 10, 1863 muster roll; Serving in Provost Guard with Company F, 1st Regiment Troops & Defenses at Camp Wight in Macon, GA Dec 1864; Married Mary Ann Johnson in Walton County, Feb 29, 1880; Filed for Confederate pension in Walton County in 1895; Reported being with wagon train in Atlanta at time of surrender.

Obediah F. Owen PVT 4 Apr 1844 Before 1875

Enlisted in Company B, 3rd Battalion, Georgia State Troops Oct 30, 1861; Mustered out April 1862; Enlisted in 42nd Georgia Sep 20, 1862; Captured at **Vicksburg**, MS July 4, 1863; Present on Dec 10, 1863 muster roll; Wounded at **Battle of Jonesborough** Aug 31, 1864; Admitted to Fairgrounds Hospital #2 in Vineville, GA with fluid in the chest Sep 3, 1864; Surrendered at **Greensboro**, NC April 26, 1865; Married Georgia Turnbull in Walton County, Oct 29, 1869; Farmer in Cut Off in 1870.

Uriah James Owen PVT Sep 1839 1862

Older brother of Obadiah Owen; PVT in Guards of the Sunny South Militia from Social Circle, GA Sep 18, 1861; Appointed 2nd SGT in Company B, 3rd Battalion, Georgia State Troops Oct 30, 1861; Mustered out April 1862; Enlisted in 42nd Georgia May 12, 1862; Died at home.

T.M. Pannell PVT After 1867

Dec 10, 1863 muster roll shows Pannell absent on Provost Guard duty in Atlanta; Took Loyalty Oath in Walton County in 1867.

W.H. Patterson PVT 30 Aug 1862

Died in Fairgrounds Hospital in Knoxville, TN.

J.H. Peters PVT 18 May 1844 Jul 1862

Died in Knoxville, TN hospital; Buried at Day Family Cemetery in Social Circle.

D.W. Phillips PVT 1863

Captured at **Vicksburg**, MS hospital July 4, 1863; Taken with sick by US Steamer *St. Maurice* from Vicksburg to Mobile & exchanged at Fort Morgan Aug 5, 1863.

Creighton C. Queen PVT 1830 20 Aug 1863

Younger brother of Edward Queen; Born in Rutherford County, NC; Captured in **Vicksburg**, MS hospital July 4, 1863; Taken by US Steamer *St. Maurice* to Mobile & exchanged at Fort Morgan Aug 5, 1863; Died at home.

Perry S. Rogers PVT 1833 After 1880

Married Margaret Ivey in Walton County, May 4, 1862; Discharged with William H. Brooks as his substitute, July 12, 1862; Farmer in Blasingame in 1880.

John N. Simmons PVT 1839 4 Jul 1892

PVT in Guards of the Sunny South Militia from Social Circle, GA Sep 18, 1861; Enlisted in Company B, 3rd Battalion, Georgia State Troops Oct 30, 1861; Mustered out April 1862; Enlisted in 42nd Georgia May 12, 1862; Captured in **Vicksburg**, MS July 4, 1863; Present on Dec 10, 1863 muster roll; Married Lucy Owen in Walton County, Aug 9, 1865; Farmer in Hempstead County, AR in 1880; Buried in Simmons Family Cemetery in Spring Hill, Arkansas.

William H. Simmons PVT 10 Sep 1835 29 Sep 1902

Older brother of John Simmons; Married Laura Swinney about 1859; PVT in Guards of the Sunny South Militia from Social Circle, GA Sep 18, 1861; Present on Dec 10, 1863 muster roll; Farmer in Homestead County, AR in 1880; Farmer in Cobb County in 1900; Buried at New Salem in Cobb County.

Martin Thomas Sims PVT 12 Apr 1831 5 Jul 1895

Born in Jackson County; Married Matilda O'Kelley in Walton County, Mar 19, 1854; 1SGT in Guards of the Sunny South Militia from Social Circle, GA Sep 18, 1861; Enlisted as 2LT in Company B, 3rd Battalion, Georgia State Troops Oct 30, 1861; Mustered out April 1862; Enlisted in 42nd Georgia May 12, 1862; Dec 10, 1863 lists Sims as AWOL since May 17, 1863; Captured at **Battle of Nashville**, TN Dec 16, 1864; Prisoner at **Camp Chase**, OH; Released May 2, 1865; Farmer in Jackson County after the war; Buried at Walnut Fork Cemetery in Braselton.

Francis Marion Smith PVT 24 Dec 1836 16 Mar 1914

Married Esther Camp Sep 7, 1861; Present on Dec 10, 1863 muster roll; Farmer in Duluth in 1880; Dry goods salesman in Athens in 1900; Buried at Old Shadowlawn Cemetery in Lawrenceville.

Richard Thomas Sorrells PVT 1842 21 Jun 1862

Appointed 3CPL; Died at home; Claim filed by his father Bennett Sorrells.
Wilson C.H. Sorrells PVT 1844 8 Apr 1865
Captured at **Vicksburg**, MS July 4, 1863; Sick at home on Dec 10, 1863 muster roll; Captured along Atlanta siege lines Aug 7, 1864; Died of pneumonia as prisoner at **Camp Chase**, OH; Buried in Confederate Cemetery, grave # 1874.
Ezekiel R. St. John PVT 1832 15 Jul 1862
Married Mahulda Mitchell in Newton County, Nov 16, 1855; Died at home on hospital furlough; Claim filed by Hulda St. John.
John W. Stark PVT 1 Sep 1833 16 Jul 1898
Married Rebecca Malcom in Walton County, Nov 16, 1857; Enlisted in Monroe Aug 15, 1862; Captured at **Vicksburg**, MS July 4, 1863; Wounded at **Battle of Resaca**, GA May 16, 1864; Admitted Madison, GA hospital May 18, 1864; Wounded in left leg at **Battle of Wise's Forks**, NC; Admitted to Pettigrew Hospital, Raleigh Mar 10, 1865; Farmer in Whitfield County after the war; Buried at West Hill Cemetery in Dalton.
Elias Jefferson Stephens PVT 15 Nov 1844 4 Mar 1903
Discharged at Bean Station, TN July 11, 1862 after accident destroyed the joint in his left wrist; Married Minerva Callahan in Walton County, Jan 31, 1867; Farmer in Paulding County after the war; Buried at Shady Grove Baptist in Dallas.
John A. Stephens PVT 1840 17 July 1863
Older brother of Jefferson Stephens; Captured in **Vicksburg**, MS hospital July 4, 1863; Died in Vicksburg.
Orion Levi Stroud PVT 31 Jul 1837 24 Nov 1911
Married Sarah Henderson in Walton County, Dec 19, 1860; Discharged upon furnishing John Lemmons as his substitute, May 12, 1862; Farmer in Broken Arrow in 1880; Farmer in Rusk County, Texas in 1900; Buried in Mount Hope Cemetery in Joinerville, Texas.
John W. Thomas PVT 1833 7 Jul 1863
Born in Clarke County; Married Nancy Clegg in Walton County, Mar 31, 1857; Farmer in Yellow River District of Gwinnett County in 1860; Enlisted in Monroe May 12, 1862; Died in Vicksburg, MS.
David Job Thompson PVT 1 Feb 1821 18 May 1888
Older brother of Nathaniel Thompson; Married Elizabeth Jane Austin Aug 12, 1849; Enlisted in Decatur Sep 22, 1863; Surrendered at **Greensboro**, NC April 26, 1865; Farmer in Walton County; Buried at Thompson Family Cemetery in Monroe.
Nathaniel R. Thompson PVT 20 Feb 1836 26 Mar 1898
Enlisted in Company B, 3rd Battalion, Georgia State Troops Jan 12, 1862; Mustered out April 1862; Enlisted in 42nd Georgia May 12, 1862; Dec 10, 1863 muster roll lists Thompson as AWOL since May 12, 1863; Wounded at **Battle of Resaca**, GA May 15, 1864; Farmer in Walton County after the war; Member of Masonic Order; Buried at Bethlehem Methodist in Barrow County.
Andrew Jackson Towler PVT 22 Nov 1835 28 Oct 1905
PVT in Guards of the Sunny South Militia from Social Circle, GA Sep 18, 1861; Married Martha Baxley in Walton County, Dec 24, 1861; Captured at **Vicksburg**, MS July 4, 1863; Present on Dec 10, 1863 muster roll; Appears on list of paroled prisoners in Augusta May 20, 1865; Farmer in Blasingame after the war; Buried at Towler Family Cemetery in Social Circle.
David Malcom Towler PVT 17 Mar 1834 10 Jul 1899

PVT in Guards of the Sunny South Militia from Social Circle, GA Sep 18, 1861; Enlisted in Monroe May 12, 1862; Present on Dec 10, 1863 muster roll; Wounded at **Battle of Resaca** May 15, 1863; Appears on list of paroled prisoners in Augusta May 20, 1865; Married Mary Frances Ivey in Walton County, Jan 7, 1866; Farmer in Blasingame after the war; Buried at Towler Family Cemetery in Social Circle; Widow's pension filed in Walton County in 1919; Reported her husband left his command near Jonesborough & was granted a medical furlough; Application supported by John J. Cox.

James Johnson Towler PVT 22 Oct 1830 6 Jan 1907

Older brother of Andrew & David Towler; Captured at **Vicksburg**, MS July 4, 1863; Present on Dec 10, 1863 muster roll; Surrendered at **Greensboro**, NC April 26, 1865; Married Nancy Ann Edmonds in 1868; Farmer in Blasingame after the war; Filed for Confederate pension in Walton County in 1889; Reported being wounded at **Battle of Resaca** May 15, 1864 resulting in 3rd finger on left hand being amputated; Buried at Towler Family Cemetery in Social Circle.

Henry Clay Tucker PVT 25 Apr 1844 2 Jun 1905

Enlisted in Monroe Mar 29, 1864; Surrendered at **Greensboro**, NC April 26, 1865; Married Francis Shelnutt in Walton County, Dec 26, 1867; Farmer in Walton County after the war; Buried at Corinth Memorial Gardens in Loganville.

James E. Tucker PVT Dec 1846 After 1900

Born in Oglethorpe County; Younger brother of Henry Tucker; Enlisted Dec 4, 1863; Married Mary Kidd in 1868; Farmer in Walton County in 1900; Buried at Tucker-Smith Family Cemetery in Walton County.

William W. Turnbull PVT 1847 20 Jul 1864

Enlisted in Monroe Sept 20, 1863; Killed in **Battle of Peachtree Creek**, GA;

Thomas J. White PVT 1832 4 May 1862

Enlisted in Company B, 3rd Battalion, Georgia State Troops Jan 12, 1862; Mustered out April 1862; Enlisted in 42nd Georgia in 1862; Died in Morristown, TN hospital.

George Whitley PVT 1845 Jun 1862

Left sick in Knoxville, TN hospital April 22, 1862; Coffin ordered Jun 5, 1862.

Andrew Jackson Williams PVT 1830 After 1870

Married Mary Ann Clegg in Walton County, Feb 15, 1852; Sutler to the 42nd Georgia; Captured at **Vicksburg**, MS July 4, 1863; Butcher in Newnan in 1870.

David Williams PVT 1834 6 Aug 1886

Married Elizabeth Willard in Morgan County, Sep 15, 1858; Enlisted in Calhoun Aug 22, 1862; Captured at **Vicksburg**, MS July 4, 1863; Present on Dec 10, 1863 muster roll; Wounded at **Battle of Atlanta** July 22, 1864; Owned dry goods store in Fair Play, Morgan County, after the war; Appointed postmaster of Fair Play in 1882.

James F. Williams PVT 1836 25 Feb 1864

PVT in Guards of the Sunny South Militia from Social Circle, GA Sep 18, 1861; Enlisted in Company B, 3rd Battalion, Georgia State Troops Oct 30, 1861; Appointed **3SGT**; Mustered out May 1862; Enlisted in Monroe May 12, 1862; Elected from ranks to 3LT Feb 24, 1863; Captured at **Vicksburg**, MS July 4, 1863; Present on Dec 10, 1863 muster roll; Killed at **Battle of Rocky Face**, GA; Buried in Confederate Section at West Hill Cemetery in Dalton, Georgia.

John Samuel Williams PVT 1832 29 Dec 1862

Older brother of James & Warren Williams; Enlisted in Company B, 3rd Battalion, Georgia State Troops Oct 30, 1861; Mustered out May 1862; Enlisted in Monroe May 12, 1862; Killed at **Battle of Chickasaw Bayou**; Claim filed by his mother Delilah Williams.

Thomas C. Williams PVT 1832 6 Dec 1862

Brother of Andrew & David Williams; Married Tinsey Clegg in Walton County, about 1855; Buried in Clegg Family Cemetery in Walton County; Claim filed by widow Tinsey Williams.

Warren H. Williams PVT 1844 18 Jul 1901

PVT in Guards of the Sunny South Militia from Social Circle, GA Sep 18, 1861; Enlisted in Company B, 3rd Battalion, Georgia State Troops Oct 30, 1861; Mustered out May 1862; Enlisted in Monroe May 12, 1862; Sick at home on Dec 10, 1863 muster roll; Appointed to regimental band; Surrendered at **Greensboro**, NC April 26, 1865; Farmer in Blasingame in 1880; Married Mary Tallulah Powell in Fulton County, Oct 29, 1885; Member of UCV Camp # 925 in Atlanta; Filed for Confederate pension in Fulton County in 1890; Application endorsed by G.W. Lyon; Obituary notes that Williams was a member of the Odd Fellows & Red Man and worked as the city produce "weigher"; Buried at Hollywood Cemetery in Atlanta.

James Miller Wilson PVT 18 May 1844 21 Jan 1900

Born in Campbell County; Wounded at **Battle of New Hope Church**, GA May 25, 1864; Appointed Ordinance Sergeant; Surrendered at **Greensboro**, NC April 25, 1865; Married Adeline Stephens Longino in 1868; Obituary notes that Captain J.M. Wilson represented Campbell County in the legislature for several terms and was senator from the 36th district for one term. He later lived in Atlanta and served several years as tax collector for Fulton County. Buried at Wilson-McLarin Cemetery in Atlanta.

Company I – Fulton Blues – 106 Men

Drawn from men living north, south & west of Atlanta. A February 20, 1864 muster roll exists for this company.

Name Rank Birth Date

John H. Barrett CPT 1835 29 Mar 1911

Born in Augusta; Married Tallulah Winningham in DeKalb County, Nov 25, 1858; Resigned Sep 26, 1862; According to Surgeon J.S. Wilson the captain was incapable of performing his duties "because of varicocele together with an affliction of the kidneys and general derangement of the urinary organ;" Auctioneer in Augusta after the war; Retired to Savannah; Buried at Laurel Grove Cemetery in Savannah.

Harvey W. Bostain 1SGT 28 Aug 1826 22 Feb 1884

Born in Jackson County; Married Mary Elizabeth Dulin in Henry County, Feb 13, 1852; Enlisted in Company F, 8th Georgia May 22, 1861 & resigned Nov 5, 1861; Received Certificate of Discharge for hemorrhoids in Normandy, TN Nov 27, 1862; Dry goods merchant in Jonesboro in 1870; Shoe manufacturer in Gainesville in 1880; Buried at Alta Vista Cemetery in Gainesville.

Wiley Franklin Baker 3SGT 26 Oct 1841 4 May 1862

Sent to Knoxville, TN hospital April 22, 1862; Buried at Mount Gilead Methodist in Ben

Hill.

William Suttles Baker 3CPL 23 Jan 1839 16 Nov 1864

Detailed as nurse in Knoxville, TN hospital April 22, 1862 by order of Colonel Henderson; Captured in **Vicksburg**, MS hospital July 4, 1863; Transported with sick from Vicksburg to Mobile on US Steamer *St. Maurice*; Exchanged off Fort Morgan Aug 5, 1863; Absent sick Feb 20, 1864 muster roll; Killed at **Battle of Nashville**, TN; Memorial marker at Mount Gilead Methodist in Ben Hill.

William F. Barrett 2CPL 1837 16 Nov 1864

Left in Knoxville, TN hospital April 22, 1862; Captured at **Vicksburg**, MS July 4, 1863; Present on Dec 10, 1863 muster roll; Captured along Atlanta siege lines Aug 7, 1864; Died of diarrhea in Camp Chase, OH prison; Buried in Confederate Cemetery Grave # 478.

Elijah M. Donehoo 2LT 30 Nov 1819 7 Mar 1877

Married Mary Jane Walker in DeKalb County, Feb 2, 1843; Father of William Donehoo; Wealthy farmer in Cooks District in 1860; Resigned April 2, 1863 upon recommendation of Assistant Surgeon J.A. Dunn; Elected Captain of Company G, 2nd Georgia Reserve in 1864; Buried at Crest Lawn Cemetery in Atlanta.

Benjamin Franklin Grubbs 4SGT 1830 1871

Born in Walton County; Married Martha Dickson Carmichael in Clarke County around 1852; Captured at **Vicksburg**, MS July 4, 1863; Present on Feb 24, 1864 muster roll; Surrendered at **Greensboro**, NC April 26, 1865; Carpenter in Atlanta in 1870.

Thomas Anderson Hill CPT 8 Dec 1841 7 Oct 1924

Born in Bastrop County, Texas; Enlisted as private in Company G, Georgia 3rd Battalion, Jan 8, 1862; Elected 2LT Feb 13, 1862; Appointed **CPT** of Company I of 42nd Georgia by General Kirby Smith Dec 18, 1862; Captured at **Vicksburg**, MS July 4, 1863; Dec 10, 1863 muster roll notes Captain was absent having gone to Bastrop County, Texas after the Vicksburg surrender; Dropped from rolls of 42nd Georgia Feb 3, 1863; Returned to Georgia as member of 8th Texas Cavalry, Terry's Texas Rangers; Wounded & captured near Atlanta Aug 7, 1864; Released from **Johnson Island**, OH prison Jun 16, 1865; Married Sallie McGehee in Hayes County, Texas, Nov 28, 1865; Ran a mercantile business in La-Grange and then Columbus, Texas. By 1900, he owned a small bank in Weimar, TX; Appears in Terry's Texas Rangers 1908 reunion panorama; Buried at Weimar Masonic Cemetery in Colorado County, Texas.

Francis C. House 1LT 1825 1883

Born with twin sister in Clarke County; Married Permelia Lee in Clarke County, Aug 17, 1843; Member of Atlanta City Council when the war began; Captured at **Vicksburg**, MS July 4, 1863; Resigned in ill-health in Decatur, GA Oct 14, 1863; House carpenter in Atlanta in 1870 and Powder Springs in 1880; Buried at Powder Springs Methodist in Cobb County.

Osborn J. Register 2LT 1826 17 Sep 1882

Resigned Jan 29, 1863 after receiving Medical Certificate for rheumatism; Vegetable dealer in Atlanta after the war; Buried at Oakland Cemetery in Atlanta.

James D. Robinson 5SGT 1842 21 Jun 1887

Born in South Carolina; Appointed regimental color bearer by order of Colonel Henderson April 1862; Wounded at **Battle of Chickasaw Bayou**, MS by bursting shell, Dec 28, 1862; Elected **2LT** Mar 8, 1863; Admitted to Fairgrounds Hospital #1 in Atlanta May 20, 1864 with dysentery; Returned to duty June 13, 1864; Resigned commission Aug 20, 1864 preferring to serve in the ranks; Married Clara Belle Neal in Fulton County, Mar 23, 1871; Pressman for the Atlanta Constitution and later promoted to manager of the Job Department; Speculator in futures on the Phelan Cotton Exchange; Buried at Oakland Cemetery in Atlanta.

James T. Smith 4CPL 1838 After 1880
Born in North Carolina; Left sick in Knoxville, TN hospital April 22, 1862; Appointed SGT Feb 1863; Captured at **Vicksburg**, MS July 4, 1863; Present on Dec 10, 1863 muster roll; Captured as a deserter along the Atlanta siege line Aug 11, 1864; Released from **Camp Chase**, OH June 26, 1865; Married Eunice Crenshaw in Fulton County, Nov 9, 1865; Working in Western & Atlantic Atlanta shop in 1870; Railroad conductor in 1880.

William Afonso Winters 1CPL 19 Jan 1834 30 Aug 1924
Born in Guilford County, NC; Captured at **Vicksburg**, MS July 4, 1863; Dec 10, 1863 muster roll reports Winters was furloughed on July 25, 1863; Present on Feb 20, 1864 muster roll; Captured near Atlanta in Sep 1864; Took Oath of Allegiance in Louisville, KY Oct 14, 1864 agreeing to stay north of the Ohio River; Bricklayer & mason in Atlanta after the war; Married Anna Veal in Fulton County, April 19, 1896; Bricklayer in Clarkston in 1900; Filed for Confederate pension in DeKalb County in 1903; Reported being on Provost Duty near Jonesborough when his detail was cut off and captured; Buried at Clarkston Cemetery in DeKalb County.

Lewis J. "Luke" Wood 2SGT 1 Aug 1835 8 Aug 1913
Married Polly Bohler in Fulton County, Oct 2, 1859; Captured at **Vicksburg**, MS July 4, 1863; Present on Feb 20, 1864 muster roll; Captured along Atlanta siege line Aug 7, 1864; Prisoner at **Camp Chase**, OH; Transferred to Point Lookout, MD for exchange Mar 18, 1865; Farmer in the Cooks District after the war; Charter member of the Bolton Lodge of Masons; Filed for Confederate pension in Cobb County in 1906; Application endorsed by L.P. Thomas & T.M. Armistead; Buried at Hollywood Cemetery in Atlanta.

John Atkinson PVT 1819 16 May 1863
Born in Greene County; Married Mary H. Landrum in Oglethorpe County, Oct 8, 1845; Carpenter in Atlanta in 1860; Appointed regimental sutler by Colonel Henderson April 1862; Captured near Lexington, KY Nov 1862; Killed at Battle of Baker's Creek, MS; Claim filed by his widow Mary H. Atkinson in Fulton County.

William L. Bates PVT 1837 24 Mar 1863
Detailed to Captain Calhoun's detachment in Knoxville, TN April 20, 1862 to guard political prisoners; Prison guard at Madison, GA Oct 31, 1862; Died in Vicksburg, MS hospital; Buried at Soldier's Rest, Cedar Hill Cemetery in Vicksburg.

Wiley F. Bell PVT 1822 29 Dec 1862
Married Sarah Elizabeth Braziel in Jackson County in 1855; Blacksmith in Gwinnett County in 1860; Killed at **Battle of Chickasaw Bayou**, MS; Claim filed by widow Sarah E. Bell in Fulton County.

James Knox Polk Berry PVT 15 May 1842 25 Feb 1914
Captured at **Vicksburg**, MS July 4, 1863; Left sick at Cato, MS hospital July 18, 1863; Returned to duty from 1st Mississippi Hospital in Jackson, MS Oct 15, 1863; Surrendered at **Greensboro**, NC April 26, 1865; Married Lina Ann Welch about 1872; Tanner in Gainesville in 1880; Working in Sugar Hill tannery in 1900; Buried at Bethlehem Methodist in Buford.

James L. Blackburn PVT 1829 Before 1867
Detailed as wagoner in Knoxville, TN by Colonel Henderson April 18, 1862; Dec 10, 1863 muster roll reports Blackburn sick in Vicksburg, MS hospital since April 27, 1863; Present on Feb 20, 1864 muster roll; Patient at Law Hospital in Eatonton, Aug 30, 1864.
Joseph Byers Blalock PVT 6 Feb 1839 27 Feb 1924
Born in Haywood County, NC; Married Betsy McWilliams in Campbell County, Jan 8, 1861; Detailed to Captain Calhoun's detachment in Knoxville, TN April 20, 1862 to guard political prisoners; Prison guard at Madison, GA Oct 31, 1862; Captured at **Vicksburg,** MS July 4, 1863; Present on Feb 20, 1864 muster roll; Captured at **Battle of Nashville,** TN Dec 16, 1864; Released from **Camp Chase,** OH June 12, 1865; Farmer in Campbell County in 1880; Filed for Confederate pension in Carroll County in 1903; Buried at Friendship Baptist in Douglasville.
Clark Heyden Bryant PVT 29 Oct 1822 19 Jun 1905
Born in South Carolina; Married Elizabeth Butler in Forsyth County, July 27, 1845; Married Martha Suggs in Fulton County, Feb 25, 1858; Detailed to Captain Calhoun's detachment in Knoxville, TN April 20, 1862 to guard political prisoners; Prison guard at Madison, GA Oct 31, 1862; Captured at **Vicksburg,** MS July 4, 1863; Sent to hospital from Decatur, GA Sep 28, 1863; Captured along Atlanta siege lines Aug 17, 1864; Discharged from **Camp Chase,** OH May 15, 1865; Married Georgia Ann Weeks in Fulton County, Dec 15, 1872; Farmer in Adamsville District in 1880; Filed for Confederate pension in Fulton County in 1895; Buried at Ramah Baptist in Palmetto.
John F. Bryant PVT 12 May 1843 14 Sep 1913
Son of Clark Bryant; Younger brother of William Bryant; Late enlistment; Captured along Atlanta siege line Aug 18, 1864; Released from **Camp Chase,** OH May 15, 1865; Married Catherine Wright in Fulton County, Nov 7, 1869; Farmer in Adamsville District in 1880; Railroad inspector in 1900; Filed for Confederate pension in Fulton County in 1905; Reported enlisting in Atlanta in fall of 1862; Buried at Sharon Baptist in Fulton County.
William G. Bryant PVT 28 Jun 1846 18 May 1924
Born in Hall County; Son of Clark Bryant; Older brother of William Bryant; Captured at **Vicksburg,** MS July 4, 1863; Present on Dec 10, 1863 muster roll; Captured in Oxford, GA by Federal cavalry July 22, 1864; Released from **Camp Chase,** OH May 15, 1865; Married Amelia Nancy Daniel in Fulton County, May 27, 1866; Farmer at Bryant's Crossroads in 1880; Married Eugenia Taylor in Fulton County, Oct 30, 1890; Member of UCV Camp # 925 in Atlanta; Filed for Confederate pension in Fulton County in 1904; Application endorsed by Colonel L.P. Thomas.
Lee Andrew Wilson Caldwell PVT 1844 19 Oct 1891
Detailed to Captain Calhoun's detachment in Knoxville, TN April 20, 1862 to guard political prisoners; Prison guard at Madison, GA Oct 31, 1862; Captured at **Vicksburg,** MS July 4, 1863; appointed **4CPL** Nov 1863; Surrendered at **Greensboro,** NC April 26, 1865; Married Lula Mason in Fulton County, Aug 14, 1870; Operated a hardware store in Austin, Texas in 1870s; Lived in Louisville, LY in 1880s; Buried at Hebron Cemetery in Anderson County, KY.
George Washington Camp PVT 1836 12 Mar 1886
Married Nancy Ann Payne in Fulton County, Sep 30, 1858; Detailed to Captain Calhoun's detachment in Knoxville, TN April 20, 1862 to guard political prisoners; Prison guard at

Madison, GA Oct 31, 1862; Captured at **Vicksburg,** MS hospital July 4, 1863; Transported from Vicksburg to Mobile on US Steamer *St. Maurice*; Exchanged off Fort Morgan Aug 5, 1863; Present Feb 20, 1864 muster roll; Captured along Atlanta siege line August 7, 1864; Prisoner at **Camp Chase,** OH; Transferred to Point Lookout, MD for release Mar 18, 1865; Among 1,088 paroled Confederate prisoners received at Boulware's Wharf Mar 27, 1865; Carpenter & builder in Atlanta after the war; Buried at Oakland Cemetery in Atlanta.

Jacob Carter PVT 2 Mar 1832 26 Sep 1907

Married Lucinda Hesterly in Fulton County, Feb 18, 1856; Left sick in Knoxville, TN hospital April 22, 1862; Captured at **Vicksburg,** MS July 4, 1863; Sick at home on Dec 10, 1863 muster roll; Present Feb 20, 1864 muster roll; Took Oath of Allegiance in Louisville, KY as a Rebel Deserter Aug 29, 1864; Agreed to remain north of the Ohio River; Filed for Confederate pension in Haralson County in 1897; Reported being captured while sick and taken to Camp Chase, OH; Farmer near Tallapoosa in 1900; Buried at Draketown Cemetery in Haralson County.

Wilson C. Carter PVT 1825 20 Mar 1865

Married Laura Hudson in DeKalb County, Oct 20, 1846; Overseer near Newnan in 1860; Detailed to Captain Calhoun's detachment in Knoxville, TN April 20, 1862 to guard political prisoners; Prison guard at Madison, GA Oct 31, 1862; Captured at **Vicksburg,** MS July 4, 1863; Captured near Atlanta Sep 23, 1864; Died of "rheumatism of the heart" at **Camp Douglas,** IL; Buried at Confederate Mound in Oak Woods Cemetery, Chicago.

David Perry Clinton PVT 3 Oct 1836 16 Jan 1913

Detailed to Captain Calhoun's detachment in Knoxville, TN April 20, 1862 to guard political prisoners; Prison guard at Madison, GA Oct 31, 1862; Captured at **Vicksburg,** MS July 4, 1863; Sent to Atlanta hospital Sep 20, 1863 & returned to duty Oct 12, 1863; Captured as deserter near Jonesborough Sep 2, 1864; Released from **Camp Douglas,** IL May 13, 1865; Married Martha Underwood in Bartow County, Nov 9, 1867; Town Marshall in Kingston, GA in 1880; Buried at Tunnel Hill Cemetery in Whitfield County.

George P. Clinton PVT 1800 28 May 1865

Father of David Clinton; Present on April 30, 1862 muster roll.

Austin Cook PVT 1837 16 May 1863

Married Martha Lee in Fulton County, Nov 13, 1857; Laborer in Black Hall District in 1860; Sent to Knoxville, TN hospital April 20, 1862 by order of Surgeon Hardeman; Killed at **Battle of Baker's Creek,** MS.

William R. Daniel PVT 1863

Died in Vicksburg, MS hospital June 20, 1863.

Henry T. Davenport PVT 25 May 1862

Left at Knoxville, TN hospital April 22, 1865; Died in Knoxville hospital; Claim filed by his father Jesse Davenport in Fulton County.

William Wilson Donehoo PVT 1844 19 Nov 1864

Son of Elijah Donehoo; Enlisted May 5, 1862 at Cumberland Gap, TN; Sent to Vicksburg, MS hospital Mar 21, 1863; Present Feb 20, 1864 muster roll; Captured at **Battle of Resaca,** GA May 15, 1864; Sent to Alton, IL prison hospital with fever Jun 23, 1864; Died as POW at **Camp Douglas,** IL; Buried at Confederate Mound in Oak Woods Cemetery, Chicago.

George Washington Drake PVT 1842 14 Dec 1864

Captured at **Vicksburg**, MS July 4, 1863; Captured near Jonesborough as deserter Sept 2, 1864; Died of smallpox as POW at **Camp Douglas**, IL; Buried at Confederate Mound in Oak Woods Cemetery, Chicago.

George Washington Elliott PVT 5 Jul 1839 25 Apr 1906

Detailed to Captain Calhoun's detachment in Knoxville, TN April 20, 1862 to guard political prisoners; Prison guard at Madison, GA Oct 31, 1862; Captured at **Vicksburg**, MS July 4, 1863; Present Feb 20, 1864 muster roll; Captured along Atlanta siege line Aug 21, 1864; Released from **Camp Chase**, OH May 15, 1864; Married Mary Melissa Brawner in Fulton County, July 29, 1866; Blacksmith & farmer in Cooks District after the war; Buried at Charles Elliott Family Cemetery in Atlanta.

David V. Faulkner PVT 1832 Before 1870

Detailed to Captain Calhoun's detachment in Knoxville, TN April 20, 1862 to guard political prisoners; Prison guard at Madison, GA Oct 31, 1862; Captured in rifle pits on Vicksburg, MS trench line June 23, 1863; Paroled July 4, 1864; Absent on medical furlough Dec 10, 1863 muster roll; Received at Atlanta distribution hospital debilitated May 7, 1864; Captured along Atlanta siege lines Aug 7, 1864; Transferred from **Camp Chase**, OH to Point Lookout, MD for release Mar 18, 1865; Paroled from Libby Prison hospital, Richmond, VA April 21, 1865; No further record.

Rowland R. Faulkner PVT 1842 1862

Younger brother of David Faulkner; Detailed to Captain Calhoun's detachment in Knoxville, TN April 20, 1862 to guard political prisoners; Death claim filed by his widow Mary Faulkner Jan 13, 1863.

Torrance Charles Gilstrap PVT 1843 Jan 1889

Detailed to Captain Calhoun's detachment in Knoxville, TN April 20, 1862 to guard political prisoners; Prison guard at Madison, GA Oct 31, 1862; Little finger shot off at **Battle of Chickasaw Bayou**; Captured in **Vicksburg**, MS hospital July 4, 1863; Transported from Vicksburg to Mobile on US Steamer *St. Maurice*; Exchanged off Fort Morgan Aug 5, 1863; Appointed musician Oct 13, 1863; Present on Feb 20, 1864 muster roll; Captured near Rough & Ready, GA Sep 12, 1864; Enlisted in Company F, 6th US Volunteers while a prisoner at **Camp Douglas**, IL Mar 26, 1865; Resident of Sacramento, CA in 1879; Buried in Mountain View Cemetery, Oakland, California.

James G. Glynch PVT 1844 Before 1867

Native of Ireland; Left in Knoxville, TN hospital April 22, 1862; Captured at **Vicksburg**, MS hospital July 16, 1863; Present Feb 20, 1864 muster roll.

John Brown Gordon PVT 4 Aug 1833 4 Sep 1898

Married Mary Richards in Cobb County in 1858; Detailed to Captain Calhoun's detachment in Knoxville, TN April 20, 1862 to guard political prisoners; Prison guard at Madison, GA Oct 31, 1862; Captured at **Vicksburg**, MS July 4, 1863; Present on Feb 20, 1864 muster roll; Surrendered at **Greensboro**, NC April 26, 1865; Railroad engineer living in Atlanta in 1880; Buried at Collins Springs Primitive Baptist in Smyrna.

Wilson Lumpkin Grubbs PVT 9 Mar 1833 30 Jun 1908

Born in Walton County; Married Seretha Hudson in Fulton County, Sep 12, 1858; Detailed to Captain Calhoun's detachment in Knoxville, TN April 20, 1862 to guard political prisoners; Prison guard at Madison, GA Oct 31, 1862; Captured at **Vicksburg**, MS July 4, 1863; Surrendered at **Greensboro**, NC April 26, 1865; Carpenter & machinist in Atlanta after the war; Member of UCV Camp # 159 in Atlanta; Filed for a Confederate pension in Fulton County in 1902; Application endorsed by John B. Jett; Buried at Oakland Cemetery in Atlanta.

Tudor Tucker Hall PVT 6 Dec 1844 28 Aug 1918
Born in McIntosh County; April 30, 1862 muster roll reports Hall had been promoted to SGT and transferred to CSA Navy; Bookkeeper in Charleston, SC in 1880; Buried at Highlands Memorial Park in Macon County, NC.

George M. Hancock PVT 24 Jul 1863
Captured at **Vicksburg**, MS hospital July 4, 1863; Died in Vicksburg hospital.

Augustus L. Harwell PVT 1843 1 Aug 1863
Captured at **Vicksburg**, MS July 4, 1863; Died at home in Stones District.

William M. Hawes PVT
Furloughed from 1st Mississippi Hospital in Jackson, MS May 1, 1863; Present on Feb 20, 1864 muster roll; Furloughed from Ocmulgee Hospital in Macon July 19, 1864.

Benjamin James Herndon PVT 22 Apr 1845 23 May 1924
Born in Alabama; Detailed to Captain Calhoun's detachment in Knoxville, TN April 20, 1862 to guard political prisoners; Prison guard at Madison, GA Oct 31, 1862; Captured at **Vicksburg**, MS July 4, 1863; Present on Dec 10, 1863 muster roll; Appointed **CPL** in 1864; Surrendered at **Greensboro**, NC April 26, 1865; Married Emma Stewart in Fulton County, Nov 5, 1871; Fireman in Atlanta after the war; Hotel clerk in Nashville, TN in 1920; Buried at Westview Cemetery in Atlanta.

David Elbert Herren PVT 17 Nov 1847 8 Jan 1925
Son of Edmund Herren; Enlisted Dec 8, 1864 at age of 17; Wounded in right foot at **Battle of Bentonville**, NC Mar 19, 1865; Foot amputated below knee at Pettigrew General Hospital #13 in Raleigh, NC; Married Elizabeth Willis in Fulton County, April 7, 1887; Farmer in Bryant's Crossroad after the war; Filed for Confederate pension for artificial leg in Fulton County in 1880; Buried at Greenwood Cemetery in Atlanta.

Edmond R. Herren PVT 1822 After 1880
Born in Marion County, SC; Married Elizabeth Winburn in DeKalb County, June 6, 1845; Father of David Herren; Enlisted in Decatur, GA Sep 23, 1863; Present Feb 20, 1864 muster roll; Admitted to Fairgrounds Hospital #2 in Atlanta April 30, 1864; Farmer in Adamsville in 1880.

James P. Hindman PVT 1843 27 Oct 1862
Born in Campbell County; Left sick in Kentucky; Captured at Chaplin Hill by Danville, KY; Buried at Belleview Cemetery in Danville.

Andrew Jackson Holbrook PVT 1830 19 Feb 1863
Farmer in Cooks District in 1860; Detailed to Captain Calhoun's detachment in Knoxville, TN April 20, 1862 to guard political prisoners; Prison guard at Madison, GA Oct 31, 1862; Died in General Hospital in Newton, MS Feb 19, 1863.

James Payton Holbrook PVT 5 Sep 1834 Jul 1875
Married Anne Coryell in Fulton County, Mar 6, 1856; No record after April 12, 1862.

Thomas James Holbrook PVT 1835 14 Nov 1913
Married Mary Toney in Fulton County, May 10, 1857; Detailed to Captain Calhoun's detachment in Knoxville, TN April 20, 1862 to guard political prisoners; Prison guard at Madison, GA Oct 31, 1862; Captured at **Vicksburg**, MS hospital July 4, 1863; Transported from Vicksburg to Mobile on US Steamer *St. Maurice*; Exchanged off Fort Morgan Aug 5, 1863; Admitted to Fairgrounds Hospital #2 in Atlanta with rubeola Jan 14, 1864;

Present Feb 20, 1864 muster roll; Surrendered at **Greensboro**, NC April 26, 1865; Farmer in Cobb County in 1900; Filed for Confederate pension in Cobb County in 1902; Reported being badly wounded by exploding shell at Battle of Nashville, TN Dec 16, 1864; Injuries have "disabled him to such an extent as to render him entirely unable to perform any sort of labor at the present time."; Buried in Mableton.

Wilson Lumpkin Holbrook PVT 20 Oct 1831 12 Feb 1893

Born in Franklin County; Married Rachel Kennedy in Forsyth County, Dec 26, 1852; Dec 10, 1863 muster roll reports that Holbrook was admitted to Knoxville, TN hospital Nov 15, 1862; Given medical furlough; Farmer in Campbell County in 1870 & Douglas County in 1880; Widows Confederate pension filed for in Fulton County in 1915; Reported leaving Command in Kentucky in Oct 1862 and carried from Cumberland Gap to hospital in Knoxville, TN; Application endorsed by Thomas Holbrook.

John W. Hornsby PVT 25 May 1846 29 Dec 1931

Transferred to Company K on Aug 1, 1862; Captured in rifle pits in front of Vicksburg, MS trenches June 16, 1863; Paroled in **Vicksburg**, MS July 4, 1863; Captured near **Jonesborough**, GA Aug 1, 1864; Discharged from **Camp Douglas**, IL June 17, 1865; Married Margaret Elizabeth Waits in Fulton County, Sep 2, 1866; Railroad section foreman after the war; Filed for Confederate pension in Fulton County in 1898; Application endorsed by Lowndes Calhoun who reported Hornsby was captured at Battle of Jonesborough; Buried at South Bend Methodist in Atlanta.

Henry Crawford House PVT 10 Mar 1846 5 Mar 1924

Born in Cobb County; Detailed to Captain Calhoun's detachment in Knoxville, TN April 20, 1862 to guard political prisoners; Prison guard at Madison, GA Oct 31, 1862; Appointed **1CPL** Feb 1863; Captured at **Vicksburg**, MS July 4, 1863; Appointed musician Oct 15, 1863; Married Bridget Murphy in Cobb County about 1866; Blacksmith in Atlanta in 1871; Blacksmith in Powder Springs in 1880; Foreman at pumping station near Birmingham, AL in 1900; Filed for Confederate pension in Jefferson County, AL in 1908; Reported being wounded at Loudon, TN (November 1863); Buried at Walker Chapel in Fultondale, Jefferson County, Alabama.

John B. Jackson PVT 1832 After 1870

Captured at **Vicksburg**, MS July 4, 1863; Admitted to Fairgrounds Hospital #1 in Atlanta Oct 16, 1863; Deserted hospital Dec 21, 1863; Present on Feb 20, 1864 muster roll; Took Reconstruction Oath in Fulton County Aug 20, 1867; Railroad mechanic in Fulton County in 1870.

John Bailey Jett PVT Apr 1837 11 Jun 1904

Detailed as teamster at Cumberland Gap, TN April 29, 1862; Captured at **Vicksburg**, MS July 4, 1863; Present on Dec 10, 1863 muster roll; Received clothing allowance at Law Hospital in Eatonton Sep 30, 1864; Married Josephine Richardson in Fulton County, May 23, 1875; Member of UCV Camp # 925 in Atlanta; Atlanta saloon keeper in 1900.

James Johns PVT 1833 Before 1867

Married Eliza Lambert in Fulton County, Dec 12, 1858; Laborer in Cooks District in 1860; Detailed to Captain Calhoun's detachment in Knoxville, TN April 20, 1862 to guard political prisoners; Prison guard at Madison, GA Oct 31, 1862; Captured at **Vicksburg**, MS July 4, 1863; Present on Feb 20, 1864 muster roll.

Elijah "Elzy" Franklin Lee PVT 3 Jan 1840 1920

Married Sarah Childers in Fulton County, Dec 13, 1860; Detailed to Captain Calhoun's detachment in Knoxville, TN April 20, 1862 to guard political prisoners; Prison guard at Madison, GA Oct 31, 1862; Captured at **Vicksburg,** MS July 4, 1863; Present on Feb 20, 1864 muster roll; Admitted to Way Hospital in Meridian, MS with debilitus Jan 19, 1865; Farmer at Bryants Crossroads in 1880; Farmer in Cullman County, AL in 1900; Filed for Confederate pension in Cullman County in 1907; Reported being wounded at **Battle of Franklin,** TN and paroled home for two month. "About ten days before Lee's surrender I left home for duty and reached Augusta, GA where I remained until discharged." Buried at Cullman City Cemetery.

Henry J. Lee PVT 7 May 1840 7 Oct 1895
Captured at **Vicksburg,** MS July 4, 1863; Admitted to Fairgrounds Hospital #1 in Atlanta Oct 2, 1863 with syphilis; Absent in hospital Feb 20, 1864 muster roll; Admitted to Ocmulgee Hospital in Macon with rheumatism May 13, 1864; Surrendered in Marshall County, TN June 9, 1865; Farmer in Cobb County in 1880; Buried at Collins Springs Primitive Baptist Church in Smyrna.

John H. Lee PVT 1844 Before 1867
Enlisted Mar 30, 1862 in Cumberland Gap, TN; Captured at **Vicksburg,** MS July 4, 1863; Present Feb 20, 1864 muster roll.

Nathaniel D. Lee PVT 9 Mar 1825 1 May 1916
Married Sallie House (Sister of Frances House) in Clarke County, Jan 9, 1837; Detailed to Captain Calhoun's detachment in Knoxville, TN April 20, 1862 to guard political prisoners; Prison guard at Madison, GA Oct 31, 1862; Captured at **Vicksburg,** MS July 4, 1863; Present on Feb 20, 1864 muster roll; Surrendered at **Greensboro,** NC April 26, 1865; Farmer in Cobb County after the war; Filed for Confederate pension in Cobb County in 1910; Reported being wounded at Battle of Resaca May 15, 1864, at Battle of New Hope Church May 25, 1864, and wounded in left leg by exploding shell at Battle of Nashville, Dec 16, 1864; Buried at Mount Harmony Baptist in Mableton.

Thomas M. Lee PVT 30 Mar 1836 21 Jul 1863
Brother of Elzy Lee; Married Nancy Hemberly about 1858; Captured in **Vicksburg,** MS hospital July 4, 1863; Died in Vicksburg hospital; Claim filed by widow Nancy Lee.

William Joe Lee PVT 18 May 1830 1911
Older brother of John & Henry Lee; Married Elizabeth Kirk in Cherokee County, Dec 17, 1850; Captured at **Vicksburg,** MS July 4, 1863; Present on Feb 20, 1863 muster roll; Farmer near Smyrna after the war; Filed for a Confederate pension in 1888; Reported being shot through the right arm at **Battle of Wyse Forks,** NC March 9, 1865 and being permanently disabled; Buried at Collins Springs Primitive Baptist in Smyrna.

John B. Linam PVT 14 Jan 1863
Detailed to Captain Calhoun's detachment in Knoxville, TN April 20, 1862 to guard political prisoners; Prison guard at Madison, GA Oct 31, 1862; Sent to Atlanta hospital from Post Hospital in Dalton Dec 1, 1862; Died in Atlanta hospital of variola.

James Edward Lynch PVT 4 Mar 1846 14 May 1911
Born in County Meath Ireland; Late enlistment; Respected contractor; Erected the Fulton County Courthouse with his brother Patrick Lynch; Filed for a Confederate pension in Harrelson County in 1907; Reported being captured near Atlanta; Application supported by Jacob Carter; Buried at Oakland Cemetery in Atlanta.

George Washington Lyon PVT 12 Aug 1838 8 Apr 1903
Married Nancy Adeline Williams in DeKalb County Nov 1, 1860; Detailed as teamster by

Colonel Henderson in Knoxville, TN April 18, 1862; Captured at **Vicksburg**, MS July 4, 1863; Detailed Oct 16, 1863 to build a bridge; Admitted to Fairgrounds Hospital #2 in Vineville, GA with kidney inflammation Dec 31, 1864; Regimental musician and band-master for the 42nd Georgia; Surrendered at **Greensboro**, NC April 26, 1865; Composer and music teacher after the war; Author of 20 sacred song books; Buried at Hollywood Cemetery in Atlanta.

William Andrew Mauldin PVT 23 Feb 1847 14 Oct 1917

Born in Greenville, SC; Detailed to Captain Calhoun's detachment in Knoxville, TN April 20, 1862 to guard political prisoners; Prison guard at Madison, GA Oct 31, 1862; Captured in rifle pits outside Vicksburg trench lines June 23, 1863; Exchanged at **Vicksburg**, MS July 4, 1863; Transported with sick from Vicksburg to Mobile on US Steamer *St. Maurice*; Exchanged off Fort Morgan Aug 5, 1863; Present Dec 10, 1863 muster roll; Admitted to Fairgrounds Hospital #2 in Atlanta with fever Feb 25, 1864; Deserted near Atlanta; Took Federal Oath of Allegiance in Chattanooga, TN Oct 26, 1864; Farmer in Cross Keys in 1900; Buried at Sardis Methodist in Atlanta.

Henry J. McDonald PVT 7 Nov 1862

Detailed to Captain Calhoun's detachment in Knoxville, TN April 20, 1862 to guard political prisoners; Died at Bean Station, TN hospital; Claim filed by his widow Jane McDonald in Fulton County.

Walter C.J. McDonald PVT 22 Jun 1830 27 Mar 1914

Born in Ireland; Detailed to Captain Calhoun's detachment in Knoxville, TN April 20, 1862 to guard political prisoners; Captured at Vicksburg, MS July 4, 1863; Detached with Captain Rowan's Artillery, Oct 1863; Surrendered at Citronelle, AL May 4, 1865; Farm laborer in Fayette County in 1870; Filed for Confederate pension in Fulton County in 1895; Application endorsed by J.B. Jett; Buried at McDonald Family Cemetery in Fulton County.

William McWilliams PVT 1833 2 Nov 1862

Born in Ireland; Married Catherine Clark in DeKalb County, Nov 2, 1852; Farmer in Panthersville, DeKalb County in 1860; Died of fever in Lexington, KY hospital; Estate probated in Fulton County in 1863.

James L. Melton PVT 1845 After 1880

Detailed to Captain Calhoun's detachment in Knoxville, TN April 20, 1862 to guard political prisoners; Captured at **Vicksburg**, MS July 4, 1863; Present on Feb 20, 1864 muster roll; Received clothing at Ocmulgee Hospital in Macon July 7, 1864; Farmer in Cobb County in 1880.

Joseph L. Melton PVT 1843 Before 1867

Older brother of James Melton; Farm family near Decatur, GA in 1860; Detailed to Captain Calhoun's detachment in Knoxville, TN April 20, 1862 to guard political prisoners; Captured at **Vicksburg**, MS July 4, 1863; Present on Feb 20, 1864 muster roll; Received clothing at Ocmulgee Hospital in Macon July 7, 1864.

Benjamin G. Neal PVT 27 Dec 1834 17 Sep 1912

Born in Butts County; Detailed to Captain Calhoun's detachment in Knoxville, TN April 20, 1862 to guard political prisoners; Dec 10, 1863 muster roll reports Neal in Atlanta hospital since Sep 24, 1863; Appointed hospital steward in Atlanta Sep 9, 1864; Surrendered at **Greensboro**, NC April 26, 1865; Took Reconstruction loyalty oath in Campbell County Aug 12, 1867; Married Sarah Matthews in Floyd County, Aug 10, 1868; Grocery clerk in Rome, GA in 1880; Filed for Confederate pension in Hall County in 1895; Reported being paralyzed in 1882 and helpless ever since; Died at Confederate Soldier's Home in Atlanta; Buried at Westview Cemetery.

Calvin Owens PVT 14 Sep 1824 28 Mar 1881

Born in South Carolina; Married Mahalia Glazier in Morgan County, Mar 21, 1847; Detailed to Captain Calhoun's detachment in Knoxville, TN April 20, 1862 to guard political prisoners; Captured at **Vicksburg**, MS July 4, 1863; Present on Feb 20, 1864 muster roll; Detailed as bridge builder on Memphis & Charleston Railroad Dec 1864; Surrendered at **Greensboro**, NC April 26, 1865; Brick mason in Atlanta in 1880; Buried at Oakland Cemetery in Atlanta.

William L. Payne PVT 1837 Before 1872

Married Margaret McCown in Fulton County, Dec 31, 1855; Captured at **Vicksburg**, MS July 4, 1863; Present on Feb 20, 1864 muster roll; Admitted to Way Hospital in Meridian, MS Jan 19, 1965; Surrendered at **Greensboro**, NC April 26, 1865.

Lewis S. Peacock PVT 1841 15 May 1864

Married Mary DeFoor in Fulton County, July 19, 1860; Detailed to Captain Calhoun's detachment in Knoxville, TN April 20, 1862 to guard political prisoners; Nurse at Catoosa Springs, GA hospital spring & summer 1863; Present on Feb 20, 1864 muster roll; Killed at **Battle of Resaca**, GA.

Thomas Jefferson Peacock PVT 6 Oct 1838 11 Dec 1898

Married Sarah Campbell in Fulton County, Dec 24, 1857; Detailed to Captain Calhoun's detachment in Knoxville, TN April 20, 1862 to guard political prisoners; Present on Feb 20, 1864 muster roll; Surrendered at **Greensboro**, NC April 26, 1865; Farmer in Campbell County after the war; Buried at Ben Hill Methodist in Fulton County.

Henry Z. Pratt PVT 1825 After 1880

Married Elizabeth Howard in Monroe County, Nov 18, 1845; Sent to Bean Station, TN hospital Jun 20, 1862; Patient at French's Division Hospital in Lauderdale, MS Aug 30, 1863; Present on Feb 20. 1864 muster roll; Admitted to Madison, GA hospital June 10, 1864; Furloughed from Blackie Hospital, Madison, GA with chronic diarrhea June 29, 1864; Captured in Egypt Station, MS hospital Dec 28, 1864; Sent to **Alton**, IL prison; Transferred to Point Lookout, MD for exchange Feb 21, 1865; Patient at Camp Winder Hospital in Richmond, VA Mar 9. 1865; Took Reconstruction Loyalty Oath in Fulton County, July 1, 1867; Laborer in Atlanta in 1880.

Preston R. Prophet PVT Before 1867

Married Sarah Ingram in Terrell County, Feb 12, 1854; Married Mary Jane Hancock in Berrien County, Jan 24, 1858; Medical furlough granted April 12, 1863; Granted medical discharge July 28, 1863 by Surgeon G.G. Roy in Atlanta, who writes "This soldier is a mere skeleton"; Dec 10, 1863 muster roll reports Prophet absent, sent to Atlanta hospital Sep 12, 1863.

Joel Queen PVT 1843 After 1880

Born in Anderson County, SC; Captured at **Vicksburg**, MS hospital July 4, 1863; Transported with sick from Vicksburg to Mobile on US Steamer *St. Maurice*; Exchanged off Fort Morgan Aug 5, 1863; Present on Feb 20, 1864 muster roll; Married Sophronia Hooper in Cobb County, Oct 23, 1866; Farmer in Cobb County in 1880; Buried at Hooper-McWilliams Family Cemetery in Cobb County.

James Edward Ramsey PVT 24 Sep 1843 8 Apr 1936

Born in Troup County; Detailed to Captain Calhoun's detachment in Knoxville, TN April 20, 1862 to guard political prisoners; Captured at Big Creek Gap near Lawrenceburg, Ky; Transported by steamboat from Louisville to Vicksburg Dec 29. 1862; Captured at **Vicksburg**, MS July 4, 1863; Present on Feb 20, 1864 muster roll; Admitted to Fairgrounds Hospital #1 in Atlanta May 15, 1864 with dysentery; Furloughed May 21, 1864; Married Sophia Robbins, July 3, 1864; Deserted along Atlanta siege line Aug 19, 1864; Released from **Camp Chase**, OH May 13, 1865; Farmer in Jackson Parish, LA in 1880; Farmer in Hopkins County, TX in 1900; Filed for Confederate pension in Anderson County, Texas in 1913; Reported moving to Texas in 1891; Buried at Shady Grove Cemetery in Shelby County, Texas.

John D. Ranton (Rantin, Routon) PVT 1824 Before 1900

Born in Spartanburg, SC; Sailor in Macon, GA in 1860; Married Julia Reeves in Fulton County, Oct 24, 1861; Detailed as signal operator for General Carter Stevenson from Jan 17, 1863 to Aug 31, 1863; Captured at **Vicksburg**, MS July 4, 1863; Appointed **1SGT** 21 Nov 1863; Captured near **Jonesborough** Sep 2, 1864; Enlisted in the 6th US Volunteers as a prisoner at **Camp Douglas**, IL April 1, 1865; Tailor in Atlanta in 1892; Filed for Confederate pension in Bibb County in 1895.

William M. Rantin PVT 24 Jun 1864

Married Eleanor Brown in Fulton County, Jan 24, 1854; Detailed to Captain Calhoun's detachment in Knoxville, TN April 20, 1862 to guard political prisoners; Ordered to perform hard labor by court martial court Oct 13, 1863; December 10, 1863 muster roll reported Rantin was under court martial at Chickamauga, GA; Present on Feb 20, 1864 muster roll; Died at Academy Hope Hospital near Marietta.

William M. Ray PVT 26 May 1864

Detailed to Captain Calhoun's detachment in Knoxville, TN April 20, 1862 to guard political prisoners; Captured at **Vicksburg**, MS July 4, 1863; Present on Feb 20, 1864 muster roll; Died in Atlanta; Buried at Oakland Cemetery in Atlanta.

Wiley Hamilton Roach PVT 1836 1862

Born in Anderson County, SC; Married Arabella Amelia Gantt in Fulton County, Sep 10, 1857; Present on April 30, 1862 muster roll; Died shortly thereafter.

Benjamin Franklin Roberts PVT 26 Apr 1846 3 Jan 1938

Enlisted in Atlanta Feb 29, 1864 at age of 18; Wounded at the thigh and shoulder at **Battle of Bentonville**, NC Mar 18, 1865; Surrendered at **Greensboro**, NC April 26, 1865; Married Cynthia Caroline Wallace Jan 20, 1869; Farmer at Bryant's Crossroads after the war; Filed for Confederate pension in Fulton County in 1905; Application endorsed by J.D. Willis & L.P. Thomas; Died at Soldier's Confederate Home in Atlanta; Buried at Ben Hill Methodist in Fitzgerald, Ben Hill County.

Samuel C. Sewell, Jr. PVT 31 Oct 1831 6 Nov 1908

Married Catherine Oliver in Fulton County, Oct 19, 1854; Married Isabella Robbins in Campbell County, Dec 23, 1859; Detailed as a teamster by order of Colonel Henderson April 20, 1862; Captured at **Vicksburg**, MS July 4, 1863; Feb 20, 1864 muster roll shows Sewell absent in hospital; Captured as a deserter along Atlanta siege line Aug 22, 1864; Oath of Allegiance taken in Louisville, KY Sep 26, 1864; Agreed to remain north of the Ohio River; Farmer in Campbell County in 1880; Filed for a Confederate pension in Fulton County in 1907; Reported being captured at Vicksburg, MS and taken north of the Ohio River for the remainder of the war; Buried at Mount Gilead Methodist at Ben Hill in Fulton County.

William N. Sewell PVT 16 Feb 1838 11 Jul 1906
Born in Franklin County; Detailed to Captain Calhoun's detachment in Knoxville, TN April 20, 1862 to guard political prisoners; Appointed **3CPL** Feb 1863; Captured at **Vicksburg,** MS July 4, 1863; Captured at **Battle of Nashville,** TN Dec 16, 1864; Released at **Camp Chase,** OH May 15, 1865; Married Susan Cornelia Diggs in Fulton County Sep 7, 1866; Farmer in Bryants Crossroads in 1880; Farmer in Bartow County in 1900; Buried at Salacoa Baptist in Cherokee County.

James T. Shierling PVT Jun 1831 After 1900
Born in DeKalb County; Married Sarah Williamson in Washington County, Oct 11, 1848; Captured at **Vicksburg,** MS July 4, 1863; Present on Feb 20, 1864 muster roll; Surrendered at **Greensboro,** NC April 26, 1865; Farmer at Bryants Crossroads in 1880; Filed for Confederate pension in Carroll County in 1898; Application endorsed by W.L. Wilson; Farmer in Flint Corner, Carroll County in 1900.

Henry L. Sims PVT 1826 1864
Born in South Carolina; Married Narcissa Webb in DeKalb County, Jan 12, 1851; Carpenter in Atlanta in 1860; Detailed to Captain Calhoun's detachment in Knoxville, TN April 20, 1862 to guard political prisoners; Captured at **Vicksburg,** MS July 4, 1863; Present on Feb 20, 1864 muster roll; Died in service.

Francis Olin Smith PVT 24 Aug 1843 20 May 1919
Born in Coweta County; Transferred to 42nd Georgia from Company K, 41st Georgia Mar 15, 1862; Detailed to Captain Calhoun's detachment in Knoxville, TN April 20, 1862 to guard political prisoners; Captured in Lawrenceville, KY Oct 15, 1862; Transported with sick on US Steamer *City of Madison* from Louisville to Vicksburg, MS for exchange Dec 22, 1862; Present on Feb 20, 1864 muster roll; Deserted along Atlanta siege line Aug 18, 1864; Released from **Camp Chase,** OH May 13, 1865; Married Emily Jane Miller in 1865; Retail grocer in Atlanta in 1880, Grocer in McGregor, Texas in 1900; Filed for Confederate pension in DeKalb County in 1911; Reports being captured near Atlanta on July 1, 1864 and taken to Camp Chase for the rest of the war; Buried at Antioch Methodist in Fairburn.

John T. Smith PVT 1838 19 Jul 1863
Detailed to Captain Calhoun's detachment in Knoxville, TN April 20, 1862 to guard political prisoners; Captured in the rifle pits in front of the Vicksburg siege lines June 23, 1863; Paroled at **Vicksburg,** MS July 4, 1863; Died in Cato, MS hospital.

Noah Benson Smith PVT 1839 9 Oct 1864
Son of Methodist minister Noah Smith; Older brother of Francis Smith; Sent to hospital Nov 23, 1863; Admitted to Fairgrounds Hospital #1 in Atlanta April 30, 1864; Furloughed June 30, 1864; Deserted along Atlanta siege line Aug 19, 1864; Died of smallpox as prisoner at **Camp Chase,** OH; Buried in Confederate Cemetery, Grave # 290.

John Andrew Jackson Spear PVT 21 Jan 1838 21 Jan 1912
Born in Jasper County; Detailed to Captain Calhoun's detachment in Knoxville, TN April 20, 1862 to guard political prisoners; Captured at **Battle of Baker's Creek,** MS May 16, 1863; Paroled at **Vicksburg,** MS July 4, 1863; Present on Feb 20, 1864 muster roll; Surrendered at **Greensboro,** NC April 26, 1865; Married Elizabeth Meeks in Fulton County, Jan 21, 1869; Farmer in Jasper County after the war; Filed for Confederate pension in Jasper County in 1900; Reported being hit in the right arm "by a minie ball fired by the

Federals while in line of battle at Baker's Creek": Spear wrote that the wound has made "the limb substantially and eventually useless"; Application endorsed by G.W. Elliott & L.J. Wood; Buried at Enon Baptist Church in Jasper County.

William M. Stringer PVT 1833 After 1870

Married Mary Green in Fulton County, May 1, 1859; Sent to Knoxville, TN hospital April 22, 1862; Feb 20, 1864 muster roll reports Stringer absent in Atlanta hospital since Nov 9, 1863; Nurse at Medical College Hospital in Atlanta Dec 1, 1862; Captured at **Battle of Resaca**, GA May 15, 1864; While prisoner at **Alton**, IL took Oath of Allegiance and joined 5th US Infantry April 14, 1865; Took Reconstruction Loyalty Oath in Campbell County Aug 17, 1867; Farmer in Shelby County, TN in 1870.

Washington William Tarnt PVT 1 May 1831 11 Nov 1905

Born in Forsyth County; Detailed to Captain Calhoun's detachment in Knoxville, TN April 20, 1862 to guard political prisoners; Appointed **5SGT** Feb 1863; Captured at **Vicksburg**, MS hospital July 4, 1863; Transported from Vicksburg to Mobile on US Steamer *St. Maurice*; Exchanged off Fort Morgan Aug 5, 1863; Present on Feb 20, 1864 muster roll; Surrendered at **Greensboro**, NC April 26, 1865; Carpenter in Atlanta in 1870; Buried at Oakland Cemetery in Atlanta.

Benjamin Franklin Tuck PVT 1843 Before 1867

Farmer in Cooks District in 1860; Captured at **Vicksburg**, MS July 4, 1863; Present on Feb 20, 1864 muster roll; Died in service.

Thomas Threwitt Underwood PVT 11 Oct 1830 10 Dec 1910

Born in Warren County; Married Susan Rutledge in Fulton County, Jun 4, 1861; Granted Certificate of Disability at Big Springs, TN Aug 10, 1862; Enlisted as 1SGT in Company E, 64th Georgia in Atlanta Mar 1, 1863; Deserted June 10, 1863; Enlisted in Company E of the 2nd (later 4th) US Volunteers Oct 14, 1864 as prisoner at **Point Lookout**, MD; Appointed CPL Nov 1, 1864 at Fort Monroe, VA; Deserted on Missouri River near Jefferson, MO while on US Steamboat *Mars* May 13, 1865; Farmer, carpenter & Methodist preacher in Hall County, TX after the war; Buried at Dreamland Cemetery in Turkey, Texas.

Daniel David Marion Waits PVT 5 Aug 1826 28 Jul 1863

Married Hannah Jane Rotusee in DeKalb County, July 15, 1846; Appointed **CPL**; Captured at **Vicksburg**, MS July 4, 1863; Died in Montgomery, AL hospital.

Thomas Kennedy Waits PVT 24 Oct 1832 10 Jul 1908

Married Mary Salena Baker in Fulton County, Dec 20, 1860; Captured at **Vicksburg**, MS July 4, 1863; Absent in hospital on Feb 20, 1864 muster roll; Military prison guard in Atlanta; Married Julia Ann Baker in Fulton County, Mar 5, 1871; Farmer in Bryants Crossroads after the war; Filed for Confederate pension in Fulton County in 1889; Reported being wounded by shell in left Leg at **Battle of Baker's Creek** May 16, 1863; Reported being wounded in left arm at **Battle of Nashville**, Dec 16, 1864 "rendering it substantially & essentially useless"; Application endorsed by Lowndes Calhoun; Member of UCV Camp #159 in Atlanta; Buried at Ben Hill Methodist in Fulton County.

John Goodman Wallace PVT 28 Mar 1835 1 Mar 1901

Married Sarah Elizabeth Waits in Campbell County, Oct 5, 1857; Detailed as teamster in Vicksburg, MS April 1863; Captured at **Vicksburg**, MS July 4, 1863; Present on Feb 20, 1864 muster roll; Farmer in Campbell County in 1880; Farmer in Atascosa County, Texas in 1900; Buried at Sand Hollow Cemetery in Frio County, Texas.

Joseph DeKalb Willis PVT May 1843 8 Mar 1907
Left in Knoxville, TN hospital April 22, 1862; Wounded in arm at **Battle of Chickasaw Bayou,** MS Dec 29, 1862; Captured at **Vicksburg,** MS July 4, 1863; Patient at Flewellen Hospital in Cassville Oct 31, 1863; Present on Dec 10, 1863 muster roll; Surrendered at **Greensboro,** NC April 26, 1865; Married Mary Childers in Fulton County, Dec 7, 1868; Farmer in Bryants Crossroads after the war; Member of UCV Camp #159 in Atlanta; Filed for Confederate pension in Fulton County in 1906; Application endorsed by Colonels W.H. Hulsey & L.P. Thomas; Buried at Utoy Primitive Baptist in Atlanta.
William Leak "Leek" Wilson PVT 28 Jan 1840 20 Aug 1915
Born in Franklin County; Younger brother of Surgeon John Wilson; Appointed **4SGT** Feb 1863; Captured at **Vicksburg,** MS July 4, 1863; Elected **2LT** Dec 26, 1863; Present on Feb 20, 1864 muster roll; Appointed **1LT**; Furloughed near Lovejoy Station, GA Aug 17, 1864; Surrendered at **Greensboro,** NC April 26, 1865; Married Kate Hornsby in Campbell County in 1871; Farmer in Sandtown, Campbell County after the war; Served as Sheriff of Campbell County; Buried at Wilson-McLarin Family Cemetery in Atlanta.
Michael Woolf PVT 15 May 1864
Detailed to Captain Calhoun's detachment in Knoxville, TN April 20, 1862 to guard political prisoners; Captured at **Vicksburg,** MS July 4, 1863; Present on Feb 20, 1863 muster roll; Killed at **Battle of Resaca,** GA.

Company K – Calhoun Guards – 104 Men

Company drawn from the city of Atlanta. Late enlistments from Newton & Walton counties were placed in this company.

Name Rank Birth Date
William Lowndes Calhoun CPT 21 Nov 1837 16 Nov 1908
Married Mary Jane Oliver in Fulton County, Sep 13, 1857; Led a 50-soldier detachment detailed to guard political prisoners from Knoxville, TN to Madison, GA April 20, 1862 through November 1862; Captured at **Vicksburg,** MS July 4, 1863; Served on Court Marshall Board Dec 1863; Seriously wounded at **Battle of Resaca,** GA May 15, 1864; Furloughed at Lovejoy Station, GA Aug 23, 1864; Furloughed with battle wound at Way Hospital, Meridian, MS Mar 27, 1865; Served four terms in the Georgia legislature representing Atlanta; Elected Mayor of Atlanta in 1878; Elected Judge of Fulton County Ordinary Court for 16 years beginning in 1884; Historian of 42nd Survivors Association; Commander of the Confederate Veterans Association of Fulton County in 1888; Commander of UCV Camp #159; Author of History of the 42nd Georgia published in 1900; Colonel & Judge Advocate of Georgia UDC; President of the Board of Trustees of the Confederate Soldiers Home in Atlanta for 14 years; President of the John B. Gordon Monument Association; Calhoun was described in his OBIT as,"Of retiring disposition, wholly unostentatious, deeply sympathetic and loveable in his nature, he was yet a man of determination and achievement."; Buried at Oakland Cemetery in Atlanta.
James H. Brown 1LT 16 Jun 1864
Detailed by Colonel Henderson to stay in Knoxville, TN April 23, 1862 as regimental business agent; Elected **1LT** in Decatur Oct 11, 1863; Captured at **Vicksburg,** MS July 4,

1863; Wounded at **Battle of Resaca** May 15, 1863; Admitted to Fairgrounds Hospital #1 in Atlanta May 16, 1864 with gunshot wound; Died in the hospital; Buried at Oakland Cemetery in Atlanta;

Giles Rufus Chapman 4CPL Jul 1840 19 Dec 1904

Married Francis Eddelman in Fulton County, 3 Sep 1861; Captured at **Vicksburg,** MS July 4, 1863; Promoted to **5SGT** Oct 1863; Sick at home on Dec 1863 muster roll; Wounded at **Battle of Atlanta** July 22, 1864; Transferred to Company D, 9th Battalion Artillery Jan 11, 1865; Running a sawmill in Telfair County in 1880; Farmer in Angelina County, Texas in 1900; Buried at Keltys Cemetery in Angelina County, Texas.

William C. Cockrell 1CPL

Detailed as musician April 3, 1863; Captured at **Vicksburg,** MS hospital July 4, 1863; Transported with sick from Vicksburg to Mobile on US Steamer *St. Maurice*; Exchanged off Fort Morgan Aug 5, 1863; Detailed as drummer Oct 8, 1863; Captured at **Battle of Jonesborough,** Aug 31, 1864.

William Thomas Farrar 4SGT 30 Mar 1834 19 Sep 1895

Born in Putnam County; Married Ellen Maria Peck in Putnam County, Sep 4, 1856; Present on Dec 1863 muster roll; Captured at **Vicksburg,** MS July 4, 1863; Promoted to **1SGT**; Surrendered at **Greensboro,** NC April 26, 1865; Clerk in Macon in 1870. Farmer in Baldwin County in 1880; Buried at Farrar-Lane-Finney Family Cemetery in Milledgeville.

William J. Frederick 5SGT 1845 After 1910

Son of Jesse Frederick; Captured in the rifle pits in front of the Vicksburg siege lines, June 23, 1863; Paroled at **Vicksburg,** MS July 4, 1863; Present on Dec 1863 muster roll; Wounded at **Pumpkin Vine Creek,** GA Jun 5, 1864; Machinist in Cairo, GA in 1880; Living in Bainbridge, GA in 1900.

Benjamin Harris 2CPL 13 Mar 1825 1906

Left in Knoxville, TN hospital April 23, 1862; Captured at **Vicksburg,** MS hospital July 4, 1863; Present on Dec 1863 muster roll; Transported with sick from Vicksburg to Mobile on US Steamer *St. Maurice*; Exchanged off Fort Morgan Aug 5, 1863; Carpenter in Atlanta in 1870; Farmer in Campbell County in 1880; Filed for Confederate pension in Fulton County in 1895; Reported being captured in Marietta, GA in June 1864.

Leonard A. Ratterree 3SGT 1 Apr 1829 29 Jun 1863

Born in Chester County, SC; Married Evaline King Bacon in Fulton County, Feb 26, 1857; Appointed **2SGT** Nov 26, 1862; Detailed as hospital guard April 1863; Died at Magnolia, Webster County, MS; Claim filed by his father Alex Ratterree.

Henry Young Snow 2SGT 27 Feb 1840 22 Dec 1924

Born in Darien, McIntosh County; Appointed **1SGT** Nov 26, 1862; Captured at **Vicksburg,** MS July 4, 1863; Elected **2LT** Jan 5, 1864; Wounded at **Battle of Wyse Forks,** NC Mar 14, 1865; Surrendered at **Greensboro,** NC April 26, 1865; Married Nancy Chaffin Hanby in Knox County, TN Oct 14, 1869; Grocery merchant & real estate investor in Atlanta after the war; Member of UCV Camp #159 in Atlanta; Moved to Palatka County, FL after the death of his wife in 1896; Became the largest stockholder in Kennerly Hardware; Died in Palatka, Florida; His OBIT noted that "the deceased was a man of culture , and his courtly manner bespoke of the type of the 'Old South,' now so fast passing." Buried at Oakland Cemetery in Atlanta.

John Thomas Ware 3CPL Jan 1844 2 Jun 1909

Born in Floyd County; Discharged with disability for chronic bronchitis at Big Spring, TN Aug 10, 1862; Conductor on the Atlanta, West Point Railroad in 1880; Retired to raise Shawl Neck fighting cocks; As manager of the Cold Spring Cue Club he became famous for his barbeque & Brunswick stew; Buried at Oakland Cemetery in Atlanta.

Thomas Agnew PVT 19 Nov 1862

Died at the Fairgrounds Hospital in Knoxville, TN; Buried at Bethel Cemetery in Knoxville; Claim filed by widow Mary Agnew.

Wiley Jackson Baggett PVT 31 Oct 1835 11 Feb 1911

Married Malinda Brand in Gwinnett County, Aug 8, 1862; Enlisted in Calhoun July 28, 1862; Nurse at Lauderdale Springs, MS hospital from April through August 1863; Present on Dec 1863 muster roll; Wounded in left leg at **Battle of Nashville**, TN Dec 16, 1864; Admitted to Ocmulgee Hospital in Macon with fractured left leg from exploding shell Mar 1, 1865; Furloughed from Ocmulgee on Mar 28, 1865; Filed for Confederate pension in Walton County in 1897; Farmer in Rockdale in 1900; Buried at Haynes Creek Primitive Baptist in Loganville.

John Bailey PVT 1834 After 1880

Married Winney Hill in Gwinnett County, Jan 23, 1859; Enlisted in Calhoun July 28, 1862; Captured at **Vicksburg**, MS July 4, 1863; AWOL on Oct 31, 1863 muster roll; Captured as a deserter near Marietta June 28, 1864; Took Oath of Allegiance in Louisville, KY July 18, 1864; Agreed to remain north of the Ohio River; Farmer in Hog Mountain in 1880.

Andrew James Boss PVT 17 Jul 1828 25 Jul 1863

Born in Walton County; Married Susan Francis Martin; Enlisted in Calhoun July 20, 1862; Captured at **Vicksburg**, MS July 4, 1863; Died in Conyers; Buried at Shiloh Primitive Baptist in Loganville.

Henry Thomas Boss PVT 1829 22 Sep 1863

Enlisted in Calhoun July 20, 1862; Admitted to Post Hospital in Dalton Nov 21, 1862.

Peter W. Boss PVT 1836 27 Nov 1862

Married Frances Peters in Walton County, Dec 20, 1859; Enlisted July 16, 1862; Died of disease at Bean Station, TN. Claim filed by his widow Francis Boss in Walton County.

Jasper N. Bruce PVT 1833 After 1880

Married Rebecca Barker in Lumpkin County, Oct 9, 1853; Captured at **Vicksburg**, MS July 4, 1863; AWOL on Dec 1863 muster roll; Teamster in St. Louis, MO in 1870; Married Elizabeth Stiles in St. Louis, MO, Mar 28, 1874; Farmer in Fannin County, Texas in 1880.

Timothy M. Butt PVT 1838 After 1900

Born in Campbell County; Left sick in Knoxville, TN hospital April 23, 1862; Detailed by General Bragg as hospital nurse Nov 7, 1863; On detail Aug 12, 1864; School teacher in Atlanta in 1875; Lawyer in Atlanta in 1885.

James T. Calhoun PVT

Enlisted in Atlanta Aug 20, 1862; Captured at **Vicksburg**, MS July 4, 1863; Present on Dec 1863 muster roll; Promoted to **5SGT**; Surrendered at **Greensboro**, NC April 26, 1865.

Frederick H. Casey PVT 6 Sep 1835 29 Aug 1863

Born in Walton County; Married Sarah Austin in Walton County, Nov 19, 1855; Laborer in Buckhead, GA in 1860; Died in Lauderdale Springs, MS hospital; Believed to be buried at Lauderdale Springs, MS Confederate Cemetery; Government marker at Oakland Cemetery in Atlanta.

Edward Thomas Chubb PVT 17 Oct 1838 29 May 1920

Born in Lexington County, SC; Married Martha Ann Turner in Newton County, Dec 19, 1860; Enlisted in Calhoun July 28, 1862; Captured at **Vicksburg,** MS July 4, 1863; Present on Dec 1863 muster roll; Wounded at **Battle of Nashville,** TN Dec 16, 1864; Wounded at **Battle of Bentonville,** NC Mar 19, 1865; Surrendered at **Greensboro,** NC April 26, 1865; Farmer in DeKalb County after the war; Filed for Confederate pension in DeKalb County in 1906; Application endorsed by E.L. Turner & W.P. Medlock; Buried at Chupp Family Cemetery in Lithonia.

James H. Clark PVT 7 Dec 1846 14 Apr 1924

Born in Newton County; Married Julia Malvina Mann in Newton County, Jan 30, 1863; Farmer in Newton County in 1880; Farmer in South Bend in 1900; Filed for Confederate pension in Fulton County in 1910; Reported enlisting in Atlanta July 20, 1864 & being captured two days later while guarding a railroad bridge; Clark reports he was released from Camp Chase, OH Jun 26, 1865; Buried at Westview Cemetery in Atlanta.

Charles Collier Clayton PVT 6 Dec 1834 11 Oct 1922

Late enlistment; Captured along Atlanta siege line Aug 7, 1864; Transferred from **Camp Chase,** OH to Point Lookout, MD for exchange Mar 18, 1865; Married Mariah Myers about 1872; Farmer in Cobb County after the war; Filed for Confederate pension in Cobb County in 1910; Application endorsed by E.R. Howell, who remembers Clayton being captured at Battle of Jonesborough; Buried at Liberty Hill Cemetery in Acworth.

James A. Collier PVT 1836

Enlisted in Calhoun Aug 1, 1862; Captured at **Vicksburg,** MS July 4, 1863; Present on Dec 1863 muster roll; Wounded at **Battle of Rocky Face,** GA Feb 25, 1864; Wounded at **Battle of New Hope Church** May 25, 1864; Admitted to Fairgrounds Hospital #1 in Atlanta May 27, 1864 with gunshot wound; Transferred to Vineville hospital July 7, 1864; Surrendered at **Greensboro,** NC April 26, 1865.

William Smith Crumley PVT 2 Jan 1829 3 Aug 1895

Married Cynthia King in Gwinnett County, May 12, 1849; Enlisted in Calhoun July 28, 1862; Admitted to Post Hospital in Dalton Dec 23, 1862; Patient at Lauderdale Springs, MS hospital April 30, 1863; AWOL on Dec 1863 muster roll; Deserted from Lauderdale Springs, MS hospital Aug 5, 1863; Gave Oath of Allegiance in Louisville, KY Sep 26, 1864 & agreed to remain north of the Ohio River; Farmer in Gwinnett County in 1880; Buried at Prospect Methodist in Lawrenceville.

Joseph J. Davis PVT 1840 11 Sep 1862

Died at Bean Station, TN hospital; Claim filed by his father Thomas W. Davis in Fulton County.

Vincent Davis PVT 21 Apr 1839 After 1900

Born in Anderson County, SC; Married Julia Ray in Fulton County, Sep 13, 1857; Captured in the rifle pits before the Vicksburg siege lines Jun 16, 1863; Paroled at **Vicksburg,** MS July 4, 1864; Transported with sick from Vicksburg to Mobile on US Steamer *St. Maurice*; Exchanged off Fort Morgan Aug 5, 1863; Appointed *3CPL* in Oct 1863; Present on Dec 1863 muster roll; Admitted to Fairgrounds Hospital #2 in Atlanta with dysentery May 16, 1864; Wounded in **Battle of Atlanta** July 22, 1864; Admitted to Pim Hospital in Griffin July 24, 1864; Surrendered at **Greensboro,** NC April 26, 1865; Brick mason in

Atlanta after the war; Filed for Confederate pension in Fulton County in 1898; Affidavit of Physician provided by W.M. Durham.

John Floyd Dennard PVT 30 May 1845 24 Feb 1926

Born in Newton County; Enlisted in Decatur Feb 10, 1864; Wounded at **Battle of Jonesborough** Aug 31, 1864; Wounded at **Battle of Nashville**, Dec 16, 1864; Surrendered at **Greensboro**, NC April 26, 1865; Married Anna Winston in Newton County, Dec 12, 1867; Farmer in Rockdale County after the war; Filed for a Confederate pension in Rockdale County in 1906; Application endorsed by E.L. Turner; Buried at Almand Cemetery in Conyers.

Malon J. Doby PVT 12 Jan 1839 29 Jul 1912

Married Mary Watkins in Newton County, Nov 3, 1859; Enlisted in Company B, 18th Georgia Feb 22, 1862; Discharged with disability Aug 5, 1862; Enlisted in 42nd Georgia in 1864; Captured near Decatur as a deserter Sep 12, 1864; Swore Oath of Allegiance in Louisville, KY Oct 11, 1864 & agreed to remain north of the Ohio River; Farmer in Gwinnett County in 1900; Filed for Confederate pension in Gwinnett County in 1907; Reported that he was not able to serve after being discharged from the 18th Georgia; Buried at Trinity Methodist in Duluth.

William Irving Farris PVT 3 Dec 1826 22 Oct 1897

Married Mary Malinda Wells in 1850; Detailed as teamster; Absent sick at home December 1863 muster roll; Admitted to Fairgrounds Hospital #2 in Atlanta with gunshot wound Feb 13, 1864; Farmer in Panthersville, DeKalb County after the war; Buried at Antioch Baptist in Atlanta.

Jesse J. Frederick PVT 1821 19 Feb 1865

Father of William Frederick; Farmer in Alpharetta in 1860; Assigned to Quartermaster Department as teamster April 1862; Listed on March 1863 roll as teamster in Vicksburg, MS; Listed on Dec 1863 muster roll as teamster; Died of pneumonia in Army of Tennessee hospital.

David D. Garrison PVT 1833 24 Jun 1863

Killed defending rifle pits in front of Vicksburg siege lines; Federals reported he was "too badly wounded to move." Buried at Cedar Hill, Vicksburg, MS.

James D. Garrison PVT 1828 16 Dec 1864

Born in South Carolina; Farmer in Cooks District in 1860; Captured at **Vicksburg**, MS July 4, 1863; Killed at **Battle of Nashville**, TN.

William J. Garrison PVT 24 Feb 1839 21 Feb 1885

Younger brother of David Garrison; Left sick at Knoxville, TN hospital April 23, 1862; Wounded at **Battle of Baker's Creek**, MS May 16, 1863; Captured at Washington Hospital in **Vicksburg**, MS July 4, 1863; Transported from Vicksburg to Mobile on US Steamer *H. Chouteau*; Married Margaret McEver in Paulding County, Jan 5, 1871; Farmer in Paulding County after the war; Buried at Crossroads Baptist in Dallas, Paulding County.

L.C. Gault PVT 1840 1864

Enlisted as 2SGT in Company F of the 23rd Georgia Aug 31, 1861; Mustered out in Yorktown, VA Jan 10, 1862; Late enlistment in 42nd Georgia; Admitted to Fairgrounds Hospital #2 in Atlanta May 20, 1864.

James H. Gossett PVT 1843 10 Sep 1894

Born in South Carolina; Married Sarah Tidwell in Fulton County Dec 31, 1863; Late enlistment; Captured in 1864 as deserter; Took Oath of Allegiance in Louisville, KY Oct

14, 1864 & agreed to stay north of the Ohio River; Butcher in Atlanta after the war; Died in Atlanta.

Andrew W. Greene PVT 9 Sep 1838 20 Feb 1920

Born in Polk County, TN; Married Sarah Cornett about 1858; Enlisted in Decatur June 1, 1863; Admitted to Griffin Hospital with typhoid fever in Mar 1864; Farmer near Toccoa after the war; Filled for Confederate pension in Fannin County in 1905; Reported rupturing back loading ammunition on railroad cars in Jackson, MS Feb 7, 1864; Also reported losing eyesight through "exposures of all kinds."; Reported receiving furlough from Surgeon Robert Foster at Griffin Hospital; Buried at Harmony Baptist in Blue Ridge, Fannin County.

John Hall PVT 1842 Jul 1863

Younger brother of William Hall; Enlisted in Calhoun July 28, 1862; Wounded along Vicksburg, MS siege lines; Captured in **Vicksburg**, MS Hospital #3 July 4, 1863; Died from his wounds; Memorial marker at Harmony Grove Methodist in Auburn, Barrow County.

William B. Hall PVT 5 Feb 1832 1 Nov 1866

Enlisted in Calhoun July 28, 1862; Wounded at **Battle of Baker's Creek**, May 16, 1863; Captured at **Missionary Ridge**, TN Nov 25, 1863; Took Oath of Allegiance at **Rock Island**, Il prison April 21, 1864; Buried at Harmony Grove Methodist in Auburn, Barrow County.

Asberry H. "Berry" Hardy PVT 1823 After 1880

Born in South Carolina; Married Jane Little in DeKalb County, Oct 12, 1845; Captured at **Vicksburg**, MS July 4, 1863; Present on Oct 31, 1863 muster roll; Took Reconstruction Oath in Fulton County, Aug 20, 1867; Farmer in Cobb County in 1870; Married Mary Posey in Marshall County, AL, Jan 15, 1880; Farmer in Marshall County, AL in 1880.

Henry A. Harris PVT 1826 After 1880

Born in Tennessee; Married Frances Redmon in Forsyth County, Feb 26, 1846; Carpenter in Atlanta in 1860; Captured at **Vicksburg**, MS July 4, 1863; Present on Dec 1863 muster roll; Appointed **3CPL** in Oct 1863; Signed Reconstruction Oath in Fulton County July 19, 1867; Carpenter in Dawson County in 1880.

William Harrison PVT

Discharged with debilitating disability at Knoxville, TN Oct 16, 1862.

Thomas E. Henderson PVT

Captured at **Vicksburg**, MS July 4, 1863; Present on Dec 1863 muster roll; Wounded on picket line at Kennesaw Mtn Jun 27, 1864;

C.A. Hendricks PVT 14 Apr 1864

Landowner in Campbell County; Late enlistment; Died at Erwin Hospital in Kingston, GA.

John H. Hilton PVT 1827 1908

Born in Walton County; Married Mary Sentell in Fulton County, Dec 12, 1858; Captured at Vicksburg, MS July 4, 1863; Present on Dec 1863 muster roll; Took an Oath of Allegiance as a deserter at Louisville, KY Sep 26, 1864; Agreed to remain north of the Ohio River; Carpenter in Atlanta in 1870; Farmer in Cross Keys in 1880; Filed for Confederate pension in DeKalb County in 1905.

James Franklin Hollingsworth PVT 21 Oct 1835 9 Mar 1914

Born in Newton County; Married Sarah Adeline Ozburn in Clayton County, Mar 1, 1860;

Farmer near Conyers in 1860; Enlisted in Calhoun July 28, 1862; Captured at **Vicksburg**, MS July 4, 1863; Present on Dec 1863 muster roll; Carpenter in Atlanta in 1880; Member of UCV Camp #159 in Atlanta; Buried at Westview Cemetery in Atlanta.

John Newton Hollingsworth PVT 9 Sep 1833 13 Nov 1862

Died at Morristown, TN hospital.

John W. Hornsby PVT 25 Mar 1846 29 Dec 1931

Enlisted in Calhoun 1 Aug 1862 at age 16; Captured in rifle pits in front of Vicksburg siege lines Jun 16, 1863; Exchanged at **Vicksburg**, MS July 4, 1863; Captured at **Battle of Jonesborough**, GA Aug 31, 1864; Released from **Camp Douglas**, IL June 17, 1865; Married Margaret Waits in Fulton County, Sep 2, 1866; Railroad section foreman after the war; Filed for Confederate pension in Fulton County in 1898; Application endorsed by Captain W.L. Calhoun; Buried at Old South Bend Methodist in Atlanta.

Elijah Richard Howell PVT 11 Oct 1845 12 Jun 1926

Born in Cherokee County; Younger brother of James Howell; Married Sarah Ann Hames in Cherokee County, Oct 11, 1868; Farmer in Cherokee County in 1880; Farmer in Cobb County in 1900; Filed for Confederate pension in Cobb County in 1900; Reported enlisting in 42nd Georgia in Jan 1864; Reported being wounded in the chest at Battle of Atlanta July 22, 1864; Reported being wounded at Battle of Bentonville, NC Mar 19, 1865; Surrendered at Greensboro, NC April 26, 1865.

James Jefferson Howell PVT Oct 1835 25 Apr 1900

Born in Cherokee County; Married Margaret Garner about 1860; Wounded along Vicksburg siege line; Captured at **Vicksburg**, MS July 4, 1863; Absent with Colonel Henderson's permission Dec 1863 muster roll; Admitted with gunshot wound Way Hospital in Meridian, MS Jan 13, 1865; Furloughed from hospital; Farmer in Cherokee County, AL in 1880; Farmer in Walker County, GA in 1900; Buried at Trinity Cemetery in Walker County.

John L. Humphrey PVT 1831 11 Jan 1891

Older brother of Richard & William Humphrey; Carpenter in Atlanta in 1860; Captured at **Battle of Baker's Creek**, MS May 16, 1863; Received as POW at Fort Delaware, DE June 9, 1863; Exchanged July 4, 1863; Paroled at Camp Lee near Richmond, V July 19, 1863; AWOL on Dec 1863 muster roll; Carpenter in Atlanta in 1870.

Richard A. Humphrey PVT 1840 20 Nov 1889

Born in Forsyth County; Captured at **Vicksburg**, MS July 4, 1863; Present on Dec 31, 1863 muster roll; Wounded at **Battle of Resaca**, GA May 15, 1864; Admitted to Pim Hospital in Griffin July 24, 1864; Farmer in Buckhead in 1880; Cleaning Pullman sleeping cars in Atlanta in 1889.

William Jasper Humphrey PVT 1836 After 1880

Born in Forsyth County; Married Martha Redmond in Fulton County, May 13, 1855; Captured at **Battle of Baker's Creek**, MS May 16, 1863; POW at **Camp Morton**, IN; Paroled at Fort Delaware, DE July 4, 1864; Exchanged at Camp Lee, Richmond, VA July 19, 1863; AWOL on Dec 31, 1863 muster roll; Brick mason in Atlanta in 1870; Brick mason in Forsyth County in 1880.

Samuel James PVT 16 May 1864

Late enlistment; Killed at **Battle of Resaca**, GA.

David Henry Johnson PVT 3 Jul 1838 14 Feb 1916

Born in Walton County; Enlisted in Calhoun July 28, 1862; Captured at **Vicksburg**, MS

July 4, 1863; Captured along Atlanta siege lines Aug 7, 1863; POW at **Camp Chase**, OH; Exchanged at Point Lookout, MD Mar 18, 1865; Blacksmith in Gwinnett County after the war; Filed for Confederate pension in Gwinnett County in 1901; Application endorsed by J.S. Davis of Milton Tigers (C); Buried at Pleasant Hill Cemetery in Dacula.

E.B. Johnson PVT

Enlisted in Dalton by Captain Calhoun Dec 31, 1863; Surrendered at **Greensboro**, NC April 26, 1865.

J.A. Johnson PVT

Late enlistment; Wounded in thigh and captured at **Battle of Resaca**, GA May 15, 1864.

Leander A. Kelley PVT 1830 Before 1867

Born in Alabama; May have previously served in Company C, 21st Georgia; Late enlistment in 42nd Georgia; Captured along Atlanta siege lines Aug 7, 1864; POW at **Camp Chase**, OH; Exchanged at Point Lookout, MD Mar 18, 1865.

Zenus S. Lunsford PVT 7 Dec 1831 After 1870

Married Lucy Estes in Newton County, Oct 28, 1852; Enlisted in Calhoun Aug 6. 1862; Present Oct 31, 1863 muster roll; Admitted to Fairgrounds Hospital #1 in Atlanta Oct 22, 1864; Returned to duty Nov 16, 1864; Farmer in Morgan County in 1870.

Joseph W. Maddox PVT

Captured in the rifle pits in front of Vicksburg siege lines June 16, 1863; Paroled at **Vicksburg**, MS July 4, 1863; Sick at home on Dec 31, 1863 muster roll; Admitted to Fairgrounds Hospital #1 in Atlanta Oct 2, 1863 with fever; Returned to duty Oct 18, 1863;

W.P. Maddox PVT

Late enlistment; Wounded at **Battle of Nashville**, Dec 16, 1864;

Lucius Maxwell PVT 25 Jan 1828 1 Oct 1892

Married Frances Smith in Henry County, July 22, 1850; Detailed on ordinance duty in Knoxville, TN May & June 1862; Detailed as harness marker in Knoxville, TN; Present on Dec 31, 1863 muster roll; Shot in head and captured at **Battle of Atlanta**, July 22, 1864; POW at **Camp Douglas**, IL; Transferred to Point Lookout, MD for release Mar 14, 1865; Patient at Jackson Hospital in Richmond, VA Mar 25, 1865; Harness maker in Atlanta in 1870 & Hampton, Henry County in 1880; Buried at McDonough City Cemetery in Henry County.

William T. McWaters PVT 1823 7 Apr 1862

Born in Newton County; Married Mary M. Mabry in Newton County, Jan 3, 1845; Carpenter in Atlanta in 1860; One of the first recruits to die in service. Buried at Oakland Cemetery in Atlanta; Claim filed by widow Mary M. McWaters in Fulton County; McWaters left a wife and seven children.

William Parks Medlock PVT 6 Jun 1837 25 Mar 1903

Married Velma Mason in Fulton County, Aug 27, 1861; Late enlistment; Admitted with wound to Way Hospital in Meridian, MS Mar 23, 1865; Farmer in Fulton County in 1870 & 1880; Married Susan Kitteridge in DeKalb County, Nov 25, 1885; Filed for Confederate pension in DeKalb County in 1889; Reported being wounded in both hands by artillery shell at Battle of Nashville, TN Dec 16, 1864; Farmer near Decatur in 1900; Buried at Decatur City Cemetery in DeKalb County.

Joel Nichols PVT 1818 30 Jul 1901

Born in Sampson County, NC; Married Margaret Smith in Fayette County, April 14,

1850; Wounded in the chest at **Battle of Chickamauga**, Sep 19, 1863; Given medical discharge on Oct 15, 1863; Filed for Confederate pension in Fulton County in 1889; Application endorsed by Captain W.L. Calhoun, who reported "in the campaign in Mississippi by reason of cold & exposure he contracted a disease which manifested itself in a large and painful swelling on his breast; that it soon broke making two large sores some distance apart; that said sores continued to discharge for over three years greatly reducing him in flesh and strength; finally undermining completely his constitution."; Buried on Confederate Knoll in Westview Cemetery in Atlanta.

Thomas O'Dillon PVT 1841 1863
Married Sarah Allen in Clarke County, Aug 3, 1862; Left sick in Knoxville, TN hospital April 23, 1862; Captured in Harrodsburg, KY Oct 10, 1862; Transferred from Louisville, KY to Vicksburg, MS on US Steamboat *City of Madison* Dec 5, 1862; Oct 31, 1863 muster roll reports, "Left sick in Versailes, KY – Not heard from since.";

W.N. Osborn PVT 31 Jan 1863
Enlisted May 15, 1862; Died in Catoosa Springs, GA hospital south of Chattanooga.

James F. Palmer PVT 1839 Before 1870
Enlisted July 28, 1862 in Calhoun; Captured at **Vicksburg**, MS July 4, 1863; Present on Oct 31, 1863 muster roll; Captured at **Battle of Nashville**, TN Dec 16, 1864; Released from Camp Chase, OH June 12, 1865; Told Federals he lived in Newton County.

John Alexander Paris PVT 21 Nov 1830 25 Aug 1905
Born in Columbia, SC; Married Mary Ann Hemphill about 1850; Sign painter in Atlanta after the war; Buried at Hollywood Cemetery in Atlanta; Widow's pension filed in Fulton County in 1905; Reported her husband served 8 months and was given medical discharge from Cumberland Gap, TN.

Dr. Anderson Marion Parker PVT 16 Jan 1822 19 Apr 1898
Born in Morgan County; Pioneer citizen & merchant of Marthasville; Married Oney Jennings in Fayette County, Aug 25, 1846; Medical doctor in Atlanta in 1860; Made independent medical physician to 42nd Georgia April 12, 1862; Physician in Canton in 1880; Became wealthy investing in Atlanta real estate; Steward at Trinity Methodist in Atlanta; Filed for Confederate pension in Fulton County in 1897; Reported being taken by severe cold at Cumberland Gap and sent home after six months of service; Application endorsed by Dr. W.M. Durham; Buried in Oakland Cemetery in Atlanta.

James B. Peavy PVT 23 Nov 1832 10 Aug 1879
Married Elizabeth Caroline Barron in Fulton County, Dec 19, 1850; Farmer in Buckhead in 1860; 32-year-old farmer in 1864 militia census for Fulton County; Enlisted in 1864; Wounded at **Battle of New Hope Church**, GA May 25, 1864; Admitted to Ocmulgee Hospital in Macon with chronic bronchitis Jan 13, 1865; Furloughed Mar 21, 1865; Buried at Oakland Cemetery in Atlanta.

George Washington Powell PVT 7 Aug 1846 11 Jan 1914
Late enlistment; Admitted to Pim Hospital in Griffin July 24, 1864; Captured at Rough & Ready, GA Aug 31, 1864; Discharged from **Camp Douglas**, IL June 17, 1865; Married Eugenia Prather in DeKalb County, Jan 28, 1869; Farmer in DeKalb County after the war; Buried at Klondike Baptist in DeKalb County.

George W. Puckett PVT 19 Aug 1833 3 Jun 1914
Born in Gwinnett County; Furloughed from Enterprise, MS hospital June 18, 1863; Sick in hospital Oct 31, 1863; Married Eliza Owen in Cherokee County, Aug 8, 1867; Farmer

in Cherokee County after the war; Member of UCV Camp # 1088 in Canton; Filed for Confederate pension in Cherokee County in 1907; Reported serving last months of the war in Johnson's Regiment of Georgia Militia; Buried at Dogwood Hills Church Cemetery in Waleska, Cherokee County.

Alexander J. P. Ratterree PVT 1847 After 1892

Son of John Ratteree; Enlisted in Vicksburg, MS Jan 8, 1863; Captured in Washington Hospital in **Vicksburg**, MS July 4, 1863; Transported from Vicksburg to Mobile on US Steamer *H. Chouteau*; Sick in hospital Cot 31, 1863; Admitted to Fairgrounds Hospital #1 in Atlanta Nov 19, 1863 with fever; Returned to duty Nov 28, 1863; Surrendered at **Greensboro**, NC April 26, 1865; Married Clementine Jordan in Fulton County, June 23, 1867; Railroad brakeman in Atlanta in 1874; Conductor in Cherokee County in 1880; Coupler for Western & Atlantic in Atlanta in 1892.

John Ratterree PVT 12 Jan 1824 7 Aug 1897

Born in York County, SC; Married Julia Ann Ratterree in DeKalb County, Jan 8, 1846; Father of Alexander Ratterree; Wood cutter in Black Hall District in 1860; Captured in rifle pits in front of Vicksburg siege lines June 16, 1863; Paroled at **Vicksburg**, MS July 4, 1863; Sick in hospital on Oct 31, 1863 muster roll; Transported from Vicksburg to Mobile on US Steamer *H. Chouteau*; Wounded near East Point, GA Aug 6, 1864; Farmer in Fulton County in 1870; Farmer in Kings Mountain, SC in 1880; Buried at Laurelwood Cemetery in Rock Hill, York County, SC.

Cornelius L. Roberts PVT 1841 After 1880

Enlisted in Madison Sep 8, 1862; Captured at **Vicksburg**, MS July 4, 1863; Wounded at **Battle of Rocky Face**, GA Feb 25, 1864; Received Atlanta Distribution hospital Feb 26, 1864; Admitted to Stout Hospital in Madison, GA hospital Feb 27, 1864; Furloughed from Madison, GA hospital with gunshot wound to left shoulder, Mar 15, 1864; Furlough extended May 24, 1864; Surrendered at Greensboro, NC April 26, 1865; Wheelwright in Cedartown in 1880.

Joseph E. Robinson PVT May 1842 24 Nov 1911

Born in Greene County; Overseer in Covington in 1860; Left sick in Knoxville, TN April 28, 1862; Present on Dec 1863 muster roll; Furloughed from Ocmulgee Hospital in Macon Nov 11, 1864; Signed Reconstruction Oath in Newton County June 26, 1867; Farm laborer in Pickneyville, Gwinnett County in 1900; Filed for Confederate pension in Gwinnett County in 1908; Reported that he was present at final surrender; Died at Confederate Soldier's Home in Atlanta.

William Jackson Scarbrough PVT May 1847 12 Sep 1923

Born in Henry County; Initially rejected by mustering officer as underaged; Wounded at **Battle of Baker's Creek**, MS May 16, 1863; Captured at **Vicksburg**, MS July 4, 1863; Present on Oct 31, 1863 muster roll; Captured at **Battle of Jonesborough** Aug 31, 1864; POW at **Camp Douglas**, IL; Transferred to New Orleans, LA for exchange May 4, 1865; Discharged June 20, 1865; Married Nancy Gizzard in Fulton County, May 13, 1866; Applied for Confederate pension in Fulton County in 1909; Application endorsed by B.F. Roberts; Farmer & dry-goods merchant in Campbell County after the war; Buried at Shadnor Baptist in Union City.

James P. Shaw PVT Oct 1832

Initially discharged April 12, 1862 as "weak eyed"; Re-enlisted in Calhoun July 28, 1862; Captured at **Vicksburg**, MS July 4, 1863; Present on Dec 1863 muster roll.

Kelly Sims PVT 7 May 1830 30 Nov 1908
Born in Laurens County, SC: Married Jane Cook in Henry County, Dec 23, 1858; Enlisted in Calhoun 28, 1862; Admitted to Fairgrounds Hospital #1 in Atlanta Oct 17, 1862 with typhoid fever; Furloughed Nov 6, 1862; Captured at **Vicksburg**, MS July 4, 1863; AWOL on Oct 31, 1863 muster roll; Farmer in Henry County after the war; Buried at Flat Shoals Primitive Baptist in Henry County; Widows pension applied for in Henry County in 1909; Reports her husband was present at surrender in Bentonville, NC.
James Samuel Skelton PVT 1832 7 Nov 1862
Married Mary Hall in Jackson County, Dec 7, 1856; Enlisted in Lawrenceville July 29, 1862; Died in Fairgrounds Hospital in Knoxville, TN; Buried at Bethel Cemetery in Knoxville, TN; Claim filed by widow Mary Skelton; Memorial stone at Harmony Grove Methodist in Auburn, Barrow County.
Absalom Franklin Smith PVT 28 Sep 1847 18 Feb 1924
Younger brother of John & Joseph Smith; 1864 enlistment; Patient at Ocmulgee Hospital in Macon July 16, 1864; Captured along Atlanta siege lines Aug 1864; Admitted to General Hospital, Camden Street, Baltimore, MD Jan 30, 1865; Escaped hospital May 23, 1865; Married Mae Steele in Copiah County, MS, Nov 5, 1870; Machinist in Tyler, Texas in 1880; Buried at Rose Hill Cemetery in Smith County, Texas; Widow's pension applied for in Smith County in 1924.
J.K. Smith PVT 22 Oct 1862
Died in Fairgrounds Hospital, TN.
John Henry Smith PVT 23 Mar 1842 10 Jan 1917
Captured at **Vicksburg**, MS July 4, 1863; Present on Dec 1863 muster roll; Captured as deserter along Atlanta siege lines Aug 4, 1864; Took Oath of Allegiance in Louisville, KY Oct 11, 1864; Agreed to remain north of the Ohio River; Married Sarah Doster in Clark County, Indiana , Aug 1, 1869; Farmer in Hall County in 1880 & Jackson County in 1900; Filed for Confederate pension in 1914; Reported escaping about Nov 17, 1864 while en route to Alton Prison. IN; Worked for Ohio Falls Car Works in New Albany, IN until end of war; Application endorsed by brother A.F. Smith; Buried at Zion Hill in Braselton.
Joseph W. Smith PVT 1844 24 Nov 1862
Brother of Absalom & John Smith; Died in Atlanta hospital; Claim filed by his father Absalom C. Smith.
Pinckney Smith PVT 1822 15 Mar 1869
Born in South Carolina; Married Louisa Gardner Wade in Hall County, Jan 21, 1844; Captured in Kentucky fall 1862; Paroled in Chattanooga, TN Jan 11, 1863; Wounded along Vicksburg siege lines June 1863; Captured at Hospital #3 in **Vicksburg**, MS July 4, 1863; Sick at home on Oct 31, 1863 muster roll; Among "retired men mustered for pay at post of Atlanta, Ga." June 30, 1864; Widow's pension filed in Fulton County in 1891; Reported that her husband died from the effects of his wound received during the Siege of Vicksburg in 1863.
Joshua Spillman PVT 1824 12 Apr 1885
Born in Louisiana; Married Frances Gurley in DeKalb County, Mar 13, 1851; Captured at **Vicksburg**, MS hospital July 4, 1863; Present on Dec 1863 muster roll; Deserted Confederate lines Aug 1864; Took Oath of Allegiance in Louisville, KY Oct 11, 1864; Agreed to remain north of the Ohio River; Carpenter in Atlanta after the war; Buried at Oakland Cemetery in Atlanta.

Frederick T. Stier PVT 8 Jan 1832 1 Nov 1903
Born in Rockingham County, NC; Captured at **Vicksburg**, MS July 4, 1863; Appointed **1CPL** Oct 1, 1863; Appointed **5SGT**; Surrendered at **Greensboro**, NC April 26, 1865; Married Susan Frances Jackson in Fulton County, Mar 1, 1866; House carpenter in Campbell County in 1870, in Newnan in 1880, & Atlanta in 1890; Member of UCV Camp #159 in Atlanta; Buried on Confederate Knoll at Westview Cemetery in Atlanta.

James Garrett Sullivan PVT 6 Mar 1837 12 Feb 1923
Born in DeKalb County; Left sick at Knoxville, TN hospital April 23, 1862; Wounded and captured at **Battle of Baker's Creek**, MS May 16, 1863; Present on Oct 31, 1863 muster roll; Captured at **Battle of New Hope Church**, GA May 25, 1864; Admitted to **Rock Island**, Il prison hospital Sep 16, 1864 with remittent fever; Enlisted in 3rd US Infantry for frontier service Oct 15, 1864; Married Mary Ellen McCutcheon about 1867; Grocer in Marietta in 1880; Produce peddler in Gainesville in 1900; Buried at Marietta City Cemetery in Cobb County.

Josiah E. Sullivan PVT 1835 8 Aug 1863
Born in Laurens County, SC; Older brother of James & Pleasant Sullivan; Captured at **Vicksburg**, MS July 4, 1863.

Pleasant Augustus Sullivan PVT 10 Apr 1843 27 Jan 1920
Captured at **Battle of Baker's Creek**, MS May 16, 1863; POW at **Camp Morton**, IN; Paroled at Fort Delaware, DE July 4, 1863; Present on Oct 31, 1863 muster roll; Captured at **Battle of Missionary Ridge**, TN Nov 25, 1863; POW at **Rock Island**, IL; Volunteered for frontier service but was rejected; Released from Rock Island Oct 27, 1864 after taking Oath of Allegiance; Married Susan Beavers in Fulton County, Dec 25, 1867; Farmer in Milton County after the war; Filed for Confederate pension in Milton County in 1904; Application endorsed by his brother J.G. Sullivan.

Moses Gerome Terry PVT 28 Jan 1842 27 Jun 1910
Born in Greene County, Alabama; Captured at **Vicksburg**, MS July 4, 1863; Present on Dec 1863 muster roll; Appointed **4CPL**; Admitted to Fairgrounds Hospital #1 in Atlanta June 24, 1864; Furloughed July 3, 1864; Wounded at **Battle of Nashville**, Dec 16, 1864; Admitted to Way Hospital in Meridian, MS with wound Feb 8, 1865; Furloughed Mar 21, 1865; Married Susan Rebecca Moore about 1870; Farmer in Moscow, MS after the war; Buried at Bethel Methodist in Kemper County, Mississippi.

Elijah Luther Turner PVT 16 May 1835 28 Apr 1919
Married Permelia Gregory in Newton County, Jan 8, 1857; Enlisted in Calhoun July 28, 1862; Present on Oct 31, 1863 muster roll; Surrendered at **Greensboro**, NC April 26, 1865; Married Martha Reagin in Rockdale County, July 18, 1877; Farmer in Rockdale County after the war; Filed for Confederate pension in Rockdale County in 1917; Application endorsed by W.P. Plunkett; Buried at McClung Family Cemetery in Conyers.

William J. Turner PVT 1825 2 Jan 1865
Born in North Carolina; Older brother of Elijah Turner; Married Jane Brawner in DeKalb County, Dec 25, 1853; Farmer near Conyers in 1860; Enlisted in Calhoun, GA July 28, 1862; Patient at General Hospital in Marion, MS June 30, 1863; Present on Dec 1863 muster roll; Wounded by shell fragment and captured at **Battle of Nashville**, TN Dec 16, 1864; Right arm amputated at shoulder Jan 2, 1865; Died at US General Hospital as a result of the surgery; Buried at Mount Olivet Cemetery in Nashville, Tennessee.

Edward W. Veal PVT 21 Mar 1821 21 Feb 1863
Married Martha Bell in Madison County, Mar 28, 1844; Miller in Hall County in 1860; Died at Roy Hospital in Atlanta; Buried at Oakland Cemetery in Atlanta; Claim filed by his widow Martha Veal.
Andrew Jackson Waldrop PVT 4 Apr 1837 24 May 1896
Captured as a deserter in Carroll County; Took Oath of Allegiance in Louisville, KY July 31, 1864; Agreed to remain north of the Ohio River; Married Elizabeth McClelland in Jasper County in 1866; Farmer in Carroll County in 1870 & Jasper County in 1880; Buried at Mount Vernon Baptist in Indian Springs, Butts County.
Benjamin Franklin Walker PVT 12 Dec 1835 1899
Married Sarah Medlock in Fulton County, Nov 22, 1855; Elected Junior **2LT** Mar 20, 1862; Captured at **Vicksburg**, MS July 4, 1863; Admitted to Fairgrounds Hospital #1 in Atlanta Oct 5, 1863 with fever; Returned to Duty Oct 19, 1863; Elected **2LT** Oct 11, 1863; Cashiered by court martial court March 29, 1864; Elected LT of Company A, 6th Georgia Militia in 1864; Paroled in Milledgeville May 25, 1865; Farmer in Bryants Crossroads in 1880; Civil engineer and surveyor for Fulton County; Left government to become important real estate developer in Atlanta; Member of UCV Camp #159 in Atlanta; Buried at Oakland Cemetery in Atlanta.
Green Jackson Watkins PVT 23 May 1835 15 Jun 1910
Born in Alabama; Married the widow Emeline Lucinda Mayson Mell in 1867; Appointed **3CPL** in 1863; Captured at **Vicksburg**, MS July 4, 1863; Present on Dec 1863 muster roll; Captured at **Battle of Bentonville**, NC Mar 19, 1865; Released from **Point Lookout**, MD June 22, 1865; Farmer & ferryman in Adamsville District after the war; Buried at Mayson Family Cemetery in Fulton County.
Jesse M. Weaver PVT 1829 19 Nov 1863
Married Martha Waters in Newton County, Feb 6, 1851; Clerk in Atlanta in 1860; Died at Institute Hospital in Atlanta.
Augustus Franklin Williams PVT 10 Mar 1821 7 Sep 1922
Lived 101 years; Born in Oglethorpe County; Married Amanda Harbor in Murray County, Feb 2, 1845; Transferred from Company D, 40th Georgia; Surrendered at **Greensboro**, NC April 25, 1865; Filed for Confederate pension in Houston County in 1905.
Joseph Ashley Williams PVT 1826 After 1880
Married Martha McGinnis in Cherokee County about 1850; Captured in Harrodsburg, KY; Taken by steamboat from Louisville to Vicksburg, MS and exchanged on Nov 29, 1862; Captured at **Vicksburg**, MS July 4, 1863; Present on Dec 1863 muster roll; Captured at **Battle of New Hope Church**, GA May 25, 1864; Released from **Rock Island**, IL prison Jun 18, 1865; Dairyman in Caseys District in 1880.
Asa "Acey" M. Willis PVT 1828 After 1870
Married Sarah Kirksey in Fulton County, Nov 20, 1858; Captured at **Vicksburg**, MS July 4, 1863; Admitted to Fairgrounds Hospital #1 in Atlanta Oct 2, 1863 with syphilis; Returned to duty Oct 16, 1863; Appointed **3SGT** Oct 1863; Sick at home on Dec 1863 muster roll; Captured at **Battle of Jonesborough**, GA Aug 31, 1864; Laborer in Paulding County in 1870.
William Joseph Wright PVT 3 May 1822 17 Feb 1910
Married Amanda Holcomb in Newton County, July 23, 1851; Discharged May 22, 1863; Farmer in Newton County after the war; Buried at Starrsville Cemetery in Newton

County.

Silas J. Yarborough PVT 1823 After 1880

Present on Dec 1863 muster roll; Captured at **Battle of Jonesborough** Aug 31, 1864; Farmer in Goshen, Clermont County, Ohio after the war.

Bibliography

PERIODICALS

Atlanta Constitution
Atlanta Georgian & News
Atlanta Historical Bulletin
Atlanta Journal
Atlantic Monthly
Blue & Gray Magazine (Columbus, OH)
Civil War Times Illustrated
Confederate Veteran
Daily Picayune (New Orleans, LA)
Daily South Carolinian (Columbia, SC)
Gwinnett County Historical Society Heritage Newsletter (GA)
Harper's Weekly
Ironton Register (OH)
Milwaukee Daily Sentinel
News & Observer (NC)
North Georgia Journal
North South Trader's Civil War (Orange, VA)
Poughkeepsie Daily Press
Register of the Kentucky Historical Society
Sacramento Steamer Union
Southern Banner (Athens, GA)
Southern Watchman (Athens, GA)f
Atlanta Southern Confederacy
Tennessee Historical Quarterly
Weekly News & Courier (Charleston, SC)
Weimar Mercury (TX)

PRIMARY SOURCES

Asbury, Rufus. War-time diary from September 17, 1862, to February 17, 1863. Part of James Blick Artrope Collection. Drawer 283, microfilm box 16. Georgia Archives, Morrow, GA. Cited as Asbury diary.

"Camp 52nd Georgia Regiment, 1864 March 30." America's Turning Point: Documenting the Civil War Experience. Box 14, Folder 8, E. Merton Coulter manuscript collection II. MS 2345. Hargrett Rare Book and Manuscript Library, University of Georgia Libraries, Athens, GA. http://dlg.galileo.usg.edu/id:dlg_turningpoint_harg2345-014-008.

Civil War Times Illustrated Collection of Soldiers' Correspondence and Diaries, 1861–1865. U.S. Army Heritage and Education Center, Carlisle, PA.

Clayton, Henry De Lamar. Military papers. W. S. Hoole Special Collections. University of Alabama Libraries, Tuscaloosa.

Confederate Prisoners Consolidated in Louisville, U.S., Civil War Prisoner of War Records, 1861–1865. National Archives and Records Administration (NARA), M598, roll 88.

Connor, Wesley Olin. Cherokee Artillery war diary. Wesley O. Connor Family Papers. Ms3102, box 1, folder 5. Hargrett Rare Books & Manuscript Library, University of Georgia Libraries, Athens, GA.

Confederate States of America prison (Madison, Ga.) records. Hargrett Rare Book and Manuscript Library, University of Georgia Libraries, as presented in the Digital Library of Georgia. http://dlg.galileo.usg.edu/harg/turningpoint/do:harg0455-001-001.

Daniel, George Hewitt. Letters. George Hewitt Daniel Collection. Ms 85f. James G. Kenan Research Center, Atlanta History Center, Atlanta, GA.

DeGress, Captain Francis. Report of Captain Francis DeGress to Major T. Maurice, July 22, 1864. Francis DeGress papers, Kenan Research Center, Atlanta History Center, Atlanta, GA.

Edgar, Hiram, to Mary Edgar. April 10, 1862. Transcribed letter posted by user n19lga on Hiram Edgar family tree on ancestry.com.

Edwards Gallery. *Dinner for President Theodore Roosevelt.* Sepia toned silver gelatin print. 10x14. Atlanta History Photograph Collection, Kenan Research Center, Atlanta History Center, http://album.atlantahistorycenter.com/cdm/singleitem/collection/athpc/id/858/rec/1 1905.

Elliott, Robert. Letters of Angus McLeod and John Nash. Private collection. Grayson, GA. Cited as Elliott collection.

Fowler, George Thornton. Letters. Davis-Quillin Family Collection. Ms 90f. James G. Kenan Research Center, Atlanta History Center, Atlanta, GA. Cited as Fowler papers.

Gourlie, John C. Letters. *Civil War Times Illustrated Collection of Soldiers' Correspondence and Diaries, 1861–1865.* U.S. Army Heritage and Education Center at Carlisle Barracks, Carlisle, PA.

Gunter, Ira Lawson. Letters. In private collection of James Latimer, Jonesboro, GA. Cited as Latimer collection.

Henderson, Robert. "Chickasaw Bayou Field Report of Col. Robert Henderson." Reproduced in Lee Alexander and James Bogle, "George Hewitt Daniel (1817–1864)." *Atlanta Historical Bulletin* 8/3 (September 1968): 29–31.

Houghton, Charles A. Letters. *Civil War Times Illustrated Collection of Soldiers' Correspondence and Diaries, 1861–1865.* U.S. Army Heritage and Education Center at Carlisle Barracks, Carlisle, PA.

Hoyle, Georgia. "An Address to the Fowler Guards." Handwritten copy made by Eliza Frances Davis Quillan. Kenan Research Library, Atlanta History Center, Atlanta, GA.

Hulsey, William. Report to General Clayton on the Operations of the 42nd Georgia, August 9, 1864. In military papers of General Henry De Lamar Clayton. W. S. Hoole Special Collections. University of Alabama Libraries.

Kellogg, Henry C., 43rd Georgia, Cassville, Georgia, to J. P. C. Whitehead, Stovall's Brigade, December 20, 1863. RG3.22 Charles Henry Hardin, 1875–1877. Missouri State Archives. http://cdm16795.contentdm.oclc.org/cdm/ref/collection/msamerge/id/12812.

Latimer, James. Letters of Ira Lawson Gunter. Private collection, Jonesboro, GA. Cited as Latimer collection.

"List of All Passengers Taken on Board the Steam Ship *Dan'l Webster.*" San Juan del Norte, June 1855. Passenger lists of vessels arriving at New Orleans, 1820–1902. M259, roll 42. National Archives and Records Administration, Washington, DC.

Maxwell, Singleton Allen. "A Journal of the 42nd GA Regt. & of S.A. Maxwell." 1862–1863. CWM, RG 57-1-1, box 20, location 3338-10. Georgia Archives, Morrow, GA. Cited as Maxwell journal.

"May 1, 1865, Final Surrender Documents of the 42nd Georgia" and "May 2, 1865, Final Surrender Documents of the 40th Georgia." Muster Rolls and Lists of Confederate Troops Paroled in North Carolina. M1781, roll 3. National Archives and Records Administration, Washington, DC.

McLeod, Angus, to Griffin Vaughn. January 20, 1863. In the private collection of Robert Elliott, Grayson, GA. Cited as Elliott collection.

McNab, Laura J. "L. P. Thomas Chapter, U.D.C.," Norcross, Ga." *Confederate Veteran* (January 1904): 15.

Nash, John T., to Robert Bolt Nash. March 25, 1863. In the private collection of Robert Elliott, Grayson, Georgia. Cited as Elliott collection.

Petty, E. T. "The Cumberland Gap Campaign." In *The* National Tribune *Scrap Book: Stories of the Camp, March, Battle, Hospital and Prison Told by Comrades,* 119–21. Washington, DC: The National Tribune, n.d. Copy available at the U. S. Army War College Library and Archives, Carlisle, PA.

Phillips, Almon. Letters. Company C, 42nd Regiment Georgia Volunteers. George Thornton Fowler Collection, ac64-125, mf 154, drawer 15, box 72. Georgia Archives, Morrow, GA.

Rainey, Harvey Asberry. Letters. Company C, 42nd Regiment Georgia Volunteers folder. CWM, R6 57-1-1, box 7, location 3342-10. Georgia Archives, Morrow, GA. Cited as Rainey letters.

Reynolds, Edward H. Diary. *Civil War Times Illustrated Collection of Soldiers' Correspondence and Diaries, 1861–1865.* U.S. Army Heritage and Education Center at Carlisle Barracks, Carlisle, PA

Robinson, Private Henry W. Letters. Company B, 42nd Georgia. In Henry W. Robinson Letters, 1862–1863, manuscript collection 392. Stuart A. Rose Manuscript, Archives, and Rare Book Library, Emory University, Atlanta, GA. Cited as Robinson letters.

Shrimp, William T. Letters. *Civil War Times Illustrated Collection of Soldiers' Correspondence and Diaries, 1861–1865*. U.S. Army Heritage and Education Center at Carlisle Barracks, Carlisle, PA.

Stovall, Anna M. Notebook. Southern Historical Collection. Papers of Marcellus Stovall. Wilson Library, University of North Carolina, Chapel Hill, NC.

Stovall, Marcellus. Marcellus Augustus Stovall scrapbooks, 1861–1863, MS 1047. 3 vols. Georgia Historical Society, Savannah, GA.

Stovall, Marcellus. Unpublished autobiography. 1868. Marcellus Stovall Papers, 85-01-A, B, C. Reese Library Archives, Augusta State University. Cited as Stovall papers, ASU.

Stovall, Lewis. Letters. 43rd Georgia. Stovall Family file. Forsyth County Historical Society, Cumming, GA.

Taylor, Thomas Thomson. Papers. 47th Ohio. 1861–1865. Manuscript Coll. No. 354. Stuart A. Rose Manuscript, Archives, and Rare Book Library, Emory University, Atlanta, GA. Cited as Taylor papers.

Thomas Family folder. Gwinnett Historical Society, Lawrenceville, GA. Cited as Thomas Family folder, GHS.

Thomas, Lovick P. Collected papers. 4 vols. Bell Research Center, Cumming, GA. Cited as Thomas collection, BRC.

Thomas, Lovick P. Papers. In the private collection of Dr. Vick Thomas, Santa Fe, New Mexico. Cited as Thomas papers, Santa Fe.

Thomas, Lovick P. Papers (includes two scrapbooks), 1901–1910. Mss 700f. Kenan Research Center, Atlanta History Center. Cited as Thomas papers, AHC.

Weaver, Ben. Letters. Company B, 42nd Georgia. Claudia Weaver Collection, drawer 56, box 64. Georgia Archives, Morrow, GA. Cited as Weaver letters.

Wilson, Billington A. Letters. Company G, 42nd Georgia. United Daughters of the Confederacy Bound Transcripts. Vol. 6. Georgia Archives, Morrow, GA.

BOOKS, DISSERTATIONS, JOURNALS, AND WEBSITES

Alexander, Lee, and James Bogle. "George Hewitt Daniel (1817–1864)." *Atlanta Historical Bulletin* 8/2 (September 1968): 19–52.

Allardice, Bruce S. *Confederate Colonels: A Biographical Register*. Columbia: University of Missouri Press, 2008.

———. *More Generals in Gray*. Baton Rouge: Louisiana State University Press, 2006.

Ambrose, Stephen. "The Battle of Nashville: Fighting Gets Hotter." *Civil War*

Times Illustrated 3/8 (December 1964): 31.

———. "Struggle for Vicksburg." *Civil War Times* (July 1967): 54.

Armistead, Zachariah Jabez. "Letters from Zachariah Armistead." In vol. 6 of *Bible Records, Military Posters & Reminiscences of Confederate Soldiers*. Atlanta: Georgia Division, United Daughters of the Confederacy, 1942.

Atlanta Businessmen's League. *The Proposed Atlanta National Military Park: A Brief Outline of the Project*. Atlanta: Businessmen's League, 1900.

Bailey, George W. *A Private Chapter of the War*. St. Louis, MO: G. I. Jones & Company, 1880.

Barnes, J. D. *What I Saw You Do: A Brief History of the Battles, Marches, and Sieges of the 20th Iowa Volunteer Infantry, during their three years' of active service in the War of the Rebellion*. Port Bryan, IL: Owen & Hall, 1896.

Barnwell, V. T. *Atlanta City Directory of 1867*. Compiled for the *Atlanta Intelligencer*, 1867.

Baxley, Neil. *Walk in the Light: The Journey of the 10th and 19th South Carolina Volunteer Infantry*. The South Carolina Regimental-Roster Set. Wilmington, NC: Broadfoot Publishing, 2013.

Bearss, Edwin Cole. *The Campaign for Vicksburg*. 3 vols. Dayton, OH: Morningside House, Inc., 1985.

Bell, Hiram Parks. *Men & Things: Being Reminiscent, Biographical and Historical*. Atlanta, GA: Foote & Davies, 1907.

Benton, Charles E. *As Seen from the Ranks: A Boy in the Civil War*. New York: G. P. Putnam's Sons, 1902.

Blount, Russell W. *The Battles of New Hope Church*. Gretna, LA: Pelican Publishing Company, 2010.

Bogle, Joseph, and William L. Calhoun. *Historical Sketches of Barton's (later Stovall's) Georgia Brigade, Army of Tennessee, C.S.A.*, introduced and edited by William Stanley Hoole and Martha DuBose Hoole. Confederate Regimental Series, no. 5. University, AL: Confederate Publishing Company, 1984.

Bohannon, Keith S. "The Northeast Georgia Mountains during the Secession Crisis and Civil War." PhD diss. Pennsylvania State University, 2001.

Bond, Judy, ed. *Civil War Letters: Written by William Trimble, George Trimble, John Trimble*, Conyers, GA: Privately printed. A copy can be found in the Conyers Public Library.

Bradley, Mark L. *Last Stand in the Carolinas: The Battle of Bentonville*. Campbell, CA: Savas Woodbury Publishers, 1996.

———. *This Astounding Close: The Road to Bennett Place*. Chapel Hill: University of North Carolina Press, 2000.

———, ed. "'Two Confederate Views of Bentonville': The Official Report of Maj. Gen. Henry D. Clayton and Reminiscences of Col. Henry D. Bunn." *Civil War Regiments: A Journal of the American Civil War* 6 (1998): 93–106.

Bragg, C. L. *Distinction in Every Service: Brigadier General Marcellus A. Stovall,*

C.S.A. Shippensburg, PA: White Mane Books, 2002.

Bragg, William Harris, and William R. Scaife. *Joe Brown's Army: The Georgia State Line, 1862–1865*. Macon: Mercer University Press, 1987.

Brewer, George E. "Incidents of the Retreat from Nashville." *Confederate Veteran* 18/7 (July 1910): 327.

Broadwater, Robert P. *Battle of Despair: Bentonville and the North Carolina Campaign*. Macon: Mercer University Press, 2004.

Brooke, Ted O. *Into the Gray Vale: Registers of the Georgia Confederate Soldier's Home, Atlanta, Georgia, 1901–1941*. Cumming, GA: Ted Brooke, 2012.

Brown, James Madison. "James Madison Brown's 1925 Recollections." In *Wayfarers in Walton: A History of Walton County, Georgia 1818–1967*, edited by Anita B. Sams. Doraville, GA: Foote & Davies, 1967.

Bull, Rice C. *Soldering: The Civil War Diary of Rice C. Bull, 123rd New York Volunteer Infantry*. Edited by Jack Bauer. San Rafael, CA: Presidio Press, 1995.

Bunch, Jack A. *Roster of the Courts-Martial in the Confederate States Armies*. Shippensburg, PA: White Mane Books, 2001.

Calhoun, William Lowndes. *History of the 42d Regiment, Georgia Volunteers, Confederate States Army, Infantry*, Atlanta, GA: 1900.

"Calhoun, W. L. Biography." Calhoun Family Papers, 1834–1960. MSS 50, 1–3. James G. Kenan Research Center Manuscript Collection, Atlanta History Center, Atlanta, GA.

Campbell, Henry. *A Hoosier's Civil War Diary, Bugler, 18th Indiana Battery*. National Park Service, 1961.

Carnes, F. G. "'We Can Hold Our Ground': Calvin Smith's Diary." *Civil War Times Illustrated* 24 (April 1985): 24–31.

Castel, Albert. *Decision in the West*. Lawrence: University Press of Kansas, 1992.

Cathey, M. Todd, and Gary W. Waddey. *"Forward My Brave Boys!" A History of the 11th Tennessee Volunteers Infantry*. Macon, GA: Mercer University Press, 2016.

Compiled Service Records of Confederate Soldiers from Georgia Units. M266. National Archives and Records Administration, Washington, D.C.

Colton, Abraham Moses. Letters. *Letters from Confederate Soldiers 1860–1865*. Bound typescripts. Vol. 2. Georgia Division, United Daughters of the Confederacy, 1940.

Comstock, Daniel W. *Ninth Cavalry: One Hundred and Twenty-first Regiment*. Richmond, IN: J. M. Coe, 1890.

Cozzens, Peter. *The Shipwreck of Their Hopes: The Battle of Chattanooga*. Urbana: University of Illinois Press, 1994.

Cunyus, Lucy Josephine. *History of Bartow County, Georgia*. 1933. Easley, SC: Southern Historical Press, 1983.

Daniel, Larry J. *Soldiering in the Army of Tennessee*. Chapel Hill: University of North Carolina Press, 1991.

Davidson, Captain J. H. "Battle of Tazewell, Tennessee, August 6, 1862, As Described by Captain J. H. Davidson, 14th Kentucky Volunteer Infantry," comp. Michael Wood, 16th Ohio Volunteer Infantry website. http://www.mkwe.com/ohio/pages/H011-04.htm.

Davis, Stephen. "Sherman in North Georgia: The Battle of Resaca." *Blue & Gray* 31/4 (Summer 2015): 6–28, 42–50.

———. *What the Yankees Did to Us: Sherman's Bombardment and Wrecking of Atlanta.* Macon: Mercer University Press, 2012.

Davis, William C. "Cumberland Gap, 1862." *Civil War Times Illustrated* 10/6 (October 1971): 26.

Dodwell, James. "Bouton's Battery." In *Events of the Civil War*, by Edward Bouton, 85–87. Los Angeles: Moles & Collins, 1906.

Dorward, L. L. *Deep Running Roots, Far-Reaching Branches: The Story of the City of Oxford, Georgia.* History pamphlet. Oxford, GA: n.p., 2014.

Douthit, John. "Johnny Reb's Impressions of Kentucky in the Fall of 1862." Edited by N. Fowler. *The Register of the Kentucky Historical Society* 48/164 (July 1950): 205–15.

Dunkerly, Robert M. *The Confederate Surrender at Greensboro: The Final Days of the Army of Tennessee, April 1865.* Jefferson, NC: McFarland & Co., 2013.

Ecelbarger, Gary. *The Day Dixie Died: The Battle of Atlanta.* New York: St. Martin's Press, 2010.

Edwards, Dan, ed. "Lt. William Drennan, Ordinance Officer of Featherston's Brigade: Letters to His Wife, Georgia Torrey Drennan." http://www.battleofchampionhill.org/drennan.htm.

Elliott, Sam Davis. "'I Regard Maj. Gen. Stewart as the Best Qualified of the Maj. Genls of This Army': Alexander P. Stewart in the First Phase of the Atlanta Campaign." In *Confederate Generals in the Western Theater*, vol. 2. Edited by Lawrence Bergeron & Arthur Hewitt, 133–57. Knoxville: University of Tennessee Press, 2010.

———. *Soldier of Tennessee: General Alexander P. Stewart and the Civil War in the West.* Baton Rouge: Louisiana State University Press, 1999.

Erquitt, Bill. "Cotton Mill Prison Canteen Revisited: Prison Log and a Site Excavation Shed New Light." *North South Trader's Civil War* 38/2 (March 2014): 26–37.

Evans, Clement Anselm. *Confederate Military History: Extended Edition* (1899). 19 vols. Wilmington, NC: Broadfoot Publishing, 1987.

Evans, David. *Sherman's Horseman: Union Cavalry Operations in the Atlanta Campaign.* Bloomington: Indiana University Press, 1999.

Force, Manning F. "Personal Recollections of the Vicksburg Campaign: A Paper Read before the Ohio Commandery of the Military Order of the Loyal Legion of the United States." In vol. 1 of *Military Order of the Loyal Legion of the United States* (MOLLUS) (1888). Wilmington, NC: Broadfoot Publishing,

1999.

Freeman, Henry V. "A Colored Brigade in the Campaign and Battle of Nashville." In vol. 2 of *Military Order of the Loyal Legion of the United States* (MOLLUS) (1888). Wilmington, NC: Broadfoot Publishing, 1999.

Garrett, Franklin M. *Atlanta and Environs: A Chronicle of Its People and Events.* 2 vols. Athens: University of Georgia Press, 1954.

Georgia Capital Museum. *Hallowed Banners: Historic Flags in the Georgia Capitol Collection.* Atlanta: Georgia Capitol Museum, 2005.

Gibson, R. L. "Report of Brigadier-General R. L. Gibson of Operations in Vicinity of Nashville, January 11, 1865." *Southern Historical Society Papers* 4/3 (July–December 1877): 142–43.

Glatthaar, Joseph T. *General Lee's Army: From Victory to Collapse.* New York: Free Press, 2008.

Goff, John H. "Edward Lloyd Thomas, Surveyor." *Emory University Quarterly* (Summer 1962).

Goodson, Gary Ray. *Georgia Confederate 7,000.* 4 vols. Shawnee, CO: Goodson Enterprises, 1997.

———. *Letters and Diaries from the Confederate Soldiers of Brigadier General Seth M. Barton & Marcellus A. Stovall's Georgia's Infantry Brigade.* Volume 2 of *The Georgia Confederate 7,000.* Shawnee, CO: Goodson Enterprises, 1997.

Governor's Log, April 8, 1862. Telegraphs between Colonel Robert Henderson at Camp McDonald and Governor Joseph E. Brown at Milledgeville. Governor's Incoming Correspondence, Civil War, RG 1-1-5. Georgia Archives, Morrow, GA.

Grabau, Warren. *Ninety-Eight Days: A Geographer's View of the Vicksburg Campaign.* Knoxville: University of Tennessee Press, 2000.

Gresham, James Monroe. "A Partial History of the Walton Tigers." In *Wayfarers in Walton: A History of Walton County, Georgia 1818–1967*, edited by Anita B. Sams, 660–64. Doraville, GA: Foote & Davies, 1967.

———. "Prison Life at Johnson's Island, Ohio." In *Wayfarers in Walton: A History of Walton County, Georgia 1818–1967*, edited by Anita B. Sams, 427–29. Doraville, GA: Foote & Davies, 1967.

Groom, Winston. *Shrouds of Glory: From Atlanta to Nashville: The Last Great Campaign of the Civil War.* New York: Atlantic Monthly Press, 1995.

Hafendorfer, Kenneth. *Perryville: Battle for Kentucky.* Louisville, KY: K. H. Press, 1991.

Hansard, Mary A. *Old Time Tazewell.* Sweetwater, TN: Mary Lorena Hansard Wilson, 1979.

Harris, Bonnie P. *The Confederate Hospitals of Madison, Georgia: Their Records & History.* Saline, MI: McNaughton & Gunn, 2014.

Hattaway, Herman. *General Stephen D. Lee.* Oxford: University Press of Mississippi, 1976.

Henderson, Lillian. *Roster of the Confederate Soldiers of Georgia, 1861–1865.* 6 vols. Hapeville, GA: Longino & Porter, 1960.

Henderson, Luana. *Inventory of United Confederate Veterans Records.* Baton Rouge: Louisiana University Press, 1996.

Herring, Dorothy Holland. *Company A of the Fortieth Georgia Infantry in the Confederate Service.* Westminster, MD: Willow Bend Books, 2000.

Hess, Earl J. *Kennesaw Mountain: Sherman, Johnston and the Atlanta Campaign.* Chapel Hill, NC: University of North Carolina Press, 2013.

———. *The Battle of Ezra Church and the Struggle for Atlanta.* Chapel Hill, NC: University of North Carolina Press, 2015.

Hewett, Janet, et al., eds. *Supplement to the Official Records of the Union and Confederate Armies.* 100 vols. Wilmington, NC: Broadfoot Pub. Co., 1994–2001.

Hoehling, A. A. *Vicksburg: 47 Days of Siege.* New York, NY: Fairfax Press, 1969.

Holman, Tyler. "Factory Town: A History of the Lawrenceville Manufacturing Company." Gwinnett Historical Society. *The Heritage* 44/3 (Fall 2015): 58–65.

Hood, John Bell. *Advance & Retreat; Personal Experiences in the United States and Confederate States Armies.* Philadelphia: Burk & Fetridge, 1880.

Hood, Stephen M. *The Lost Papers of Confederate General John Bell Hood.* El Dorado Hills, CA: Savas Beatie LLC, 2015.

Hopkins, Owen J. *Under the Flag of the Nation: Diaries and Letters of a Yankee Volunteer in the Civil War.* Edited by Otto Bond. Columbus: Ohio State University Press, 1961.

Horn, Stanley F. *Tennessee's War: 1861–1865 Described by Participants.* Nashville: Tennessee Civil War Centennial Commission, 1965.

———. *The Decisive Battle of Nashville.* Baton Rouge: University of Louisiana Press, 1956.

Howard, Oliver Otis. "The Battles about Atlanta." *Atlantic Monthly* 37 (October 1876): 395.

Hoyle, Georgia. "An Address to the Fowler Guards." Handwritten copy made by Eliza Frances Davis Quillan. Kenan Research Library, Atlanta History Center, Atlanta, GA.

Huff, Frederick Ware, comp. *Four Families: Winn, Thomas, Ware, Garrett of the Southern United States from 1600s to 1993.* Kennesaw, GA: F. W. Huff, 1993.

Hughes, Nathaniel C., and Gordon D. Whitney. *Jefferson Davis in Blue: The Life of Sherman's Relentless Warrior.* Baton Rouge: Louisiana State Press, 2006.

———. *Bentonville: The Final Battle of Sherman & Johnston.* Chapel Hill: University of North Carolina Press, 1996.

Ingmire, Frances, and Carolyn Ericson. *Confederate P.O.W's Buried in Northern Cemeteries* (Washington: National Archives, 1984).

International Cotton Exposition of Products, Machinery, and Manufacturers. Prospectus. Atlanta, GA: Jas. P. Harrison & Co., 1881.

Jenkins, Robert D. *The Battle of Peach Tree Creek: Hood's First Sortie, 20 July 1864.* Macon, GA: Mercer University Press, 2013.

Johnson, David E. *Douglas Southall Freeman.* Gretna, LA: Pelican Publishing, 2002.

Johnston, John. "Cavalry of Hood's Left at Nashville." *Confederate Veteran* 13/1 (January 1905): 28–30.

Johnston, John Warfield. "Joseph W. Anderson, B.L." *The University Memorial: Biographical Sketches of Alumni of the University of Virginia Who Fell in the Confederate War.* Edited by Rev. John Lipscomb Johnson. Baltimore, MD: Turnbull Brothers, 1871.

Jones, George H. "George H. Jones Recollections, Submitted by L. P. Thomas Chapter 320, UDC, Norcross, Georgia." In vol. 18 of *Confederate Reminiscences and Letters.* Atlanta: Georgia Division of United Daughters of the Confederacy, 2002.

Kibler, James Everett Jr. *The Classical Origins of Southern Literature.* McClellanville: Abbeville Institute Press, 2017.

Kimball, Charles B. *History of Battery "A" First Illinois Light Artillery Volunteers.* Chicago, IL: Cushing Printing Company, 1899.

King, Augusta Wylie. "International Cotton Exposition." *Atlanta Historical Bulletin* 4/18 (July 1939): 186–88.

Kinsland, William S. "A Band of Brothers; The Men and the Legends of the 52nd Georgia Regiment." *North Georgia Journal* 2/2 (Summer 1985): 8–30.

Knight, James R. *Hood's Tennessee Campaign: The Desperate Venture of a Desperate Man.* Charleston, SC: History Press, 2014.

Lane, Mills, ed. *"Dear Mother: Don't Grieve about Me. If I Get killed, I'll Only Be Dead": Letters from Georgia Soldiers in the Civil War.* Savannah, GA: Beehive Press, 1990.

Lee, General Stephen D. "From Palmetto, GA, to Defeat at Nashville." 1876 account. *Confederate Veteran* 16/6 (June 1908): 257.

"Lieutenant S. E. M. Underhill, Volunteer Aide-de-camp of Brig. Gen. S. D. Lee, Account of the Vicksburg Campaign, May–June 1863, Letter to Mother." *Guernsey Times* (Cambridge, OH), August 20, 1863, 1. Courtesy Dan Masters. http://www.battleofchampionhill.org/underhill-2.htm.

Losson, Christopher. *Tennessee's Forgotten Warriors: Frank Cheatham and His Confederate Division.* Knoxville: University of Tennessee Press, 1989.

Manigault, Arthur Middleton. *A Carolinian Goes to War: The Civil War Narrative of Arthur Middleton Manigault.* Edited by R. Lockwood Tower. Columbia: University of South Carolina Press, 1983.

Massey, R. J. "William Lowndes Calhoun." In volume 3 of *Men of Mark in Georgia,* edited by William F. Northen. Atlanta, GA: A. B. Caldwell, 1911.

Mathisen, Robert R. *The Routledge Sourcebook of Religion and the American Civil War: A History in Documents.* New York: Routledge, 2015.

Maxwell, Singleton Allen. "A Journal of the 42nd GA Regt. & of S.A. Maxwell." 1862–1863. Georgia Archives, CWM, RG 57-1-1, box 20, location: 3338-10. Cited as Maxwell journal.

McCurry, Richard M. *Two Great Rebel Armies: An Essay in Confederate Military History*. Chapel Hill: University of North Carolina Press, 1989.

McDaniel, Henry. *With Unabated Trust: Major Henry McDaniel's Love Letters from Confederate Battlefields as Treasured in Hester McDaniel's Bonnet Box*. Edited by Anita B. Sams. Monroe, GA: Historical Society of Walton County, 1977.

McDonough, James Lee. *Nashville: The Western Confederacy's Final Gamble*. Nashville: University of Tennessee Press, 2014.

———. *William Tecumseh Sherman: In the Service of my Country*. New York: W. W. Norton & Company, 2016.

McInvale, Morton R. *The Battle of Pickett's Mill: "Foredoomed to Oblivion."* Georgia Department of Natural Resources, 1977.

McKnight, Brian D. *Contested Borderland: The Civil War in Appalachian Kentucky and Virginia*. Lexington: University of Kentucky Press, 2006.

McMorries, Edward Young. *History of the First Regiment, Alabama Volunteer Infantry, C.S.A.* Montgomery, AL: Brown Printing Company, 1904.

McMurry, Richard M. *Atlanta 1864: Last Chance for the Confederacy*. Lincoln: University of Nebraska Press, 2000.

McNab, Laura J. "L. P. Thomas Chapter, U.D.C., Norcross, GA." *Confederate Veteran* 12/1 (January 1904): 15.

Meeks, Beth Thomas, and Bonnie Speer. *Heck Thomas, My Pap*. Norman, OK: Levite of Apache, 1988.

Memoirs of Georgia: Containing Historical Accounts of the State's Civil, Military, Industrial, and Professional Interests, and Personal Sketches of Many of its People. 2 vols. Atlanta: Southern Historical Association, 1895.

Military Order of the Loyal Legion of the United States (MOLLUS). 70 vols. Wilmington, NC: Broadfoot Pub. Co., 1991–1997.

Minutes of the Eighth Annual Meeting & Reunion of the United Confederate Veterans (1899). New Orleans: Hopkins' Printing Office, 1899.

Minutes of the Seventh Annual Meeting and Reunion of the United Confederate Veterans (1898). New Orleans: Hopkins' Printing Office, 1899.

Morgan, Thomas J. "Reminiscences of Service with Colored Troops in the Army of the Cumberland." In vol. 5 of *Military Order of the Loyal Legion of the United States* (MOLLUS) (1885). Wilmington, NC: Broadfoot Publishing, 1999.

Morhous, Henry C. *Reminiscences of the 123rd New York: Giving a Complete History of its Three Years Service in the War*. Greenwich, NY: Peoples Journal Book & Job Office, 1879.

Mullins, Michael A. *The Fremont Rifles, A History of the 37th Illinois Veteran Volunteer Infantry*. Wilmington, NC: Broadfoot Publishing, 1990.

Nash, Francis H. *Civil War Diary of Francis H. Nash*. Groesbeck, TX: Frank Bryan,

1956.

Newton County Historical Society. *History of Newton County, Georgia*. Covington: Newton County Historical Society, 1988.

Nisbet, James Cooper. *Four Years on the Firing Line*. Chattanooga, TN: The Imperial Press, 1914.

Nourse, Captain Henry. "The Atlanta Campaign." In *The Story of the Fifty-Fifth Regiment Illinois Volunteer Infantry in the Civil War 1861–1865*. Compiled by a Committee of the Regiment. Clinton, MA: W. J. Coulter, 1887.

O'Kelley, Harold Ernest. *Dahlonega's Blue Ridge Rangers in the Civil War*. Gainesville: Georgia Printing Company, 2002.

"Passengers: San Francisco: 1800s." Maritime Heritage Project—San Francisco 1846–1899. https://www.maritimeheritage.org/passengers/Clarissa-Andrews-22May1852.html.

Patton, J. T. "Personal Recollections of Four Years in Dixie." In vol. 1 of *Military Order of the Loyal Legion of the United States* (MOLLUS) (1893). Michigan War Papers, vol. 1. Wilmington, NC: Broadfoot Publishing, 1999.

Pierce, J. Kingston. *America's Historical Trails with Tom Bodett*. San Francisco: kQED Books, 1997.

Pioneer Citizen's Society of Atlanta. *Pioneer Citizen's History of Atlanta*. Atlanta: Byrd Printing Co., 1902.

Pochem, David. "Ain't No Company 'J' in This Man's Army." Pochefamily.org/Books/No%20Company%20J.html. Accessed November 3, 2019.

Provine, William B. "The Legend of 'Long Tom' at Cumberland Gap." *Tennessee Historical Quarterly* 24/3 (Fall 1965): 256–64.

Reagan, Alice E. *H. I. Kimball: Entrepreneur*. Atlanta: Cherokee Publishing Company, 1983.

Richards, Leonard. *The California Gold Rush and the Coming of the Civil War*. New York: Alfred Knopf, 2007.

Ridley, Bromfield L. *Battles and Sketches of the Army of Tennessee* (1906). Dayton, OH: Morningside Bookshop, 1978.

Robertson, Jno., comp. *Michigan in the War*. Lansing, MI: W. S. George & Co., State Printers & Binders, 1882.

Rodgers, Robert L. *History of the Confederate Veterans Association of Fulton County*. Atlanta: V. P. Sisson, 1890.

———. *Who Captured the DeGress Battery? Report of Robert L. Rodgers, Historian to Atlanta Camp No. 159, U. C. V. on the Capture of the DeGress Battery and Battery A, 1st Ill. Light Artillery, in the Battle of Atlanta, July 22d, 1864, with Other Papers Bearing Thereon*. Atlanta, GA: Some Survivors of Manigault's Brigade, 1898. Copy can be found in the archives of Kenan Research Center, Atlanta History Center, Atlanta, Georgia.

Rogers, Ian Michael W. "A Missed Opportunity: Preservation at Peachtree Creek."

American Battlefield Trust, https://www.civilwar.org/learn/articles/missed-opportunity.

Rowland, Grattan Whitehead. *Charles Alden Rowland 1833–1902: Diary for the Year 1859 and Series of Letters to His Wife Catharine Barnes Whitehead May 2, 1861 to May 3, 1865.* Augusta, GA: Privately published, 1975.

Russell, James M. "Atlanta, Gate City of the South, 1847 to 1885." PhD diss. Princeton University, 1971.

Salling, Stuart. *Louisianians in the Western Confederacy: The Adams-Gibson Brigade in the Civil War.* Jefferson, NC: McFarland & Company, 2010.

Sams, Anita B., ed. *Wayfarers in Walton: A History of Walton County, Georgia 1818–1967.* Doraville, GA: Foote & Davies, 1967.

———. *With Unabated Trust: Major Henry McDaniel's Love Letters from Confederate Battlefields as Treasured in Hester McDaniel's Bonnet Box.* Monroe: Historical Society of Walton County, 1977.

Sanders, Stuart W. *Maney's Confederate Brigade at the Battle of Perryville.* Charleston, SC: The History Press, 2014.

Scribner, Theo T. *Indiana's Roll of Honor.* Volume 2. Indianapolis: A. D. Streight Publishing, 1867.

Secrist, Philip L. *The Battle of Resaca: Atlanta Campaign 1864.* Macon, GA: Mercer University Press, 1998.

Shaffer, Michael K. "History of Camp McDonald." Friends of Camp McDonald Park. http://www.campmcdonaldpark.org/history.html.

Shavin, Norman, and Bruce Galphin. *Atlanta: Triumph of a People.* Atlanta: Capricorn Corp., 1982.

Sherman, William T. *Memoirs of William T. Sherman.* 2nd ed., revised and corrected. 2 vols. New York: D. Appleton and Company, 1886.

Sholes, A. E. *Sholes' Directory of the City of Atlanta for 1877.* Atlanta, GA: Sunny South Publishing, 1877.

Shoup, Francis Asbury. "Works at the Chattahoochee River." *Confederate Veteran* 3 (September 1895): 262–65.

Smedland, William S. *Camp Fires of Georgia's Troops, 1861–1865.* Kennesaw: Kennesaw Mountain Press, 1994.

Smith, Calvin Morgan. *A Smith in Service: Diaries of Calvin Morgan Smith, 1847–1864.* Edited by Leland L. Smith. Rogersville, TN: Hawkins County Genealogical Society, 2000.

Smith, Derek. *In the Lion's Mouth: Hood's Tragic Retreat from Nashville, 1864.* Mechanicsburg, PA: Stackpole Military History Series, 2011.

Smith, Timothy B. *Champion Hill: Decisive Battle for Vicksburg.* New York: Savas Beatie, 2004.

Sokolosky, Wade, and Mark A. Smith. *"To Prepare for Sherman's Coming": The Battle of Wise's Forks, March 1865.* El Dorado Hills, CA: Savas Beatie, 2015.

Stevenson, B. F. *"Cumberland Gap," A Paper Read before the Ohio Commandery of the Military Order of the Loyal Legion of the United States*. Cincinnati: H. C. Sherick & Co., 1885.

Strayer, Larry M., and Richard A. Baumgartner, eds. *Echoes of Battle: The Atlanta Campaign: An Illustrated collection of Union and Confederate Narratives*. Huntington, WV: Blue Acorn Press, 1991.

Taylor, William D. *"A fit representation of pandemonium": East Tennessee Confederate Soldiers in the Campaign for Vicksburg*. Macon, GA: Mercer University Press, 2008.

Thomas, Lovick Pierce. "The Fight at Bentonville." In *Southern Historical Society Papers* 29 (1901): 215–22. Edited by Robert Alonzo Brock. Millwood, NY: Kraus Reprint Company, 1978.

———. "'Reminiscent': An Interesting Letter from Col. L. P. Thomas to Capt. Mitchell." Green County newspaper clipping, 1896. MSS 240F. Joshua M. Mitchell Collection, James G. Kenan Research Center Manuscript Collection, Atlanta History Center, Atlanta, GA.

———. "Reminiscences of Forty-Second Georgia." *Confederate Veteran* 12/1 (January 1904): 14–15.

Thornton, G. B., comp. "List of casualties in Clayton's Division, Army of Tennessee, in the Battle before Franklin, Tenn., November 30, 1864." Loose report in Tennessee Archives of Surgeon Gustavus Brown Thornton of Clayton's Division.

United Daughters of the Confederacy, Georgia Division. *Bible Records, Military Posters & Reminiscences of Confederate Soldiers*. 14 volumes of bound typescripts. 1940. Georgia Department of Archives and History, Morrow, GA. Cited as UDC collection.

United States. War Department. *The War of the Rebellion: A Compilation of the Official Records of the Union and Confederate Armies*. 128 volumes. Washington, D.C., 1880–1901. Cited as *OR*.

Van Wyck, Richard T. *A War to Petrify the Heart: The Civil War Letters of a Dutchess County, N.Y. Volunteer*. Edited by Virginia Hughes Kaminsky. Hensonville, NY: Black Dome Press, 1997.

Wade, Gregory L. "Historian recounts Hood's retreat on 149th anniversary." Franklin (TN) Home Page. https://franklinhomepage.com/historian-recounts-hoods-retreat-on-149th-anniversary/.

Walker, Cornelius Irvine. "1898 Report submitted by Col. C.I. Walker." In Rodgers, *Who Captured the DeGress Battery?* Pages 6–11. Atlanta, GA: Some Survivors of Manigault's Brigade, 1898. Copy can be found in the archives of Kenan Research Center, Atlanta History Center, Atlanta, GA.

———. *Rolls & Historical Sketches of the Tenth Regiment, South Carolina Volunteers*. Charleston, SC: Walker, Evans, & Gogswell, 1881.

Walker, Henry James, Jr. "Henry Delamar Clayton: Secessionist, Redeemer, Soldier, Redeemer." PhD diss. University of Alabama, 1995.

Warner, Ezra J. *Generals in Blue: Lives of the Union Commanders*. Baton Rouge: Louisiana State University, 1964.

———. *Generals in Gray: Lives of the Confederate Commanders*. Baton Rouge: Louisiana State University, 1959.

Watkins, Sam R. *"Co. Aytch": A Side Show of the Big Show* (1882). Wilmington, NC: Broadfoot Publishing, 1994.

———. *"Co. Aytch," Maury Grays, First Tennessee Regiment* (1882). Wilmington, NC: Broadfoot Publishing, 1992.

Welsh, Jack D. *Two Confederate Hospitals and Their Patients: Atlanta to Opelika*. Macon: Mercer University Press, 2005.

Werstein, Irving. *Kearny the Magnificent: The Story of General Philip Kearny*. New York: John Day Company, 1962.

Whatley, John C. *The Typical Georgia Confederate: Compiled from His Written Correspondence to the Folks Back Home and from His Post-war Reminiscences*. Norcross, GA: Saint Brendan Press, 2007.

Wildman, David. *Iowa's Martyr Regiment: The Story of the 38th Iowa Infantry*. Iowa City, IA: Camp Pope, 2010.

Williams, Thomas J. *An Historical Sketch of the 56th Ohio Volunteer Infantry: During the Great Civil War From 1861 to 1866*. Columbus, OH: Lawrence Press, 1899.

Winn, Fannie Thomas, "A Sketch in the Life and War Record of Colonel Lovick Pierce Thomas II by his Daughter (Ca. 1930)." In *Col. Lovick Pierce Thomas, II, CSA (1835–1910) of Forty-Second Infantry Regiment Georgia Volunteers: A Brief Biography*, edited by Frederick Ware Huff, 49–61. Kennesaw, GA: F. W. Huff, 2000.

Winsor, Mark. Letters of the Temple Franklin Cooper Family. Clayton, GA: Camp #1929 S.C.V. n.d.

"W. L. Calhoun Biography." Calhoun Family Papers, 1834–1960. MSS 50, 1–3. James G. Kenan Research Center Manuscript Collection, Atlanta History Center, Atlanta, GA.

Wolbach, Theodore. "Battle of Tazewell, Tennessee, August 6, 1862, As Described by a Soldier from Company E, 16th Ohio," comp. Michael Wood, 16th Ohio Volunteer Infantry website. http://www.mkwe.com/ohio/pages/H011-15.htm.

Woodbury, Henry H. *Complete History of the 46th Illinois Veteran Volunteer Infantry, from the date of its organization in 1861, to its final discharge, February 1st, 1866*. Freeport, IL: Bailey & Ankeny, 1866.

Index